Mechanisms of Plant Growth and Improved Productivity

BOOKS IN SOILS, PLANTS, AND THE ENVIRONMENT

Mechanisms of Plant Growth and Improved Productivity

Modern Approaches

edited by

Amarjit S. Basra
Punjab Agricultural University
Ludhiana, India

Marcel Dekker, Inc.   New York•Basel•Hong Kong

Library of Congress Cataloging-in-Publication Data

Mechanisms of plant growth and improved productivity : modern approaches/
edited by Amarjit S. Basra.
    p. cm. — (Books in soils, plants, and the environment)
    Includes bibliographical references and index.
    ISBN 0-8247-9192-4 (acid-free paper)
    1. Crops—Growth. 2. Crops—Genetic engineering. 3. Growth (Plants)
    4. Plant genetic engineering. 5. Crop yields. 6. Primary productivity
    (Biology) I. Basra, Amarjit S. II. Series.
    SB112.5.M43 1994
    581.3'1—dc20   94-12080
                                                         CIP

The publisher offers discounts on this book when ordered in bulk quantities. For more information, write to Special Sales/Professional Marketing at the address below.

This book is printed on acid-free paper.

Marcel Dekker, Inc.
270 Madison Avenue, New York, New York 10016

Current printing (last digit):
10 9 8 7 6 5 4 3 2 1

PRINTED IN THE UNITED STATES OF AMERICA

Preface

The fascinating challenge of understanding the mechanisms of plant growth has gained a new urgency as we strive to exploit agricultural crops for improved productivity. At no other time has this understanding been more relevant nor the field more stimulating or provocative. Increasingly refined tools of biochemistry, biophysics, and molecular biology now allow us to approach these mechanisms at various levels of plant organization. The time is ripe to bring together new knowledge arising from this field, giving it a clear perspective and identifying research priorities for the future.

Plant productivity is simply a measure of the total photosynthesis of the plant less any respiration that has occurred during its growth. Thus, the factors limiting plant productivity are factors limiting net photosynthesis. Since most of the dry weight of plants consists of carbon compounds, the increase in harvested yield is intimately linked to changes in the photosynthetic fixation of carbon dioxide per unit of land area and the subsequent partitioning of this assimilate between the harvested parts and the rest of the plant. With this in mind, Chapter 1 discusses the mechanisms and factors influencing and controlling the photoassimilate transport in plants.

Conservation of respiratory carbon by nonphotosynthetic refixation is considerable in certain plant parts and may contribute to yield. Efficiency in nutrient uptake and utilization is a desirable goal for improved plant productivity. Considerable research has focused on roles of mineral nutrient elements and effects of nutrient deficiencies on plant growth. All these aspects along with accounts of factors governing nutrient uptake and transport and practical aspects of plant nutrition are dealt with in Chapters 2 and 5.

Symbiotic nitrogen fixation (Chapter 3) continues to be an area of great interest and intensive study. The current status of research is assessed, especially in regard to regulation, factors affecting nitrogen fixation, energy sources, exchange of metabolites between the symbionts, and genetics and molecular biology.

Efficient nitrate metabolism is vital for healthy plant growth. Chapter 4, on nitrate assimilation in plants, discusses current investigations into the physiology, genetics, and molecular biology of nitrate uptake, reduction, and regulation.

Plant growth regulators (PGRs) play a very important role in the regulation of plant growth and development. The diversity of phenomena involving PGRs has puzzled many researchers regarding the fundamental mechanism of hormone action in plants. Basic research also has much to gain from novel chemical tools derived from industrial research to better understand hormone synthesis, metabolism, and mode of action as well as the developmental phenomena they direct. Agriculture has recently seen great strides in the effective use of this knowledge for chemical regulation of crop growth and productivity. Chapter 6 deals with the various aspects of plant growth regulators.

Plant growth and productivity are limited by a large number of adverse conditions of the ambient environment. We must better understand mechanisms of plant resistance to these stresses for achieving improved productivity under such environments. Chapters 11 and 12 assess the progress of research in plant resistance to abiotic and biotic stresses.

Because of the increased burning of fossil fuels, the atmospheric carbon dioxide concentration has increased significantly. The atmospheric concentrations of other gases such as methane, nitrous oxides, and chlorofluorocarbons are also increasing at an alarmingly rapid rate. The rising levels of these gases will have global climatic and biological effects on plant productivity. In view of this, the implications of the greenhouse effect on plant growth and productivity are discussed (Chapter 7).

Biotechnology makes it possible to isolate, characterize, and manipulate specific genes. This new technology provides a powerful tool to understand plant growth and development and offers a way to directly manipulate the processes leading to improved productivity. Its potential is enormous and the technology for gene transfer in plants is advancing rapidly. Three chapters (Chapters 8–10) cover this important topic.

Obviously, a book on a topic of such wide scope cannot be all-inclusive. My hope is that this book will provide its readers with a strong awareness of current approaches being followed in the field and an appreciation of the potential of biotechnological tools to achieve improved productivity.

Amarjit S. Basra

Contents

Contributors

Michael T. Abberton, Ph.D.   Genetics Group, AFRC Institute for Grasslands and Environmental Research, Aberystywth, Dyfed, Wales

Craig A. Atkins, Ph.D., D.Sc.   Professor, Department of Botany, The University of Western Australia, Nedlands, Western Australia, Australia

Dennis A. Baker, Ph.D., F.I.Biol.   Professor, Department of Biological Sciences, Wye College, University of London, Ashford, Kent, England

Dorothea Bartels, Ph.D.   Department for Plant Breeding and Yield Physiology, Max-Planck-Institut für Züchtungsforschung, Cologne, Germany

Amarjit S. Basra, Ph.D.   Department of Botany, Punjab Agricultural University, Ludhiana, India

D. S. Brar, Ph.D.   Division of Plant Breeding, Genetics, and Biochemistry, International Rice Research Institute, Manila, Philippines

David B. Collinge, Ph.D.   Associate Professor, Department of Plant Biology, The Royal Veterinary and Agricultural University, Frederiksberg, Denmark

Per L. Gregersen, Ph.D.   Senior Research Fellow, Department of Plant Biology, The Royal Veterinary and Agricultural University, Frederiksberg, Denmark

Peter Hedden, Ph.D. Department of Agricultural Sciences, University of Bristol, and Long Ashton Research Station, AFRC Institute of Arable Crops Research, Long Ashton, Bristol, England

Gordon Victor Hoad, Ph.D.* UG6 Head, Department of Plant Sciences, University of Bristol, and Long Ashton Research Station, AFRC Institute of Arable Crops Research, Long Ashton, Bristol, England

Gabriel Iturriaga, Ph.D. Instituto de Biotecnologia, Universidad Nacional Autonoma de Mexico, Cuernavaca, Mexico

Allison R. Kermode, Ph.D. Assistant Professor, Department of Biological Sciences, Simon Fraser University, Burnaby, British Columbia, Canada

Gurdev S. Khush, Ph.D. Principal Plant Breeder and Head, Division of Plant Breeding, Genetics, and Biochemistry, International Rice Research Institute, Manila, Philippines

Ernest A. Kirkby, C.Chem., FRSC, C.Biol., F.I.Biol. Department of Pure and Applied Biology, The University of Leeds, Leeds, England

Jacques Le Bot, Ph.D.[†] The University of Leeds, Leeds, England

Nancy Longnecker, Ph.D. Department of Soil Science and Plant Nutrition, School of Agriculture, The University of Western Australia, Nedlands, Western Australia, Australia

John Anthony Milburn, Ph.D., F.I.Biol. Professor, Department of Botany, University of New England, Armidale, New South Wales, Australia

David J. Pilbeam, Ph.D. Department of Pure and Applied Biology, The University of Leeds, Leeds, England

Hans Thordal-Christensen, Ph.D. Senior Research Fellow, Department of Plant Biology, The Royal Veterinary and Agricultural University, Frederiksberg, Denmark

Marc Van den Bulcke, Ph.D. Laboratorium Voor Genetica, Universiteit Gent, Ghent, Belgium

Sylvan H. Wittwer, Ph.D. Director Emeritus, Agricultural Experiment Station, Michigan State University, East Lansing, Michigan

John L. Wray, Ph.D. Plant Sciences Laboratory, School of Biological and Medical Sciences, University of St. Andrews, St. Andrews, Scotland

Current affiliations:
*Department of Biology, College of Science, University of Bahrain, Bahrain
[†]INRA Station d'Agronomie, Centre de Recherches Agronomiques, Montfavet, France

Mechanisms of Plant Growth and Improved Productivity

1

Photoassimilate Transport

Dennis A. Baker

Wye College
University of London
Ashford, Kent, England

John Anthony Milburn

University of New England
Armidale, New South Wales, Australia

PATHWAYS DEVELOPMENT AND STRUCTURE

Transport in Sieve Tubes

Evidence from bark-girdling experiments for the operation of sieve tubes predated their discovery. The roots, when detached from their supply of nutrients, died. This "ring-barking" has provided a simple yet effective method of forest clearance used since antiquity. Anatomically, however, sieve tubes are difficult to discern, even with a modern optical microscope. Also, despite the fact that it is now known that their sugary contents are under considerable positive pressure, evidence of their bleeding in response to injury was very scanty. The exceptions lay in the tropics where there was a widespread array of techniques for tapping large quantities of sap from several palms. This phenomenon was only poorly understood, however, and indeed was largely ignored in temperate latitudes where the critical questions were eventually asked about the function of sieve tubes.

The discovery of sieve tubes by the German forester, Theodore Hartig in 1837, and the subsequent observation of exudation in minute amounts from forest trees by his son, Robert Hartig, confirmed the potential role of these transport conduits. The amount of exudates, though small, was considerably greater than could flow from a single punctured sieve-tube element: hence longitudinal flow was an obvious deduction. We now understand that the sieve plates, the porosity of which has been a major cause of controversy, usually restrict the loss of sap. This is an ecological mechanism which protects plants from excessive predation by phloem-feeding animals. Though doubts persisted into the 1920s and even later, about the possible role of xylem as a possible *major* conductor of organic nutrients, these were progressively eliminated, e.g., by the work of Mason and Maskell [1], so that Dixon could write in 1933 [2], "It is possible to show that the movement of bast sap takes place in the sieve tubes" and go on to describe many ingenious experiments on exudation and injection. Perhaps because this paper was neglected, much of his message remained to be rediscovered in the 1970s.

It is now established beyond doubt that sieve tubes (and their numerous but tiny sieve-plate pores) transport the major quantities of elaborated nutrients within plants. All sinks (roots, fruits, buds, and secondary meristems) are supplied by the phloem sieve tubes which, therefore, collectively represent a pathway of enormous significance in the world of biology.

An interesting feature of sieve tubes is that they are laid down in close proximity to xylem conduits, which has made it difficult to distinguish between their respective roles in conduction. There may be a good reason for this situation, and it possibly reflects that sieve tubes consume and also exude considerable quantities of water from, and into, the adjacent tissues of the plant. These tissues are discussed more extensively below (see pp. 8–16).

It is known that sieve tubes originate from a series of longitudinally elongated cells (the sieve elements of the final sieve tube). The cell contents are degraded to the extent that the nucleus and tonoplast disappear, but nonetheless seem to remain "alive." Sieve elements are partially fused at their end walls. In this region the junction wall is then locally digested to produce the characteristic "sieve plate." This plate usually contains 50–150 pores each of which is lined with callose, a carbohydrate used in many regions of the plant as a sealant (e.g., in pollen tubes, where it helps maintain turgor). These pores can seal with callose on a seasonal basis as in temperate trees, or as a result of wounding [3]. However, on wounding the main mechanism for the sealing operation is the almost instantaneous deposition of plugs of protein in and over the pores (originally called slime; now called "P protein" for "phloem protein"). Unless special precautions are taken when microscopic sections are cut, the pores seal automatically. This phenomenon caused a long-standing controversy between anatomists, who said it was impossible for sieve tubes to function as simple open pipes, and the transport physiologists, many of whom said that they must.

Transport in Xylem Conduits

As noted above, it is quite possible for xylem conduits to transport significant quantities of photoassimilates. These long tubes are wide and water filled, when functional, and easily permeated by dissolved inorganic and organic solutes. However, on account of the direction of the transpiration stream, such a pathway for the transport of photoassimilates is of limited usefulness.

Xylem vessels are the most efficient sap transporting units, consisting of large numbers of cells fused end-to-end by *perforation plates.* Despite their name, perforation plates may be reduced to a mere rim with little evidence of a plate: others have rows of relatively huge pores, which may be useful in identifying wood specimens. A vessel consists of several cells linked by perforation plates. At its ends, however, sap must pass through the much more finely porous pit membranes, which act as filters, removing both suspended particles and bubbles. This gives a clue as to the major function of pit membranes, which is to prevent the spread of bubbles and thus prevent catastrophic embolization. Tracheids and fibers operate as small, short, vessels differing in that they originate from single cells. All of these units behave similarly in conducting sap, often being called collectively *conduits.* Although pit membranes can filter out very small particles and certain colloids, solutes in solution can pass through with relative ease.

Xylem conduits may become directly involved in long-distance transport, when they are immature. Before the walls of conduits are strengthened by lignin deposition, they are voluminous and thin walled, maintaining their shape through high turgor pressure. In this condition they strongly resemble sieve tubes, being filled with concentrated solutes, including sugars. From time to time there have been suggestions that these xylem initials may be engaged in assimilate transport, and indeed, if severed in the early spring they sometimes exude like certain sieve tubes.

Another suggestion was that these *immature* xylem initials might be involved in ion accumulation in roots. Through the death of the cells, the loss of membrane control would automatically release the ions into the transpiration stream, by the so-called "test-tube" hypothesis [4]. There is no firm evidence, however, that this similarity with sieve tubes is more than coincidental, arising from their mode of development, which requires vigorous cell expansion to achieve sufficiently large elements to make correspondingly large vessels. Any long-distance transport of photoassimilates in this manner seems to be insignificant.

There is, however, a major transport of organic chemicals, which are synthesized in roots, toward the aerial organs. Xylem sap has been found to convey significant quantities of organic amides and ureides in this way [5]. Another example which is very well documented, is the transportation of nicotine, an alkaloid, from the roots where it is synthesized to the leaves where it normally would protect the plant from insect predation. It is harvested by mankind to make tobacco products, which include useful insecticides [6].

Transport and Function of Rays

An important transport system in plants, which is often ignored, is the ray system. As the name suggests, rays are radial in distribution and are prominent in the medullary region: as a consequence those rays are termed *medullary rays*. In fact, rays extend from the medulla outwards across the cambial region into the phloem tissues where they become progressively harder to recognize in microscopic transverse sections.

Close examination of the ray cells reveals the living contents; organelles such as plastids are common, and the cells are often packed with starch or other reserve products. Between the cells are plentiful plasmodesmata giving an important clue to their physiological role. The fact that these living cells have thick cell walls in the xylem, but not the phloem, seems an essential requirement to withstand the considerable mechanical pressures that develop within growing tree trunks, which would otherwise crush the cells.

The tissue organization of rays varies from plant to plant. In longitudinal transverse section, ray cells are normally grouped into vertical ellipses (Figure 1), the size and number of cells in each group varying from species to species. This indicates that ray transport consists of a series of similar pathways connected laterally with one another, but each capable of promoting transport along the cells in a radial direction. Because the rays interconnect

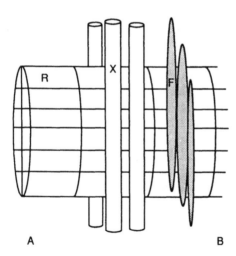

Figure 1   Diagram of the way in which xylem conduits (X) and fibers (F) make contact with ray cells in wood of a tree such as *Acer*. The ray cells (R) run radially in groups like the spokes of a wooden wheel from the center of the tree trunk (B) toward the periphery (A).

sieve tubes and the xylem conduits, they are in an ideal position to be able to extract or secrete solutes into the vertically oriented transpiration stream. *Acer* is a species which has been studied extensively in this respect [7,8,9, 10], and this work will be described here in more detail as a good example of ray and xylem transport.

Acer has been used for many centuries in Europe and Northern America as a source of enriched xylem sap. This sap contains a high proportion of sugars. Collection is made from trees in the early spring when they are still leafless. Boreholes are made in the trunks and metal tubes are hammered in to a depth of about 2 cm. In the simplest arrangement up to four such tubes are fitted per tree (depending on its size). Buckets are suspended from them to collect the sap. Flow, which is most intense on warm sunny days, only occurs to any significant extent following sharp frosts. These have been proved to provide the mechanism driving, first a suction in the tree at night, when it becomes imbibed with water; then a positive pressurization by day, which expels the sap if the trunks have been wounded.

In *Acer saccharum* the concentration of solutes is typically up to 4%, most of which is sugars. Boiling the sap removes the water and simultaneously caramelizes the concentrated sap. The product is the well-known maple syrup. All members of the genus *Acer* are capable of this type of flow, but the solute concentrations and flow are generally better from *A. saccharum*, the famed "sugar-bush" which supports a minor industry.

What is the source of the solutes in the sap and what is their function in the intact tree? The flow of solutes arises from mobilization of organic reserves, especially starch, from the ray tissue. Rising spring temperatures induce secretion of sugars from the rays into the adjacent xylem conduits. This flow of solutes is not entirely diffusion dependent, because it is augmented by the pressurizing mechanism. Pressure changes are generated by gas bubbles entrained in the fibers (which can expand and contract in response to temperature and ice invasion) producing an internal ebb and flow within the tissues of the trunk (see Figure 2).

By taking samples from a tree trunk at different heights, it can be shown that the solute concentration increases as the sap ascends the trunk, carried passively by the weak transpiration stream within the leafless tree. It appears that the sap stream is enriched by an action akin to an ion-exchange column, until it reaches the finer twigs when the concentration falls. The twigs are, apparently, utilizing solutes for the spring growth of leaves within buds and also new vascular tissue. The transpiration stream provides a convenient transport system for this supply of nutrients (see Figure 3).

Why do the organic stores, which were produced by the foliage the previous summer, need to be retransported in this manner? By what mechanism, and for what reason, have they been retained over the winter within the tree

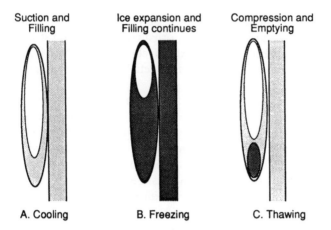

Suction and Filling	Ice expansion and Filling continues	Compression and Emptying
A. Cooling	B. Freezing	C. Thawing

Figure 2   Diagram of how fibers subtract or contribute water to the conduits as they freeze or thaw. The gas phase acts as a spring and contracts on cooling (A) but the effect is especially strong when it is invaded by ice formation (B). This cycle is reversed when thawing occurs producing a flow of sap or pressurization (C). Liquid sap is horizontal-hatched: ice is shown as a diagonal-hatch. (Adapted from [9].)

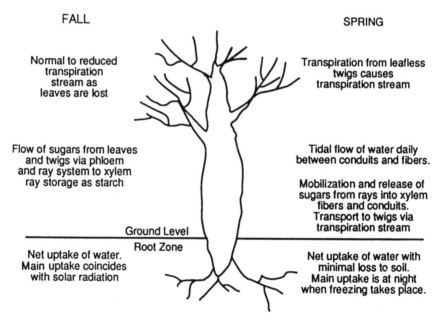

FALL   SPRING

Normal to reduced transpiration stream as leaves are lost

Transpiration from leafless twigs causes transpiration stream

Flow of sugars from leaves and twigs via phloem and ray system to xylem ray storage as starch

Tidal flow of water daily between conduits and fibers.

Mobilization and release of sugars from rays into xylem fibers and conduits. Transport to twigs via transpiration stream

Ground Level

Root Zone

Net uptake of water. Main uptake coincides with solar radiation

Net uptake of water with minimal loss to soil. Main uptake is at night when freezing takes place.

Figure 3   Diagram indicating how a maple (*Acer* spp.) tree stores its carbohydrates in the wooden trunk for overwintering. It then utilizes wood fiber pressurization to transport the mobilized sugars to the transpiration stream where they become available for new cambial tissues and especially new bud growth.

trunks? The pathway of the initial storage in maple has not been studied in detail; but it can be assumed with confidence to be the sieve tubes of the phloem within the bark. From sieve-tube exudation experiments on many other species, they have been established to transport a rich supply of photo-assimilates down tree trunks (see Figure 3). This flow is strongest in fall of each year and hence we can safely deduce that some of this flow is intercepted by the rays for storage in the trunk over the winter.

The question remains: why store in tree trunks? The reason seems to be more ecological than physiological. A tree trunk is a potentially secure storage place, much safer from predation by insects, birds, and small mammals than would be the case if stored materials were deposited in the delicate and easily accessible twigs. It appears that this whole circulatory mechanism, involving rays, xylem conduits, and phloem sieve tubes, has been evolved through specialization of those processes normally required in very many plants for secondary thickening and root maintenance.

In this section we have concentrated on ray transport in the genus *Acer* but it can be assumed with confidence that similar circulation occurs in a wide range of woody plants, possibly on a smaller scale. Much of this work has still to be documented in other trees as part of the annual cycle in the perennial habit. Similar solute fluxes have been shown to occur in herbaceous plants [11], leaving no doubt of the important role of ray transport throughout terrestrial plants in general.

TRANSFER OF DRY MATTER

The most important function of sieve tubes is the transportation of dry matter, a subject which has been investigated for about 70 years. Münch [12] investigated the transport of dry matter into sinks, i.e., potato tubers and annual rings of wood in trees, respectively. Curiously, Dixon did so while believing that the main transport channel was the xylem; however, Münch had no similar misconceptions. Dixon [2] calculated that the rate of flow through the sieve tubes would have to be 49 cm/h to deposit the amount of dry matter found in a potato after 100 days growth. Münch made similar calculations on the deposition of wood and, by measuring the cross section of the phloem, was able to calculate the "specific mass transfer" (SMT) of dry matter in phloem.

This approach has been extended by many workers subsequently, with ever increasing precision, so that corrections could be added for losses in dry matter from respiration. Sinks seldom gain dry matter from photosynthesis, even though they often possess chloroplasts and stomata. Most sinks grow in a sigmoid pattern, meaning that dry matter transfer is slower both initially and finally [13]. Nevertheless the data collected by Münch and Dixon have been confirmed for a large range of intact plants. Canny [14] has drawn attention to the fact that such tissues have typically SMTs around 4 g/cm² phloem/h.

Another interesting way to consider dry matter transfer is from wounded exuding plants. The amount of dry matter in the sieve-tube sap is easily found by drying it. The SMTs can be calculated by measuring the cross section of phloem, or even better the sieve tubes, and the time taken for quantities of exudate to collect. Some enormous values have been measured, i.e., up to about 1500 g/cm²/h. These indicate that the sieve tubes have an enormous capacity to transport dry matter, which is seldom fully exploited by sink uptake in the intact, unwounded condition. This type of investigation shows that sieve plates are far from being the near impenetrable barriers to the longitudinal flow of sap once thought. Instead it seems that in normal undisturbed plants they have a remarkably low hydraulic resistance, with the result that phloem sap is supplied to sinks at a considerable hydrostatic pressure.

A constant problem in increasing the precision of SMT measurements has been the dimensions of sieve tubes. They are difficult to identify in cross section, even if stained with fluorochromes such as aniline blue. The frequent assumption that sieve tubes can be regarded as 20% of the nonfibrous phloem is no longer acceptable for serious work because the proportion has been found to vary from about 8 to 60%. Precise work necessitates many careful measurements to find the correct proportion.

MECHANISM: PRESSURE FLOW

Historical Synopsis

Though sieve tubes and their osmotic contents were discovered by the Hartigs around 1850, there was no clear concept of a mechanism beyond the assumption that osmotic pressure was likely to be involved. A mechanism was clearly formulated in 1930 by Münch [13] after extensive studies on the transport of photoassimilate, mainly through bark girdling experiments and studying the annual depositions of wood.

Another important element of Münch's hypothesis was that there *must* be a circulation of water. This he predicted would be secreted at the zone of deposition of solutes, and he claimed to collect this water from tongues of tree bark. Unfortunately others were unable to repeat this experiment causing doubt about his hypothesis.

In the course of time Münch's mechanism has become almost universally accepted. However, though in its essentials it is fairly simple, in its operation within the tissues of the plant, there are many modifying factors which have caused some researchers to question his hypothesis. To understand these misgivings it is appropriate to study the pressure flow mechanism as postulated by Münch in more detail.

Münch's Pressure Flow Mechanism

Münch envisioned a pair of opposed osmometers connected so that an increase in solute concentration in one, would induce a localized influx of water. This increased hydrostatic pressure would force water from the other osmometer by *reverse osmosis*. In this way water would enter sieve tubes at sinks, along with the loaded solutes. Conversely, water would be expelled in sink tissues in parallel with the removal of solutes by sink tissues. It will be noted that there is a mass flow of water carrying the solutes passively along with it, from source to sink. Furthermore, though pressure from osmosis is held responsible for the flow, no pressure gradient could be demonstrated in the model introduced by Münch to illustrate his hypothesis. On the other hand, a solute concentration gradient could be demonstrated: this situation had led others [1,14] to suggest that a mechanism akin to diffusion rather than pressure might account for the transport. Of course diffusion is far too slow over these distances, often measured in meters, requiring "facilitated diffusion" by some mysterious mechanism.

The most vexed issue was, of course, the sieve plates. Münch assumed that they would not prevent the flow of solution, to such an extent that they were not even included in his model. If there is no hydraulic resistance there cannot be a pressure gradient, as indicated above. However, anatomists had found that sieve plates seemed to be blocked by phloem [P] protein, which would prevent flow. Various pumping hypotheses were advanced according to which the proteins might promote flow either by electroosmosis or peristalsis [15]. In fact as fixation techniques have improved, more and more sieve plate pores have been found to be open. This evidence, in conjunction with the fact that many plants can be induced to exude phloem sap, has now laid the foundations of modern understanding. According to this view the primary function of sieve plates is to seal and so prevent the flow of sap when the tubes are damaged, to prevent excessive loss of plant nutrients to a potential predator [16]. Concentration gradients have been found to be quite small in short cereal stems which indicates that in the intact plant, sieve plates cannot possibly represent a major hydraulic resistance [17].

A great deal of controversy has centered around the overliteral application of Münch's model, which represents sieve tubes as simple glass tubes. If, however, radioactive solutes travel down the tube they tend to become static and concentrated near the point of entry into sieve tubes. We now interpret this observation as indicating that the tubes are in fact effectively permeable. Solutes are exchanged and stored in parenchyma cells along the sieve tube giving a misleading impression of the concentration gradients within the sieve tubes themselves.

The reason why Münch's evidence for his theory in collecting water from reverse osmosis could not be repeated by others has also become clearer.

The design of his experiment was not ideal: water could have been collected from transpired water rather than reverse osmosis because his collection system would allow distillation to occur. Mason and Maskell, who failed to repeat his observations, probably failed for another reason, however. For water to be exuded a barkflap *must unload materials* (as commonly happens when a stem is bark-girdled, causing a swollen callus), and it seems possible that the bark of cotton may have been too badly damaged while being separated from the xylem to unload successfully. It has been demonstrated, however [18], that if bark flaps of *Salix* are kept under paraffin oil, droplets of liquid water exude, which is exactly what Münch sought to demonstrate.

ANALYSIS OF SIEVE TUBE CONTENTS

Analysis is usually conducted on those phloem saps which can be obtained using various collecting methods. A few plants bleed phloem sap when incisions are made into the bark tissues. These include a number of dicotyledonous trees [19], some cucurbits [20], the castor bean [21,22], *Yucca* [23] and the fruits of some legumes [24]. The majority of plants do not bleed from the phloem and investigators have developed other techniques to sample the phloem sap. These include severing phloem-feeding aphids from their inserted stylets. The stylets then continue to exude phloem sap, due to the high turgor within the sieve elements, and this can be collected for analysis. The methods employed for stylectomy range from simple cutting with a sharp blade [25] to surgical lasers. Another method, which has been developed recently, involves *cryopuncture*, whereby a needle, cooled in liquid nitrogen, induces bleeding when inserted into the phloem tissues [26]. Other investigators have used 5–20 mM solutions of EDTA into which cut stems or petioles are placed. The destruction of membranes by the EDTA induces bleeding from the phloem [27], but contamination from the contents of other tissues is an unavoidable defect of this technique.

Notwithstanding the problems of obtaining sap for analysis, several investigators have reported on the composition of phloem sap [28]. Such reports indicate a remarkable consistency in that all phloem saps are characterized by their high sugar and organic nitrogen content and their high K^+/Na^+ and Mg^{2+}/Ca^{2+} ratios. An alkaline pH of between 7.2 and 8.5 is also a common characteristic. The analysis presented in Table 1 for phloem sap from *Ricinus communis*, the castor bean, shows a typical composition from a herbaceous plant.

The high dry matter content of phloem saps (between 10 and 25%) contains 90% or more of sugar. In many plants sucrose is the only sugar present, but other sugars, namely raffinose, stachyose, and verbascose, and

Table 1 The Composition of the Exudate Obtained from Incisions Made in the Bark of *Ricinus* Plants. Some Values for Xylem Exudate Are Included for Comparison (concentrations are expressed in mg cm^{-3} and also in eq m^{-3} or mol m^{-3} where relevant)

	Phloem sap		Xylem sap
Dry matter	100–125		2 mg cm^{-3}
Sucrose	80–106		0
Reducing sugars		Absent	0
Protein	1.45–2.20		
Amino acids	5.2 (as glutamic acid)	37.4 mol m^{-3}	
Keto acids	2.0–3.2 (as malic acid)	30–47 eq m^{-3}	
Phosphate	0.35–0.55	7.4–11.4 eq m^{-3}	
Sulfate	0.024–0.048	0.5–1.0 eq m^{-3}	
Chloride	0.355–0.675	10–19 eq m^{-3}	
Nitrate	Absent		
Bicarbonate	0.010	1.7 eq m^{-3}	
Potassium	2.3–4.4	60–112 eq m^{-3}	
Sodium	0.046–0.092	1.0–4.6 eq m^{-3}	7–8 eq m^{-3}
Calcium	0.020–0.092	1.0–4.6 eq m^{-3}	3–4 eq m^{-3}
Magnesium	0.109–0.122	9–10 eq m^{-3}	
Ammonium	0.029	1.6 eq m^{-3}	
Auxin	10.5 × 10^{-6}	0.60 × 10^{-4} mol m^{-3}	
Gibberellin	2.3 × 10^{-6}	0.67 × 10^{-5} mol m^{-3}	
Cytokinin	10.8 × 10^{-6}	0.52 × 10^{-4} mol m^{-3}	
ATP	0.24–0.36	0.40–0.60 mol m^{-3}	
pH	8.0–8.2		6.0
Conductance	1.3 mS m^{-1} at 18°C		
Solute potential	−1.42 to −1.52 MPa		
Viscosity	1.34 × 10^{-3} N s m^{-2} at 20°C		

Source: Ref. 28.

occasionally the sugar alcohols mannitol and sorbital, may occur. Amino acids and amides are found in phloem saps at between 0.2 and 0.5% normally, although during leaf senescence the level may rise to 5%. A wide variety of organic and inorganic compounds, including growth substances and enzymes, are also present.

The translocated sugars are all similar, consisting of glucose and fructose units, joined by a glycosidic bond, as in sucrose, with one or more D-galactose units attached (Figure 4). The resultant sugars, raffinose, stachyose, verbascose, and ajugose are sometimes found in small amounts with sucrose as the major component, but in some plant families the longer oligosaccharides

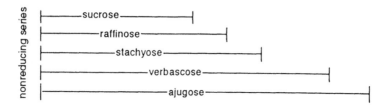

fructosyl-β, α-glucosyl-α-galactosyl-α-galactosyl-α-galactosyl-α-galactose

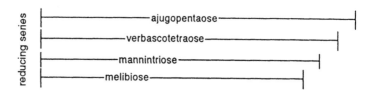

Figure 4   The carbohydrates transported in the phloem are usually nonreducing members of the raffinose family of oligosaccharides. Each higher member is formed by the addition of one 1,6-galactosyl residue.

are more abundant (e.g., Cucurbitaceae, where stachyose is the major sugar translocated). The sugar alcohols D-mannitol and D-sorbitol are also found in some tree species, the most intensively studied genus being *Fraxinus*.

Hexoses are not normally found in phloem exudates, although glucose and fructose are usually present in the nonconducting cells of phloem tissues as the result of sucrose hydrolysis. There has been a recent report of glucose in phloem sap from *Ricinus* seedlings under special conditions [29]. The physiological significance of the fact that translocated sugars are normally nonreducing seems to be that these sugars function as protected derivatives of glucose which are not so readily metabolized and thus provide more stable material for long-distance transport. In addition, weight for weight, these nonreducing sugars contain additional energy because of the glycosidic bond. A further possible advantage is that more carbon can be transported in the form of oligosaccharides than in the form of hexoses for the same osmotic value. With respect to this latter point, it is generally observed that herbaceous species transport mainly sucrose, while trees and climbing plants, with longer translocation pathways, favor the higher oligosaccharides such as raffinose and stachyose.

The organic nitrogen component of phloem sap comprises a wide variety of amino acids and amides in addition to a relatively high protein content (see Table 2). The amino acid and amide contents vary with the age and species of plant. A selective uptake of amino acids and amides into the phloem has been proposed [30] and the observation that the phloem sap amino acid composition does not merely reflect that of the source leaf lends support to this proposal (see Table 2). The protein content of phloem sap is mainly the phloem or P protein which seals the sieve plate pores of damaged sieve tubes (see p. 2). Additionally a large number of enzymes, particularly those associated with carbohydrate and nitrogen metabolism, are also found. Whether these enzymes are all functioning in a metabolic sense is debatable. During its differentiation the sieve tube loses not only nuclear information but also ribosomal translation sites and secretory dictyosomes (see p. 2). Thus any enzymes present in phloem sap *must* originate from the adjacent companion cells and their presence may be fortuitous or may indicate a key role, as in the utilization of sucrose in the formation of sieve-tube starch and of callose, which is made possible by the presence of UDPG-fructose glycosyl transferase in the phloem sap.

The inorganic ions present in the phloem sap betray its cytoplasmic origin (Table 1). Potassium is the major cation, with concentrations of up

Table 2 The Amino Acid Composition of Phloem Exudate Obtained from *Ricinus* Plants

Amino acid	Concentration	
	(mol m^{-3})	(%)
Glutamic acid	13.0	34.76
Aspartic acid	8.8	23.53
Threonine	5.4	14.44
Glycine	2.4	6.42
Alanine	2.0	5.35
Serine	1.6	4.28
Valine	1.6	4.28
Isoleucine	1.0	2.67
Phenylalanine	0.6	1.60
Histidine	0.4	1.07
Leucine	0.4	1.07
Arginine	Trace	Trace
Methionine	Trace	Trace
Total amino acids	37.4	100.00

Source: Ref. 28.

to 112 mM(mol m^{-3}), and has been proposed to play a role in stimulating the plasma membrane H$^+$-ATPase [31]. The high Mg^{2+}/Ca^{2+} ratio, referred to above, is typical for cytoplasmic compartments, the low calcium concentration being reflected in the deficiencies of this element which sometimes occur in developing storage sinks. Inorganic anions are present in much lower amounts than the cations, the charge balance being provided by organic anions. Chloride and phosphate are the most common inorganic anions; traces of sulphate and bicarbonate are often present. Nitrate is not found in phloem sap although some early analyses record its presence (see [32]). In those species which transport nitrate in the xylem, such as *Ricinus* [33], the absence of nitrate in the phloem sap argues against any *direct* circulation of nitrogenous material between the two conducting pathways. In many species the concentration of these phloem mobile ions increases markedly in the phloem sap during leaf senescence, when they are exported through the sieve tubes prior to leaf fall. In addition to the major inorganic ions mentioned above, traces of molybdenum, copper, iron, manganese, and zinc have been reported in phloem exudate from *Yucca* inflorescence stalks [34].

In addition to the sugars and other substances described above a wide range of other physiologically important materials are found in phloem saps. Organic acids, such as malate, are found in substantial amounts in phloem exudates and translocated malate has been implicated in the regulation of nitrate uptake [35]. Certainly the presence of a dicarboxylic acid, such as malate, provides a portion of the charge balance required for the large amount of potassium present in phloem saps, as mentioned above. Adenosine triphosphate (ATP) has been identified in phloem saps at relatively high concentrations and is utilized to fuel the H$^+$-ATPase on the sieve element plasma membrane. Nucleic acids and the vitamins biotin, nicotinic acid, pantothenic acid, folic acid, pyridoxine, riboflavin and thiamine are also found [32].

The presence of growth substances in phloem saps at physiologically active concentrations has been the subject of speculation as to their role. All of the major plant hormones, auxin, cytokinins, gibberellins, abscisic acid, and ACC (the precursor of ethylene) have been found [36]. The polyamines, putrescine, spermidine, and spermine, which are believed to have a growth regulatory role, are also present in phloem sap [37].

Phloem saps can thus be seen to contain, at least in small amounts, a considerable number of the myriad of substances which are involved in plant metabolism. The list of substances identified in phloem exudates continues to lengthen as more sophisticated analytical techniques become available, and it seems probable that at least traces of all the water-soluble substances naturally occurring within plants will be found. In addition to these endogenous substances many compounds applied exogenously will be translocated to some extent within the phloem. The subject of xenobiotic transport and its relevance to pesticide application is discussed on p. 19.

TRANSPORT WITHIN PHOTOSYNTHETIC CELLS

The transport of photoassimilates commences within the chloroplast of photosynthetic cells. This transporting organelle must strike a balance between exporting elaborated carbon and chemical energy, recycling newly synthesized intermediates, and internal storage of excess products.

The production of triose phosphate by the Calvin cycle within the chloroplast stroma can be considered as the starting point for photoassimilate transport. The triose phosphate can then either be exported through the phosphate:triose phosphate translocator located in the inner chloroplast envelope, condensed into starch within the stroma, or recycled in the Calvin cycle. The control of these processes is affected by the availability of inorganic phosphate P_i, which is counterexchanged for triose phosphate. The triose phosphate subsequently provides the substrate for sucrose synthesis within the cytosol, the fine control of which is affected by fructose 2,6-biphosphate [38].

Thus, during the day, the rates of sucrose and starch synthesis and the rate of photosynthetic carbon assimilation are coordinated to maintain a balance between the sucrose level temporarily stored within the vacuole, in the cytoplasm available for export, the starch accumulated in the chloroplast, and the amount of assimilated carbon returning to the Calvin cycle. At night, the sucrose stored in the vacuole during the day and the starch accumulated in the chloroplast stroma are remobilized and utilized for leaf metabolism or export, providing an around-the-clock availability of photoassimilates [39].

The above system is considerably modified in C_4 plants where there is compartmentation of sucrose and starch formation in the bundle sheath and sucrose synthesis in the mesophyll. This is the result of the export of triose phosphate from the mesophyll chloroplasts and its import by the bundle sheath chloroplasts, in exchange for phosphoglycerate in each case [39]. These diffusional movements of triose phosphate and phosphoglycerate are in addition to the transfer of C_4 acids between the mesophyll and bundle sheath cells and are dependent upon the generation of relatively large intercellular concentration gradients within the symplast.

In plants with a Crassulacean acid metabolism (CAM), where dark fixation of CO_2 with malate occurs, the malate is invariably stored in the vacuole during the nocturnal acidification and transported to the cytosol during the deacidification in the light. Carbohydrate is stored either in the chloroplast as starch or extrachloroplastically as sugars (as in pineapple). This transfer of fixed carbon involves glycolytic breakdown in the dark and gluconeogenic synthesis in the light of the carbohydrates involved. Large amounts of carbon must therefore enter the chloroplast in starch-forming species when malate is decarboxylated during deacidification.

It will be apparent from the above description that inorganic phosphate recycling or supply plays a key role in the production of photoassimilates.

If a limited inorganic phosphate supply limits photosynthesis and growth of cereals at relatively low temperatures or in plants growing in an enriched CO_2 environment, an understanding of phososynthetic regulation could be used for the genetic or chemical improvement of crop productivity under these conditions.

A transport of photoassimilates occurs between kinetically distinct pools within the leaf. These pools are predominantly of sucrose, referred to above, but often other photosynthetic products are involved, representing distinct storage and export pools. The site of the soluble carbohydrate storage pool is generally thought to be the vacuoles of the mesophyll cells, but little is known of how transport across the tonoplast of these cells is regulated and what control mechanisms operate in terms of providing carbohydrates for export to the phloem.

The soluble carbohydrates destined for export, mainly sucrose (see p. 10), must travel from a producer cell to a sieve element. These two compartments may be contiguous but some mesophyll cells are distanced from the sieve elements by two or three adjacent cells. The prevailing view is that assimilate movement between the mseophyll cells follows a route through the symplast.

LOADING OF PHOTOASSIMILATES

Introduction

Partitioning of photoassimilates involves the concentration of sugars in the sieve tubes from the sites of production, i.e., the chloroplasts in the leaf mesophyll cells. Evidence was presented by Geiger [40] that, because photoassimilates synthesized from radioactive CO_2 escaped from lightly abraded leaves into a bathing solution, the photoassimilates must be free to move via the apoplast toward the sieve tubes by diffusion. At the apoplast evidence has accumulated that there is a marked increase in concentration within both companion cells and sieve tubes: these are now called the "cc-se complex." At the membranes of these cells the sharp jump in concentration is held to imply a vigorous loading step from mesophyll apoplast into cc-se complex symplast.

More recent evidence has somewhat obscured this straightforward picture. Madore et al. [41] have shown that if certain fluorescent dyes (Lucifer Yellow) are injected into the symplast of mesophyll cells they are transferred to the sieve tubes without escaping into the apoplast (see p. 17). Thus it seems there are two possible pathways for photoassimilate loading, and the possible roles of each pathway in different species of plant is still hotly debated. A useful review of partitioning in a wide range of situations in plants has been published recently [42].

The various materials found in phloem saps have been discussed in the preceding section. These substances must enter the phloem sieve elements at or near the source tissues in the process of phloem loading. This process may be regarded as the terminal step in the lateral transport of photoassimilates within the leaf and involves a selective accumulation of the major solutes destined for long-distance transport to the sink tissues. The topic has been the subject of a number of reviews in recent years [43], and is currently the subject of intensive study in a number of research programs. A major difficulty encountered is in interpreting a number of recent reports which indicate a symplastic pathway, as an alternative or in addition to the more generally accepted apoplastic route for phloem loading.

Symplast versus Apoplast

It has long been recognized that both apoplastic and symplastic pathways were potentially available for phloem loading [43] but the high concentration of sucrose in the phloem sap relative to that of the mesophyll cells, coupled with a paucity of plasmodesmatal connections between these two compartments in C_4 grasses such as maize [44], have been taken as circumstantial evidence for a membrane-mediated selective uptake of this solute from the apoplastic interface between these producer and translocator cells. Studies in a number of laboratories have indicated that sucrose uptake from the apoplast is in response to a proton gradient created by a proton pumping ATPase located on the sieve element plasma membrane. The essentials of this process are illustrated in Figure 5 where it can be seen that the inward movement of protons, in response to the electrochemical gradient ($\Delta\mu H^+$) generated by the proton pump, is coupled to sucrose movement across the membrane. This proton:sucrose cotransport results in the sucrose becoming concentrated within the sieve element, resulting in an influx of water which raises the internal turgor pressure. Thus loading provides the driving force for the long-distance transport of sucrose by pressure driven mass flow (see p. 9).

There have been a number of reports, based on the microinjection of fluorescent dyes, that a symplastic continuity exists between the mesophyll and the sieve elements. Lucifer Yellow CH, a fluorescent dye readily visible in living cells at nontoxic concentrations, has been injected into the mesophyll cells of a source leaf of *Ipomea* and extensive movement observed into the minor veins [45]. As this dye is unable to cross cell membranes, the above result implies symplastic continuity between the mesophyll cells and phloem sieve elements, a situation which questions one of the tenets of apoplastic loading. It has been questioned whether the movement of Lucifer Yellow CH, a charged dye (MW457), along its concentration gradient will reflect the movement of uncharged sucrose (MW 352) against a concentration gradient,

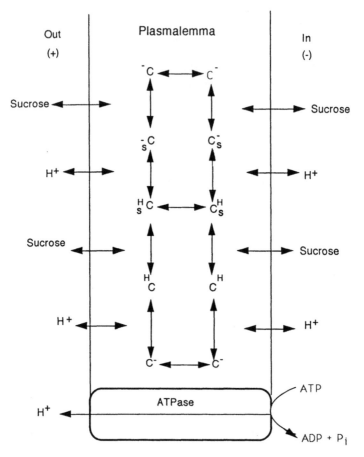

Figure 5 A model for sucrose-proton cotransport across the plasmalemma.

or how hexoses of MW180 are excluded from the sieve element [46]. Treatment with plasmolyzing osmotica, which would reduce or eliminate any symplastic pathway, have been shown to generally promote phloem loading of both $^{14}CO_2$ derived assimilates and exogenously applied [^{14}C] sucrose [40].

On the basis of the above observations it is apparent that the active and selective nature of the loading process appears more compatible with an apoplastic pathway. This may however reflect our lack of understanding of the functioning of plasmodesmata, which could conceivably function as one-way valves, and we must await the production of unequivocal evidence before dismissing one or other of these, not necessarily mutually incompatible, pathways.

A feature of phloem loading is the high substrate specificity of the sucrose carrier [47]. When presented with a range of other sugars and compounds only nitrophenyl glucosides and galactosides are readily recognized by the carrier, with maltose and trehalose also being recognized, but to a lesser extent [43].

Loading of Xenobiotics

In recent years increased attention has been paid to the systemicity of pesticides with a view to improving their efficacy. The target tissues for many insecticides, fungicides, and herbicides are often the various meristems of the shoot and root systems and the use of systemic, phloem mobile compounds would allow lower, and thus more environmentally acceptable, application rates. The xenobiotic compound can be carried to all parts of the plant connected by the phloem providing the compound is able to access the sieve elements and be transported. Two possible mechanisms have been postulated to explain the phloem systemicity of certain pesticides, namely the "acid trap" and "intermediate permeability" hypotheses.

The acid trap hypothesis is based on the well-known behavior of weak acids which, dependent on the pKa values, are undissociated in the acidic apoplast of the cell wall and are dissociated in the alkaline symplast compartment of the cytoplasm. The undissociated form has a high lipid permeability, and moves easily across the cell membrane, whereas the dissociated anion is much less permeable. As a result the anionic form of the compound will become accumulated within cytoplasmic compartments, including the sieve elements. This hypothesis is also applicable to endogenous substances, such as IAA and ABA, which are phloem mobile [48] and is the basis of the systemic behavior of many of the phenoxyacetic acids such as 2,4-D and its analogs.

The phloem mobility of pesticides which are neutral or zwitterionic is explained by their intermediate permeability. This hypothesis is based on the lipid permeability of compounds and their residence time within the phloem. Compounds which are very lipid permeable will move in and out of the phloem so readily that they will not be translocated. Conversely, compounds with a low permeability will not enter in sufficient amounts to have an effect. Those compounds with an intermediate permeability will enter the sieve elements and be translocated in sufficient amounts to reach sink organs in an effective amount.

A diffusion model unifying the acid trap and intermediate permeability hypotheses has been developed [49]. This model predicts that membrane permeability is the overriding factor which determines phloem mobility of xenobiotics and that for a given acid dissociation there is an optimum permeability which determines the systemicity or otherwise of the solute [50].

The possibility that xenobiotics may be modified so that they could access the phloem via the sucrose carrier ("piggyback" entry) has been mooted [47], but the high specificity of this carrier for sucrose, referred to earlier in this section, has severely restricted this mode of entry. It does, however, indicate a model for a potential herbicide activity based on sucrose analogs. If an analog of sucrose could be recognized by and bind to the sucrose carrier without being loaded, this would effectively limit the supply of sucrose and other phloem mobile solutes to developing sink tissues. Such a herbicide would have the added advantage of having a low toxicity to animals, which do not possess sucrose carrier systems in their membranes, and thus be more environmentally friendly than most herbicides in current usage.

UNLOADING: SYMPLAST VERSUS APOPLAST

The unloading of photoassimilates is the term given to the events occurring in sink tissues by which the solutes, which have undergone long-distance transport, are transferred from the conducting to the nonconducting cells.

As with the loading process, discussed in the previous section, there are two possible routes for this transfer, the apoplast and the symplast. Whereas in vegetative sinks, apical meristems, and some storage organs, both apoplastic and symplastic pathways are theoretically available, in reproductive sinks, where the developing endosperm or cotyledons are the storage organs, no symplastic continuity occurs between these sporophyte (parent plant) and gametophyte (developing seed) generations. Thus in such reproductive sinks an apoplastic step is an inevitable consequence of this alternation of generations and their associated genetic isolation.

The alternative pathways, apoplastic and symplastic, for unloading are illustrated in Figure 6 where it can be seen that apoplastic unloading may be subdivided according to the site of sucrose hydrolysis. In some instances, such as in the maize kernel, the sucrose is hydrolyzed in the apoplast by an acid invertase and the resultant hexoses are accumulated by the sink tissues, a process presumably mediated by hexose carriers on the plasma membranes of these cells [51]. In other cases, such as in the wheat grain, the sucrose is not hydrolyzed in the apoplast and is accumulated in the endosperm where direct synthesis to starch occurs with no intervening hydrolysis [53]. However, in all sink tissues which have been studied, including sucrose storage tissues, there is always a step involving sucrose hydrolysis or starch synthesis which results in the sucrose moving down a gradient from the sieve elements, where its concentration is high, into the sink tissues where its concentration is low. Subsequent reactions may include resynthesis of sucrose within a storage compartment, such as the vacuole, as occurs in sugar beet and cane [53].

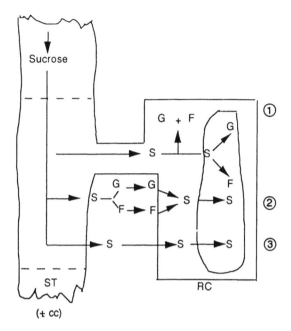

Figure 6 Stylized summary of representative unloading mechanisms. ST, sieve tube; RC, receiver cell or sink cell; CC, companion cell; S, sucrose; G, glucose; F, fructose. The receiver cell's vacuole is represented by an oval, and plasmodesma is shown connecting the sieve tube and receiver cell in type 1. Small circles represent solute carriers. (1) Sugarbeet sink leaves, corn root tips, and bean endocarp, (2) Sugarcane stalks, corn kernels, and sorghum grain, (3) Sugarbeet taproot, legume seeds, and wheat grains.

As will be discussed in the following section, these hydrolytic and synthetic events are key regulatory activities which will affect the strength of particular sinks and thus play a major role in determining crop productivity.

Little is known about the initial release of sucrose from the sieve elements. It is generally assumed that in symplastic unloading, and in instances where a symplastic pathway precedes an apoplastic step as, for instance, in the wheat grain, the sucrose moves out across the plasma membrane in response to the downhill gradient. It is unlikely that a molecule of the size and hydrophilicity of sucrose is able to cross the membrane without the mediation of a sucrose carrier molecule. The presence of an ATPase on the plasma membrane of sieve elements in sink tissues has led to speculation that sucrose unloading may involve a countertransport with protons in response to a gradient of electrochemical potential for protons established by the proton-extruding activity of this pump [54]. An alternative idea is that abscisic acid may reduce ATPase activity in the sieve elements of the sink and that protons cotransport sucrose down a reduced pH gradient [55]. However, as the phloem sap is normally about pH 8 and the pH of the sink apoplast is around pH 6, the large pH gradient reversal required seems unlikely. However, to date such systems are based purely on speculation and there is a dearth of information on the nature of the sucrose efflux from the sieve elements, both in terms of the location of this event and the driving forces involved.

The regulatory processes which govern unloading have been the subject of considerable interest with the realization that manipulation of the key enzymes involved offers a means to influence the quantity and, possibly, the quality of crop yields. From a biochemical viewpoint, unloading commences at the chloroplast envelope, and all the other compartments within the plant, which are unable to carry out photosynthesis, are therefore sinks [56]. Thus the enzymes of the biochemical pathways involved in sucrose synthesis, as outlined in the section "Transport within Photosynthetic Cells" (p. 15), are potential targets for genetic manipulations, as are those involved with phloem loading per se.

Physiologically defined sinks also present a complement of potential target enzymes. The acid invertase found in the sink apoplast is widely distributed throughout the vascular tissues of plants. The ectopic expression of yeast derived invertase in transgenic tobacco plants resulted in a very stunted growth and limited development of sink organs, presumably as a result of the reduced availability of sucrose for translocation [57].

Sucrose synthease, which catalyzes the reversible catabolism of sucrose plus UDP to fructose plus UDP glucose, appears to be active in sucrose hydrolysis in meristematic sink tissues, where symplastic unloading is postulated to occur. This reaction is optimal at pH 7.6 and UDP is restricted to cytoplasmic compartments of the cell. Subsequent catabolism of UDP-glucose by UDP-glucose pyrophosphorylase requires pyrophosphate and it is probable that the supply of pyrophosphate may limit sucrose breakdown by sucrose synthase, thus providing a possible metabolic control of this process. As the subsequent synthesis of cell walls, starch, proteins, and nucleic acids all produce pyrophosphate, it is possible that this supply of pyrophosphate may provide a feedback regulation for further sucrose breakdown [58].

In the wheat grain, and presumably the grains of other C3 grasses, imported sucrose is converted directly into starch by the activity of the enzyme starch synthase within the cytoplasm of the endosperm cells. As the activity of this enzyme is reported to decline as the grain approaches maturity, and the synthesis of starch cannot be enhanced by enhancing the sucrose supply [59], it is probable that the conversion of sucrose to starch, rather than sucrose supply, is the rate-limiting process for grain filling in wheat and related cereals. In maize kernels the hydrolysis of sucrose in the pedicel apoplast precedes hexose uptake and starch synthesis by the endosperm. In this case it is the activity of multimeric enzymes such as ADP-glucose pyrophosphorylase which catalyzes a key regulatory step in starch biosynthesis, and is likely to be the rate-limiting factor for kernel filling.

SOURCE/SINK REGULATION: HORMONAL CONTROL OR TURGOR REGULATION?

Increases in crop productivity can be achieved by improving the partitioning of assimilates, but such developments require a better understanding of the integration of source/sink interactions. The concept of sources and sinks connected by a vascular continuity is illustrated in Figure 7. It must be emphasized however that the source/sink status within a plant is a dynamic one, since a net importing organ may convert to a net exporting one or (rarely) vice versa.

The potential source strength of a leaf will be determined by its photosynthetic capacity as affected by environmental conditions and its developmental stage. The triose phosphates exported from the chloroplast and synthesized into sucrose in the cytoplasm in association with starch synthesis and breakdown can partially account for a short-term control of carbon allocation in the leaf. Subsequently, storage of sugars in the vacuole, or

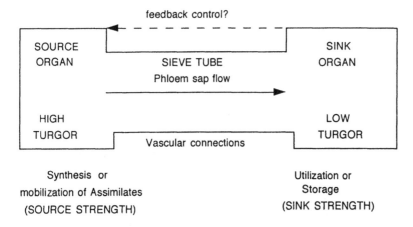

Figure 7 Source/sink relationships for photoassimilates in a transport system. Both source and sink organs interact through the pathway to regulate the balance between supply and demand [61].

utilization in growth and maintenance or export provide control points for longer-term carbon allocation within the plant.

Starch metabolism, its synthesis and breakdown, can affect export but is unlikely to regulate this process except in extreme circumstances, such as when starch is the major carbon source for export during low light or darkness. Some of the carbon exported from source leaves during the night comes from starch breakdown but no suitable mechanism has been identified to explain the regulation of this process [60] and further work is needed to resolve the implications for the control of export. Mechanisms for controlling starch synthesis are known and starch content may be regulated by changes in the synthesis rate with a continuous and constant degradation [60]. However, the weight of experimental evidence indicates an absence of degradation during starch synthesis in the light and favors some control of starch degradation [61]. The production of starch within the chloroplast limits the transfer of a proportion of the fixed carbon for export and thus these two processes, starch synthesis and export, are in competition. Factors affecting starch synthesis are mainly environmental signals, such as day length, which will affect the synthetic process directly or indirectly by changing the activity of sinks or the processes of export [62].

Sucrose synthesis is dependent on a number of processes and will be controlled in the light, when fixation exceeds export, by the fructose 2,6-bisphosphate effector described by Stitt [63]. The synthesis of sucrose is important in controlling export rates, although compartmentalization of the sucrose between vacuole and cytoplasm makes it difficult to accurately measure the amount available for export.

Sucrose loading, discussed earlier in this chapter, is an important determinant of the export rate from source tissues. This process may be controlled at two points, during sucrose release from mesophyll cells into the cell walls and by uptake from this apoplastic compartment across the sieve element plasma membrane. Where symplastic loading occurs some other, and currently unknown, control presumably exists. Of these potential control points there is little evidence for a regulatory role being played by the release into the mesophyll apoplast. The loading of sucrose into the phloem by proton-sucrose cotransport is known to be stimulated by endogenous hormones with an increase in sucrose loading [64]. The potassium status of the source leaf is also a critical determinant, affecting sucrose efflux from the mesophyll and enhancing uptake of sucrose into the sieve elements [65]. However, it is unlikely that the loading process per se limits phloem transport as it is clearly capable of responding to a wide range of available sucrose concentrations and to rapid changes in sink demand [66].

Dark respiration can account for an appreciable proportion of fixed carbon, being as high as 10% of that fixed during the previous light period

[67]. It has been reported that whereas some of this respiratory loss can be accounted for by maintenance respiration, sustaining vital metabolism and energizing the loading process, a portion may be "wasted" by poorly controlled metabolism when some soluble metabolites are in excess [69]. Such losses will affect the export potential of source leaves when available fixed carbon exceeds the current sink demand.

The diurnal production of organic and amino acids, which are themselves exported from source leaves, does not greatly influence carbon export rates [69]. Organic acids, such as malate, play an important role in balancing the accumulated potassium and maintaining the osmotic and pH balance in the leaf mesophyll and in the sieve elements, while the production of amino acids is important metabolically for maintenance and protein synthesis.

Carbon export from a source leaf evidences a number of controls which provide rapid and sensitive responses matching supply to export. These controls maintain the sucrose supply during short-term environmental fluctuations but more persistent environmental changes require longer-term responses and appear to be regulated by the loading process. Such changes will then feed back to affect sucrose and starch production which are then modified to reflect any alteration in export rate.

Many investigators have observed that a proportionality exists between photosynthetic rate and export rate [70]. This relationship between fixation and export is a result of the short-term regulation referred to above and breaks down when fixation is low and reserves mobilized, or when fixation is suddenly increased by environmental changes.

Although most crop plants are source limited there are situations in which carbon fixation exceeds the demands of the major sink organs and secondary storage sinks are utilized [71]. The demand of the sink organs influences export from source leaves by a feedback mechanism operating through the various regulatory steps outlined above. Such a system is very plastic, and when the demand from one sink is altered the redistribution of available assimilates to others may result in overall export from the source leaves remaining constant. Thus it is only in the longer term that any changes in export can be observed. This slow response by the export rate to changes in sink demand presumably prevents overload of the export capacity and allows the various systems within the source leaves to adjust [72].

Plants have adopted widely different strategies of export and storage. Some, such as the C_4 grasses, immediately export most of their recently fixed carbon, while others store a large proportion as starch or soluble sugars for export during the night. Such differences probably reflect differing ecological demands. Storing sugars helps to maintain turgor and could be advantageous to plants exposed to cold or arid environments, or avoid any reduction of starch degradation during cold nights. Plants growing in more temperate regions have maximum growth at night when any water loss is minimized and consequently benefit from a large store of fixed carbon in the form of starch. The use of starch as a storage form is energetically and osmotically more efficient than sugars and reduces the potential exploitation of these plants by pests or pathogens [61].

Throughout the preceding discussion, which has focused on the regulation of export from source tissues, reference has been made to the major role of sink organs in the partitioning of photoassimilates within plants. Individual sink organs may receive assimilates due to a "push" from the source or by a "pull" to meet their demands due to their metabolic activity. The product of this sink activity and sink size gives rise to sink strength which can be equated with the absolute growth rate [70]. This relationship ignores the respiratory losses in the sink tissues, and the term *gross sink strength*, which includes this loss of carbon, has been introduced to compensate for this underestimate [71]. However, measurements of import rates or absolute growth rates reveal nothing of the mechanisms that regulate partitioning, which involve both morphological and biochemical determinants. Sink activity must be estimated in terms of the rate-limiting steps which include phloem unloading, transport of assimilates within sink tissues (across plasma membranes and tonoplasts where this occurs), and chemical conversion and utilization [61].

Sink organs may be classified into either those involved with utilization or those involved with storage (Table 3). It can be seen that there is considerable variety in terms of the possible control points for the import of assimilates and the enzymes involved in hydrolysis, transport, and synthesis will undoubtedly play major, but sometimes diverse, roles in this process.

The various sink organs within a plant compete for assimilates. The resultant partitioning usually follows an order of priority between the sink organs which often undergoes change during plant development. It is the regulation of this sink competition which determines the partitioning of photoassimilates within plants.

The strength of individual competing sinks reflects both genetic determinants, in terms of cell number and capacity, and regulation by environmental factors, which may be expressed through changes in the levels of endogenous growth regulators which are known to influence sink strength. When a greater understanding of these processes is achieved the potential for improving crop productivity by the regulation of partitioning may be realized.

CONCLUSIONS: FUTURE RESEARCH

In the preceding sections, attention has been focused on the key events associated with the loading and unloading of photoassimilates. An understanding

Table 3 Classification of Sink Organs in Plants According to the Route of Unloading and the Related Metabolic Processes

Sink organs	Routes	Processes
A. Utilization sinks (meristematic tissues)	Symplastic	Hydrolysis of sucrose prior to synthetic and respiratory utilization of hexoses
B. Storage sinks	Symplastic and/or apoplastic	Regulation of unloading and compartmentation of assimilates.
1. *Sugar sinks*		
(a) Sucrose sinks (sugar beet tap root)		H^+/sucrose antiport across tonoplast
(sugar cane stalk)		Sucrose hydrolysis in cell wall and hexose transport across tonoplast, followed by sucrose storage in vacuole.
(b) Hexose sinks		Sucrose hydrolysis prior to hexose storage in vacuole.
(c) Fructan sinks		Sucrose hydrolysis followed by hexose uptake and fructan storage.
2. *Sugar/starch sinks* (tomato)		Hydrolysis of sucrose and accumulation of hexoses and starch.
3. *Starch sinks* (wheat grain)		Sucrose uptake into endosperm and starch synthesis.
(corn kernel)		Sucrose hydrolysis, hexose uptake into endosperm and starch synthesis.

Source: Ref. 61.

of how these processes are regulated must feature high on the list of priorities for future research into assimilate transport within plants. As with many other areas of biological research, our knowledge and understanding of the processes involved is relatively minimal, but, with the development of powerful new techniques, answers can be obtained provided we pose the right questions.

Over the past decade approaches have been focused within the fields of research on photoassimilate transport. These encompass various areas, ranging from agronomy to molecular biology, where the integrative role played by phloem transport within the major processes of the plant has become increasingly recognized. This plethora of information emanating from a wide range of disciplines, now requires careful evaluation to provide an understanding of the key processes involved in the partitioning and compartmentation between source and sink tissues within the plant. Thus the subject of phloem transport has undergone major shifts in emphasis from the time when a few researchers hotly debated the pros and cons of hypothetical long-distance transport mechanisms in a manner reminiscent of the scientists in *Gulliver's Travels* who sought to trap sunlight in bottles when surrounded by photosynthesizing plants.

There are several specific areas in assimilate transport where future research should now appropriately be concentrated. These include the major enzyme-mediated processes which regulate the export and import of assimilates by source and sink tissues, respectively. Here the identification, isolation, and purification of key enzymes, utilizing biochemical techniques, will allow the genes to be identified and cloned, using molecular genetics. Ectopic expression of these enzymes in transgenic plants then provides a powerful means of resolving the question of whether the enzymes which play the major roles in these regulatory processes can themselves be manipulated to adjust the source:sink balance, both in a qualitative and a quantitative sense.

However, a number of features of the long-distance transport system still need to be elucidated. In this sense "long-distance" refers to cell to cell distances and greater. We do not yet understand the regulation of plasmodesmata: are there real pores from cell to cell which can be opened and closed on demand? It is also pertinent to reflect on the fact that sieve-plate pores of the sieve tubes, which seem to be under somewhat similar control, are very poorly understood. We know that their "porosity" seems to increase in response to demand under stable conditions favorable to growth. However, many factors ranging from temperature and alien chemicals to mechanical shock seem to precipitate sealing.

We still do not yet understand how to control the sealing systems in the phloem and this must be one goal for future long-distance transport research. If it can be mastered, not only will it be possible to learn a great deal more about the internal condition of plants in general, but it may even be possible to harvest photoassimilates in a nondestructive manner. We already understand many of the gross aspects of loading and unloading sieve tubes, but much of the fine detail remains to be clarified. A good deal is yet to be discovered about how the contending pathways in apoplast and symplast are utilized under appropriate conditions and in different species. The chemical control of loading and unloading clearly has enzymic, physical, and hormonal aspects which have yet to be defined. In future plants may well be engineered to be susceptible to artificial chemicals, which will be able to enter the plants systemically, to assist in growth control and more efficient harvesting.

There are two auxiliary systems which can at times become very important in the life of the plant: the rays and the xylem. Ray tissue seems to operate simultaneously as a circulatory link between xylem and phloem and as a temporary and seasonal store for photoassimilates. The xylem seems to be utilized as a passive "piggyback" transport system when the transpiration stream is appropriately directed toward target tissues, especially when growth starts to accelerate at the beginning of the growing season. Up to now these transport systems have only been examined in a relatively few species. Until many more have been studied it will be impossible to gauge their importance in plants generally; nevertheless the importance of such studies for understanding the growth of timber in forestry cannot be underestimated.

Obviously, the transport of photoassimilates is of immense importance in controlling crop yield. Potentially large gains in yield should flow from a better understanding and manipulation of the processes involved. An unavoidable by-product of such research is the better understanding of the physiology and biochemistry of higher plants, on which civilization is so heavily dependent.

REFERENCES

1. T. G. Mason and E. J. Maskell, *Annals of Botany, 50*: 23 (1931).
2. H. H. Dixon, Scientific Proceedings of the Royal Dublin Society, *20*: 487 (1933).
3. K. Esau, *Encyclopedia of Plant Anatomy, 5,* Gebruder Borntraeger, Berlin and Stuttgart, p. 38 (1969).
4. B. Hylmo, *Physiol. Plant., 6*: 333 (1953).
5. J. S. Pate, *Encyclopedia of Plant Physiology, 1* (M. H. Zimmermann and J. A. Milburn, eds.), Springer Verlag, Berlin, Heidelberg, and New York, p. 451 (1975).
6. J. S. Bonner, *Plant Biochemistry,* Academic Press, New York, p. 328 (1950).
7. J. W. Marvin, *The Physiology of Forest Trees* (K. V. Thimann, ed.), Ronald Press, New York, p. 95 (1958).
8. J. J. Sauter, W. Iten, and M. H. Zimmermann, *Can. J. Bot., 51*: 1 (1973).
9. J. A. Milburn and P. E. O'Malley, *Can. J. Bot., 61*: 3100 (1984).
10. M. T. Tyree, *Plant Physiol., 73*: 277 (1983).
11. J. S. Pate, *Transport of Photoassimilates* (D. A. Baker and J. A. Milburn, eds.), Longman, Harlow, Essex, U.K., p. 138 (1989).
12. E. Münch, *Die Stoffbewegungen in der Pflanze*, Fischer, Jena (1930).
13. J. Kallarackal and J. A. Milburn, *Austral. J. Plant Physiol., 56*: 211 (1984).
14. M. T. Canny, *Phloem Transport,* Cambridge University Press, Cambridge, U.K. (1973).
15. E. MacRobbie, *Biol. Rev., 46*: 429 (1971).
16. J. A. Milburn, *Encyclopedia of Plant Physiology, 1* (M. H. Zimmermann and J. A. Milburn, eds.), Springer Verlag, Berlin, Heidelberg, and New York, p. 328 (1975).
17. J. Passioura and A. E. Ashford, *Austral. J. Plant Physiol., 1*: 521 (1974).

18. J. A. Milburn, *Water Flow in Plants,* Longmans, Harlow, Essex, U.K., p. 168 (1979).
19. M. H. Zimmermann and H. Ziegler, *Encyclopedia of Plant Physiology, 1* (M. H. Zimmermann and J. A. Milburn, eds.), Springer Verlag, Berlin, Heidelberg, and New York, p. 480 (1975).
20. A. S. Crafts, *Plant Physiol., 7*: 183 (1932).
21. J. A. Milburn, *Planta, 95*: 272 (1970).
22. S. M. Hall, D. A. Baker, and J. A. Milburn, *Planta, 100*: 200 (1971).
23. J. Van Die and P. M. L. Tammes, *Encyclopedia of Plant Physiology, 1* (M. H. Zimmermann and J. A. Milburn, eds.), Springer Verlag, Berlin, Heidelberg, and New York, p. 196 (1975).
24. J. S. Pate, P. J. Sharkey, and O. A. M. Lewis, *Planta, 120*: 229 (1974).
25. J. S. Kennedy and T. E. Mittler, *Nature, 171*: 528 (1953).
26. J. S. Pate, M. B. Peoples, and C. A. Atkins, *Plant Physiol., 74*: 499 (1984).
27. R. J. Simpson and M. J. Dalling, *Planta, 151*: 447 (1981).
28. J. A. Milburn and D. A. Baker, *Transport of Photoassimilates* (D. A. Baker and J. A. Milburn, eds.), Longman, Harlow, Essex, U.K., p. 345 (1989).
29. J. Kallarackal and E. Komor, *Planta, 177*: 327 (1989).
30. A. L. Kursanov and M. I. Brovchenko, *Fiziol. Rast., 8*: 270 (1966).
31. D. A. Baker, British Plant Regulator Group, Monograph 12 (B. Jeffcoat, A. F. Hawkins, and A. D. Stead, eds.), p. 163 (1985).
32. H. Ziegler, *Transport in Plants I. Phloem Transport Encyclopedia of Plant Physiology* (M. H. Zimmermann and J. A. Milburn, eds.), Springer Verlag, Berlin, Heidelberg, and New York, p. 59 (1975).
33. D. J. F. Bowling, A. E. S. Macklon, and R. M. Spanswick, *J. Exp. Bot., 17*: 410 (1966).
34. J. Van Die and P. M. L. Tammes, *Transport in Plants I. Phloem Transport, Encyclopedia of Plant Physiology* (M. H. Zimmermann and J. A. Milburn, eds.), Springer Verlag, Berlin, Heidelberg, and New York, p. 196 (1975).
35. A. Ben-Zioni, Y. Vaadia, and W. Lips, *Plant Physiol., 24*: 288 (1971).
36. S. M. Hall and D. A. Baker, *Planta, 106*: 131 (1972).
37. R. Friedman, N. Levin, and A. Altman, *Plant Physiol., 82*: 1154 (1986).
38. M. N. Stitt and H. W. Heldt, *Planta, 164*: 179 (1985).
39. M. N. Sivak, R. C. Leegood, and D. A. Walker, *Transport of Assimilates* (D. A. Baker and J. A. Milburn, eds.), Longman, Harlow, p. 1 (1989).
40. D. R. Geiger, *Transport in Plants 1. Phloem Transport, Encyclopedia of Plant Physiology* (M. H. Zimmermann and J. A. Milburn, eds.), Springer Verlag, Berlin, Heidelberg, New York, p. 395 (1975).
41. M. A. Madore, J. W. Oross, and W. J. Lucas, *Plant Physiol., 82*: 432 (1986).
42. I. F. Wardlaw, *New Phytol., 116*: 341 (1990).
43. S. Delrot, *Transport of Photoassimilates* (D. A. Baker and J. A. Milburn, eds.), Longman, Harlow, Essex, U.K., p. 167 (1989).
44. S. H. Russell and R. F. Evert, *Planta, 164*: 448 (1985).
45. M. A. Madore and W. J. Lucas, *Planta, 171*: 197 (1987).
46. J. Daie and R. E. Wyse, *Physiol. Plant., 64*: 547 (1985).
47. D. A. Baker, British Plant Growth Regulator Group Monograph *18*: 71 (1989).

48. S. M. Hall and D. A. Baker, *Planta, 106*: 131 (1972).
49. D. A. Kleier, *Plant. Physiol., 86*: 303 (1988).
50. F. C. Hsu, D. A. Kleier, and W. R. Melander, *Plant Physiol., 86*: 811 (1988).
51. F. C. Felker and J. C. Shannon, *Plant Physiol., 65*: 864 (1980).
52. A. H. Rijven and R. M. Gifford, *Plant Cell Environ., 6*: 417 (1983).
53. K. T. Glaszion and K. R. Gayler, *Plant Physiol., 49*: 912 (1972).
54. D. A. Baker, K. Poustini, and F. Didehvar, *Recent Advances in Phloem Transport and Assimilate Compartmentation* (J. L. Bonnemain, S. Delrot, J. Dainty, and W. J. Lucas, eds.), Ouest Editions, Nantes, p. 286 (1991).
55. W. Tanner, *Phloem Loading and Related Processes* (W. Eschrich and H. Lorenzen, eds.), Gustav Fischer Verlag, Stuttgart and New York, p. 349 (1980).
56. A. Herold, *New Phytol., 86*: 131 (1980).
57. L. Willmitzer, *Recent Advances in Phloem Transport and Assimilate Compartmentation* (J. L. Bonnemain, S. Delrot, J. Dainty, and W. J. Lucas, eds.), Ouest Editions, Nantes, p. 10 (1991).
58. W. Claussen, B. R. Loveys, and J. S. Hawker, *Physiol. Plant., 65*: 275 (1985).
59. E. W. R. Barlow, G. R. Donovan, and J. W. Lee, *Austral. J. Plant Physiol., 10*: 99 (1983).
60. M. N. Stitt, *Storage Carbohydrates in Vascular Plants: Distribution, Physiology and Metabolism* (D. H. Lewis, ed.), SEB Seminar Ser. 19, p. 205 (1984).
61. L. C. Ho, R. I. Grange, and A. F. Shaw, *Transport of Photoassimilates* (D. A. Baker and J. A. Milburn, eds.), Longman, Harlow, Essex, U.K., p. 306 (1989).
62. S. J. Britz, *Phloem Transport* (J. Cronshaw, W. J. Lucas, and R. T. Giaquinta, eds.), Alan R. Liss, New York, p. 527 (1986).
63. M. N. Stitt, *Plant Physiol., 84*: 201 (1987).
64. F. Malek and D. A. Baker, *Plant Sci. Lett., 11*: 233 (1978).
65. D. A. Baker and S. Chaudry, *Phloem Transport* (J. Cronshaw, W. J. Lucas, and R. T. Giaquinta, eds.), Alan R. Liss, New York, p. 77 (1986).
66. R. I. Grange, *J. Exp. Bot., 36*: 1749 (1985).
67. J. Azcon-Bieto, *Advances in Photosynthesis Research* (C. Sybesma, ed.), Nijhoff/Junk, the Hague, The Netherlands, p. 905 (1984).
68. L. C. Ho, *Ann. Bot., 43*: 437 (1979).
69. J. H. Thorne and H. R. Koller, *Plant Physiol., 54*: 201 (1974).
70. J. Warren Wilson, *Crop Processes in Controlled Environments* (A. R. Rees, K. E. Cuckshull, D. W. Hand, and R. G. Hurd, eds.), Academic Press, London and New York, p. 7 (1972).
71. A. J. Walker and L. C. Ho, *Ann. Bot., 41*: 813 (1977).
72. L. C. Ho, *Phloem Transport* (J. Cronshaw, W. J. Lucas, and R. T. Giaquinta, eds.), Alan R. Liss, New York, p. 317 (1986).

2

Plant Mineral Nutrition in Crop Production

Jacques Le Bot,* David J. Pilbeam, and Ernest A. Kirkby

The University of Leeds
Leeds, England

INTRODUCTION

The tissues of higher plants are made up largely of organic compounds which contain carbon, hydrogen, oxygen, and nitrogen. Carbon and oxygen atoms in the form of CO_2 are taken up from the atmosphere and fixed, together with hydrogen and oxygen from water, in the dark reactions of photosynthesis. Nitrogen is absorbed as ions (NO_3^- or NH_4^+) from the soil solution, or in the case of nodulated legumes may be fixed as molecular N_2 from the atmosphere. In addition to these elements plants also contain a number of other mineral elements which are present in lower concentration but which are essential for normal plant growth. These mineral elements include potassium, phosphorus, sulfur, calcium, magnesium, iron, manganese, copper, zinc, molybdenum, boron, and chlorine (Table 1).

These elements are taken up from the soil mainly as inorganic ions. Some are assimilated (e.g., SO_4^{2-}), or incorporated into organic molecules in essentially unchanged form (e.g., $H_2PO_4^-$ and Mg^{2+}). Others remain in cells to activate enzymes, as osmotica, or simply as counterions to neutralize inorganic or organic ions with the opposite charge. The subject of plant mineral nutrition covers the uptake, assimilation, and transport of all the mineral elements both in natural and agricultural ecosystems, and has been

---

Current affiliation: Centre de Recherches Agronomiques, Montfavet, France

Table 1 Average Composition of Plant Tissue (% in Dry Matter)

Carbon (C)	42%	Nitrogen (N)	2%	Calcium (Ca)	1%
Oxygen (O)	44%	Phosphorus (P)	0.4%	Magnesium (Mg)	0.5%
Hydrogen (H)	6%	Potassium (K)	2%	Sulfur (S)	0.5%

Micronutrients: Fe, Zn, Mn, B, Cu, Mo, Cl in total 1%

Other elements ("beneficial") Na, Co, I, Al, Si, . . .

an important area of interest since Liebig first showed that plants assimilate inorganic compounds and do not require organic molecules and "life force."

From our understanding of plant nutrition has come the widespread application of fertilizers to field crops, where particular nutrients are known to be in comparatively short supply or where fertilizer application can enhance crop yield or quality. It has also led to the development of hydroponics, where plants are grown in aqueous media containing dissolved mineral salts.

There are two particularly important concepts that have governed the use of fertilizers and hydroponics: (1) Sprengel's Law of the Minimum states that the growth of a plant is limited by the supply of whichever element is least available in relation to the plant's demand for it. This has led to the formulation of "ideal" mixtures of elements for different crops, mixtures that are approximately the proportions in which the elements would be present in each crop species if it was grown in unlimiting concentrations of all the essential elements. (2) The Law of Diminishing Returns, attributed to Mitscherlich [1], demonstrates that for a low rate of supply of a nutrient, addition of an increment can give a big increase in yield, but for every additional increment supplied the increase in yield is progressively smaller. This shows both that the biggest increase in yields can be obtained by supplying fertilizers to crops that would otherwise be growing in nutrient-poor soils and also that supplying fertilizers to nutrient-rich soils can give rise to wastefully high concentrations of nutrients in the soil.

From these two concepts it can be seen that there is a demand for knowledge about the optimum rates of supply of different nutrients that give maximum plant yields without undue waste of fertilizers. This is especially important now that an excess of some nutrients in soils has been recognized as causing environmental problems. Furthermore, there is considerable interest both in how a shortage of one or more nutrients affects the growth of plants when other nutrients are adequately supplied, and how excesses of nutrients influence plant growth. Given that there is now the means to study plant physiology at the molecular level it is hardly surprising that these topics are considered in a range of organizational levels, from ecosystems to organelles. In this review we will concentrate on three key aspects of plant mineral nutrition as the subject is currently being studied:

1. The physiology of nutrient uptake, translocation, and assimilation, and how this relates to the growth of whole plants.
2. How plants respond to imbalances in the supply of elements and how this affects their growth.
3. How conditions in the rhizosphere affect nutrient availability, and how plants themselves influence these conditions.

PHYSIOLOGY OF NUTRIENT ASSIMILATION

Uptake Mechanisms

Nutrient ions are taken up by plants from hydroponic solutions or soils, where they occur in a wide range of concentrations. Some gaseous compounds, or nutrients supplied in liquid fertilizers, may be taken up through the foliage, but it is the plant root that provides the usual site for entry of mineral nutrients.

It was the work of Hoagland and co-workers [2] on the uptake of mineral ions by algae that laid the foundations for understanding ion uptake, both in lower and higher plants. The important conclusions from their work were that plant cells can accumulate nutrient ions to a higher concentration than they are present in the external medium, that the accumulation requires the expenditure of metabolic energy, and that the uptake is selective between ions [3].

Accumulation of ions against an electrochemical gradient is energy-dependent, and is thought to be facilitated by carriers. These carriers are believed to be proteins that change their conformation in the presence of an energy source, and transfer an ion from one side to the other of the plasmalemma of a root epidermal cell in a manner that can be described by the Michaelis-Menten kinetics used for enzyme-catalyzed reactions. Living cells are negatively charged as compared with the outer medium, so that anions entering cells must do so by an active mechanism, regardless of their external concentration. Conversely, cations may be taken up as a result of electrostatic attraction, although in the case of K^+ both passive and active uptakes are involved. Diffusion of ions through the plasmalemma of root epidermal cells from high external concentration, or down an electrochemical gradient, may occur through channels in the membrane, and it is currently thought that carriers may occur in these channels so that both active and passive uptake occur at the same site [3,4].

There often appear to be at least two, and possibly more, phases to uptake. The rate of uptake increases with increasing concentrations of nutrient supplied to a plant, until a saturating concentration is reached and the rate of uptake reaches a V_{max} value. At higher concentration of the nutrient in the medium, however, another saturation curve occurs and another V_{max}

is reached. These dual systems of ion uptake were first described by Epstein [5], and a good example is given by nitrate uptake. In *Arabidopsis thaliana* (L.) Heynh. one system operates between 0.01–0.05 mol m^{-3} nitrate (V_{max} = 4.0 μmol NO$_3^-$g^{-1}h^{-1}; K_m = 0.04 mol m^{-3} nitrate), with another operating at higher concentrations of nitrate (V_{max} = 700 μmol NO$_3^-$g^{-1}h^{-1}; K_m = 25 mol m^{-3} nitrate) [6]. The uptake of nitrate is also an example of where multiple uptake systems are claimed to exist. Such a multiple system has been reported in *Phaseolus vulgaris* L. [7], but the possible existence of such systems is still a matter of controversy.

In the same way that enzyme reactions can be inhibited by analogs of the substrate, carriers can often by inhibited by ions similar in size and charge to those normally transported across membranes. Each ion appears to have its own site of uptake and may interfere with the uptake of other ions at different sites. For example, Cl$^-$ and NO$_3^-$ ions are generally antagonistic to each other [8]. These antagonistic effects may be due either to nonspecific replacement effects or to direct competition for carrier sites (e.g., K$^+$/Rb$^+$).

Rate of uptake of ions is largely dependent on the concentration of the ions at the root surface. For example, barley (*Hordeum vulgare* L.) seedlings transferred from a range of nitrate concentrations from 10 to 250 mmol m^{-3} to a range of concentrations up to three times higher, showed big increases in the rate of nitrate uptake [9]. It is probable that uptake of nitrate from low concentrations occurs by means of a constitutive carrier, and when plants are transferred to high concentrations of nitrate the increasing amounts of nitrate entering the root by this carrier induce a high capacity system [10]. In barley roots induced by nitrate there is a nitrate uptake system with a K_m for nitrate of 7 mmol m^{-3} that saturates at 40 mmol m^{-3}, another system with a K_m of 34–36 mmol m^{-3} that saturates at 200 mmol m^{-3}, and possibly a linear system above 500 mmol m^{-3} [11].

Plants show increases in the rates of uptake of ions following nutrient deprivation. Barley seedlings deprived of phosphorus and then resupplied after several days showed higher rates of uptake of phosphate than control plants continuously supplied. Similar observations were made for nitrogen, sulfur, and chlorine [12]. This higher uptake results from either an increase in the number of binding sites or a rise in activity of sites, or both these factors. This short-term response enables plants to compensate for temporary shortages of individual nutrients.

Acquisition, Uptake, and Assimilation of N, P, and K

The most important three elements in determining the potential yields of crops are nitrogen, phosphorus, and potassium, and there is a strong positive correlation between a country's average cereal yield and its average rate of application of N, P, and K fertilizers [13]. Because these three components are the major constituents of fertilizers worldwide it is worth considering the uptake of N, P, and K by plants in some detail.

Nitrogen is taken up by the roots of plants as the nitrate (NO_3^-) ion, the ammonium (NH_4^+) ion, or, in the case of plants that form root nodules, as atmospheric dinitrogen (N_2). Root nodules are considered elsewhere in this volume, and so the uptake of N_2 will not be covered here.

Nitrogen

Nitrogen is a very mobile element that circulates between the atmosphere, the soil, and living organisms. In soil it is present mainly as organic molecules in living organisms or dead matter, but it also occurs as ammonium, nitrite, and nitrate ions. In the nitrogen cycle amino N is released from organic molecules by proteolysis, and is converted to NH_3 by ammonification. The NH_3 formed is oxidized to NO_2^-, and then to NO_3^-, in the reactions of nitrification, reactions which like those of ammonification are carried out by microorganisms. Nitrate and ammonium may also arise from the addition of fertilizers and from lightning discharges and vehicle emissions. Nitrate may be lost from the soil by the action of denitrifying bacteria or by leaching, and both ammonium and nitrate are lost from the soil by being taken up by plants. Further details of the nitrogen cycle are given elsewhere [14], but it is important to note that the key reactions are carried out by microorganisms.

Environmental conditions that affect these microorganisms thus have a strong impact on the availability of NH_4^+ and NO_3^- ions. Nitrification reactions require oxygen, and so aerobic conditions are required for the oxidation of ammonium to nitrate. Low soil pH inhibits microbial nitrification [15], and low temperatures inhibit both ammonification and nitrification. Denitrification is promoted by high soil moisture, neutral pH, high soil temperatures, and a low rate of oxygen diffusion so that obligately aerobic bacteria use NO_3^- as the terminal electron acceptor of electron transport during respiration [14]. From these differing responses to environmental conditions it can be seen that both low and high soil temperatures can give rise to soils with low concentrations of NO_3^- ions. Nitrate is easily washed from soils because the NO_3^- ion is highly mobile. On the other hand NH_4^+ ions are attracted to negative charges of the clay and humus colloids so that ammonium is much less mobile.

The concentrations of both ions are obviously greatly increased by application of ammonium or nitrate fertilizers, and also by application of organic fertilizers that slowly break down under the action of the bacteria operative in the nitrogen cycle. An additional source of fixed nitrogen arises from nitrifying free-living microorganisms, which can convert atmospheric

N_2 into organic forms of N. Under temperate conditions these organisms make little contribution to N supply for crop growth. However, under tropical conditions as much as 60–90 kgN ha^{-1} can be fixed, and it has been suggested that nitrogen supply to crop plants can be enhanced by inoculation with nitrifying bacteria [16]. In wetland cultivation of rice it is now standard practice in many countries to grow the aquatic fern *Azolla pinniata* in paddy fields, so that the organic N compounds (up to 30 kgN ha^{-1} year^{-1}) formed from N_2 by the blue/green alga *Anabaena azollae* that lives in association with the fern are released into the water when the fronds die [17].

For crop plants that grow in agricultural soils the NO_3^- ion is the form of nitrogen most commonly taken up, although there is also some uptake of NH_4^+ ions. In plants adapted to growing in acid soils, where NH_4^+ is predominant, the NH_4^+ ion is the preferred N source. The comparative uptake and assimilation of these two ions has recently been discussed elsewhere [18], and the physiology of nitrate uptake and assimilation has also been the subject of recent reviews [19].

Nitrate in the soil solution is taken up by plants through the carrier system already described, with either accompanying uptake of a proton (NO_3^-/H$^+$ symport) [20] or extrusion of an OH$^-$ ion (NO_3^-/H$^-$ antiport) [21]. Current views favor the NO_3^-/H$^+$ symport for both the high– and low–affinity uptake systems, with 2 protons taken up for every nitrate ion [22]. The uptake of ammonium occurs by means of an electrogenic transporter, with the exchange of protons for NH_4^+ ions [23]. Like NO_3^- uptake, there may be biphasic or even multiphasic uptake systems. For soybean (*Glycine max* [L.] Merr.), for example, there appear to be three phases [24]. Unlike the nitrate carrier, the NH_4^+ carrier does not seem to be inducible.

After uptake ammonium is assimilated by the same reactions irrespective of whether it is taken up from the soil or arises from the internal reduction of nitrate. Whereas the shoot is often the major site of nitrate reduction [25], the roots appear to be more important in ammonium assimilation. From experiments on *Beta vulgaris* L. grown at a range of pH values it seems likely that NH_4^+ ions taken up from the soil are mainly assimilated in the roots with the shoots only being a significant site of assimilation in plants grown in media with pH values greater than 6 [26]. However, even where NH_4^+ is mainly assimilated in the roots some release and reassimilation of NH_3 occurs during photorespiration in the leaves [27].

The complexity of the reactions of uptake and assimilation of ammonium and nitrate gives rise to many sites where regulation of nitrogen assimilation can occur (Figure 1). The two enzymes of conversion of nitrate to ammonium, nitrate reductase, and nitrite reductase, may be subjected to feedback inhibition by ammonium or amino acids (the end products of ammonium assimilation) or both [28–30], although amino acids have very little

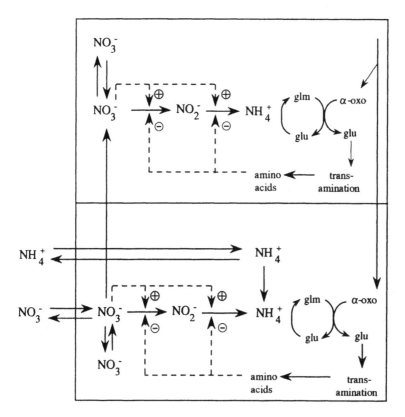

Figure 1 The assimilation of nitrate and ammonium in plants.

effect on in vitro nitrate reductase activity [31]. Both reductases are induced by nitrate in the rooting medium [32], and so too is the nitrate carrier.

This combination of controls ensures that plants optimize nitrate use as follows: (1) nitrate is actively "pumped" into the roots and "leaks" out [33], so that the rate of influx is proportional to the external concentration [34]. (2) With higher concentrations of nitrate in the rooting medium more nitrate is taken up as the carrier is induced, but the efflux component does not necessarily increase as nitrate and nitrite reductases are both inducible, particularly in the shoot. In plants grown in low concentrations of nitrate a large proportion of nitrate assimilation occurs in the roots, but for plants grown in high concentrations of nitrate a larger proportion of the assimilation occurs in the shoots [35,36]. The activity of nitrate reductase in leaves appears to be proportional to the flux of nitrate in the xylem [37], so that the additional nitrate taken up can be rapidly assimilated. (3) The rate of nitrate assimilation slows if the amino acids formed are not incorporated into proteins, and accumulate at their sites of synthesis.

A further factor controlling the assimilation of nitrogen is the availability of energy, because uptake and the assimilation stages all require energy. The reduction of nitrate to nitrite requires NADH and the reduction of nitrite to ammonium requires reduced ferredoxin. The synthesis of glutamine from glutamic acid and ammonium requires ATP, and the subsequent formation of two molecules of glutamate from 2-oxoglutarate and glutamine requires either NADH or reduced ferredoxin. Furthermore, the supply of 2-oxoglutarate itself represents a drain on the carbon economy of the plant, and so requires energy. Therefore, it is not surprising that both assimilation and uptake of nitrate show strong diurnality, with most assimilation occurring during the light period [36,38]. The control of this diurnal rhythm is even more precise, however, than would be brought about by changes in the availability of energy sources within plants. During a light/dark cycle, nitrate reductase mRNA increases rapidly before the end of the dark period, reaching a maximum at the start of the light period. The amount of the nitrate reductase protein reaches a maximum approximately 2 h after this [39]. Light also markedly enhances the accumulation of nitrate reductase mRNA that occurs during induction of the enzyme by nitrate [40] and there is a requirement for both light and nitrate in order for active nitrate reductase to be synthesized [41].

Even though uptake and assimilation of nitrate appear to be regulated in such a way that they ought to be perfectly in balance, this is not always the case. Under some circumstances "luxury consumption" of nitrate occurs, where the rate of uptake exceeds the rate of assimilation, and the excess nitrate is sequestered in storage pools [42]. When plants grow with a limited supply of nitrogen they have more capacity for the production of carbon skeletons than is required, and nonstructural carbon compounds accumulate [43].

The accumulation of carbon compounds cannot occur indefinitely because although nitrogen assimilation is dependent upon photosynthesis, the photosynthetic apparatus is the site of a large proportion of the total protein content of a plant, and thus requires reduced nitrogen compounds. In particular the enzyme ribulose bisphosphate carboxylase/oxygenase is a major site of reduced N in plants, accounting for 40–80% of the total soluble leaf protein in a C3 plant [44]. It is considered to be the rate-limiting enzyme in C3 photosynthesis, and its concentration was shown to increase by approximately 120% in response to nitrogen application to potato (*Solanum tuberosum* L.) [45]. It has also been shown that expression of genes coding for some enzymes of C4 photosynthesis is regulated by nitrogen availability [46].

In discussing the molecular biology and biochemistry of nitrogen assimilation we have not dealt with the relationships between the acquisition, uptake, and assimilation of nitrogen and the growth of whole plants. It has already been mentioned that the assimilation of nitrate occurs predominantly in the shoots of plants, so that the rate of leaf development has a big effect on the capacity of a plant to assimilate nitrate. The availability of nitrogen strongly influences leaf area, so that plants with insufficient supply of nitrogen have small leaves [47]. An important reason for supplying adequate nitrogen to plants is to lengthen the life of the leaves, so that leaf area duration is maintained [48]. This gives rise to a greater production of the products of photosynthesis, and over a longer period, but is also important for metabolism of nitrogen itself: nitrate reductase activity increases in young leaves, and then gradually declines as leaves age. In a plant in the exponential phase of growth the increase in the number of new leaves developing more than offsets the senescence of older leaves, so that the total capacity of the leaf system to reduce nitrate also increases exponentially [49].

Because the supply of nitrogen to plants maintains the leaf area, plants well supplied with nitrogen have higher shoot:root ratios than plants poorly supplied [50]. If the nitrogen supply is interrupted, the rate of shoot growth quickly decreases, whereas the rate of root growth remains constant [51]. In a split-root experiment on wheat, Lambers et al. (52) showed that for optimal or supraoptimal supply of nitrate to one half of the root system, if the other half of the root system received no nitrate the length of its main axis was considerably greater. For supply of nitrate at one-quarter of the optimal rate the lengths of the main axes of both halves were greater. At the low rate of supply the fed half of the root system had more laterals but the number of laterals on the starved half showed very little difference between treatments.

From these findings it can be seen that where parts of root systems are in areas of soil where nitrate concentrations are very low, growth of the main axis fueled by uptake of nitrate from other parts of the root system can carry the roots into unexplored areas of soil [53,54]. This is obviously an important mechanism for maximizing the uptake of nutrients because the rate of nitrogen uptake is proportional to root length. At high rates of N supply there is an almost constant relationship between nitrogen uptake and root length per plant, but at low rates of N supply there is a progressively higher rate of uptake with increasing length [50].

Increases in root:shoot ratio with lower rate of supply of N can be seen from experiments of Larsson and co-workers in which barley plants were grown with different rates of relative addition of nitrate [55]. The use of relative addition is derived from the formula that describes the change in the nitrogen content of plants with time, $N_t = N_o e^{RAt}$ where N_t and N_o are the nitrogen contents of plants at harvest and t = o respectively, t is the time to harvest and RA is the relative rate of nitrogen addition to the plants. By choosing different values of RA, plants can be grown with different rates of supply of nitrogen, although for each value of RA ever-increasing amounts of nitrogen are supplied each day during the exponential phase of growth, and so steady-state conditions should ideally occur. At low rates of RA the relative growth rates of the barley roots were lower than at higher rates of supply, but at the low RA, values were higher than RA itself. This is counter to the theory of the relative addition technique, so that these plants were not in the state of balance expected and a greater proportion of plant growth occurred in the roots. V_{max} of nitrate uptake increased with increasing RA up to approximately 50% of the maximum RA value, and then decreased (Figure 2A). This could be construed as being caused by induction of the nitrate carrier, but when V_{max} was plotted against nitrogen concentration in the root there was a very similar relationship to root RGR plotted against root N concentration at the lower values, and so V_{max} must have been strongly related to total root protein rather than specific carriers (Figure 2B). At higher values of root N concentration (brought about by higher values of RA), V_{max} decreased, and so here the accumulation of nitrate in the root presumably caused feedback inhibition of uptake [55].

When V_{max} was transformed into relative units (nitrogen taken up per unit nitrogen in the plant and unit time), at low values of RGR of the barley plants, the values of relative V_{max} were up to 30-fold higher than RGR (Figure 2C). Therefore, although absolute V_{max} (uptake of N per gram root-dry weight per day) may have been comparatively low, the increased relative size of the root system more than compensated for this, and low internal nitrogen concentrations allowed the uptake of large amounts of nitrate [55]. Relative V_{max} at growth limiting rates of nitrate supply is considerably in excess of the demand set by growth, whereas when nitrate supply is saturating, relative V_{max} is lower and nearly equals demand [56]. This shows that at low rates of nitrate supply the slow growth of the plants limits uptake and not vice versa.

As has already been pointed out, one of the short-term effects of fluctuations of supply of nitrate to plants is luxury consumption of the ion. When there is a rapid increase in the supply of nitrate to plants the ion may accumulate in storage pools, and these storage pools can then buffer the plants against decreases in the rate of supply of nitrate. In barley plants transferred to a nitrogen-free medium from a nitrate-containing medium, both the concentrations of nitrate in the plants and the rate of nitrate reductase activity decreased, but the rate of decrease of enzyme activity was slower in plants that had larger storage pools of nitrate at the time of nitrate withdrawal [57]. Even when the rate of reduction of nitrate has declined to almost zero, however, plants appear to maintain a large amount of potentially active nitrate

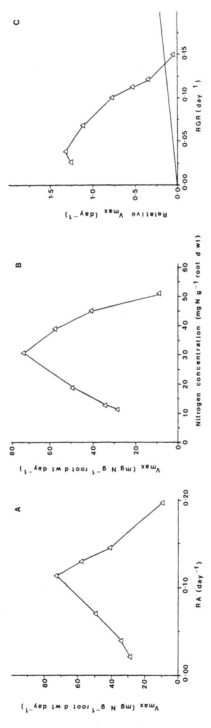

Figure 2 Kinetic properties of nitrate uptake by barley (cv Golf). (A) Relationship between V_{max} and relative addition rate. (B) Relationship between V_{max} and nitrogen concentration in root. (C) Relationship between relative V_{max} and relative growth rate (straight line represents relative V_{max} = RGR). Redrawn from [55].

reductase enzyme in the leaves, so that when nitrate is resupplied (as roots grow into areas of soil where nitrate is present) assimilation of the ion can commence almost immediately [58].

If this luxury consumption serves to protect plants from short-term deficiencies in the supply of nitrate, it can be seen that there is a critical rate of supply of nitrate, above which value there is no long-term effect on plant yield. It has been shown that there is a relationship between nitrate concentration in a plant, nitrogen concentration, and shoot fresh weight [59] (Figure 3).

From the above discussion it can be seen that shoot:root ratio is directly related to the nitrogen content of a plant. It has been suggested that there is a simple equilibrium in distribution of assimilates, particularly carbon and nitrogen compounds [51,60,61]. Plant growth regulators may control shoot:root ratio [62], or it may be regulated by changes in the equilibrium of distribution of assimilates. When there is a low rate of supply of nitrate to the plants, and most of the assimilation of nitrogen occurs in the roots, there will be less metabolic activity in the shoot. The consequent decrease in demand for carbohydrates will lead to an imbalance in the distribution of carbon compounds between shoot and root. According to the Münch theory of carbon translocation in phloem, movements are mainly due to the gradient in concentration between shoots and roots, and so under N deficiency more carbohydrates would be allocated to the roots.

The assimilation of nitrate (or ammonium) in the roots enables amino acids and proteins to be synthesized from these carbohydrates, thereby maintaining root growth. This is evident from experiments on *Zea mays* L. fed on $^{15}NO_3$, where a constant proportion of the ^{15}N taken up was shown to be incorporated into root proteins regardless of the concentration of NO_3 supplied [63]. Under low rates of nitrogen supply the rate of production of proteins and structural carbohydrates in the roots eventually declines as the decrease in metabolic activity in the shoots causes a reduction in photosynthesis and the export of carbon compounds to the roots is lowered. When there is an uneven distribution of supply of nitrate to different parts of a root system, although there is different dry matter allocation to different parts of the system its overall size is unaffected [56].

The growth of crops in the field is directly proportional to the amount of radiation that the crop receives; i.e., the intensity of light seems to be of no direct importance in the field, but the yield of C3 crops in Northern Europe is strongly correlated with the fraction of radiation that is intercepted. Slight increases in the fraction of light intercepted give rise to big increases in yield at low values, but as the factor rises above 0.6 toward 1.0 the increments of extra yield decrease in size [64]. One might expect that the relationship between nitrogen supply and the development of leaves referred to earlier could

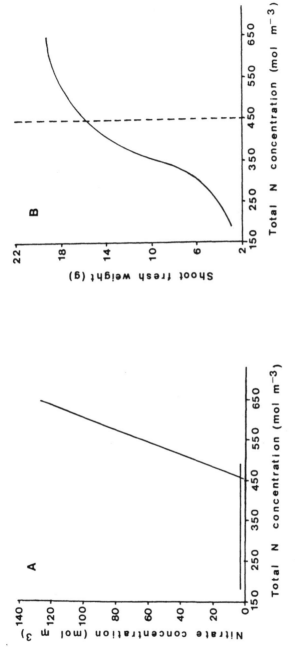

Figure 3 Relationships between nitrate concentration, total N concentration, and fresh weight of shoots of wheat. (A) Relationship between nitrate concentration and total N concentration in wheat shoots. (B) Relationship between shoot fresh weight and total N concentration in shoot. Redrawn from [59].

be of importance in this context, with increased supply of nitrogen leading to enhanced interception of radiation. In experiments where leaves are removed from plants, however, increased photosynthesis in the remaining leaves compensates for their absence (see references in [64]). The relationship between crop yield (W) and fraction of radiation intercepted (F) is mirrored by an identical relationship between W and $W/(1 + W)$; i.e., at any one time the final yield of a crop depends on its weight at that time.

If plants maintain an approximately constant C:N ratio during vegetative growth there will be a direct relationship between the total nitrogen content and the weight of the plant at any time. It has been demonstrated that the rate of photosynthesis of individual leaves shows a positive correlation with the concentration of nitrogen in the leaf, with a minimum value of N concentration below which photosynthesis does not occur [64]. Therefore, if the concentration of nitrogen in leaves is directly proportional to the nitrogen concentration in the whole plant, there will be a positive relationship between the rate of photosynthesis (and growth) and the nitrogen concentration in the plant, at least up to the concentration at which free nitrate accumulates. Despite the differences in growth rates of shoots and roots that occur when plants become nitrogen-deficient, over a growing season growth rates of different plant parts are more in balance. Greenwood et al. [64] give examples of four European crops where the rate of photosynthesis is positively correlated with the nitrogen concentration in the plant.

From experiments in which crop plants are grown with different levels of N fertilizer applications, and yields at different times of the growing season are matched with %N in the plants, it can be seen that there is a critical %N value above which no further increases in yield occur (presumably the concentration above which free nitrate accumulates in storage pools). It has been shown that the relationship between critical %N in the dry matter and yield is that in larger plants the value of critical %N is much lower than in younger plants [64]. This is because as plants get bigger there is an increase in the proportion of structural carbohydrate and a decrease in the proportion of protein-containing cytoplasm, and so in fact there is not a constancy in the C:N ratio over the entire growth of a plant.

Greenwood et al. were able to produce a dynamic model for C3 crops in Northern Europe, that relates changes in yield with time of growth to the weight and the critical nitrogen concentration at each time. Even with adequate supplies of nutrients and water the relative growth rate of plants declines as absolute growth rate increases, and both are affected by the plant mass itself. However, this model produces a growth rate coefficient that is independent of plant mass, and which is linearly related to the ratio of %N in the plant: critical %N for plants with limited N supply.

Greenwood and co-workers have subsequently shown that the critical %N is strongly related to plant mass per unit area in a wide variety of species [65]. Only very small differences have been found in the relationship between C3 crops, with a bigger difference between C3 and C4 crops. As the rate of N uptake per unit of radiation intercepted by both C3 and C4 crops was shown to be similar for crops of the same weight per unit area, values of lower %N seen in the dry matter of C4 crops can be attributed to C4 photosynthesis being more efficient than C3 photosynthesis.

The similarities between C3 crops presumably arise because the component of nitrogen in the plant that is essential for photosynthesis can be related to plant mass per unit area to virtually the same extent in different species [66]. The rate of photosynthesis in individual leaves is proportional to the nitrogen content of each leaf, as already discussed, and if rates of photosynthesis increase in unshaded leaves to compensate for decreases in rate in shaded leaves (see earlier discussion) it is reasonable to assume that organic nitrogen is distributed in a plant to maximize the rate of photosynthesis of the whole canopy [66]. Once the canopy has reached maximum size, plants continue to grow at a constant rate, so that the proportion of photosynthetic material decreases, and the proportion of structural material increases. As the structural material is mainly carbohydrate, the %N in the plant decreases with age, and the small differences between species in the relationship between rate of dry matter production and mass per unit area are presumably due to differences in the distribution of nitrogen between metabolic and structural components of dry matter [66].

Phosphorus

In general N, P, and K can all produce large increases in crop yields [13], but larger increases in yield can be brought about by applications of smaller amounts of P than K, and by smaller amounts of K than N, although at higher concentrations all produce similar increases. The absorption of phosphorus by plants continues until late stages of growth, when the uptake of nitrogen and potassium is declining, and so the use of P fertilizers gives increased crop yields almost to the end of the growing season.

Phosphorus is taken up as phosphate ($H_2PO_4^-$) by means of an active mechanism with H^+ cotransport or HCO_3^- antiport properties [67]. Within 10 minutes of uptake 80% of the phosphate absorbed has been shown to be incorporated into organic compounds, mainly hexose phosphates and uridine diphosphate [68]. In plant cells the phosphorus occurs as orthophosphate, and to a minor extent pyrophosphate, and is not further metabolized but is transferred to different molecules essentially unaltered. This gives rise to at least two pools of phosphorus in plants, organic and inorganic phosphorus, and it has been suggested that the partitioning between these pools is a major determinant of the phosphorus efficiency of different plants [69].

In order to be available to plants, phosphate has to be dissolved in the soil solution. There is also a fraction of phosphate that is solid, and can rapidly exchange with the solution phosphate (the labile phosphate fraction) and there is an insoluble fraction that only exchanges with the labile fraction very slowly [14]. In some soils phosphate is adsorbed very strongly, and the availability of phosphate to plants is limited. Anions are adsorbed more strongly at low pH than at high pH, and thus phosphate in low supply in acid soils becomes more available following liming. The composition of the soil also affects the degree of adsorption, with aluminium and iron oxides being strongly adsorptive. In general oxisols (red tropical soils) adsorb large quantities of phosphate, whereas podzols have a lower capacity for adsorption. Therefore, oxisols require larger amounts of P fertilizers in order for the concentration of phosphate in the soil solution to be adequate for vigorous plant growth, but they are better buffered against P deficiency. Rapidly growing crops absorb approximately 1 kg P ha^{-1} day^{-1}, and as soil P concentration is typically between 0.3 and 3.0 kg P ha^{-1}, the solution phosphate has to be rapidly replenished [14]. As the concentration of phosphate in the soil solution drops due to uptake of phosphate by plants, desorption of phosphate occurs and so the immobilized phosphate becomes an important source of phosphorus for plant growth. In order to compensate for the adsorption of phosphate into this immobilized fraction, however, the application rate of P needs to be 10–15% higher than crop demand for phosphorus [14].

The availability of phosphorus in soils is a crucial factor in determining the rate of growth of plants, and in an experiment on lettuce (*Lactuca sativa* L.) plants grown in 13 different U.K. soils there were up to threefold differences in the rates of growth of the young plants due to differences in phosphate availability [70]. Differences in the growth rates of the young plants gave rise to big differences in the yields at harvest, and so it was suggested that phosphate should be provided as a "starter" fertilizer to vegetable crops to accelerate early growth of the seedlings [70].

When plants are deprived of phosphorus, changes occur in the root system to enable more uptake to occur. As already mentioned, phosphorus-deprived plants can take up phosphate at a faster rate after resupply [12], and as with nitrogen deprivation, roots grow longer so that a larger volume of soil is penetrated [71]. The effects of phosphorus nutrition on plant growth and shoot:root ratio have recently been discussed by Sanders [72].

The enhanced growth of plants with P fertilization is brought about by a promotion of metabolic activity. In *Phaseolus vulgaris* there is a positive correlation between the rate of photosynthesis per unit leaf area and the concentration of phosphorus in the primary leaves [73], so that when plants are well supplied with phosphate their growth rate is higher because photosynthesis is more efficient.

In a comparison of phosphorus-efficient and phosphorus-inefficient lines of *Phaseolus vulgaris* (i.e., a line with a high dry weight yield and a line with a low dry weight yield with low P supply, respectively) the phosphorus-efficient line showed a bigger increase in photosynthesis in response to increased P concentration in the primary leaves than the phosphorus-inefficient line [73]. When the plants were grown with a low rate of P supply, however, both shoots and roots of the inefficient line contained a much higher concentration of P than the efficient line. If the concentrations of P in the primary leaves were the same as the values reported for the whole tops, the higher concentration of P in the leaves of the inefficient line combined with the lower photosynthetic efficiency would have given approximately the same rate of photosynthesis per unit area of leaf in the two lines. As the plants of the phosphorus-efficient line were much larger, the total rate of photosynthesis per plant would have been higher at any time, but the initial cause of the increased growth is less easy to identify. As efficient plants contained a lower concentration of phosphorus in their larger mass the rate of absorption of phosphate appeared to be less important than the efficiency of its use in contributing to the growth of the plants [73].

In another study on *P. vulgaris* [74], lines that produced only low yields with P deprivation had higher leaf area indices and higher leaf weights at 21 days after germination than phosphorus-efficient lines, although these differences had disappeared by flowering. In this study, however, efficiency related to yield of fruits at harvest, and as the inefficient lines had produced more stem weight than the efficient lines by flowering, and they accumulated more phosphorus in the roots, the allocation of phosphorus and photosynthetic products to different plants parts may be an essential ingredient of phosphorus efficiency.

Potassium

Like phosphate, potassium ions are adsorbed to soil particles, in this case by ion exchange reactions, and the amount of adsorbed K^+ determines the buffer capacity of the soil. K^+ ions readily pass through plant membranes, and uptake occurs largely by facilitated diffusion with exchange of protons into the soil solution [75]. At low K^+ concentrations in the soil solution an active uptake process that shows Michaelis-Menten kinetics occurs [76].

Within plant cells K^+ ions are held by the negative electrical potential that exists. It is a very mobile ion, however, presumably because of the permeability of membranes. Most potassium is taken up during vegetative growth, and it has particularly important roles as a counterion in the transport of anions in xylem and phloem, in water relations of plants, in meristematic growth, and in enzyme activation [14].

As potassium has been shown to cause turgor in young leaves of *Phaseolus vulgaris* [77] and to regulate the opening of stomata [78], it is hardly surprising that one of the first symptoms of potassium deficiency in plants is decrease in turgor. Other solutes can contribute to turgor, however, and during potassium deficiency other cations or sugars and amino acids may accumulate in vacuoles [79]. The vacuole is the main site of K^+ storage in plant cells, and the main role of this K^+ pool is as an osmoticum. Cytoplasmic potassium also contributes to the generation of osmotic potential, but in addition is essential for the activity of some cytoplasmic enzymes, and it is the concentration of K^+ ions in the cytoplasm that determines the rate of growth of plants [80]. During potassium deficiency the K^+ concentration in the vacuole decreases to a minimum possible value without any effect of growth, but then growth is inhibited as the cytoplasmic concentration subsequently decreases. The availability of other cations can affect the critical K^+ concentration for growth as they can substitute for K^+ ions in the vacuole, but in the absence of other cations organic solutes must substitute for K^+ and so are not available for biosynthetic pathways [80].

Some of the main effects of potassium are on photosynthesis, and on the translocation of photosynthates. Where K^+ concentrations in leaves are high there is enhanced photosynthesis, more translocation of photosynthates to the storage organs, more mobilization of leaf proteins into the seeds, and longer duration of leaf area [81]. In plants with a high dry matter production, e.g., sugar beet or potatoes, there is generally a high concentration of K^+ ions in the leaves [82].

The concentrations of K^+ ions in a range of crop plants growing under optimal conditions have been examined in detail, and so it is possible to analyze plant material to check that the critical K^+ concentration is exceeded. However, %K^+ in dry matter of crop plants declines sharply during the growing season, and so even if regular measurements of K^+ concentration are made, prediction of plant potassium requirements needs very precise knowledge of what the ideal concentration is at a particular time of the season. Leigh [79] has shown that the concentration of K^+ ions in tissue water remains relatively constant throughout vegetative and early reproductive growth in a number of crops, and so once the critical K^+ concentration in tissue water has been determined for any crop it becomes comparatively simple to check regularly to see whether this value is exceeded. As soil K^+ supply strongly influences tissue water K^+ concentration, any deficiency can be alleviated by application of fertilizer [79].

As with nitrogen and phosphorus, different species and cultivars show different levels of efficiency of use of potassium, as measured by amount of dry matter produced per unit of K^+ taken up. For example, in barley varieties grown with a low rate of K^+ supply, the varieties with the highest absolute and relative growth rates not only had higher rates of K^+ uptake by the roots, but also had higher efficiencies of K^+ utilization than the slower growing varieties [83]. The higher rates of uptake of K^+ ions were associated with a decrease in shoot:root ratio, and those varieties that were efficient at low rates of K^+ supply had larger and more efficient root systems.

In experiments on *Phaseolus vulgaris* [84] discussed by Mengel [82], there were big differences in dry matter yield of the plants of four varieties across a range of potassium concentrations, although the K^+ concentrations in the dry matter at maximum yields varied very little between the varieties. The efficiency of use of potassium by the plants varied very little, therefore, although some were more efficient at taking up potassium than others. In cereals, where some varieties are thought to have more efficient use of K^+ within the plant than others, the efficiency is expressed in terms of yield of grain per unit of K^+ taken up. In the efficient varieties there is a higher harvest index [82], as is also the case with nitrogen-efficient and phosphorus-efficient varieties.

Potential Improvements in Crop Production from an Understanding of the Physiology of Uptake and Assimilation of N, P, and K

Prediction of Nitrogen Requirement of Cereals by Measuring Nitrate Reductase Activity

Following the work of Hageman and co-workers [85] who demonstrated positive correlations between nitrate reductase activities in leaves of cereals and both grain yield and grain protein content, attempts have been made to use measurements of the activity at different times in the growing season to assess the nitrogen requirement of cereal crops [86]. This will probably prove to be too cumbersome a procedure to be adopted by many farmers.

Prediction of Nitrogen Requirement of Crops by Measuring Internal Nitrate Concentration in Plants

Although it is time-consuming to measure the activity of nitrate reductase it is much simpler to measure the concentration of its substrate, nitrate. Measurements of internal nitrate concentration have successfully been used on cereal and vegetable crops as a means of assessing nitrogen requirement of the crops [87,88]. Although there are simple and rapid tests for the measurement of nitrate it may prove difficult to persuade farmers and growers to adopt this technique in large numbers. However, it is particularly valuable for crops that tend to accumulate high concentrations of nitrate (e.g., spinach) not as a means of assessing fertilizer requirement, but as a means of ensuring that acceptable concentrations of nitrate are not exceeded [89]. Although evidence of risks to the health of adults from nitrate in the diet is not conclusive, the risks of methemoglobinemia in infants from exposure to nitrate in the diet make these tests invaluable in crops that will be sold for use in baby food.

Avoidance of High Concentrations of Nitrate in Plants

Although it is useful for farmers and growers to be able to determine the concentration of nitrate in crop plants before harvest, it is even more useful to be able to minimize any possible accumulation of the ion. The reactions of nitrate reduction occur most rapidly during the light period and it has been shown that illuminating spinach (*Spinacea oleracea* L.) plants at low intensity during the dark period immediately prior to harvest significantly reduces the nitrate content of the leaves [90].

Competition Between Ions for Uptake Sites

The competition for uptake between similar ions has been used to advantage following the disaster at the Chernobyl nuclear reactor in 1986. Application of K-fertilizers to soils of low potassium content has been recommended to lower the amount of ^{137}Cs ions taken up by plants [91].

Measurements of Nutrient Solution pH as Simple Tests for Nutrient Uptake Efficiency

The uptake of K^+ ions is accompanied by extrusion of H^+ ions, and as there is a strong positive correlation between the rate of K^+ influx and the rate of H^+ efflux, it has been suggested that noting the color change of a redox indicator included in a nutrient solution could be a simple means of identifying plant varieties with a fast rate of K^+ uptake [75].

Use of Plant Growth Regulators to Improve the Performance of Enzymes of Nitrogen Assimilation

There are large numbers of plant growth regulators used in agriculture, and many of these may affect enzymes of nitrogen assimilation and other processes. However, "improvement" in the performance of one enzyme may be of no benefit, or even be detrimental, to the growth of plants. At the current time it is difficult to predict what would be the effects on the nutrition of plants of altering the activities of any individual enzyme or carrier, and the development of new uses for plant growth regulators is still very much a "trial and error" process.

Genetic Engineering of Crop Plants Containing Enzymes of Carriers with Desired Activities

A considerable amount of work has been carried out on the genetics of plants in relation to uptake and assimilation of N, P, and K. However, although it is comparatively straightforward to insert a gene into a plant species, adding desirable nutritional abilities is much more difficult. As shown here, the interaction between different aspects of plant growth is so complex that it would be impossible to predict the effects of transferring a carrier protein or assimilatory enzyme into another species. Genetic engineering will not enable the production of nutrient-efficient plants in the near future, but the ability to alter the characters of enzymes and carriers could prove to be an invaluable tool in research on plant nutrition.

Regulation of Plant Growth by Plant Growth Regulators (PGRs)

Although the use of PGRs to affect specific enzymes and carriers has already been discounted above, it may be questioned whether properties of plants at the gross morphological level, such as root:shoot ratio could be beneficially altered. Deficiency of N, P, or K naturally increases root:shoot ratio, and thus any use of PGRs to bring this about would result in an even bigger root:shoot ratio under deficiency. This could limit photosynthesis and so greatly decrease plant growth that such manipulation would probably not have any beneficial effects. Findenegg [50] has shown that root: shoot ratio is usually optimum for plant growth at all rates of N supply, and he concludes that manipulation of root:shoot ratio by PGRs would be unsuitable.

Use of Models to Predict Nutrient Requirement of Crops

The model of Greenwood and co-workers [64–66] gives a good understanding of how climate and nitrogen concentration in crop plants influence their growth. When fully refined these sorts of models will allow agronomists to match the supply of nitrogen to the climatic conditions at any one time to guarantee optimum yields of crops. However, these models are only half of what is required because the availability of nitrogen in the soil also has to be considered. Nitrate in soils mainly flows to the roots of plants in the water that ultimately becomes the transpiration stream, and the concentration of nitrate is largely determined by the rate at which plants take it up. However, the initial concentration and the rate of resupply by mineralization or fertilizer application are also crucially important.

In parts of the world where climatic conditions are regularly repeated each year it is possible to measure nitrate concentration in the soil at a few different times in spring with a simple colorimetric test, and then to predict the rate of mineralization in that soil and add just enough fertilizer N to give optimum growth of crops [92]. A more accurate approach, but one that is more time-consuming and requires access to more complex equipment, is to measure the nitrate and ammonium content of the soil by Electro-Ultrafiltration (EUF). Surveys of different farm soils by EUF show that farmers invariably overestimate the requirement of their soils for N, and use of such techniques could be invaluable in conjunction with models of plant growth to optimize nitrogen fertilizer use [93]. However, farmers have to be persuaded to use these new techniques and as it appears that only small numbers of them use tests for determining nitrate concentrations in plants (see earlier), this is probably also true for soil testing.

Plant Breeding

It ought to be possible to breed plants for enhanced nutrition, but the difficulty here is in knowing what characters to select for. Root:shoot ratio could theoretically be altered, but the objections to this are the same as for PGRs. Short-term responses of plants to deficiency or excess of nitrogen, for example, may be more suitably altered.

There is some evidence that in soils where nitrate concentration is low, plants bred to grow in high concentrations of nitrate are at a selective disadvantage, and do not grow efficiently [94]. It is certainly possible to breed plants for enhanced uptake of nutrients, but this does not always lead to an improved rate of growth of plants. In a study of 25 varieties of barley, a ranking of the varieties in order of rate of K^+ influx did not match the ranking in order of growth rates [95].

Characters of phosphorus efficiency in *Phaseolus vulgaris* are heritable [73], and it would be sensible to breed lines that use phosphorus in such a way as to give maximum yields in soils with low availability of P. In a study of a large number of lines of *P. vulgaris*, plants were classified as being either efficient or inefficient with low P supply, and additionally were subdivided into two groups according to whether or not they were responsive to P fertilization [74]. With such a classification it should be possible to breed plants that are not only efficient at low P availability, but that also respond to added P.

With breeding programs already established to improve the nutrition of crop plants, it is useful to consider how breeding of wheat, the plants that have been longest in cultivation, has improved their efficiency of nutrient utilization.

The phosphorus efficiency of wheat, measured as yield of grain per mg P in the shoot of plants at a low rate of P supply, has been shown to follow the order hexaploid species > tetraploid species > diploid species, whereas the concentration of P in the grains follows the reverse order [96].

Therefore, the efficiency of the hexaploid species is brought about by the fact that they can produce high yields with low concentrations of P in the grain, not because there is more uptake of P. Grain weight has increased with evolution, mostly due to an increase in the number of endosperm cells, and there has been a decrease in the %N in the grains [97].

The increases in the size of grains were paralleled by increases in leaf area, reduction in the rate of photosynthesis per unit leaf area, and reduction in the rate of photorespiration in plants grown with nonlimiting supplies of nutrients. There is more movement of assimilates to the grains of modern wheats, with a greater loss in dry weight from the stems during grain filling and less movement of assimilates to roots, shoots, and tillers [98]. Breeding (and accidental evolution) have not so much increased the characters of the plants directly associated with nutrition, but have affected morphology and photosynthesis and have caused the increase in Harvest Index already referred to [82]. These factors together may then have affected nutrient demand.

MAINTENANCE OF BALANCED NUTRITION

Ionic Balance in Plants

Plants are dependent upon the uptake in defined amounts of those ionic species that are required in order to sustain growth. Although there is a high selectivity in ion uptake by roots, the uptake of one ion species can influence the rate of uptake of another. Nutrients are mainly taken up from the soil solution as charged ions, either as cations or anions, and it is important to consider the ionic charge balance of uptake and to relate this balance to plant metabolism. Since some of the nutrient ions are transported to the shoots, it is necessary also to consider the ionic balance between mineral ions in translocation as well as in uptake.

It is possible to use microelectrodes, indicator dyes, or NMR spectroscopy to measure differences in potential between the inside and the outside of a cell. Such measurements have established that potential differences occur across the plasmamembrane, and if the outer medium is taken as the reference having a potential of 0, the inside of the cell is negatively charged. The difference of potential is in the order of -100 mV. Ionic species respond to this difference of potential by being either attracted to (cations) or repulsed (anions) from the inner surface of the membrane. The transport of anions across the plasmalemma, a process clearly occurring against an electrochemical gradient (uphill transport), requires energy (ATP) and specific membrane carriers. The transport of cations across the membrane is more complex. In addition to the existence of specific carriers at the plasmalemma (active uptake), part of the uptake can be considered as a diffusion process along the electrochemical gradient (downhill transport). In recent work on *Arabidopsis thaliana* root hairs it appears that the working of a plasma membrane proton pump, a constant current source with a voltage range from 0 to -200 mV, maintains the negative potential of the cell and this potential is sufficient to cause enough inflow of K^+ ions for cellular expansion [99].

It is also well established that tissues have a cation exchange capacity (CEC) brought about by the existence of electronegative sites on cell walls. Epstein and Leggett [100] showed that barley roots immersed in a solution of labeled strontium took up *Sr rapidly at first and more slowly afterwards. As discussed earlier, this type of kinetics results from two simultaneous processes, an active uptake constant over a limited period of time and a passive diffusion which is not limited in time. When the roots of barley were removed from the radioactive strontium solution and placed in a solution containing calcium, there was an efflux of *Sr into the Sr-free solution. This was interpreted as the result of exchange of *Sr ions held on the electronegative sites in the Donnan free-space by Ca^{2+} ions. This result clearly indicates that there is competition between cations for the binding sites on cell walls. The divalent cations such as Ca^{2+} and Mg^{2+} are mainly involved with this exchange capacity while monovalent cations such as K^+ or Na^+ are not adsorbed to any measurable extent. This may be of great importance in soils where the concentration of certain cation species is extremely low (like Fe^{3+} in alkaline soils) or extremely high (e.g., toxic soils high in Mn^{2+}), especially for the uptake of divalent or trivalent cations.

In the roots, uptake of nutrients takes place at very different rates. Some ions such as NO_3^-, Cl^-, K^+, or Na^+ are taken up rapidly while uptake rates of SO_4^{2-}, Ca^{2+}, or Mg^{2+} are lower. In terms of the charge balance of nutrient uptake, uptake of anions may be greater, lower, or equal to that of cations. Plants, however, maintain electroneutrality in their tissues by movements of positive (H^+) or negative (OH^-/HCO_3^-) charges across the plasmamembrane. The use of microelectrodes has shown, however, that intracellular pH (or at least that in the cytoplasm) is stable and controlled within narrow limits. This indicates that in maintaining electroneutrality in their tissue in response to a differential uptake of anions and cations, plants do not accumulate H^+ or OH^- ion species.

The mechanisms involved in intracellular pH regulation have been reviewed recently [101]. These possibly include cytoplasmic buffers, plasmalemma H^+-ATPase, H^+/Na^+ and H^+/K^+ antiports, HCO_3^- exchange systems (Cl^-/OH^- assimilation), and the production and consumption of organic acids. This last system for intracellular pH control is thought to be of prime importance in plants, and its mechanism as proposed by Davies [102] is known as the "biochemical pH-stat" (Figure 4). In this model, pH control in plant tissues is achieved by the net production or consumption

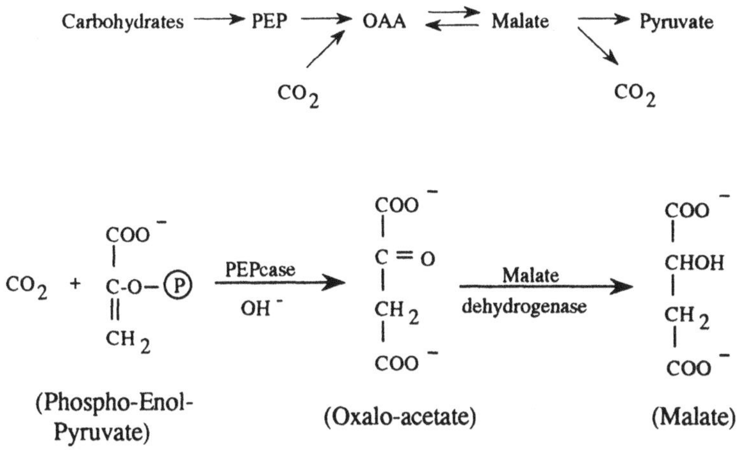

Carbohydrates $\longrightarrow$ PEP $\longrightarrow$ OAA $\rightleftharpoons$ Malate $\longrightarrow$ Pyruvate

CO_2 $\qquad\qquad\qquad\qquad\qquad$ CO_2

$$CO_2 \;+\; \begin{array}{c} COO^- \\ | \\ C\text{-}O\text{-}\textcircled{P} \\ || \\ CH_2 \end{array} \xrightarrow[OH^-]{PEPcase} \begin{array}{c} COO^- \\ | \\ C=O \\ | \\ CH_2 \\ | \\ COO^- \end{array} \xrightarrow[\text{dehydrogenase}]{\text{Malate}} \begin{array}{c} COO^- \\ | \\ CHOH \\ | \\ CH_2 \\ | \\ COO^- \end{array}$$

(Phospho-Enol-Pyruvate) $\qquad\qquad$ (Oxalo-acetate) $\qquad\qquad$ (Malate)

Figure 4   The biochemical pH stat. From [102].

of organic acids, with emphasis given to the key role of the enzyme PEP carboxylase (PEPcase). When intracellular pH increases, the activity of PEPcase is enhanced and results in net synthesis or organic acid (malate) which can be stored as an anion (malate$^-$) in the vacuole, K$^+$ being its counterion. On the other hand when intracellular pH decreases, PEPcase activity is depressed and net malate consumption takes place (decarboxylation), the $-COO^-$ anion charge of the organic acid being lost as CO_2.

Experimental evidence of changes in the pool of organic acids in response to differential uptake of cations and anions is given from data of Hiatt [103]. In his experiments nutrient uptake was measured in relation to changes in the pool of organic acids of plants fed with simple salts in which ionic constituents were taken up by plants at different uptake rates (cation >anion, cation <anion, or cation = anion).

1. When plants were fed with KCl, uptake rates of both K$^+$ and Cl$^-$ were equal and no change in the pool of organic acids in the plant was observed.

2. When plants received $CaCl_2$, uptake of Cl$^-$ was greater than that of Ca^{2+} and as a response to this differential uptake of anions over cations (A>C) there was a net decrease in the pool of organic anions. This can be interpreted as the result of a shift in intracellular pH towards acidification (H$^+$ uptake or OH$^-$/HCO$_3^-$ extrusion in order to maintain electroneutrality) and subsequent decrease in PEPcase activity (= net malate consumption) in accordance with the "pH stat" theory.

3. When plants were fed with K_2SO_4, uptake of K^+ exceeded that of SO_4^{2-} and as a result of this differential uptake of cations over anions ($C > A$) there was a net increase in the pool of organic anions in the plant tissues. This again is in accordance with the theory since the excess uptake of K^+ over SO_4^{2-} had to be balanced by a corresponding uptake of OH^- charges (or H^+ extrusion) in order to maintain electroneutrality, the shift in intracellular pH toward alkalinization resulting in a stimulation of PEPcase activity and thus in a net increase in organic anion synthesis.

Since electroneutrality in plant tissues results from the uptake (or extrusion) of H^+ or OH^- (HCO_3^-) charges in or outside the cell, pH changes in the outer nutrient medium are to be expected in response to the uptake of cations and anions. This indeed is frequently observed and a typical illustration of this phenomenon can be seen from plants growing in full nutrient solution with either NH_4^+ or NO_3^- as the nitrogen source. When plants are supplied with $NH_4 - N$, cation uptake generally exceeds anion uptake and the pH of the rooting medium decreases (net H^+ efflux or OH^- uptake in order to compensate for the greater uptake of cations over anions). On the other hand, most of the plant species supplied with $NO_3 - N$ take up an excess of anions over cations and the pH of the rooting medium increases (net OH^- extrusion or H^+ uptake) [104–106].

Nutrient Transport in Plants

Once taken up, nutrients are either stored or assimilated in the roots or are transported to the tops where they are assimilated, stored, or recirculated to the roots. Upward transport of nutrients to the shoot mainly occurs through the xylem vessels and follows water movements from roots to shoot (mass flow). These movements are mainly brought about by the difference in water potential between the atmosphere and the soil, and water moves along the soil–plant–atmosphere continuum according to the rate of plant transpiration. The active secretion of ions into the xylem vessels generates differences in osmotic potential between the xylem vessels and the adjacent cells and creates water movements into the vessels. This root pressure may be of importance for nutrient transport to the shoot, especially in young plants and plants having low rates of transpiration. Movements of divalent cations can also occur by exchange reactions in the xylem walls.

In the xylem sap nutrients are transported either as ions or in reduced forms (uncharged). The site in which nitrogen is assimilated (in the roots or in the tops) determines whether it is transported as an ion (NO_3^-) or in neutral forms. This is of importance for the movements of other nutrients which are mainly transported to the shoot in dissociated form (e.g., K^+) since from the viewpoint of ionic balance, the electroneutrality of the sap has to be maintained with pH values often close to neutrality. For most dicots, however, the main site of NO_3^- reduction is in the shoot, and NO_3^- is transported as an anion in the xylem sap.

Movement of NO_3^- from root to shoot is generally rapid although this varies depending on the accompanying cation. The uptake and xylem transport of NO_3^- by maize plants fed on a variety of nitrate salts is shown in Table 2 [107].

Uptake and transport of NO_3^- in the sap was higher when K^+ rather than Ca^{2+} acted as the counterion for NO_3^- translocation. This indicates that K^+ moves more rapidly in the xylem than Ca^{2+}, possibly because Ca^{2+} is involved in exchange reactions in the xylem walls. For plants supplied with $NaNO_3$ and KNO_3, the uptake rate of nitrate was similar but NO_3^- transport to the tops was depressed when Na^+ rather than K^+ was the counterion. This emphasises the fact that although Na^+ can replace K^+ in its role in counterbalancing NO_3^- uptake, it cannot replace the role of K^+ transport in maize plants. In fact this is due to specific resorption of Na^+ from the xylem into the xylem parenchyma cells of the roots or lower stem [108].

Plants, however, vary markedly in their ability to transport Na^+ to the tops and this property determines whether they can grow in saline conditions (natrophilic species) or not (natrophobic species). This specificity is genetically controlled and can be used for the selection of cultivars adapted to saline conditions. The important role of K^+ as counterion for NO_3^- translocation to the tops is emphasized in the model proposed by Ben-Zioni et al. [109] for NO_3^- uptake and transport from roots to shoot. In this model NO_3^- taken up by the roots is transported in the xylem sap where K^+ plays the role of counterbalancing ion. In the tops, NO_3^- reduction leads to a charge transfer to OH^-/HCO_3^- radicals and the consequent increase in cytoplasmic pH results in activation of PEPcase and net synthesis of organic anion (malate$^-$) in balance with K^+. A downward movement of [malate$^-$ K^+] can then take

Table 2   Uptake and Xylem Transport of Nitrate by Maize Fed on a Variety of Nitrate Salts in Two Separate Experiments [from 107].

Nitrogen Source	Uptake (μmol g^{-1} 6 h^{-1})	Transport (% of uptake)
Ca(NO$_3$)$_2$	102.6	12.5
KNO$_3$	123.2	37.3
NaNO$_3$	75.5	10.3
KNO$_3$	78.1	28.2

place after phloem loading. On arrival in the roots the organic anion undergoes a decarboxylation reaction and the HCO_3^- produced is then exchanged in the outer medium for the uptake of NO_3^-. This model, based on the internal recycling of K^+ in the plant, has been studied in a variety of plant species (see [110] for references) but cannot be applied to all plants since the model requires two main conditions: (1) an excess anion over cation uptake with NO_3^- reduction taking place mainly in the tops, and (2) a downward movement of malate from the shoot to the roots. Plants, however, vary markedly in their ability to accumulate organic anions in the vacuole of leaf tissues (see [19] for further discussion) so that K^+ recycling may not be of great importance in all species.

Response of Plants to Imbalanced Nutrient Supply

The interaction between nutrients during the process of ion uptake is an important factor to consider in mineral nutrition of plants. Although plants have a remarkable selectivity in the uptake of mineral species, growth and plant mineral composition can be easily altered by changes in the supply of nutrients. Studies relating growth of plants to increasing doses of a given nutrient are often made while keeping the supply of other nutrients constant. In such studies, plants respond to the increasing concentration of the nutrient but also to the changes in the ratios between nutrients available to the roots.

It has been proposed that the selectivity in ion uptake by plants is better characterized by the ratios of ions taken up rather than the concentration found in the tissues [111]. For instance in the case of cation uptake it is possible to assess the selectivity of plant species in cation uptake by comparing the ratios between cations in the substrate and in the plant tissues. In order to compare the response of plants to varying K, Ca, and Mg supply, Braakhekke [111] proposed representing the data in a triaxial diagram in which the three ratios K:Ca, Ca:Mg, and Mg:K are plotted in three log-scale axes intersecting at angles of 60°. This author compared plant tissue composition of several species grown in nutrient solutions differing in their relative supply of K, Ca, and Mg (Figure 5). The response of *Trifolium repens* L. and *Rumex acetosa* L. grown at three different cation ratios (a, b, and c) illustrates the differences in selectivity in cation uptake by these two plant species.

Both *Trifolium* and *Rumex* responded to changes in cation ratio supply with varied tissue cation composition. Changes were much more limited in the plant tissue, however, than in the nutrient solution. This indicates a high selectivity in cation uptake for both species. *Trifolium* obviously has a preference for calcium over magnesium since a low Ca:Mg supply ratio resulted in a high Ca:Mg uptake ratio. *Rumex*, on the other hand, exhibited a net

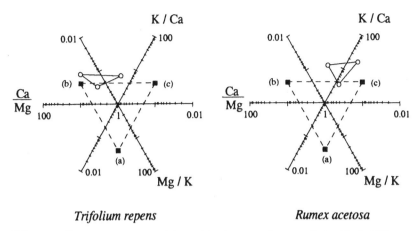

Trifolium repens   Rumex acetosa

Figure 5   Cation composition of 2 plant species supplied with 3 different nutrient solutions. Results are plotted on the triaxial ratio diagram (open symbols). The cation composition of the three nutrient solutions (closed symbols) was (meq L^{-1}): (a) K 0.2, Ca 2, and Mg 2; (b) K 2, Ca 0.2, and Mg 2; (c) K 2, Ca 2, and Mg 0.2. Redrawn from [111].

selectivity for magnesium over calcium uptake with the Ca:Mg uptake ratio remaining low even with high Ca supply. For both plant species K:Ca and K:Mg uptake ratios were high even at low K supply in the nutrient solution, indicating a high efficiency of both plants to take up potassium. Such differences in cation uptake selectivity characterize ecological differences between plant species and adaptation to particular habitats.

Thus, interactions between nutrients can be approached by the definition of optimal ratios between nutrients supplied. Such ratios should be carefully considered, however, since it is possible to maintain ratios with deficient or toxic supply of nutrients. Bearing this in mind, an optimal ratio between nitrogen and sulfur of about 17 was found to hold for several plant species [112,113]. For tomato and wheat plants grown at toxic levels of Mn, the Mg:Mn ratio in plant tissues was found to be a better indicator of toxicity than Mn concentration alone [114]. In wheat a Mg:Mn ratio lower than 20 was associated with significant yield decrease. For tomato this ratio was lower, indicating a greater tolerance of this crop to Mn.

Potassium/Sodium Selectivity

Sodium is classified as a beneficial nutrient, stimulating plant growth at low concentration but being essential to only some plant species ("halophytes"). High sodium supply may, however, be detrimental for the growth of "glycophyte" plants. In most soils, sodium concentration is similar to or higher than that of potassium and both ions are readily taken up by the roots. Plants, however, have a high selectivity in potassium uptake as compared with sodium, and under normal conditions K concentration in plant tissues largely exceeds Na concentration.

Plants are classified as "natrophilic" or "natrophobic" species based on their growth response to sodium and capacity to transport Na to the shoot [108]. Natrophilic species can readily transport Na to the shoot where it is stored in cell vacuoles. Natrophobic species, in contrast, have exclusion mechanisms by which Na is actively kept in the roots or lower stem parts by resorption of Na from the xylem into the living cells bordering the xylem vessels (K/Na antiport system). For such plant species, however, the high K:Na ratio in leaf tissues may be disturbed by saline soil conditions. Salinity is estimated to occur over 1000 million ha worldwide and to affect about one-third of irrigated crops at one stage of their growing season [see 115].

Although salinity is not exclusively ascribed to sodium salts, in a large number of cases Na is at very high concentration in saline soils. The response of plants to sodium concentration in the soil solution differs, however, between plant species. Marschner [116] proposed a classification of plants within 4 groups (A, B, C, and D) depending on growth stimulation by sodium. In group A are found the plants that respond to Na by a growth stimulation that would not occur with similar K supply (halophytes). In such plants, Na can replace a high proportion of K in the plant tissues. In contrast, group D (glycophytes) comprises plants in which K replacement by Na is not possible. These plants do not respond to Na supply by growth enhancement. Groups B and C show intermediate response to Na. Large supply of Na to plants of group C and D results in toxicity brought about by high Na concentration in the leaves (Na transport to the tops exceeding the resorption capacity) where it is detrimental, e.g., for the structure of chloroplasts [117]. In group A and B, however, Na transport to the shoot is not detrimental and Na is mainly restricted to the vacuole, where it replaces K in its role as an osmoticum. Na is highly efficient in this function and offers the plant protection against water stress by maintaining high osmotic pressure in the cells.

Salt tolerance in plants is a major issue for the development of new agricultural resources in several countries, especially developing countries. Plant response to salinity varies greatly between species and between genotypes. For tomato, wild varieties are more salt tolerant than the varieties used for fruit production. That salt tolerance is genetically controlled has led to breeding programs to develop new varieties better adapted to saline conditions. This approach is promising even if so far no commercial variety selected or bred for salt tolerance seems to have been released to growers [118].

Experiments have been carried out in order to identify the chromosomal location of the gene conferring high K:Na ratios in wheat grown in saline conditions. For modern hexaploid breadwheat containing the A, B, and D genomes, it is the D genome that carries this gene and its exact location is on the long arm of the 4D chromosome [119]. A similar trait is also found in *Aegilops* grasses that carry the D genome, and in rye (*Secale cereale* L.) in the R genome. Such findings may be of great importance for genetic manipulation techniques leading to the development of varieties better adapted to saline habitats. As emphasized by Vose [115], however, there is more than one mechanism operating to confer salt tolerance to plants. These include control of uptake, transport, and utilization of nutrients in the plant cells. The phenotypic expression of the genotype is the result of all the separate mechanisms, which have their own genetic control. The complexity of the control mechanisms, and their localization at a cellular and tissue level renders the genetic approach complicated and may explain why a certain number of plant cell lines selected for tolerance to salinity have failed to express this character in regenerated plants tested in saline soil conditions.

Plant Growth in Calcareous Soils as Related to Fe and P Shortages

In the world about 25% of soils are classified as calcareous soils [115] and in these soils, crops are at risk from iron deficiency. In alkaline soils there is also a strong fixation of P rendering its availability to plants low. In these soils, P can easily be the limiting factor for plant growth. Research on Fe and P deficiencies is an important area in plant mineral nutrition, and the study of the response mechanisms of plants grown under Fe or P limitation may identify important markers for the genetic selection of plants better adapted to alkaline soils. The effects of nutrient deficiencies on crop yields are dealt with elsewhere in this volume, and the responses of plants to deficiencies are covered later in this chapter, and so only brief mention is made of the subject here.

Fe Deficiency

Iron is present in soils in large quantities but mainly in forms unavailable to plants. The equilibrium concentration of Fe^{3+} in calcareous soil solution at pH 8.3 is in the order of 10^{-19} mol m^{-3} [120] and this low concentration together with high HCO_3^- concentration explains why certain crop species suffer Fe deficiency when grown in such soils ("lime-induced chlorosis"). Plants are called "Fe efficient" when they are capable of inducing changes in their rhizosphere to make more iron available to the roots, thus masking any visual symptoms of Fe deficiency. In contrast, "Fe inefficient" plants cannot respond to Fe shortage by such changes in the rhizosphere. The response mechanisms to Fe shortage seem to be genetically controlled since variations between plant species or even between cultivars have been observed.

P. Deficiency

As discussed earlier, the availability of phosphorus in soils is low because of adsorption. At low pH phosphate ions are adsorbed strongly, but at higher pH adsorption is less and more phosphate is available to plants. A high soil pH can decrease P availability under some circumstances, however, because phosphates can be precipitated as insoluble calcium phosphates if the concentration of Ca^{2+} ions in the soil solution is high [14]. Phosphorus deficiency can therefore occur in alkaline soils, especially when low moisture levels limit root growth and possibly reduce P diffusion [108].

INFLUENCE OF PLANTS ON RHIZOSPHERE CONDITIONS AND NUTRIENT AVAILABILITY

It has already been shown that when plants are poorly supplied with some nutrients the growth of roots is favored at the expense of shoots, so that capacity to explore new areas of soil remains high. There are many different responses of plants to nutrient deficiency that improve their prospects of making good any shortfall of nutrients, and some of these responses include changing the conditions in the rhizosphere. One of the most important rhizosphere responses to nutrient deficiency is the formation of symbiotic associations between plant roots and mycorrhizal fungi under P deficiency, and these mycorrhizal associations are considered elsewhere in the volume. Other responses, which have recently been reviewed [116], include:

1. Change in rhizosphere pH
2. Change in rhizosphere redox potential
3. Release of phytosiderophores by roots
4. Release of organic substrates for microorganisms by roots.

Good examples of these responses are given by plants subjected to Fe and P deficiency.

Rhizosphere Responses to Fe Deficiency

For monocots it seems that release of siderophores is the most important response to Fe-deficiency, but for dicots there is a wide range of responses [108]. One of the most important is a decrease in rhizosphere pH, which has been attributed to several factors. These include differential cation-anion uptake [122], the release of dissociable reductants [123], and proton pumps located at the plasmamembrane [124]. In Fe-efficient (*Cucumis sativus* L.)

plants iron deficiency stimulates an ATPase, which causes a higher trans-membrane potential and a greater capacity to acidify the external medium [125]. In all cases the acidification of the rhizosphere increases the acquisition of Fe from inorganic Fe^{3+} compounds [126].

For chickpea plants (*Cicer arietinum* L.) grown with nitrate and subjected to Fe stress, Alloush et al. [127] showed that both anion and cation uptakes were depressed compared with nonstressed plants. Anion uptake, and mainly NO_3^- as the major anion, was depressed more than cation uptake, leading to excess cation uptake. The value of (C-A) and H^+ released by the roots appeared to be in fairly close agreement, indicating that much of the induced acidification in the rhizosphere of the Fe-deficient chickpea plants was accounted for by the difference between cation and anion uptake. The acidification of the rhizosphere was concomitant with the accumulation of organic acids (citric and malic acids) in the roots [128], indicating that the two processes are linked as proposed by Landsberg [124]. Accumulation of organic acids may arise from the specific effect of Fe deficiency on the enzyme aconitase [129] and possibly enhanced rates of nonphotosynthetic CO_2 fixation via PEPcase [130]. These organic acids may provide the source of H^+ ions for extrusion (Figure 6) as well as a considerable internal anion charge possibly acting as counterion in cation uptake (thus depressing NO_3^- uptake) and transport from roots to shoot.

Enhanced reduction of Fe^{3+} to Fe^{2+} by roots of Fe-deficient plants is an important mechanism developed by Fe-efficient plants to increase iron availability to the roots. The reduction mechanism has been attributed to the release of "reductants" such as caffeic acid [131,132] that may reduce

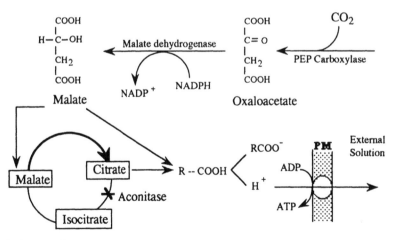

Figure 6   Response of chickpea plants to Fe deficiency. Redrawn from [128].

Fe^{3+} in the free space of the root system. An alternative hypothesis of enzymatic reduction at the plasmamembrane of rhizodermal cells has been proposed [133] but Römheld and Marschner [134] suggested that a combination of both the release of "reductants" and enzymatic reduction may account for the enhanced Fe^{3+} reduction capacity in the roots of Fe-deficient dicot species grown in field conditions. Such a reduction method has been seen in tomato, where Fe-deficiency leads to increased expression of constitutive Fe-chelate reductase isoforms in the root plasmamembrane [135].

The siderophores that have a particularly important role in mobilizing iron in the rhizosphere of monocot species are mainly nonprotein amino acids that solubilize unavailable forms of iron at high pH and chelate Fe^{3+} before its uptake by roots [108]. The release of these siderophores is under genetic control, and so it may be possible to breed for resistance to lime chlorosis in crop plants such as sorghum [136], but the control is probably multigenic and so breeding may prove to be difficult.

Rhizosphere Responses to P Deficiency

When plants are grown with $NO_3 - N$, the pH of the rhizosphere generally increases in response to the differential uptake of anions over cations and the maintenance of electroneutrality both in plant tissues and in the outer medium. In the case of P-deficient plants, the rhizosphere pH decreases even with $NO_3 - N$ supply [137–140]. The reason for the decrease in rhizosphere pH has been attributed to an excess cation over anion uptake [138, 140] or the release of organic acids (malic and citric acids) from the roots [137, 139]. Such acids may act as chelating agents in the soil, exchanging with bound phosphate and thus rendering P available to plants.

Acid conditions in the rhizosphere of P-deficient plants grown in calcareous soils enable greater solubilization of phosphorus to take place, thus increasing P availability to the plants. Evidence from work on cell suspensions of *Brassica nigra* [141] and with rape plants *Brassica napus* [142] indicates that under conditions of P-deficiency, nonphotosynthetic CO_2 fixation by the enzyme PEPcase is higher than in P-sufficient cells or plants, thus resulting in enhanced production of organic acids such as malate. In the "pH-stat" model of Davies presented earlier, such an increase in the organic acid production would lead to H^+ excretion in the outer medium in order to maintain internal pH in the cells at set values. In nitrate-fed tomato plants PEPcase activity increases in the roots within 2 days of withdrawal of P, by which time the rate of alkalinization of the growth medium is already decreasing. By 8 days after withdrawal PEPcase activity in the roots can be up to four times higher than in P-sufficient plants, and the growth medium is noticeably acidifed by this time [143].

ACKNOWLEDGMENTS

We wish to thank G. Alloush, F. E. Sanders, and R. Schulz for valuable discussions and Miss S. Hunter for preparation of the manuscript.

REFERENCES

1. E. A. Mitscherlich, *Soil Science for Farmers, Foresters and Gardeners*, 7th ed., Verlag P. Parey, Berlin, Hamburg. (1954).
2. D. R. Hoagland, *Lectures on the Inorganic Nutrition of Plants*, Chronica Botanica Company, Waltham, MA, p. 48 (1948).
3. P. Nissen, *Internat. Rev. Cytol., 126*: 89 (1991).
4. P. Nissen, *Plant Roots: The Hidden Half* (Y. Waisel, A. Eshel, and U. Kafkafi, eds.), Marcel Dekker, New York, p. 483 (1991).
5. E. Epstein, D. W. Rains, and O. E. Elzam, Proceedings National Academy Science (U.S.) *53*: 1320 (1965).
6. H. Doddema and G. P. Telkamp, *Physiol. Plant., 45*: 332 (1979).
7. H. Breteler and P. Nissen, *Plant Physiol., 70*: 754 (1982).
8. J. Wehrmann and R. Hahndel, VIth International Colloquium for the Optimization of Plant Nutrition, Montpellier, France, *Proceedings,* Vol 2, pp 679–685 (1984).
9. C. E. Deane-Drummond, *Plant Sci. Lett., 24*: 79 (1982).
10. R. Behl, R. Tischner, and K. Raschke, *Planta, 176*: 235 (1988).
11. M. Aslam, R. L. Travis, and R. C. Huffaker, *Plant Physiol., 99*: 1124 (1992).
12. R. B. Lee, *Ann. Bot., 50*: 429 (1982).
13. D. J. Greenwood, "Plant Nutrition 1982," Proceedings of the 9th International Plant Nutrition Colloquium (A. Scaife, ed.), CAB, Slough, U. K., pp. 1–6 (1982).
14. K. Mengel and E. A. Kirkby, *Principles of Plant Nutrition*, 4th ed., International Potash Institute, Berne (1987).
15. H. Munk, *Landw. Forsch., 50*: 150 (1958).
16. J. Döbereiner, *Inorganic Plant Nutrition, Encyclopedia Plant Physiology New Series, 15A* (A. Lauchli and R. L. Bielski, eds.), Springer Verlag, Berlin, p. 330 (1983).
17. I. Watanabe, C. R. Espinas, N. S. Berja, and B. V. Alimagno, *Utilization of the Azolla-Anabaena Complex as a Nitrogen Fertilizer for Rice,* IRRI *Research Paper Series No. 11*, p. 3 (1977).
18. D. J. Pilbeam and E. A. Kirkby, *Nitrogen Metabolism of Plants, Proceedings of the Phytochemical Society of Europe No. 33* (K. Mengel and D. J. Pilbeam, eds.) Oxford University Press, Oxford, pp. 55–70 (1992).
19. D. J. Pilbeam and E. A. Kirkby, *Nitrogen in Plants* (Y. P. Abrol, ed.), Research Studies Press, Taunton, U. K., p. 39 (1990).
20. W. R. Ullrich, *Nitrogen Metabolism of Plants, Proceedings of the Phytochemical Society of Europe No. 33* (K. Mengel and D. J. Pilbeam, eds.), Oxford University Press, Oxford, pp. 121–137 (1992).
21. J. B. Thibaud and C. Grignon, *Plant Sci. Lett., 22*: 279. (1981).

22. A. D. M. Glass, J. E. Shaff, and L. V. Kochian, *Plant Physiol.*, *99*: 456 (1992).
23. A. D. M. Glass, *ISI Atlas of Science*: *Animal and Plant Sciences*, *1*: 151 (1988).
24. R. Antoni Joseph, Tang Van Hai and J. Lambert, *Physiol. Plant.*, *34*: 321 (1975).
25. J. S. Pate, *Soil Biol. Biochem.*, *5*: 109 (1973).
26. G. R. Findenegg, J. A. Nelemans, and P. A. Arnozis, *J. Plant Nutr.*, *12*: 593 (1989).
27. A. J. Keys, I. F. Bird, M. J. Cornelius, P. J. Lea, R. M. Wallsgrove, and B. J. Miflin, *Nature*, *275*: 741 (1978).
28. H. Breteler and P. A. Arnozis, *Phytochemistry*, *24*: 653. (1985).
29. H. Breteler and M. Siegerist, *Plant Physiol.*, *75*: 1099. (1984).
30. A. Marion-Poll, J. C. Huet, and M. Caboche, *Plant Sci. Lett.*, *34*: 61 (1984).
31. L. P. Solomonson and M. J. Barber, *Annu. Rev. Plant Physiol. Plant Mol. Biol.*, *41*: 225 (1990).
32. L. Beevers and R. H. Hageman, *Annu. Rev. Plant Physiol.*, *20*: 495 (1969).
33. A. Scaife, *Plant and Soil*, *114*: 139 (1989).
34. C. E. Deane-Drummond and A. D. M. Glass, *Plant Physiol.*, *73*: 100 (1983).
35. J. M. Sutherland, M. Andrews, S. McInroy, and J. I. Sprent, *Ann. Bot.*, *56*: 259 (1985).
36. W. Wallace and J. S. Pate, *Ann. Bot.*, *29*: 655. (1965).
37. D. L. Shaner and J. S. Boyer, *Plant Physiol.*, *58*: 499 (1976).
38. B. T. Steer, *Plant Physiol.*, *51*: 744 (1973).
39. F. Galangau, F. Daniel-Vedele, T. Moureaux, M.-F. Dorbe, M.-T. Leydecker, and M. Caboche, *Plant Physiol.*, *88*: 383 (1988).
40. J. M. Melzer, A. Kleinhofs, and R. L. Warner, *Mol. Gen. Genet.*, *217*: 341 (1989).
41. A. Oaks, M. Poulle, V. I. Goodfellow, L. A. Class, and H. Deising, *Plant Physiol.*, *88*: 1067 (1988).
42. T. F. Ferrari, D. C. Yoder, and P. Filner, *Plant Physiol.*, *51*: 423 (1973).
43. G. Hehl and K. Mengel, *Landw. Forsch.*, *25/I*: 73 (1970).
44. R. C. Huffaker, *Encyclopedia of Plant Physiology*, *Vol. 14A* (D. Boulter and B. Parthier, eds.), Springer Verlag, Berlin, p. 370 (1982).
45. P. Millard and J. W. Catt, *J. Exp. Bot.*, *39*: 1 (1988).
46. M. Yamazaki, A. Watanabe, and T. Sugiyama, *Plant Cell Physiol.*, *27*: 443 (1986).
47. L. T. Evans, I. F. Wardlaw, and R. A. Fischer, *Crop Physiology* (L. T. Evans, ed.), Cambridge University Press, Cambridge, U.K., p. 101 (1975).
48. H. W. Woolhouse, *Physiological Processes Limiting Plant Productivity* (C. B. Johnson, ed.), Butterworth, London, p. 1 (1981).
49. N. Bellaloui and D. J. Pilbeam, *J. Plant Nutr.*, *13*: 39 (1990).
50. G. R. Findenegg, *Plant Nutrition*: *Physiology and Applications* (M. L. van Beusichem, ed.), Kluwer, Dordrecht, p. 21 (1990).
51. R. Brouwer, *Neth. J. Agric. Sci.*, *10*: 399 (1962).
52. H. Lambers, R. J. Simpson, V. C. Beilharz, and M. J. Dalling, *Physiol. Plant.*, *56*: 421 (1982).
53. M. C. Drew, L. R. Saker, and T. W. Ashley, *J. Exp. Bot.*, *24*: 1189. (1973).

54. M. C. Drew and L. R. Saker, *J. Exp. Bot.*, *26*: 79 (1975).
55. M. Mattsson, E. Johansson, T. Lundborg, M. Larsson, and C.-M. Larsson, *J. Exp. Bot.*, *42*: 197 (1991).
56. E. Öhlen and C. -M. Larsson, *Physiol. Plant.*, *85*: 9 (1992).
57. J. Barneix, D. M. James, E. F. Watson, and E. J. Hewitt, *Planta*, *162*: 469 (1984).
58. N. Bellaloui and D. J. Pilbeam, *J. Exp. Bot.*, *42*: 81 (1991).
59. R. G. Zhen and R. A. Leigh, *Plant and Soil*, *124*: 157 (1990).
60. J. H. M. Thornley, *Ann. Bot.*, *36*: 431 (1972).
61. J. B. Wilson, *Ann. Bot.*, *61*: 433 (1988).
62. P. B. Goodwin, B. I. Gollnow, and D. S. Leatham, *Phytohormones and Related Compounds—A Comprehensive Treatise, II*, (D. S. Leatham, P. B. Goodwin and T. J. V. Higgins, eds), Biomedical Press, Amsterdam, p. 215 (1978).
63. M. A. Morgan, W. A. Jackson, W. L. Pan, and R. J. Volk, *Plant and Soil, 91*: 343 (1986).
64. D. J. Greenwood, J. J. Neeteson, and A. Draycott, *Plant and Soil, 91*: 281 (1986).
65. D. J. Greenwood, G. Lemaire, G. Gosse, P. Cruz, A. Draycott, and J. J. Neeteson, *Ann. Bot.*, *66*: 425 (1990).
66. D. J. Greenwood, F. Gastal, G. Lemaire, A. Draycott, P. Millard, and J. J. Neeteson, *Ann. Bot.*, *67*: 181 (1991).
67. C. I. Ullrich-Eberius, A. Novacky, E. Fischer, and U. Luttge, *Plant Physiol.*, *67*: 797 (1981).
68. P. C. Jackson and C. E. Hagen, *Plant Physiol.*, *35*: 326 (1960).
69. G. C. Elliott and A. Lauchli, *Agron. J.*, *77*: 399 (1985).
70. P. A. Costigan, *J. Plant Nutr.*, *10*: 1523 (1987).
71. P. A. Nye and W. N. M. Foster, *J. Agric. Sci.*, *56*: 299 (1961).
72. F. E. Sanders, *Advances in Phytopathology, 9*: 135 (1993).
73. G. Whiteaker, G. C. Gerloff, W.H. Gabelman, and D. Lindgren, *J. Amer. Hort. Sci.*, *101*: 472 (1976).
74. M. Thung, *Genetic Aspects of Plant Mineral Nutrition* (N. El Bassam, M. Dambroth, and B. C. Loughman, eds.) Kluwer, Dordrecht, The Netherlands, p. 501 (1990).
75. A. D. M. Glass, M. Y. Siddiqi, and K. I. Gilesi, *Plant Physiol.*, *68*: 457. (1981).
76. J. M. Cheeseman and J. B. Hanson, *Plant Physiol.*, *64*: 842 (1979).
77. K. Mengel and W. W. Arneke, *Physiol. Plant.*, *54*: 402 (1982).
78. G. D. Humble and K. Raschke, *Plant Physiol.*, *48*: 447 (1971).
79. R. A. Leigh, "Methods of Potassium Research in Plants" Proceedings of the 21st Colloquium of the International Potash Institute, Berne, Switzerland, p. 117–126 (1989).
80. R. A. Leigh and R. G. Wyn-Jones, *New Phytol.*, *97*: 1 (1984).
81. K. Mengel and E. A. Kirkby, *Advances in Agronomy*, *33*: 59 (1980).
82. K. Mengel, "Methods of Potassium Research in Plants," Proceedings of the 21st Colloquium of the International Potash Institute, Berne, Switzerland, p. 67 (1989).

83. M. Y. Siddiqi and A. D. M. Glass, *Can. J. Bot.*, *61*: 671 (1983).
84. P. F. Shea, G. C. Gerloff, and W.H. Gabelman, *Plant and Soil*, *28*: 337 (1968).
85. E. L. Deckard, R. J. Lambert, and R. H. Hageman, *Crop Sci.*, *17*: 293 (1973).
86. V. Wojcieska, E. Wolska, and M. Ruszkowska, VIth International Colloquium for the Optimization of Plant Nutrition, Montpellier, France, Proceedings p. 1429–1433 (1984).
87. P. W. Syltie, S. W. Melstead, and W. M. Walker, *Commun. Soil Sci. Plant Anal.*, *3*: 38 (1972).
88. A. Scaife, Personal communication in *Diagnosis of Mineral Disorders in Plants I* (C. Bould, E. J. Hewitt, and P. Needham) HMSO, London, p. 113 (1983).
89. R. Schulz, personal communication.
90. E. Steingröver, J. Siesling, and P. Ratering, *Physiol. Plant.*, *66*: 557 (1986).
91. E. Haunold, "Methods of Potassium Research in Plants," Proceedings of the 21st Colloquium of the International Potash Institute, Berne, Switzerland, p. 49–65 (1989).
92. J. Wehrmann and H. C. Scharpf, *Plant and Soil*, *52*: 109 (1979).
93. K. Mengel, *Nitrogen Metabolism of Plants, Proceedings of the Phytochemical Society of Europe No. 33* (K. Mengel and D. J. Pilbeam, eds.), Oxford University Press, Oxford, p. 1 (1992).
94. C. O. Rodgers and A. J. Barneix, *Plant Physiol. Biochem.*, *27*: 387 (1989).
95. M. Y. Siddiqi, and A. D. M. Glass, *Can. J. Bot.*, *61*: 1551 (1983).
96. G. D. Batten, *Ann. Bot.*, *58*: 49 (1986).
97. R. L. Dunstone and L. T. Evans, *Austral. J. Plant Physiol.*, *1*: 157 (1974).
98. L. T. Evans and R. L. Dunstone, *Austral. J. Biol. Sci.*, *23*: 725 (1970).
99. R. R. Lew, *Plant Physiol.*, *97*: 1527 (1991).
100. E. Epstein and J. E. Leggett, *Amer. J. Bot.*, *41*: 785 (1954).
101. A. Kurdjian and J. Guern, *Ann. Rev. Plant Physiol.*, *40*: 271 (1989).
102. D. D. Davies, *Biosynthesis and Its Control in Plants* (B. V. Millborrow, ed.), Academic Press, London, p. 1 (1973).
103. A. J. Hiatt, *Z. Pflanzenphysiol.*, *56*: 233 (1967).
104. E. A. Kirkby and K. Mengel, *Plant Physiol.*, *42*: 6 (1967).
105. R. J. Haynes and K. M. Goh, *Biol. Rev.*, *53*: 465. (1978).
106. M. L. van Beusichem, E. A. Kirkby, and R. Baas, *Plant Physiol.*, *86*: 914 (1988).
107. T. W. Rufty Jr., W. A. Jackson, and C. D. Raper, *Plant Physiol.*, *68*: 605 (1981).
108. H. Marschner, *Mineral Nutrition of Higher Plants*, Academic Press, London (1986).
109. A. Ben-Zioni, Y. Vaadia, and S. H. Lips, *Physiol. Plant.*, *24*: 288 (1971).
110. B. Touraine, N. Grignon, and C. Grignon, *Plant Physiol.*, *88*: 605 (1988).
111. W. G. Braakhekke, *Genetic Aspects of Plant Nutrition* (M. R. Saric and B. C. Loughman, eds.), Martinus Nijhoff, Dordrecht, The Netherlands, p. 331 (1983).
112. P. E. Rasmussen, R. E. Ramig, L. G. Ekin, and C. R. Rhode, *Plant and Soil*, *46*: 153 (1977).

113. K. N. Bansal, D. P. Motiramani, and A. R. Pal, *Plant and Soil, 70*: 133 (1983).
114. J. Le Bot, M. J. Goss, M. J. G. P. R. Carvalho, and E. A. Kirkby, *Plant and Soil, 124*: 205 (1990).
115. P. B. Vose, *Genetic Aspects of Plant Mineral Nutrition* (W. H. Gabelman and B. C. Loughman, eds.), Martinus Nijhoff, Dordrecht, The Netherlands, p. 3 (1987).
116. H. Marschner, Proceedings of the 8th Colloquium of the International Potash Institute, Berne, Switzerland, p. 50–63 (1971).
117. C. Hecht-Buchholz, G. Mix, and H. Marschner, *Plant Analysis and Fertilizer Problems* (J. Wehrmann, ed.), German Society of Plant Nutrition, Hannover (1974).
118. E. Epstein and D. W. Rains, *Genetic Aspects of Plant Mineral Nutrition* (W. H. Gabelman and B. C. Loughman, eds.), Martinus Nijhoff, Dordrecht, The Netherlands, p. 113–125 (1987).
119. R. G. Wyn-Jones and J. Gorham, "Methods of Potassium Research in Plants" Proceedings of the 21st Colloquium of the International Potash Institute, Berne, Switzerland p. 27–36 (1989).
120. G. Julian, H. J. Cameron, and R. A. Olsen, *J. Plant Nutr., 6*: 163 (1983).
121. H. Marschner, *Plant Roots, The Hidden Half* (Y. Waisel, A. Eshel, and U. Kafkafi, eds.), Marcel Dekker, New York, p. 503 (1991).
122. K. Venkat-Raju, H. Marschner, and V. Romheld, *Z. Pflanz. Boden., 132*: 177 (1972).
123. J. C. Brown and W. E. Jones, *Physiol. Plant., 30*: 148 (1974).
124. E. C. Landsberg. *J. Plant Nutr., 3*: 579 (1981).
125. G. Zocchi and S. Cocucci, *Plant Physiol., 92*: 908 (1990).
126. V. Römheld, *Iron Transport in Microbes, Plants and Animals* (G. Winkelmann, D. van der Helm and J. B. Neilands, eds.), VCH Verlagsges, Weinheim, p. 353 (1987).
127. G. A. Alloush, J. Le Bot, F. E. Sanders, and E. A. Kirkby, *J. Plant Nutr., 13*: 1575 (1990).
128. G. A. Alloush, The Mechanism of Mobilization of Iron from Soil Minerals in the Rhizosphere of *Cicer arietinum* L., Ph.D. Thesis, The University of Leeds (1990).
129. J. S. D. Bacon, P. C. De Kock, and M. J. Palmer, *Biochem. J., 80*: 64 (1961).
130. C. R. Stocking, *Plant Physiol., 55*: 626 (1975).
131. R. A. Olsen, J. H. Bennett, D. Blune, and J. C. Brown, *J. Plant Nutr., 3*: 905 (1981).
132. N. H. Hether, N. R. Olsen, and L. L. Jackson, *J. Plant Nutr., 7*: 667 (1984).
133. R. L. Chaney, J. C. Brown, and L. O. Tiffin, *Plant Physiol., 56*: 208 (1972).
134. V. Römheld and H. Marschner, *Advances in Plant Nutrition, Volume 2* (B. Tinker and A. Lauchli, eds.), Praeger, New York, p. 155 (1986).
135. M. J. Holden, D. G. Luster, R. L. Chaney, T. J. Buckhout, and C. Robinson, *Plant Physiol., 97*: 537 (1991).
136. V. Römheld and H. Marschner, *Plant and Soil, 123*: 147 (1990).
137. M. J. Grinsted, M. J. Hedley, R. E. White, and P. H. Nye, *New Phytol., 91*: 19 (1982).

138. M. J. Hedley, P. H. Nye, and R. E. White, *New Phytol.*, *91*: 31 (1982).
139. E. Hoffland, G. R. Findenegg, and J. A. Nelemans, *Plant and Soil*, *113*: 161 (1989).
140. J. Le Bot, G. A. Alloush, E. A. Kirkby, and F. E. Sanders, *J. Plant Nutr.*, *13*: 1591 (1990).
141. S. M. G. Duff, G. B. G. Moorhead, D. D. Lefebvre, and W. C. Plaxton, *Plant Physiol.*, *90*: 1275 (1989).
142. E. Hoffland, J. A. Nelemans, and G. R. Findenegg, *Plant Nutrition: Physiology and Applications* (M. L. van Beusichem, ed.), Kluwer Academic Publishers, Dordrecht, p. 171 (1990).
143. D. J. Pilbeam, I. Cakmak, H. Marschner, and E. A. Kirkby, *Plant and Soil*, *154*: 111 (1993).

3

Symbiotic Nitrogen Fixation

Craig A. Atkins

The University of Western Australia
Nedlands, Western Australia, Australia

Legumes have been exploited widely in agricultural production systems, both in highly developed countries and, increasingly, in less well-developed and often less fertile regions of the world. Their uses range from monocultures of grain legumes, in which high-protein seed yields a valuable source of food for human or livestock nutrition, through mixed cropping systems where legumes provide green manure or offer alternative rotational cash crops to cereals. Legumes are also widely used in improved pastures which incorporate, with grasses, clovers, and medics and in agroforestry enterprises where leguminous trees provide fuel wood or timber and serve to stabilize soils and increase their fertility. The advantages of using legumes are many. Generally their high-quality grains, especially those like soybean which contain substantial levels of edible oils as well as a high level of protein, bring a good financial return to the farmer, while their use in rotations and mixed cropping systems enhance control of pests and disease as well as increasing the yield of cereals or other nonlegumes. It is their ability to fix atmospheric dinitrogen, at rates sufficiently high to provide the plant with a source of N additional to that in soil, and reduce the farmer's dependence on fertilizer N, which has always been regarded as a major attribute in their utility for agriculture.

The high cost of manufacture, transport, and application of fertilizer N remains a significant component in the efficiency of crop production, and, in less well-developed countries, constitutes a serious constraint to productivity. It is not too surprising then, that many national agricultural development programs as well as those of the major international agricultural research institutes, the agencies of the United Nations which deal with agriculture, and, to some extent, the aid organizations of the world have fostered and supported research into improving legume productivity and maximizing the benefits of symbiotic nitrogen fixation. These efforts have seen a substantial increase in the body of knowledge about legumes, their agronomy, pests, diseases, nutritional requirements, and especially in the nature of their symbiosis with *Rhizobium*. We now have a good understanding of the biochemistry and physiology of infection of roots, nodule development, and nodule functioning. The nature of some of the signal compounds involved in determining specificity in the association, in regulating expression of genes associated with infection, and the early steps in nodule differentiation (*nod* genes) have been elucidated. The rhizobial genes which code for the proteins required for nitrogen fixation (*nif* and *fix* genes) have likewise been isolated, described, and their regulation understood to the point where it is possible for the property of N_2 fixation to be transferred to nondiazotrophic bacteria (e.g., *E. coli*). The form and nature of plant metabolic pathways, which assimilate ammonia produced by nitrogenase, and the sort of nitrogenous solutes transferred from nodules to the host plant as a source of N for protein synthesis, have also been studied and are now documented in considerable detail. Given that this quite substantial body of knowledge has been accumulated it is appropriate and timely that its use and value be evaluated in relation to efforts aimed at improving the productivity of crop plants and enhancing the value of nitrogen fixation in realizing these gains. Accordingly, this chapter will address the overall aim of the volume, namely, to examine modern approaches to improving plant productivity, from the perspective of exploiting the value of N_2 fixation.

As a process, N_2 fixation occurs throughout nature in a variety of plant–microbe interactions, with a range of levels of interdependency; however, this chapter will deal almost exclusively with the legume, *Rhizobium* symbiosis. Furthermore only two main areas of research will be considered. The first of these is an examination of our understanding of factors intrinsic to the symbiosis which limit N_2 fixation and which might be altered to increase the value of legumes. The second will consider the approaches, research progress, and restrictions involved in conferring the symbiotic traits of legumes on nonleguminous plants, especially cereals. The reader is referred to the excellent review volumes which have appeared in recent years dealing with many aspects of the legume symbiosis [1–3] and, in particular, reviews dealing with nodule development [4–7] and the biochemistry and genetics of N_2 fixation [7–12]. It has become clear that rhizobia (i.e., nodule-forming bacteria) are a heterogeneous group which contain at least three distinct genera; *Rhizobium*, *Bradyrhizobium*, and *Azorhizobium* [5,11]. The species within each of these have been named largely on the basis of their legume host species (e.g., *Rhizobium lupinii* or *Bradyrhizobium japonicum*, etc.). In this review these binominals are used where appropriate and the term *rhizobia* is used as a general description of nodule-forming organisms which may include any or all of the genera.

IMPROVING THE VALUE AND UTILITY OF N_2 FIXATION BY THE LEGUME–RHIZOBIA SYMBIOSIS

Legumes generally, and grain legume species especially, frequently fail to achieve maximum N_2 fixation under field conditions. Any one of many factors which simply limit plant growth could be responsible for yield depression. These include inadequate supply of soil nutrients and water, losses due to pests and disease, or the use of species or cultivars unsuited to the soil or climatic conditions in areas where legumes are grown. There are a number of limitations which are intrinsic to legumes, however, and which are imposed by the nature of the symbiosis itself. The most important of these are:

1. A need for infection by a rhizobial strain which will result in development of an effective symbiosis.
2. A high requirement, in terms of plant nutrients, to support multiplication of rhizobia and nodule development.
3. A high cost, in terms of photosynthate use, for nitrogenase functioning and nodule maintenance.
4. A need to protect nitrogenase from inactivation by ambient O_2.
5. An inhibition of both nodulation and nodule functioning by soil and fertilizer N.
6. The premature senescence of nodules, especially during reproductive development.

Research aimed at understanding each of these limitations has received considerable attention in recent years. Guiding this research is the premise that knowledge of the mechanisms which underlie these limitations will allow manipulation of either the plant or the rhizobia to overcome their negative effects. In the context of modern biological research, "manipulation" has come to mean genetic manipulation and especially that using recombinant DNA technology. Other, more traditional, forms of manipulation, however, based on conventional plant breeding or modification of agronomic practices, will continue to have a prominent place in legume improvement.

Infection with an "Effective Rhizobial Strain"

Numerous examples have been described where inoculation of legume crops at sowing with a compatible strain of rhizobia results in enhanced N_2 fixation and yield. Although techniques of inoculant preparation and application have been developed which have led to successful establishment of symbioses under otherwise difficult environmental conditions, a number of authors have highlighted the need for selection of locally adapted strains of rhizobia [13,14] chosen for their tolerance to extreme soil conditions (e.g., acidity, salinity, alkalinity) or unfavorable climates [15]. On the other hand there have been a number of proposals for the use of novel, genetically engineered strains of rhizobia which will form more effective symbioses. These include strains which contain a unidirectional uptake hydrogenase [16,17] or in which regulation of cytochrome expression is altered [18]. Strains have also been developed which express insecticidal proteins to control specific nodule pests (e.g., control of *Rivellia angulata* larvae by the endotoxin from *Bacillus thuringiensis* [19]), or in which host plant specificity is altered [see 20]. Whether genetically engineered or naturally occurring "elite" strains are used as inocula, their success in improving legume productivity will depend, first, on their ability to survive in soils and to compete successfully against indigenous bacteria for nodule occupancy and, second, on the stability of the engineered traits and their sustained expression in the soil.

These considerations have prompted a resurgence of interest in the nature and mechanisms of "competitiveness" of bacteria introduced into soil [21] and, more specifically, into the population genetics of rhizobia [11]. In the past the lack of sensitive techniques to unambiguously identify bacterial strains in soils has been a major impediment to studies of the ecology of rhizobia. Such techniques are also necessary if the fate of engineered bacteria in soil is to be followed and an assessment made of the risk which their introduction might impose on an ecosystem. Methods involving identification with antibodies, the use of antibiotic resistance markers, or total protein profiles have been used with some success. More recently developed methods, however, which involve recognition of specific membrane proteins or lipopolysaccharide profiles [22], plasmid profiles [23], or which use specific gene probes [23–27], should allow more rapid progress [28].

A number of approaches have been used or have been proposed to increase the competitive advantage of a rhizobial strain introduced into the soil. These include the use of inoculation techniques which enhance the proportion of the introduced organisms on seed or in soil [29]; the use of highly competitive strains of rhizobia which will survive and multiply more readily than those which are indigenous [21,30]; or exploitation of legume hosts which do not nodulate with indigenous rhizobia or which have greater preference for the introduced strains [13,31,32]. Recently, Li and Alexander [33] have demonstrated that the competitiveness of both *R. meliloti* and *B. japonicum* inoculants was increased in soils by including, in the inoculum, antibiotic-producing strains of *Bacillus* sp. or *Streptomyces griseus*. While the precise roles of each partner in such coinoculations have not been defined, significant increases in yield of alfalfa and soybean have been obtained [33].

Apparently rhizobia move only short distances in soils, and do so largely due to water flow [34–36]. Consequently, seed inoculation results in a predominance of crown nodulation, while indigenous rhizobia, which are more widely distributed in the soil profile, form a more uniformly nodulated root system. Hardarson et al. [36] conducted both greenhouse and field trials with soybean in which either predominately crown nodulation or more uniform lateral root nodulation patterns were generated using seed or soil inoculation respectively. Nodule position affected N_2 fixation, with those formed on lateral roots providing a higher proportion of the plants' N than those on the crown. In view of the inhibitory effect which initially formed crown nodules are likely to have on subsequent lateral root nodulation [37,38], one consequence of seed inoculation is that later-formed nodules will be largely the result of infection with indigenous rhizobia. If these bacteria form less effective or ineffective symbioses, inadequate rates of N_2 fixation, especially late in plant growth and possibly during seed filling, are likely to occur. Furthermore, low mobility of seed inoculated rhizobia is likely to limit their migration so that in subsequent years they become restricted to the upper soil layers. Hardarson et al. [36] have suggested therefore that rhizobia should be selected, not only for effectiveness in N_2 fixation, but also for their mobility in soil. Indeed in a strain comparison with *B. japonicum*, Wadisinisuk et al. [39] were able to show that one which was apparently more competitive migrated further in soil than did a less competitive strain. Such differences are possibly related to motility attributes, cell surface characteristics [40,41], or to differences in chemotactic response [42].

Three genetic approaches to enhance competition in favor of introduced strains have been used:

1. Legume genotypes with very restricted susceptibility to infection.
2. Altering the host range of the inoculated strains so that it is very narrow.
3. Constructing strains of rhizobia with an increased ability to compete for nodule occupancy.

There are a number of examples where legume host cultivars are not infected by indigenous groups of rhizobia in the soil [see 20]. The best documented case comes from the work of Cregan et al. [43] or of Sadowsky et al. [44] in which the dominant group of *B. japonicum* (serocluster 123) in a range of soils in the U.S. Midwest has been found to infect only some soybean varieties. Similar situations have been demonstrated for *R. meliloti* and R. *leguminosarum* in specific areas [45,46]. A related, though somewhat different, situation has been shown for tropical soils of West Africa in which many strains of the "cowpea" group of rhizobia are ubiquituous. These fail to effectively nodulate the improved U.S. soybean varieties [13], thereby offering the possibility of promoting efficient nodulation by introduction of B. *japonicum* strains. In this case, however, Asian lines of soybean have been introduced which form effective symbiosis with the indigenous "cowpea" rhizobia [13] and eliminate the need for inoculation. A number of plant genes have been identified which offer the potential to alter host physiology in such a manner as to exclude infection by some bacteria and/or promote infection by others. These include a series of *rj* alleles [47] which confer non-nodulation or restricted nodulation in soybean [48] as well as traits in alfalfa, pea, and in a number of clovers in which selective strain dependence has been demonstrated [see 20]. Generally, the mechanism for strain selection is not known, but in one case, that of transgenic white clover, in which the roots were infected by R. *leguminosarum*, the change was due to specific transfer of the pea lectin gene *psl* [49].

Although selection of strains with altered host plant specificity from naturally occuring rhizobial populations in soil using host plants to "trap" organisms with the desired features has been used extensively in the search for more effective organisms, direct genetic manipulation of already isolated effective symbionts is now feasible. This has become possible as a result of recognition and isolation of a number of the bacterial genes involved in nodulation, some of which determine host range [10,11]. The nodulation (*nod*) genes in *Rhizobium* species are largely located on high molecular weight, symbiotic (*sym*) plasmids, whereas in *Bradyrhizobium* and *Azorhizobium* species these genes reside on the chromosome [11,50,51,52]. For this reason much of the work to date has used R. *meliloti* and R. *leguminosarum* biovars *trifolii* and *viciae* [10]. However, information on *nod* genes and their regulation in a number of *Bradyrhizobium* spp. has also slowly accumulated [11]. Three of the nodulation genes (*nod A, B,* and *C,* in the *nod ABC* operon) are common (common *nod* genes) to all rhizobia [53] while others (such as *nod L* and *nod MN* in R. *leguminosarum* and *nod H, G,* and *P* in R. *meliloti* [11]); are specific nodulation genes. Both the common *nod* genes as well as some host specific *nod* genes are regulated by a common *nod D* gene (or a number of copies of *nod D*) which in turn is directly influenced by the concentration of phenylpropanoid molecules (flavones or isoflavones) present in exudates produced by plant roots [4]. In this way the *nod* genes are coordinately expressed in the presence of the host. Specificity appears to be conferred in part by the *nod D* product/flavone interaction (some flavones induce *nod D* while others competitively inhibit *nod* gene activation [54,55]), and by the nature of *nod D* gene products. For example the three *nod D* genes of *R. meliloti* produce three gene products which have different inducer preferences [56].

A number of chimeric *nod D* genes involving sequences from *R. leguminosarum* and *R. meliloti* have now been constructed in which host specificity has been modified [57]. Furthermore transfer of a *nod D* allele has been effected from the broad host-range symbiont of the nonlegume *Parasponia andersonii* (NGR 234) to a *Rhizobium* strain in which normally unexpressed *nod* genes have been induced by signal compounds produced by rice roots [58]. These genetic approaches to the production of nodulated cereals will be discussed in a later section but they serve as examples of the current level of manipulation of "specificity" which has been achieved. The products of the common *nod* genes are necessary for a number of the events in early nodule development to proceed. These include the curling and deformation of root hairs [59] as well as induction of cortical cell division at specific sites in the root [60] at or prior to infection. One of these *nod* signnals, specific to alfalfa (*Nod* RM-1), has recently been isolated and identified as a sulfated β-1,4-tetrasaccharide of D-glucosamine in which three of the amino groups are acetylated while the fourth carries a C16 bis unsaturated fatty acid [61]. This signal compound specifically elicited root hair deformation at nanomolar concentrations on the host [61]. Its identification opens the way for more incisive studies of specific recognition phenomena in the development of the symbiosis and for the identification of other nodulation specific gene products which might be exploited to enhance or narrow host range preferences.

Another potential means to enhance competitiveness involves the use of strains which produce toxins or antibiotics inhibitory to infection by other strains. *R. leguminosarum* bv. *trifolii* strain T24 produces the antibiotic trifolitoxin which inhibits the growth of other pea strains [62]. The gene or genes responsible for the synthesis of this toxin have been successfully cloned (as cosmid pTFX1) and transferred to other rhizobia without compromising their symbiotic effectiveness [63]. More recently the trifolitoxin genes have been coupled with promoters which enhance toxin synthesis [21]; however, the consequences of this modification for competitiveness in soil have not been determined.

Despite the considerable progress in our knowledge of the regulation of *nod* genes and in the recognition of the diversity of approaches which might be employed to enhance the competitiveness of a rhizobial strain introduced into soil, one critical area of understanding needs to be tackled before the use of genetically engineered rhizobia can be justified. This relates to the behavior of the organism in soil and includes its population genetics, to the stability of traits introduced through genetic manipulation, and to factors which govern root colonization. The prospect of introducing genetically engineered bacteria in soil has raised the specter of irreversible change to the soil flora and the possibility of the unwitting transfer of undesirable genetic traits into the environment. As noted earlier, recently introduced techniques to recognize bacteria in soil are now available and more definitive studies in this area have been initiated. These include systematic studies of genetic diversity [64–66] and stability of rhizobia in the field [67], of recombination between *Rhizobium* spp. [68], plasmid transfer in the rhizosphere [69] and the use of Tn5 mutants with altered capacity for root colonization [28]. Triplett [21] has suggested that the genetic strategies being developed for release of *Pseudomonas* strains involving environmentally triggered promoters and "suicide" genes [70,71] might be useful for incorporation in rhizobia. Such strategies could allow selective elimination of introduced rhizobia from soil and their replacement with more desirable types.

Nutritional Requirements for Nodule Establishment

Estimates of the number of bacteroids which occupy the infected cells of a functional and mature cowpea nodule range from 8 to around 18×10^4 per cell with 1.5 to more than 13×10^8 per nodule [72]. This bacterial population was derived from a small number of infecting organisms whose division was largely dependent on the host plant for essential nutrients. It is perhaps not surprising then that seedlings of many grain and pasture legumes exhibit a transient inhibition of growth and yellowing of their foliage during the period of initial nodule formation [73,74]. The yellowing has been attributed to a "N-hunger" and can be largely overcome by providing a small amount of starter N (3–5 kg $N \cdot ha^{-1}$ [75,76]). Sprent and Thomas [77] have suggested that the N-hunger state exhibited by many legume species of epigeal germination habit (e.g., common bean or soybean) is largely due to the first fixed N being sequestered by bacteriods for their own multiplication and growth. As a consequence the first-formed foliage leaves are N deficient and their photosynthetic capacity restricted. The relative partitioning of seed N reserves to nodule development, and particularly to bacteroids, has been measured in cowpea using ^{15}N-labeled seed [74,78,79]. These experiments demonstrated the autocatalytic nature of the initial fixed N which, while used preferentially to form bacteriod protein, at the same time supplemented seed N for leaf development. At the time at which nitrogenase was first detected by H_2 evolution, or acetylene reduction, however, more than half of the N in bacteriods had come from plant seed reserves [79].

While substantial differences between species with epigeal or hypogeal germination are likely [77], cultivars of a single species which have different seed sizes also differ in the extent to which N-hunger conditions early seedling vigor. In the case of two cowpea varieties (Vita 3 and Caloona), which differed in seed dry weight and total N by twofold, only the smaller (Caloona) showed N hunger [74]. In Caloona, however, this was not due to fixation of N_2 being inhibited relative to the larger seeded variety. In fact, fixation in Caloona began earlier and at higher rates than in Vita 3. This was achieved by the plant investing 20–25% of its cotyledon reserve of N in nodules in Caloona but only 7% in Vita 3. Because of the difference in the proportional investment of N in nodules a comparison of the two varieties on the basis of cotyledon N use showed that in the first 4 weeks Caloona seedlings fixed 10.6 mg N·(mg seed N)$^{-1}$ while in Vita 3 only 5.3 (same units) were fixed. The host plant thus exercises considerable control over the rate at which the symbiotic partners reach nutritional complementation. Furthermore, although a fully functional symbiosis may be limited by the rate at which C is utilized by the respiratory processes supporting nitrogenase [80], during the period before complementation is achieved nodules are likely to be N limited [81].

Developing symbiotic bacteria are not only dependent on the plant for N and C but also for other nutrients needed in growth. In some cases, particular nutrients appear to be required in amounts additional to those for the host plant [15,82]. Other nutrients obtained from the host are sequestered by bacteroids in amounts no greater than an equivalent mass of non-nodule legume tissue. The former group, i.e., those required more specifically for symbiotic development, include Mo, Ca, Co, and Cu [82]. There have been suggestions, however, that P is also required specifically for nodule functioning both in soybean [83] and in peas [84], and recent data indicate that this may also be the case for Fe [85,86].

There is some evidence that the concentration of P is higher in nodules than in other parts of the legume plant [83,87,88], and this is not too surprising perhaps given the number of bacteria in a nodule and their content of nucleic acids and membrane phospholipids. Increases in nitrogenase activity, which have been shown for a number of species (common bean [88]; clover [87]; soybean [83,89]; lablab [90]; and pea [84]) with improved phosphorus nutrition, have generally been explained in terms of host plant growth stimulation. Israel and Rufty [91] have found that in soybean the concentration of soluble reduced-N in tissues of fully symbiotic plants was low in those under P-stress conditions and increased with P-nutrition. These changes in soluble-N were not seen in NO_3-dependent plants and they argued that N input from N_2 fixation is more sensitive to P-stress than is the subsequent use of N in forming protein and nucleic acids. The mechanism for enhanced sensitivity of symboitic plants to P deficiency is not known. In a recent review of the subjects, however, Robson and Bottomley [15] have pointed out that generally there is a strong, positive correlation between nodule weight and shoot weight with increasing P supply. This is not the case for nutrients like Ca, Mo, and Co which are clearly required specifically for symbiotic development. Keyser and Munns [92] have demonstrated strain differences among rhizobia for P in liquid culture. These differences were not reflected in the response of nodulation and nodule activity to P supply [90].

Calcium is involved in both nodulation and nodule functioning [93]. However, the most Ca-sensitive stage appears to occur quite early in association with infection and nodule initiation [93]. Adsorption of rhizobial cells to roots of alfalfa [94] and to root hairs of pea [95] have been shown to be Ca-dependent, while the ion also appears to be required for exudation of signal compounds which induce *nod* gene expression in *R. trifolii* [96, 97]. Once nodules begin to form, their development is not impeded by Ca levels even lower than those required for host growth [98]. Ca deficiency in soils is frequently associated with acidity and with, in many cases, high levels of aluminum, so that in field experiments it may be difficult to separate effects due to Ca from inhibition caused by other factors.

The nature of elevated requirements for Cu and Co by nodules has not been clearly defined. Cu appears to be required for nodule functioning rather than for nodule initiation and development [82]. Cartwright and Hallsworth [99] have noted that Cu-deficient nodules contain less cytochrome *c* oxidase than those from Cu-sufficient plants, possibly restricting their catalytic capacity for oxidative phosphorylation. Co is a constituent of the cobamide (vitamin B_{12}) coenzymes which function in a number of important enzyme-catalyzed reactions. In bacteria one of these (ribonucleotide reductase) is required for the incorporation of ribonucleotides into the deoxy derivatives in DNA [100], and so in Co-deficient nodules bacteroid multiplication is likely to be limited. However, the effects of Co level in nodules may well be due to reactions which do not involve cobalamin [101].

Molybdenum is a functional component of the Mo–Fe protein of nitrogenase [102] and, in ureide-forming nodules, also of xanthine dehydrogenase. It is perhaps not too surprising then that nodulated plants have a greatly elevated requirement for this cation. Functional nitrogenases which contain no Mo but have V or which contain neither Mo nor V have been described [see 103, 104]. These two nitrogenases as well as the normal enzyme containing Fe and Mo, which have been found in strains of *Azotobacter* species are genetically distinct [104]. While there is some evidence that non-Mo enzymes might occur in other diazotrophs [103], including *Azorhizobium caulinodans*, there have been no reports of their expression in symbioses. However, just as the *nif* structural gene probes have proved so useful in examining the distribution of the Mo-nitrogenase, so too will those (*vnf* and *anf* [105]) which are now available for the alternate enzymes.

Iron-containing proteins constitute a large proportion of the essential components for symbiotic N_2 fixation [102]. These include the Fe- and Mo

Fe-proteins of nitrogenase, ferredoxin, and, in the infected plant cells, the heme moeity of leghemoglobin. This latter may constitute as much as 30% of the soluble protein of the nodule [106] and it is not surprising that rhizobia possess high affinity transport systems (siderophores) for Fe [107]. Experiments using TN_5 insertion mutants of *R. meliloti*, defective in these transport systems, indicate that Fe translocation from the host to the bacteroids is essential for full symbiotic development [108]. Indeed Johnson and Youngblood [85] have demonstrated that alfalfa which is dependent on N_2 fixation is more sensitive to Fe limitation than is alfalfa supplied with adequate N as NO_3, while mutants of *R. leguminsarum* defective for Fe uptake are ineffective on peas [109].

Clearly the nutritional burden placed on the young legume plant, especially in the case of N, but possibly also for the catalytically essential metals (Mo, Co, and Fe), could be substantial, and, under soil deficiency conditions, a potent limitation to symbiotic development. Rosendahl et al. [110] have stressed the role of the plant-derived peribacteroid membrane (PBM) in effective transfer of nutrients from the host to the bacteroids. Considerable specificity in solute transfer across this barrier has been described and the question of the nature of the controlling steps at this level may be posed [110]. Effects of rhizobial genotype in determining the concentrations of key limiting nutrients in nodules of a number of legumes are known [111–113], so that an interaction with factors of bacterial origin seems likely. Thus selection and development of legume cultivars with increased N_2-fixing potential might well utilize genetic manipulation of rhizobia to optimize this interaction as well as to alter the plant genome to influence permeability of the PBM.

Costs of Nodule Functioning

Nitrogenase is an unusually expensive enzyme in terms of its consumption of ATP and reductant. These are generated in nodules by bacterial respiration of reduced C substrates supplied by the host. Because of the relatively high requirement for nitrogenase there have been suggestions that fixed N is more expensive for the plant than is assimilation of soil or fertilizer N. Furthermore the reduction of N_2 is accompanied by H^+ reduction to form H_2 gas, increasing further the costs of fixation; i.e.,

$$N_2 + 6e^- + 12\,ATP \rightarrow 2NH_3 + 12\,ATP + 12\,PO_4^-$$

and

$$2H^+ + 2e^- + 4\,ATP \rightarrow H_2 + ATP + PO_4^-$$

A similar $6e^-$ reduction is achieved by the enzyme nitrite reductase (NiR) but without hydrolysis of ATP. Together with nitrate reductase (NR) these two enzymes achieve reduction of soil or fertilizer NO_3^-, i.e.,

$$\text{NO}_3^- + 2e^- \xrightarrow{\text{NR}} \text{NO}_2^- + 6e^- \xrightarrow{\text{NiR}} \text{NH}_3$$

The theoretical costing of these two forms of N assimilation may be compared by equating their need for $2e^-$ (reductant) with ATP, and assuming a $P/2e^-$ ratio of 3 for oxidative phosphorylation. On this basis N_2 fixation requires 14ATP.N^{-1} and NO_3^- reduction 12 ATP·N^{-1}. Although this difference is not great, N_2 fixation is totally dependent on respiration of translocated C substrates, while NO_3^- reduction may be directly coupled, in part or in whole, to photosynthetic generation of reductant [see 9]. In addition to nitrogenase functioning, nodules have specific costs associated with N assimilation and transport and with the establishment and maintenance of the specialized structure required to house the mechanism (i.e., the nodule). Estimates of the total costs of nodule functioning range from 3 to 6 g C·g N fixed compared with 1 to 2.5 g C·g N assimilated from the soil or from fertilizer.

Experimental determination of the C cost for N_2 fixation has shown a wide range of values [114,115] varying between symbioses and at different stages of symbiotic development [79]. However, generally these fall within the theoretical estimates of 3 to 6 g C·g N fixed. In terms of the total photosynthate generated during the growth cycle of a grain legume species (e.g., cowpea, mung bean, or lupin) this C cost accounts for roughly 10% [79]. While this may seem a relatively small drain on the plants' resources it must be remembered that grain legumes generally have a low harvest index (0.3 to 0.4) for dry matter so that in terms of a competitive sink for assimilates it is significant when compared with that allocated to grain development.

A wide range of experimental evidence has shown that factors which reduce or limit photosynthesis (e.g., light, CO_2 level, photosynthetic area, drought) also restrict nodule functioning [116–119]. This sort of effect indicates what might be expected from a heterotrophic process located in the root system; a close dependence on photosynthesis. These sorts of observations have led to the proposal that N_2 fixation is strictly regulated by photosynthesis. However, in a number of cases no correlation between rates of N_2 fixation and photosynthesis could be demonstrated and attempts to select for enhanced fixation, on the basis of genetic variation in leaf gas exchange, have not been successful [120,121]. Fixation is dependent on the extent to which translocated carbon substrates are partitioned in phloem to nodules [122,123]. Thus it is the "source–sink" relations of the plant, rather than the maximum rates of photosynthesis achieved, which are critical in determining the rate at which these substrates are made available to bacteroid respiration. The nature of this question has been resolved more clearly by recent evidence for the interaction between oxygen supply and respiration in nodules [80] and will be considered in more detail in a later section. Significant differences between symbioses in the efficiency with which their nodules use oxidizable C substrates have been demonstrated [79,114]. The reasons for these differences, which may be as much as threefold [79], are not yet resolved but apparently involve both host and microsymbiont genomes.

Rhizobia selected for enhanced fixation [124] or engineered to reduce nitrogen at higher rates [125], or hydrogen at lower rates [126], offer potential to generate more efficient symbionts. The apparently wasteful H_2 production, which accompanies N_2 fixation in all diazotrophs, results in H_2 evolution in some legume symbiosis but not others. In those with little or no H_2 evolution the H_2 is reoxidized by a bacteroid-localized, unidirectional, uptake hydrogenase [16,127]. There is some experimental evidence that soybean plants formed with *Hup*$^+$ rhizobia out yield those which lack the hydrogenase (*Hup*$^-$). A closer examination of C use by nodules of these two sorts of symbioses in cowpea show that under conditions of nonlimiting light, the differences in nodule C economy were compensated for by the plant. As a consequence no yield penalty due to the lack of *Hup* could be detected [128].

Legume symbioses also differ in the nature of the nitrogenous solutes formed from fixed N in the nodule and translocated in xylem to the host plant [129]. Furthermore there have been suggestions that those that form principally ureides (allantoin and allantoic acid), compounds with a C:N ratio of 1, use less photosynthate than do those forming principally asparagine (C:N = 2). While substantial differences in the C economy of nodules of "ureide"- and "amide"-forming symbioses have been demonstrated [114, 130], it is not clear the extent to which these can be ascribed to the form of N transport which is utilized. At present there are no mutants available which vary in the nature of their translocated nitrogenous solutes. However, three lines of evidence suggest that variation of this sort might be possible. First, both asparagine- and ureide-forming plants produce, and effectively export glutamine for 2 or 3 days following exposure of the root system to 80% Ar:20% O_2 in the gas phase [78,131]. Second, treatment of nodulated root systems of both cowpea and soybean with the hypoxanthine analogue allopurinol blocks ureide synthesis in nodules (by inhibiting xanthine dehydrogenase [132]) and, as a consequence, the translocation stream carries xanthine and glutamine [132]. Thus for a short time at least, these compounds apparently serve as suitable N sources for plant growth. Finally, infection of roots of alfalfa by specific strains of *Pseudomonas syringae pv tabaci*, or their treatment with the toxin (tabtoxinine-β-lactam) formed by these bacteria, results in substantial changes in the nature of translocated N solutes in xylem [133, 134]. The effect is related specifically to nodular ammonia assimilation pathways and results in significant yield increases, greater nodulation, and elevated N_2 fixation [133].

While these results have yet to be translated into strategies for the selection of plants with an altered spectrum of translocated N solutes, they indicate that the usual pathways for ammonia assimilation in nodules may in fact restrict bacteroid functioning. Recent evidence (C. A. Atkins, M. Fernando, S. Hunt, and D. B. Layzell, unpublished) has associated the synthesis of ureides in nodules of soybean with the provision of dicarboxylic acid substrates for bacteroid respiration and, as a consequence, nitrogenase activity. This apparent interdependence could stem from the generation of reductant by reactions of purine oxidation and might also be associated with O_2 supply in the infected cells. Whether or not this sort of interdependence between C and N metabolism is a feature of other ureide-forming symbiosis or of those forming asparagine, is not known, but it seems clear that a search for plants which translocate other N solutes to the host may be rewarding.

Role of O_2 in Regulating N_2 Fixation in Nodules

N_2 fixation in the legume nodule requires a high rate of oxidative phosphorylation in the bacteroid to generate ATP for nitrogenase. Thus nitrogenase activity is directly dependent on a sustained O_2 supply. Despite this the nitrogenase proteins of rhizobia, and indeed of almost all diazotrophs, are extremely sensitive to O_2 [135] and are irreversibly inactivated in quite low subambient levels of the gas.

This apparent paradox is resolved by three key attributes of the legume symbiosis:

1. Expression of a respiratory terminal oxidase in bacteroids which has an exceptionally high affinity for O_2 (K_m = 3 to 5 nM [136]).
2. Maintenance of a high O_2 flux in infected cells by the reversible oxygenation of leghemoglobin.
3. Operation in the outer nodule tissues of a reversibly variable barrier to gaseous diffusion which functions in response to internal O_2 level.

The intracellular concentration of free O_2 close to the sites of bacteroid respiration is maintained at a low level by a high rate of O_2 consumption and at a high flux by the reversible oxygenation of leghemoglobin. In highly active N_2-fixing soybean nodules 20 to 25% of the physiologically functional leghemoglobin is oxygenated, resulting in an estimated free O_2 concentration of about 10 mM [136,137]. These estimates have been confirmed by direct spectroscopic measurements of leghemoglobin oxygenation in intact soybean root systems by Layzell et al. [138]. Thus free O_2 does not accumulate in the functioning nodule to the extent that nitrogenase is irreversibly inactivated; the rate of diffusion of O_2 into the organ being balanced by the rate of its consumption by bacteroids. However, if rhizosphere pO_2 were increased or bacteroid respiration lowered for any reason this balance would be destroyed and O_2 levels in the infected cells would rise.

In fact neither of these events leads to nitrogenase inactivation; at least in the short term. A range of studies with different symbioses [139–142] has demonstrated that stepwise increases in external pO_2 result in an increase in gaseous diffusive resistance of nodules such that this O_2 influx is controlled. Furthermore these changes in gaseous permeability are reversible [139] and have been found to be associated with a number of treatments which either reduce bacteroid respiration or inhibit nitrogenase. These include drought stress [143], application of fertilizer NO_3^- [144], inhibition of phloem transport to the nodule [145], and treatments which block N assimilation, such as exposure to acetylene or atmospheres devoid of N_2 [146] and to inhibitors, like allopurinol, which block ureide synthesis [132].

While the precise mechanism for sensing internal O_2 level and transduction of the sensor's signal into altered diffusional resistance in nodules has not been defined [72,147], there is considerable evidence that the variable barrier is located in the inner cortex [142,148,149]. The cells in this region, which lies just outside the infected cell tissue of the nodule, are comprised of two types [see 72]. The innermost 3 or 4 files appear to have large extracellular voids, when viewed by high resolution light or transmission electron microscopy of fixed material [72,150,151], while the outer 3–5 files are much smaller and have extremely small extracellular voids [150]. Although nodules from plants cultured with their whole root systems maintained in subambient pO_2 (1–10% O_2) showed a number of different morphological and fine structural adaptations, the radial dimension of the inner cortical layer of cells with little or no extracellular void was substantially reduced from one 45–50 μm thick in nodules grown in air to one only 10–20 μm thick in nodules grown in subambient pO_2 [72,150]. Theoretical considerations, based on a gas exchange model for soybean nodules developed by Hunt et al. [152], also indicate that it is the inner cortex which is most likely to provide a major barrier to gas movement.

There have been proposals that osmotic changes in cells of the cortex are likely to alter the relative extent to which the extracellular voids in the tissue contain water, and that this might constitute the basis of a "variable" barrier [153]; the water being pumped into or out of the voids. Certainly, changes in cortical cell turgor have been shown to alter gas movement in nodules [154]; although the putative site for the alteration in this case was the lenticel rather than the inner cortex. The fact that phloem supply to the nodule appears to be closely associated with diffusional control [145] might indicate a role for phloem-borne solutes, especially sucrose, in the osmotic relations of cortical cells. However, other solutes, such as the products of N metabolism, might also be involved in determining osmotic potentials

[147]. James et al. [155] have proposed that variable resistance may be a function of the water-retaining capacity of a glycoprotein which appears to be localized in cell walls of the inner cortical cell voids and has considerable antigenic homology with a component of the infection thread matrix in pea and soybean nodules [156, 157]. Van der Wiel et al. [158] have demonstrated that the cell walls of the inner cortex of both pea and soybean nodules contain a glycoprotein which is "nodule specific" (coded by the early nodulin gene ENOD2). A role for this protein in diffusional control has not been demonstrated, but these observations suggest that its early formation in inner cortical cells is a significant symbiotic feature. The relationship between the ENOD2 gene product and the glycoprotein described by James et al. [155] has not been established.

Although the means by which nodules alter their gas exchange features remains obscure it is clear that this ability is central to the functioning of an effective symbiosis. The mechanism is not restricted to determinate, spherical nodules, such as those from soybean or cowpea, but has been described for the indeterminate, branched nodules of species like pea and clover [139], while Tjepkema [159] has described, for *Parasponia* nodules, a layer of inner cortical cells corresponding to those of soybean. In nodules of *Frankia*: *Casuarina* and *Alnus* symbioses [160] there is some evidence for a plant-based barrier to O_2 permeability, but as yet there is no clear evidence for variable diffusional control in these nonlegume symbioses.

Layzell and Hunt [147] have proposed the existence of a number of sensors within legume nodules which can alter diffusion barrier resistance. One of these is envisaged as sensing the O_2 concentration within the central infected tissue. The second senses phloem translocation to the nodule, and a third the rate of N assimilation. It is equally likely, however, that levels of O_2, of phloem delivered solutes, or of products of ammonia assimilation are reflected in the output of a single sensor (e.g., adenylate charge, reductant poise, or the gradient generated by a membrane-based chemiosmotic mechanism [74]), which is in turn transduced as changes in the water/solute relations of the inner cortex. Of course these hypotheses suggest that the metabolic processes which underlie the sensor or sensors might well be located in the inner cortex. In this regard it is interesting to note that a major proportion of the enzymes of sugar utilization, dark CO_2 fixation, amino N synthesis, and ureide formation in cowpea or soybean nodules have been recovered in extracts of separated cortical tissues [161,162]. Thus a more detailed examination of the metabolic activity of these cells, particularly in respect of the processing of both phloem-delivered solutes into and of the products of N fixation out of the nodule, would seem to be justified.

The realization of the importance of O_2 diffusional control in nodules, and the linking of its function to the diverse factors, which have been described over many years as limiting fixation, represents a real breakthrough in our understanding of the legume symbiosis. As yet there has been no research on the role of the infecting rhizobia, or their gene products, on the establishment of the features of diffusional control. Similarly, although there is some evidence for differences between legume species in the rapidity or extent to which their nodules respond to the need for diffusional adjustment [163], genotypic variation in this respect has not been explored.

Effect of Combined N on N_2 Fixation

Generally, nodulation of legumes is delayed or prevented completely and N_2 fixation inhibited in N-rich soils or following application of nitrogenous fertilizer. Consequently, a roughly reciprocal relationship between nodule activity and the uptake and assimilation of combined N has been described for a wide range of symbioses [see 9]. From an agronomic point of view soil N reserves are utilized by legumes in which nodulation is restricted, just as they are by non-leguminous plants, thus reducing the potential benefit from N_2 fixation.

The interaction varies between species, with rhizobial strain, plant age, form, and site of application of combined N and with environmental conditions [see 9]. However, two separate components of the response have been clearly identified; although in practical terms, and particularly in field situations, it may be difficult to separate the two [9,164,165]. These are (1) inhibition of nodulation, and (2) inhibition of the functioning of already formed nodules. Explanations for these effects are many: They range from direct inhibition by nitrate (or products of its reduction) of infection, nodule initiation, nodule tissue differentiation, nitrogenase activity, or on the redox state of leghemoglobin or the transfer of solutes to bacteroids, to indirect effects resulting from changes in hormone balance and the partitioning of assimilates to nodules. None of these has successfully explained all aspects of the interaction and it seems likely that there is not really a single consequence of exposing a potentially nodulated plant or a functioning symbiosis to combined N, but rather a complex of effects which together serve to favor the use of soil N over that derived from fixation. From a teleological viewpoint, the uptake and assimilation of combined N is simpler than N_2 fixation in that the latter requires diversion of resources to the development of special organs, the nodules. Furthermore, the biochemistry of fixation, which is entirely dependent on heterotrophic processes, is likely to be more expensive (in photosynthate terms) than is the assimilation of NO_3^- [166]. In fact quantitative measures of the C and N economy of nodulated and nonnodulated legumes (lupins and cowpeas) indicate that the root systems of the latter are intrinsically more economic in their use of C than those which support nodules [167,168]. While the difference is greater for cowpea, a species which reduces NO_3^- principally in the shoot [168], compared to lupin, where a large proportion of NO_3^- is reduced by the root system [169], the increased respiration of the nodulated root in both species is associated not only with the activity of nodules but also of the supporting roots.

Recently, mutants of pea [170], common bean [171], and soybean [172] have been generated in which the inhibitory effects of NO_3^- on nodulation have been effectively overcome. These mutants, which have been termed *supernodulators* [172], form many more nodules than their parent varieties, apparently as a consequence of changes to the normal "autoregulatory" response of legumes to infection by rhizobia [173]. The trait of nitrate tolerant symbiosis (*nts*) appears to be simply inherited, and, in soybean [173] and some pea mutants [174], is controlled through the shoot, possibly through basipetal translocation of specific growth-regulating substances. Whether or not this is the case, the *nts* phenotype offers significant potential, not only to increase the nodulation of legumes in N-rich soils, but also to manipulate genetically the intensity of nodulation as yield level in legumes is raised.

At present there is no clear indication that the "supernodulation" feature itself leads to increased yield or N_2 fixation. Significantly the nitrogenase activities of the *nts* lines of soybean are little different to those of their parent line in tolerating NO_3^- application [175]. This is also true of the partially NO_3^--tolerant soybean mutants isolated by Gremaud and Harper [176]. While some of the less severe *nts* soybean lines appear to offer more potential in this regard than the extreme supernodulators (e.g., *nts* 1116 versus *nts* 1007 [175]), it is clear that truly isogenic materials are required before a proper evaluation of the agronomic value of the character can be made.

There have been a number of suggestions that efforts to improve the tolerance of symbioses to combined N should concentrate on means to limit uptake and/or assimilation of NO_3^- [165,177,178]. Progress in this area has not been rapid but the nitrate reductase-deficient mutants of pea [178,179] or soybean [180] may provide a means to test the hypothesis. Needed equally is a closer examination of the tolerance to soil N of less "domesticated" legume species, especially those adapted to conditions of low soil fertility [181]. In this respect, two indigenous West African legumes (Kersting's bean, *Macrotyloma geocarpum* and Bambara groundnut, *Vigna subterranea*) which are virtually unimproved, show considerable tolerance of nitrogenase activity to NO_3^- compared with the more common West African crop, cowpea (F. D. Dakora, C. A. Atkins, and J. S. Pate, unpublished data). Nodulation of both these geocarpic species is reduced by NO_3^-, but the ability of their nodules to fix N_2, especially in the case of Kersting's bean, is unaffected by levels of NO_3^- up to 15 mM. Kersting's bean is a crop grown essentially for family use in widely scattered small plots in village compounds [182] and as such has been subjected to virtually no selection for improved agronomic features.

Part of the problem in mounting a concerted effort to select or engineer symbioses in which N_2 fixation is tolerant of soil N is a lack of understanding about precisely how NO_3^- inhibits nitrogenase activity. Most hypothesis are based either on direct toxicity of NO_2^- formed at or close to bacteroids as a result of plant or rhizobial nitrate reductase activity, or on an indirect effect of NO_3^- assimilation elsewhere in the plant resulting in reduced assimilate supply to nodules [76,165,183]. As noted earlier, recent studies indicate that NO_3^- treatment results in increases in gas diffusion resistance in nodules [144, 184], with the consequence that O_2 supply to bacteroids becomes limiting. While the connection between diffusion resistance adjustment and the effect of NO_3^- is not yet clear (see p. 87), Vessey et al. [145] have demonstrated that with NO_3^- application phloem supply to nodules is sharply reduced. They [145] have argued that, in addition to restricting the supply of assimilates, phloem insufficiency results in too little water being available in the nodule for export of the products of N_2 fixation in xylem [185]. Furthermore, although NO_2^- formed close to bacteroids could potentially inhibit nitrogenase or interfere with leghemoglobin functioning [186], there is good evidence to suggest that almost all NO_3^- entering a nodule is reduced and assimilated outside the bacteroid-containing cells [9,187] and that any NO_2^- which is recovered in nodule extracts is an artifact of their extraction. Taken together these data would seem to support the "assimilate diversion" hypothesis but do not provide any evidence for the nature of the "sensory" mechanism of control.

Premature Senescence of Nodules at the Onset of Reproductive Growth in the Legume

Sinclair and de Wit [188] proposed that because grain legumes form protein-rich seed they are forced to undergo "self-destruction" of their photosynthetic potential during reproductive growth with leaf N providing a significant proportion of the seed N. As a consequence legumes suffer premature leaf senescence, and because of reduced assimilate supply, nodules might be expected to senesce and N_2 fixation to fall to a low level or cease altogether [79].

While there are undoubtedly a number of correlative controls, especially those of nutrient fluxes, which are involved in the course of leaf senescence, there is strong experimental evidence that the process is triggered by a "senescence signal" translocated from the developing seeds to the rest of the plant [189,190]. The nature of this signal is unknown, and, although all known plant growth regulators (IAA, gibberelins, ABA, cytokinins, and C_2H_4) have been implicated in the course of leaf senescence [191,192], none has been identified as causative. Recent studies of C, N, and H_2O economy of cowpea fruits have demonstrated a significant reversed xylem flow of water, and possibly solutes, from the developing embryo to the leaves sub-tending the fruit [193,194]. Furthermore this flow carries the excess water delivered in phloem so that its intensity is likely to increase with greater flow of assimilates to the fruit. If this stream turns out to carry the senescence factor then selection of genotypes for delayed leaf senescence may involve traits which enhance fruit transpiration as much as for those which maintain leaf functioning [195–197].

In species such as cowpea [198], soybean [199,200], and lupin [201], around half of the seed N at harvest is derived from mobilized foliar N. Much of this N is released from leaves following flowering when there is a decline in photosynthesis [202,203], chlorophyll, and Rubisco [198], and a steady increase in leaf proteolytic enzymes [198,204]. Generally also rates of N_2 fixation decline after anthesis [79,198,201], roughly in parallel with the decline in photosynthesis. While the major competitive sink for assim-ilates appears to be the growing seeds and fruits in species like cowpea this is not the case in those like lupin, where the demands of the parent root for respiratory substrates appear to be a more significant negative influence on nodule functioning [201].

However, despite the likely impairment of symbiotic activity during reproductive development, application of fertilizer N seldom leads to in-creased seed yield. In fact Imsande [205] has presented convincing evidence that net photosynthetic output and seed yield of soybean grown in both chamber and field conditions are greater in effectively nodulated plants com-pared to those relying solely on soil and/or fertilizer N. In addition to leaf senescence there is good evidence to show that root growth and functioning decline sharply following anthesis in grain legumes [206,207]. In soybean this results in declining nitrate uptake and the capacity for its assimilation during pod fill [205,208], so that nodulated plants actually achieve a higher rate of N input during this period than those without nodules [209]. The general applicability of these findings to legumes other than soybean will need to be established. In many field situations effective crown root nodula-tion is achieved by seed inoculation with an introduced rhizobial strain. However, the initial crop of nodules is frequently senescent by early pod fill [200], and subsequent nodulation, apparently extremely critical for en-hanced seed yield [205], is due to less or ineffective indigenous strains [36].

The precise events which lead to nodule senescence have not been clearly defined [164] and the role of translocated plant growth regulators in main-taining nodule function or, indeed, in triggering their decline, is poorly under-stood. Nodules which are formed by *fix⁻* rhizobia invariably senesce pre-maturely, and there is clear evidence that this can be a direct consequence of the lack of N [78,114,209]. Growing plants with their roots and nodules con-tinuously maintained in atmospheres devoid of N_2 (80% Ar:20% O_2 [78,114])

results in a normal symbiosis in which high levels of nitrogenase and leg-hemoglobin are expressed together with low levels of the plant enzymes of C and N metabolism essential for symbiotic functioning. These nodules also senesce prematurely, however, and their structural disintegration cannot be prevented by low levels of combined N [114]. These results emphasize the all-important role of ammonia production by the microsymbiont and suggest that maintenance of its (or a product of its assimilation) flux across the peribacteroid membrane may be the single most sensitive step in determining the longevity of nodule function. As noted earlier (p. 83), some of the unique properties of the peribacteroid membrane have been recently described [110], including the presence of transporters for C4 dicarboxylic acids. Akkermans et al. [210] and Kahn et al. [211] have proposed the activity of shuttle mechanisms (e.g., involving malate and aspartate [212]) which, if located in the peribacteroid membrane, could provide the molecular mechanism essential for C and N nutritional complementation of the symbiotic partners.

IMPROVING THE UTILITY OF N_2 FIXATION BY TRANSFERRING SYMBIOTIC TRAITS TO NONLEGUMES

The practical achievement of a nodulated cereal plant has been the "holy grail" of research in symbiotic N_2 fixation for many years. Progress has been slow, and, although there may be a good argument against the utility of nodulated cereals, until an example is found or engineered their real potential and value remains unknown.

Research has proceeded from two fairly distinct approaches: (1) nongenetic means for induction of nodulation and (2) genetic manipulation to transfer traits for nodulation or to engineer bacterial symbionts which will induce nodulation.

The basic premise supporting both of these approaches is that most, if not all, the traits required in a plant genome for the development of nodules are present in nonlegumes, just as they are in legumes, but that they might require a unique set of "triggers" from the potential bacterial partner for their ordered expression. These approaches presuppose that an infective prokaryote is the only means of introducing functional *nif* genes into the plant. Incorporating *nif* genes into the eukaryote genome, possibly into an organelle like the mitochondrion or plastid, offers another, probably more difficult, and certainly more challenging, strategy. Phillips [20], in a recent review has pointed out that the structure of the chloroplast genome is probably more suited to this task than is the peculiar genetic organization of the mitochondrion. The potentially toxic O_2 formed in photosynthesis probably means, however, that root plastids are likely to be the most hospitable host. To date there have been no reports of attempts, successful or otherwise, to transfer *nif* genes in to plant organelles.

Nongenetic Manipulation Approach

In the pioneering work of Cocking and his associates at Nottingham (reviewed recently in [212]), hydrolytic enzymes were used to remove or reduce the physical barriers to infection of nonlegume roots by rhizobia. Using this technology associations between *Rhizobium* or *Bradyrhizobium* spp. and a range of nonlegume crop plants (rice, wheat, and oil-seed rape) have been generated. While nodular structures have been recorded, in which bacteria were found both within and between plant cells, the levels of nitrogenase activity were extremely low; typically 0.1% or less of that found for legume nodules [213]. In the case of rape seed, nodulation was induced by *Bradyrhizobium parasponium* without the need for enzyme treatment of the roots [213]; but again nitrogenase was barely detectable.

Early studies on the effects of hormonal herbicides, particularly 2,4-D, have described the frequent distortion and formation of nodulelike protrusions on roots of both legumes and nonlegumes [214,215]. These observations have been extended by a number of researchers in China (reviewed by Tchan and Kennedy [216]), and, more recently, by Kennedy et al. [217,218], to achieve a high frequency of nodulelike structures (termed *para* nodules) on wheat roots. Inoculation of the plants with one or a number of 15 diverse rhizobial strains, at the same time as the roots were exposed to 2,4-D, did not result in an association which showed nitrogenase (C_2H_2 reduction) activity. However, inoculation with strains of *Azospirillum* (*A. brasilense* Sp7) has resulted in the formation of nodulelike structures which habor bacteria and which show consistently high rates of C_2H_2 reduction in atmospheres of 1 or 2% O_2 [219]. The bacteria appear to be localized toward the basal part of the *para* nodule, but there is no evidence for an intracellular location and it is not yet clear whether the azospirilla in these structures colonize live or dead cells. Preliminary experiments [219,220] have shown that application of the photosystem II inhibitor DCMU to leaves abolishes C_2H_2 reduction by *para* nodulated wheat roots suggesting that their nitrogenase activity is closely coupled to the translocation of photosynthate. Measurable ^{15}N enrichment of *para* nodules following their exposure to $^{15}N_2$ has been recorded [219], but there is no evidence that fixed N is transferred to the host. This is perhaps not too surprising because the *para* nodules, which apparently form from root initials, have a central stele of vascular tissue rather than the peripheral (cortical) network of vascular bundles more typical of legume nodules.

Undoubtedly the *para* nodules formed on wheat plants offer a useful model for further investigation of novel associations with diazotrophic bacteria. The major restriction to their utility appears to be the lack of suitable structures to restrict O_2 diffusion and to enhance the exchange of solutes. Other *Azosprillum* strains, or perhaps those of *Frankia*, might offer greater potential. While to date the only confirmed nonlegume symbiosis with *Rhizobium* is that of *Parasponia* spp. (family Ulmaceae [221]), *Frankia* symbioses occur in 17 genera from 8 families [222]. Furthermore frankiae fix N_2 over a wide range of pO_2 [223] and in some of their symbioses (e.g., those with *Coriaria* and *Datisca* spp.), the infected cells appear to be highly aerated [224].

A recent report [225] has described nodulation of barley roots with *Rhizobium astragali* under the influence of a permanent magnetic field. The "pseudo nodules" [225], which formed with rather low frequency, showed infection threads and typical legume-type infected cells, with the individual bacteria enclosed within membrane-bound vacuoles. However, nitrogenase (C_2H_2 reduction) activity could not be detected in detached nodules. The authors [225] suggest that the magnetic field might act through slow removal of barriers to recognition between the heterologous partners to the symbiosis, but rather more extensive quantitative data are needed to establish the phenomenon more clearly.

Genetic Manipulation Approach

In theory at least, traits which promote bacterial infection and lead to the nodulation response of the legume reside in the genomes of both partners to a symbiosis. While direct manipulation of the host plant genome by transferring traits from legumes to nonlegumes might seem an obvious line of investigation, it has in fact received little attention. Recently, Kjine et al. [226] have transferred a pea lectin gene into white clover with the result that the ability to be infected and nodulated by *R. trifolii* had been transferred. However, there has been little other work in this area. Perhaps the major reason for the lack of attention to the plant is that a number of the early events in the legume response to bacterial infection, and some of the early stages of nodule development, are largely controlled by the products of bacterial genes [4,10]. These events include attachment of bacteria, root hair curling, plant cell wall breakdown, cortical cell division, and the establishment of a primary nodule meristem. While there is interaction with the activity of the plant genome, for example in the production of flavonoid signals which enhance *nod* gene expression (see p. 78), or in the proliferation of infection threads and the formation of early nodule-specific proteins (early nodulins [4]), there has been remarkable progress in nodulation of cereals simply by using engineered bacteria to elicit a basic "nodulation response."

Genes associated with root hair curling (*hac*) in *R. meliloti* have been transferred to suitable plasmids and overexpressed in a *R. trifolii* strain or *Azospirillum brasilense* which then induced expression of hair curling on rice and maize roots [227,228]. More significantly Rolfe and his colleagues [229] have used the rhizobial strain isolated originally from *Parasponia* nodules (NGR 234) to form nodulelike structures on rice roots without the need to provide high levels of phytohormones. The bacterium was constructed to carry many copies of an allelic form of the regulatory *nod* D gene and this apparently allowed expression of normally unexpressed *Rhizobium* nodulation genes in the presence of rice rather than legume roots. The nodules which were formed contained membrane-encapsulated bacteria within cells of the central tissue and showed a peripheral cortical vasculature [229]. However, nitrogenase activities were negligible or even zero and the frequency of nodulation was very low (1 in 400–1000 plants).

Jing et al. [230] have used mutant strains of *Rhizobium sesbania*, isolated from *Sesbania cannabina*, which also formed nodules on the roots of particular varieties of rice. Preliminary data indicate some expression of nitrogenase but at levels only 1 or 2% of those in *Sesbania* nodules. There has been one report of high rates of N_2 fixation by a symbiosis formed between rhizobia and cereals [231]. In this report (actually a patent application), an F_1 transconjugant between two different rhizobial strains (*R. phaseoli* and *R. leucaena*) was crossed with *R. trifolii* and, in the presence of a variety of plant extracts, formed nodules on five different wheat varieties. However, these data have not been confirmed in a properly controlled study.

Clearly the next stage in forming "functional" root nodules in cereals requires expression of the bacterial genes which lead to sustained symbiosis. These include the nitrogenase (*nif*) genes themselves as well as *fix* genes (the products of which are required for nitrogenase function) and genes for the high O_2 affinity terminal oxidase and other respiratory catalysts needed for ATP synthesis under microaerobic conditions. Similarly, expression of regulatory genes involved in nitrogen metabolism (e.g., *ntr* genes in the expression of *nif* as well as those for bacterial glutamine synthetase, see [232]) is also required under symbiotic conditions. Studies with free-living symbiotic diazotrophs indicate that all of these genes are regulated predominantly, if not exclusively by cellular pO_2 [232,233] and regulatory genes have been identified in both *R. meliloti* (*fix* L [234]) and *B. japonicum* (*fix* R *nif* [235]), which either produce an O_2 sensing protein (in *R. meliloti*) or which are themselves an O_2 sensor (in *B. japonicum*).

Thus the key to symbiotic N_2 fixation, once a nodule structure has been achieved on nonlegume roots, is the establishment of an internal microaerobic environment. While the essential features which sustain or regulate low pO_2 in the infected cells of the legume nodule are now quite well understood (see p. 93), the events which lead to its initial establishment are not clear. One event of critical significance to symbiotic functioning is the exchange of solutes in a specific fashion across the peribacteroid membrane ([110]; see p. 0). Some of the nodule-specific proteins which form as the symbiosis becomes fully functional [236,237,238] are expressed in its structure, and it seems likely that their regulation is related to the establishment of a low O_2 environment. The membrane itself is of plant origin [239,240]; however, its formation appears to require at least four separate signals from the bacteria [238]. These involve a number of *fix* genes which have now been identified in *R. meliloti* [241] and which are likely to be regulated through the O_2 sensory product of *fix* L [234]. Important though the release of bacteria into PBM-bounded vesicles appears to be in many legume symbiosis, it does not occur in *Parasponia* nodules. Here the bacteria remain confined to an infected thread [242] but are still able to achieve high rates of N_2 fixation [243]. Recently Faria et al. [244,245] have shown that similar structures are also formed in nodules from at least 12 legume genera. Presumably in all these cases the plasma membrane of the infected cell provides the exchange barrier between the bacteria and the host, functioning as does the PBM in more typical legume nodules.

In legume nodules the onset of nitrogen fixation coincides with expression of a series of late nodulin genes [246] some of which undoubtedly function in maintaining the microenvironment which allows sustained fixation. The most obvious of these is leghemoglobin which plays a crucial role in O_2 transport and which, in legume nodules at least, is formed in very substantial amounts (3 mM in soybean cell cytosol [247]). The relationship between the appearance of leghemoglobin and the creation of a microaerobic environment is not clear. While the heme is obviously required for assembly of functional leghemoglobin, the formation of the apoprotein does not require the onset of prior heme synthesis [248], nor in fact does it require bacteroid differentiation [249]. The appearance of leghemoglobin and nitrogenase activity in nodules is essentially coincidental, and initiation of their synthesis does not require N_2 fixation [114]. Furthermore in functional non-legume symbioses, such as that with *Parasponia* or those involving *Frankia*, the levels of hemoglobin may be much lower than those in the typical legume nodule [222] suggesting that mechanisms other than, or additional to, the massive synthesis of heme are responsible for creating the microaerobic environment needed for *nif* and *fix* gene expression.

In soybean nodules two other nodule-specific proteins, nodulin-23 and nodulin-24, appear in the infected cells at about the same time as leghemoglobin [237,250]. Nodulin-24 is a PBM located protein [251]. The genes for each of these show considerable 5' end sequence homology and Verma et al. [251] have postulated that their coordinate regulation involves binding of a common *trans* activator. Whether such activators are of plant or bacterial origin remains to be determined. The major pathways of N assimilation and C metabolism in nodules also utilize nodule-specific isozymes (e.g., glutamine synthetase, glutamate synthase, uricase, and sucrose synthase, etc. [251,252]) which are highly expressed and which appear to be regulated in their expression by symbiotic development [131,253,254]. There is some evidence indicating that these particular pathways are essential to effective nodule functioning [132], but there is also some evidence that the nature of the overall products of NH_3 assimilation may be changed [133] without a negative effect on fixation. Thus, while induction of expression of nodulin genes coding for PBM-based proteins or leghemoglobin may be necessary for successful symbiosis in legumes, it is not yet clear whether or not this requirement is also true for the pathways of C metabolism which fashion the C_4 acids, or other substrates required for bacteroid respiration, or for the enzymes of amino acid or ureide synthesis which fashion the particular solutes which are transported to the host plant. At this stage a role for the genes encoding enzymes for these pathways in nodules formed on nonlegumes cannot even be proposed.

CONCLUDING REMARKS

The relationship between basic and applied research is frequently difficult to define, and, although fundamental research may unearth a multitude of new facts or change our interpretation of the meaning and significance of older information, the transfer of new knowledge from the bench of the researcher to the farmers' field is extraordinarily tortuous. Henzell [255] has examined this problem with respect to research priorities for biological nitrogen fixation in the context of its likely benefits to world agriculture in 1988. His overall conclusion was that no important agricultural innovations have yet emerged from the impressive recent advances in basic research in this area and, further, that the pace of application of these advances, particularly those which relate to the burgeoning knowledge of the molecular biology of the symbiosis, should not be forced.

These conclusions are probably more justified for the legume–rhizobia symbiosis than for many other agricultural enterprises. This is, in part, because our knowledge of the bacterium, and especially the genetics and biochemistry of its symbiosis, has far outstripped our knowledge of the physiological, biochemical, and genetic aspects of the legume. Compared to most cereal crops, the grain legumes (except perhaps for soybean) are not improved to the point where they approach their yield potential, and, though their utility in providing their own N is obvious, their potential in this regard is also rarely realized. Clearly our knowledge of those factors intrinsic to the symbiosis which limit its value has expanded enormously in recent years. The tools of molecular biology have yielded new information at a level of detail which is adequate for serious genetic engineering of the rhizobial symbionts to be undertaken. Factors in the symbiosis which relate to specificity of infection and, perhaps also to "competitiveness" of introduced organisms in soil, appear to be especially amenable to these efforts. However, the regulatory systems of the legume which govern nodule development and the nutritional complementation of the symbiotic partners remain to be described. The recent understanding of the plant's ability to control symbiotic activity by altering, on an hour by hour basis, the gaseous diffusional properties of nodules, is a most significant development. Knowledge of this mechanism should provide a basis for the direct assessment of how effectively the nodules on a plant in the field operate and how closely they approach their potential for N_2 fixation. It also offers a framework within which the genetics of the legume's contribution to the symbiosis can be approached and in nonlegumes, possibly modified so that new novel symbioses can be developed.

There is no doubt that research into the N_2 fixing symbiosis in legumes will continue to offer new challenges to the microbiologist and plant scientist alike. However, it also seems likely that the two, often separate, threads of research in this area will need to be integrated more effectively before the progress in basic research can be translated into improved crop yields.

REFERENCES

1. W. J. Broughton, *Nigrogen Fixation*, Vol. 3, *Legumes* (W. J. Broughton, ed.), Clarendon Press, Oxford (1983).
2. M. J. Dilworth and A. R. Glenn, *Biology and Biochemistry of Nitrogen Fixation* (M. J. Dilworth and A. R. Glenn, eds.), Elsevier, New York (1991).
3. G. Hardarson and T. A. Lie, *Breeding Legumes for Enhanced Symbiotic Nitrogen Fixation* (G. Hardarson and T. A. Lie, eds.), Martinus Nijhoff/Junk, Dordrecht, The Netherlands (1983).
4. B. G. Rolfe and P. M. Gresshoff, *Ann. Rev. Plant Physiol. Plant Mol. Biol., 39*: 297 (1988).
5. J. I. Sprent, *New Phytol., 111*: 129 (1989).
6. N. T. Keen and B. Staskawicz, *Ann. Rev. Microbiol., 42*: 42 (1988).
7. K. R. Schubert, *Ann. Rev. Plant Physiol., 37*: 539 (1986).
8. C. A. Atkins and L. Beevers, *Nitrogen in Higher Plants* (Y. P. Abrol, ed.), Research Studies Press Ltd., U.K., p. 223 (1990).
9. J. S. Pate and C. A. Atkins, *Nitrogen Fixation*, Vol. 3, *Legumes* (W. J. Broughton, ed.), Clarendon Press, Oxford, p. 245 (1983).
10. J. A. Downie and A. W. B. Johnston, *Plant, Cell Environ., 11*: 403 (1988).
11. E. Martinez, D. Romero, and R. Palacios, *Plant Sci., 9*: 59 (1990).
12. S. R. Long and J. Cooper, *Molecular Plant—Microbe Interactions* (D. P. S. Verma and R. Palacios, eds.), APS Press, St. Paul, MN, p. 163 (1988).

13. E. A. Kueneman, W. R. Root, K. E. Dashiell, and J. Hohenberg, *Breeding Legumes for Enhanced Symbiotic Nitrogen Fixation* (G. Hardarson and T. A. Lie, eds.), Martinus Nijhoff/Junk, Dordrecht, The Netherlands, p. 115 (1984).
14. P. H. Graham and S. R. Temple, *Breeding Legumes for Enhanced Symbiotic Nitrogen Fixation* (G. Hardarson and T. A. Lie, eds.), Martinus Nijhoff/Junk, Dordrecht, The Netherlands, p. 43 (1984).
15. A. D. Robson and P. J. Bottomley, *Biology and Biochemistry of Nitrogen Fixation* (M. J. Dilworth and A. R. Glenn, eds.), Elsevier, New York, p. 320 (1991).
16. K. R. Schubert and H. J. Evans, *Proceedings National Academy of Sciences (U.S.), 73*: 1207 (1976).
17. H. J. Evans, F. J. Hanus, R. A. Haugland, M. A. Cantrell, L. S. Xu, S. A. Russell, G. R. Lambert, and A. R. Harker, *World Soybean Research Conference III. Proceedings* (R. Shibles, ed.), Westview Press, Boulder, CO, p. 935 (1985).
18. M. Soberon, H. D. Williams, R. K. Poole, and E. Escamilla, *J. Bacteriol., 171*: 465 (1989).
19. P. T. C. Nambia, S. W. Ma, and V. N. Iyer, *Appl. Environ. Microbiol., 56*: 2866 (1990).
20. D. A. Phillips, *Biology and Biochemistry of Nitrogen Fixation* (M. J. Dilworth and A. R. Glenn, eds.), Elsevier, New York, p. 408 (1991).
21. E. W. Triplett, *Molecular Plant—Microbe Interactions, 3*: 199 (1990).
22. R. A. de Maagd, C. van Rossum, and B. J. J. Lugtenberg, *J. Bacteriol., 170*: 3782 (1988).
23. I. L. Pepper, K. L. Josephson, C. S. Nautiyal, and D. P. Bourque, *Soil Biol. Biochem., 21*: 749 (1989).
24. M. J. Sadowsky, P. B. Cregan, and H. H. Keyser, *Appl. Environ. Microbiol., 56*: 1768 (1990).
25. J. E. Cooper, A. J. Bjourson, and J. K. Thompson, *Appl. Environ. Microbiol., 53*: 1705 (1987).
26. P. R. Schofield, A. H. Gibson, W. F. Dudman, and J. M. Watson, *Appl. Environ. Microbiol., 53*: 2942 (1987).
27. R. Wheatcroft and R. J. Watson, *Appl. Environ. Microbiol., 54*: 574 (1988).
28. S. T. Lam, D. M. Ellis, and J. M. Ligon, *Plant and Soil, 129*: 11 (1990).
29. J. Brockwell, R. J. Roughley, and D. F. Herridge, *Austral. J. Agric. Res., 38*: 61 (1987).
30. G. Hardarson, G. H. Heichel, D. K. Barnes, and C. P. Vance, *Crop Sci., 24*: 895 (1982).
31. P. B. Cregan and H. H. Keyser, *Crop Sci., 26*: 911 (1986).
32. D. K. Barnes, G. H. Heichel, C. P. Vance, and W. R. Ellis, *Breeding Legumes for Enhanced Symbiotic Nitrogen Fixation* (G. Hardarson and T. A. Lie, eds.), Martinus Nijhoff/Junk, Dordrecht, The Netherlands, p. 31 (1984).
33. D.-M. Li and M. Alexander, *Plant and Soil, 129*: 195 (1990).
34. G. Ciafardini and C. Barbieri, *Agron. J., 79*: 645 (1987).
35. Y. A. Hamdi, *Soil Biol. Biochem., 3*: 121 (1971).
36. G. Hardarson, M. Golbs, and S. K. A. Danso, *Soil Biol. Biochem., 21*: 783 (1989).

37. R. M. Kosslak and B. Bohlool, *Plant Physiol., 75*: 125 (1984).
38. M. Pierce and W. D. Bauer, *Plant Physiol., 73*: 286 (1983).
39. P. Wadisinisuk, S. K. A. Danso, G. Hardarson, and G. D. Bowen, *Appl. Environ. Microbiol., 55*: 1711 (1989).
40. P. Ames and K. Bergman, *J. Bacteriol., 148*: 728 (1981).
41. R. F. Zdor and S. G. Pueppke, *Can. J. Microbiol., 37*: 52 (1991).
42. W. D. Bauer and G. Caetano-Anollés, *Plant and Soil, 129*: 45 (1990).
43. P. B. Cregan, H. H. Keyser, and M. J. Sadowsky, *Appl. Environ. Microbiol., 55*: 2532 (1989).
44. M. J. Sadowsky, R. E. Tully, P. B. Cregan, and H. H. Keyser, *Appl. Environ. Microbiol., 53*: 2624 (1987).
45. D. H. Demazas and P. J. Bottomley, *Soil Sci. Soc. Amer. J., 48*: 1067 (1984).
46. M. B. Jenkins and P. J. Bottomley, *Soil Sci. Soc. Amer. J., 49*: 326 (1985).
47. L. F. Williams and D. L. Lynch, *Agron. J., 46*: 28 (1954).
48. L. A. Materon and J. M. Vincent, *Field Crops Res., 3*: 215 (1980).
49. C. L. Diaz, L. S. Melchers, P. J. J. Hooykaas, B. J. J. Lugtenberg, and J. W. Kijne, *Nature, 338*: 579 (1989).
50. A. Quespel, *Physiol. Plant., 74*: 783 (1988).
51. A. W. B. Johnston, J. L. Beynon, A. V. Buchanan-Wollaston, S. M. Setchell, P. R. Hirsch, and J. E. Beringer, *Nature, 276*: 634 (1978).
52. H. Hennecke, H. M. Fisher, S. Ebeling, M. Gubler, B. Thony, M. Gottfert, J. Lamb, M. Hahn, T. Ramseier, B. Regensburger, A. A. Ivarez-Morales, and D. Studer, *Molecular Genetics of Plant — Microbe Interactions* (D. P. S. Verma and N. Brisson, eds.), Martinus Nijhoff, Dordrecht, The Netherlands, p. 191 (1987).
53. J. R. Gallon and A. E. Chaplin, *An Introduction to Nitrogen Fixation*, Cassell, London (1987).
54. J. L. Firmin, K. E. Wilson, L. Rossen, and A. W. B. Johnston, *Nature, 324*: 90 (1986).
55. M. A. Djordjevic, J. W. Redmond, M. Batley, and B. G. Rolfe, *EMBO. J., 6*: 1173 (1987).
56. Z. Gyorgypal, N. Iyer, and A. Kondorosi, *Mol. Gen. Genet., 212*: 85 (1988).
57. B. Horvath, C. W. B. Bachem, J. Schell, and A. Kondorosi, *EMBO J., 6*: 841 (1987).
58. B. G. Rolfe, D. Barnard, L. Preston, and G. L. Bender, Proceedings of the 9th Australian Nitrogen Fixation Conference, Canberra (A. E. Richardon and M. B. Peoples, eds.), p. 71 (1991).
59. G. Truchet, F. Debelle, J. Vasse, B. Terzaghi, A. M. Garnerone, C. Rosenberg, J. Batut, F. Maillet, and J. Dénarié, *J. Bacteriol., 164*: 1200 (1985).
60. M. E. Dudley, T. W. Jacobs, and S. R. Long, *Planta, 171*: 289 (1987).
61. P. Lerouge, P. Roche, C. Faucher, F. Maillet, G. Truchet, J. C. Promé, and J. Dénarié, *Nature, 344*: 781 (1990).
62. E. A. Schwinghamer and R. P. Belkengren, *Arch. Mikrobiol., 64*: 130 (1968).
63. E. W. Triplett, Proceedings National Academy of Sciences (U.S.), 85: 3810 (1988).
64. J. P. W. Young, *J. Gen. Microbiol., 131*: 2399 (1985).

65. D. Pinero, E. Martinez, and R. K. Selander, *Appl. Environ. Microbiol., 54*: 2825 (1988).
66. J. P. W. Young and M. Wexler, *J. Gen. Microbiol., 134*: 2731 (1988).
67. A. H. Gibson, D. H. Demezas, R. R. Gault, T. V. Bhuvaneswari, and J. Brockwell, *Plant and Soil, 129*: 37 (1990).
68. A. W. B. Johnston and J. E. Beringer, *J. Appl. Bacteriol., 40*: 375 (1976).
69. W. J. Broughton, V. Samrey, and J. Stanley, *FEMS Microbiol. Lett., 40*: 251 (1987).
70. J. L. Fox, *ASM News, 55*: 259 (1989).
71. L. K. Poulsen, N. W. Larsen, S. Molin, and P. Andersson, *Mol. Microbiol., 3*: 1463 (1989).
72. F. D. Dakora and C. A. Atkins, *Planta, 182*: 572 (1990).
73. J. I. Sprent and J. A. Raven, Proceedings of the Royal Society, Edinburgh, 85B: 215 (1985).
74. C. A. Atkins, J. S. Pate, P. J. Sanford, F. D. Dakora, and I. Matthews, *Plant Physiol., 90*: 1644 (1989).
75. P. J. Dart and F. V. Mercer, *Arch. Microbiol., 51*: 233 (1965).
76. C. G. O. Oghoghorie and J. S. Pate, *Proceedings of the Technical Meeting on Biological Nitrogen Fixation of the International Biological Programme* (T. A. Lie and E. G. Mulder, eds.), p. 185. *Plant and Soil* (Special Volume), Martinus Nijhoff, The Netherlands (1971).
77. J. I. Sprent and R. J. Thomas, *Plant Cell Environ., 7*: 637 (1984).
78. C. A. Atkins, B. J. Shelp, J. Kuo, M. B. Peoples, and J. S. Pate, *Planta, 162*: 316 (1984).
79. C. A. Atkins, *Outlook on Agric., 15*: 128 (1986).
80. F. D. Dakora and C. A. Atkins, *Austral. J. Plant Physiol., 16*: 131 (1989).
81. L. E. Williams, T. B. Dejong, and D. A. Phillips, *Plant Physiol., 68*: 1206 (1981).
82. A. D. Robson, *Nitrogen Fixation*, Vol. 3, *Legumes* (W. J. Broughton, ed.), Clarendon Press, Oxford, p. 36 (1983).
83. D. W. Israel, *Plant Physiol., 84*: 835 (1987).
84. E. Jakobsen, *Breeding Legumes for Enhanced Symbiotic Nitrogen Fixation* (G. Hardarson and T. A. Lie, eds.), Martinus Nijhoff/Junk, Dordrecht, The Netherlands, p. 155 (1984).
85. G. V. Johnson and G. J. Youngblood, *Plant and Soil, 130*: 219 (1991).
86. R. E. Terry, K. U. Soerensen, V. D. Jolley, and J. C. Brown, *Plant and Soil, 130*: 225 (1991).
87. A. D. Robson, G. W. O'Hara, and L. K. Abbott, *Austral. J. Plant Physiol., 8*: 427 (1981).
88. Reference to be provided.
89. K. G. Cossman, A. S. Whitney, and R. L. Fox, *Agron. J., 73*: 17 (1981).
90. M. G. Zaroug and D. N. Munns, *Plant and Soil, 53*: 329 (1979).
91. D. W. Israel and T. W. Rufty, Jr., *Crop Sci., 28*: 954 (1988).
92. H. H. Keyser and D. N. Munns, *Soil Sci. Soc. Amer. J., 43*: 519 (1979).
93. D. N. Munns, *Plant and Soil, 32*: 90 (1970).
94. G. Caetono-Annollés, A. Lagares, and G. Favelukes, *Plant and Soil, 117*: 67 (1989).
95. G. Smit, J. W. Kijne, and B. J. J. Lugtenberg, *J. Bacteriol., 169*: 4294 (1987).

96. A. E. Richardson, M. A. Djordjevic, B. G. Rolfe, and R. J. Simpson, *Soil Biol. Biochem., 20*: 431 (1988).
97. A. E. Richardson, R. J. Simpson, M. A. Djordjevic, and B. G. Rolfe, *Appl. Environ. Microbiol., 54*: 2541 (1988).
98. W. L. Lowther and J. F. Loneragan, *Plant Physiol., 43*: 1362 (1968).
99. B. Cartwright and E. G. Hallsworth, *Plant and Soil, 33*: 685 (1970).
100. J. R. Cowles, H. J. Evans, and S. A. Russell, *J. Bacteriol., 97*: 1460 (1969).
101. M. J. Dilworth, A. D. Robson, and D. L. Chatel, *New Phytol., 83*: 63 (1979).
102. R. W. Miller, *Biology and Biochemistry of Nitrogen Fixation* (M. J. Dilworth and A. R. Glenn, eds.), Elsevier, New York, p. 9 (1991).
103. R. N. Pau, *TIBS, 14*: 183 (1989).
104. R. N. Pau, *Biology and Biochemistry of Nitrogen Fixation* (M. J. Dilworth and A. R. Glenn, eds.), Elsevier, New York, p. 37 (1991).
105. R. L. Robson, *Biology and Biochemistry of Nitrogen Fixation* (M. J. Dilworth and A. R. Glenn, eds.), Elsevier, New York, p. 142 (1991).
106. D. P. S. Verma and S. Long, *International Review of Cytology, Suppl. 14* (K. Jeon, ed.), Academic Press, New York, p. 211 (1983).
107. M. L. Guerinot, *Plant and Soil, 130*: 199 (1991).
108. P. R. Gill, Jr., L. L. Barton, M. D. Scoble, and J. B. Neilands, *Plant and Soil, 130*: 211 (1991).
109. K. D. Nadler, A. W. B. Johnston, J. W. Chen, and T. R. John, *J. Bacteriol., 172*: 670 (1990).
110. L. Rosendahl, A. R. Glenn, and M. J. Dilworth, *Biology and Biochemistry of Nitrogen Fixation* (M. J. Dilworth and A. R. Glenn, eds.), Elsevier, New York, p. 259 (1991).
111. D. L. Chatel, A. D. Robson, J. W. Gartrell, and M. J. Dilworth, *Austral. J. Agric. Sci., 29*: 1191 (1978).
112. M. J. Dilworth, A. D. Robson, and D. L. Chatel, *New Phytol., 83*: 63 (1979).
113. R. K. Howell, *J. Plant Nutr., 10*: 1297 (1987).
114. C. A. Atkins, *Plant and Soil, 82*: 273 (1984).
115. K. R. Schubert, *The Energetics of Biological Nitrogen Fixation Workshop Summaries—I, Amer. Soc. Plant Physiol.* (1982).
116. R. W. F. Hardy and U. D. Havelka, *Science, 188*: 633 (1975).
117. G. J. Bethlenfalvay and D. A. Phillips, *Plant Physiol., 60*: 868 (1977).
118. R. J. Lawn and W. A. Brun, *Crop Sci., 14*: 11 (1974).
119. C. Y. Huang, J. S. Boyer, and L. N. Vanderhoef, *Plant Physiol., 56*: 228 (1975).
120. J. D. Mahon, *Can. J. Plant Sci., 62*: 5 (1982).
121. S. L. A. Hobbs and J. D. Mahon, *Can. J. Bot., 60*: 2594 (1982).
122. R. M. Rainbird, C. A. Atkins, and J. S. Pate, *Plant Physiol., 72*: 308 (1983).
123. R. M. Rainbird, C. A. Atkins, and J. S. Pate, *Plant Physiol., 73*: 392 (1983).
124. L. E. Williams, *Crop Sci., 23*: 246 (1983).
125. T. M. DeJong, N. J. Brewin, and D. A. Phillips, *J. Gen. Microbiol., 124*: 1 (1981).
126. T. M. DeJong, M. J. Brewin, A. W. B. Johnson, and D. A. Phillips, *J. Gen. Microbiol., 128*: 1829 (1982).

127. K. R. Schubert and H. J. Evans, *Recent Developments in Nitrogen Fixation* (W. E. Newton, G. R. Postgate, and C. Rodriquez-Barrueco, eds.), Academic Press, New York, p. 469 (1977).

128. R. M. Rainbird, C. A. Atkins, J. S. Pate, and P. Sanford, *Plant Physiol., 71*: 122 (1983).

129. C. A. Atkins, *Biology and Biochemistry of Nitrogen Fixation* (M. J. Dilworth and A. R. Glenn, eds.), Elsevier, New York, p. 293 (1991).

130. D. B. Layzell, R. M. Rainbird, C. A. Atkins, and J. S. Pate, *Plant Physiol., 64*: 888 (1979).

131. C. A. Atkins, B. J. Shelp, P. J. Storer, and J. S. Pate, *Planta, 162*: 327 (1984).

132. C. A. Atkins, P. J. Sanford, P. J. Storer, and J. S. Pate, *Plant Physiol., 88*: 1229 (1989).

133. T. J. Knight and P. J. Langston-Unkefer, *Science, 241*: 951 (1988).

134. P. L. Langston-Unkefer, P. A. Macy, and R. D. Durbin, *Plant Physiol., 76*: 71 (1984).

135. R. L. Robson and J. R. Postgate, *Ann. Rev. Microbiol., 34*: 183 (1980).

136. C. A. Appleby, *Ann. Rev. Plant Physiol., 35*: 443 (1984).

137. C. A. Appleby, *Biochim. Biophys. Acta, 188*: 222 (1969).

138. D. B. Layzell, S. Hunt, and G. R. Palmer, *Plant Physiol., 92*: 1101 (1990).

139. J. E. Sheehy, F. R. Minchin, and J. F. Witty, *Ann. Bot., 53*: 13 (1983).

140. S. Hunt, B. J. King, and D. B. Layzell, *Plant Physiol., 91*: 315 (1989).

141. P. R. Weisz and T. R. Sinclair, *Plant Physiol., 84*: 906 (1987).

142. J. F. Witty, J. Skøt, and N. P. Revsbech, *J. Exp. Bot., 38*: 1129 (1987).

143. P. R. Weisz, R. F. Denison, and T. R. Sinclair, *Plant Physiol., 78*: 525 (1985).

144. J. K. Vessey, K. B. Walsh, and D. B. Layzell, *Physiol. Plant., 73*: 113 (1988).

145. J. K. Vessey, K. B. Walsh, and D. B. Layzell, *Physiol. Plant., 74*: 137 (1988).

146. F. R. Minchin, J. R. Witty, J. E. Sheehy, and M. Müller, *J. Exp. Bot., 34*: 641 (1983).

147. D. B. Layzell and S. Hunt, *Physiol. Plant., 80*: 322 (1990).

148. J. D. Tjepkema and C. S. Yocum, *Planta, 119*: 351 (1974).

149. F. D. Dakora and C. A. Atkins, *Planta, 182*: 572 (1990).

150. F. D. Dakora and C. A. Atkins, *Plant Physiol., 96*: 728 (1991).

151. E. H. Newcomb, Y. Kaneko, and K. A. van der Bosch, *Protoplasma, 150*: 150 (1989).

152. S. Hunt, S. Gaito, and D. B. Layzell, *Planta, 173*: 128 (1988).

153. J. F. Witty, F. R. Minchin, L. Skøt, and J. E. Sheehy, *Oxford Survey Plant Mol. Cell Biol., 3*: 275 (1986).

154. E. J. Ralston and J. Imsande, *J. Exp. Bot., 33*: 208 (1982).

155. E. K. James, N. J. Brewin, J. A. Chudek, F. R. Minchin, and J. I. Sprent, *Proceedings AFRC Meeting on Plant and Soil Nitrogen Metabolism* (P. J. Lea, ed.), Lancaster University, Agricultural and Food Research Council, Swindon, U.K., p. 29 (1989).

156. D. J. Bradley, E. A. Wood, A. P. Larkins, G. Galfre, G. W. Butcher, and N. J. Brewin, *Planta, 173*: 149 (1988).

157. K. A. van der Bosch, D. J. Bradley, J. P. Know, S. Perotto, G. W. Butcher, and N. J. Brewin, *EMBO J., 8*: 335 (1989).

158. C. van der Wiel, B. Scheres, H. Franseen, M-J van Lierop, A. van Lammeren, A. van Kammen, and T. Bisseling, *EMBO J., 9*: 1 (1990).
159. J. D. Tjepkema, *Current Perspectives in Nitrogen Fixation* (A. H. Gibson and W. E. Newton, eds.), Australian Academy of Sciences, Canberra, Australia, p. 386 (1981).
160. J. D. Tjepkema and M. A. Murray, *Plant and Soil, 118*: 111 (1989).
161. C. A. ATkins, R. M. Rainbird, and J. S. Pate, *Z. Pflanzenphysiol., 97*: 249 (1980).
162. L. Copeland, J. Vella, and Z. Q. Hong, *Phytochem., 28*: 57 (1989).
163. F. R. Minchin, J. E. Sheehy, and J. F. Witty, *Nitrogen Fixation Research Progress* (H. J. Evans, P. J. Bottomley, W. E. Newton, eds.), Martinus Nijhoff, Dordrecht, The Netherlands, p. 285 (1985).
164. W. D. Sutton, *Nitrogen Fixation Vol. 3: Legumes* (W. J. Broughton, ed.), Clarendon Press, Oxford, p. 144 (1983).
165. J. E. Harper and A. H. Gibson, *Crop Sci., 224*: 797 (1984).
166. J. S. Pate, C. A. Atkins, and R. M. Rainbird, *Current Perspectives in Nitrogen Fixation* (A. H. Gibson and W. E. Newton, eds.), Australian Academy of Sciences, Canberra, Australia, p. 105 (1981).
167. J. S. Pate, D. B. Layzell, and C. A. Atkins, *Plant Physiol., 64*: 1083 (1979).
168. C. A. Atkins, J. S. Pate, G. J. Griffiths, and S. T. White, *Plant Physiol., 66*: 978 (1980).
169. C. A. Atkins, J. S. Pate, and D. B. Layzell, *Plant Physiol., 64*: 1078 (1979).
170. E. Jacobsen and W. J. Feenstra, *Plant Sci. Lett., 33*: 337 (1984).
171. B. R. Buttery and S. J. Park, *Can. J. Plant Sci., 70*: 375 (1989).
172. B. J. Carroll, D. L. McNeil, and P. M. Gresshoff, *Proceedings National Academy of Sciences (U.S.), 32*: 4162 (1985).
173. A. C. Delves, A. Matthews, D. A. Day, A. S. Carter, B. J. Carroll, and P. M. Gresshoff, *Plant Physiol., 82*: 588 (1986).
174. G. Duc and A. Messager, *Plant Sci., 60*: 207 (1989).
175. A. P. Hansen, M. B. Peoples, P. M. Gresshoff, C. A. Atkins, J. S. Pate, and B. J. Carroll, *J. Exp. Bot., 40*: 715 (1989).
176. M. F. Gremaud and J. E. Harper, *Plant Physiol., 89*: 169 (1989).
177. A. H. Gibson and J. E. Harper, *Crop Sci., 25*: 497 (1985).
178. E. Jacobsen, *Plant and Soil, 82*: 427 (1984).
179. W. J. Feenstra, E. Jacobson, A. C. P. M. van Sway, and A. J. C. de Visser, *Z. Pflanzenphysiol., 105*: 471 (1982).
180. S. A. Ryan, R. S. Nelson, and J. E. Harper, *Plant Physiol., 72*: 510 (1983).
181. P. H. Graham and S. R. Temple, *Plant and Soil, 82*: 43 (1984).
182. D. E. Kay, *TPI Crop and Product Digest No. 3*, Tropical Products Institute Ministry of Overseas Development, London, U.K., pp. 17, 205 (1979).
183. J. G. C. Small and O. A. Leonard, *Amer. J. Bot., 56*: 187 (1969).
184. F. R. Minchin, M. I. Minguez, J. E. Sheehy, J. F. Witty, and L. L. Skøt, *J. Exp. Bot., 37*: 1103 (1986).
185. K. B. Walsh, M. J. Canny, and D. B. Layzell, *Plant Cell Environ., 12*: 713 (1989).
186. J. Rigaud and A. Puppo, *Biochim. Biophys. Acta, 497*: 702 (1977).

187. J. I. Sprent, C. Giannakis, and W. Wallace, *J. Exp. Bot., 38*: 1121 (1987).
188. T. R. Sinclair and C. T. de Wit, *Science, 189*: 565 (1975).
189. S. J. Lindoo and L. D. Noodén, *Plant Physiol., 59*: 1136 (1977).
190. L. D. Noodén and C. S. Mauk, *Physiol. Plant., 70*: 735 (1987).
191. H. W. Woolhouse, *The Molecular Biology of Plant Development* (H. Smith and D. Grierson, eds.), Blackwells, Oxford, U.K., p. 256 (1982).
192. K. V. Thimann, *Biol. Plant., 27*: 83 (1985).
193. M. B. Peoples, J. S. Pate, C. A. Atkins, D. R. Murray, *Plant Physiol., 77*: 142 (1985).
194. J. S. Pate, M. B. Peoples, A. van Bel, J. Kuo, and C. A. Atkins, *Plant Physiol., 77*: 148 (1985).
195. S. S. Abu-Shakra, D. A. Phillips, R. C. Huffaker, *Science, 199*: 973 (1978).
196. D. A. Phillips, R. O. Pierce, S. A. Edie, K. W. Foster, and P. F. Knowles, *Crop Sci., 24*: 518 (1984).
197. G. L. Loberg, R. Shibles, D. E. Green, and J. J. Hanway, *J. Plant Nutr., 7*: 1311 (1984).
198. M. B. Peoples, J. S. Pate, and C. A. Atkins, *J. Exp. Bot., 34*: 563 (1983).
199. J. J. Hanway and C. R. Weber, *Agron. J., 63*: 406 (1971).
200. F. Zapata, S. K. A. Danso, G. Hardarson, and M. Fried, *Agron. J., 79*: 172 (1987).
201. C. A. Atkins, D. F. Herridge, and J. S. Pate, "Isotopes in Biological Dinitrogen Fixation" (C. N. Welsh, ed.), Proceedings IAEA Meeting on N_2 Fixation, Vienna, p. 221 (1979).
202. A. Sesay and R. Shibles, *Ann. Bot., 45*: 47 (1980).
203. J. S. Pate, M. B. Peoples, and C. A. Atkins, *J. Exp. Bot., 34*: 544 (1983).
204. V. A. Wittenbach, R. C. Ackerson, R. T. Giaquinta, and R. R. Hebert, *Crop Sci., 20*: 225 (1980).
205. J. Imsande, *Agron. J., 81*: 549 (1989).
206. W. C. Mayaki, I. D. Teare, and L. R. Stone, *Crop Sci., 16*: 92 (1976).
207. S. A. Barber, *Agron. J., 70*: 457 (1978).
208. J. Imsande and D. G. Edwards, *Agron. J., 80*: 789 (1988).
209. J. Imsande, *J. Exp. Bot., 39*: 1313 (1988).
210. A. D. L. Akkermans, K. Huss-Danell, and W. Roelofson, *Physiol. Plant., 53*: 298 (1981).
211. M. L. Kahn, J. Kraus, and J. E. Somerville, *Nitrogen Fixation Research Progress* (H. J. Evans, P. J. Bottomley, and W. E. Newton, eds.), Martinus Nijhoff, Dordrecht, The Netherlands, p. 193 (1985).
212. M. A. Appels and H. Haaker, *Plant Physiol., 95*: 740 (1991).
213. M. K. Al-Mallah, M. R. Davey, and E. C. Cocking, *J. Exp. Bot., 41*: 1567 (1990).
214. M. H. Wilde, *Amer. J. Bot., 38*: 79 (1951).
215. K. A. Allen, O. N. Allen, and A. S. Newman, *Amer. J. Bot., 40*: 429 (1953).
216. Y. T. Tchan and I. R. Kennedy, *Agric. Sci. (AIAS Melbourne), 2*: 47 (1989).
217. I. R. Kennedy, S. Sriskandarajah, D. Yu, Y. F. Nie, and Y. T. Tchan, Proceedings of the 9th Australian Nitrogen Fixation Conference, Canberra (A. E. Richardson and M. B. Peoples, eds.), p. 78 (1991).

218. I. R. Kennedy, A. Zeman, Y. T. Tchan, P. B. New, S. Sriskandarajah, and Y. F. Nie, *Transactions 14th International Congress Soil Science, 3*: 146 (1990).
219. Y. T. Tchan, A. M. M. Zeman, and I. R. Kennedy, *Plant Soil, 137*: 43 (1991).
220. A. M. Zeman, Y. T. Tchan, and I. R. Kennedy, Proceedings of the 9th Australian Nitrogen Fixation Conference, Canberra (A. E. Richardson and M. B. Peoples, eds.), p. 80 (1991).
221. M. J. Trinick, *Nature, 144*: 459 (1973).
222. C. S. Schwintzer and J. D. Tjepkema, *Biology and Biochemistry of Nitrogen Fixation* (M. J. Dilworth and A. R. Glenn, eds.), Elsevier, Amsterdam, p. 350 (1991).
223. R. Parsons, W. B. Silvester, S. Harris, W. T. M. Gruijters, and S. Bullivant, *Plant Physiol., 83*: 728 (1987).
224. W. B. Silvester, S. L. Harris, and J. D. Tjepkema, *The Biology of Frankia and Actinorhizal Plants* (C. R. Schwintzer and J. D. Tjepkema, eds.), Academic Press, New York, p. 157 (1990).
225. Y. Jing, B. T. Zhang, and X. Q. Shan, *FEMS Microbiol. Lett., 69*: 123 (1990).
226. J. W. Kjine, C. L. Diaz, and B. J. J. Lugtenberg, *Signal Molecules in Plant and Plant-Microbe Interactions* (B. J. J. Lugtenberg, ed.), Springer, Berlin, Heidelberg, p. 251 (1989).
227. J. Plazinski, R. W. Innes, and B. G. Rolfe, *J. Bacteriol., 163*: 812 (1985).
228. L. Piana, M. Delledone, M. N. Antonelli, and C. Fogher, *Azospirillum IV, Genetics, Physiology, Ecology* (W. Klingmuller, ed.), p. 83 (1988).
229. B. G. Rolfe, D. Barnard, L. Preston, and G. L. Bender, Proceedings 9th Australian Nitrogen Fixation Conference, Canberra (A. E. Richardon and M. B. Peoples, eds.), p. 71 (1991).
230. Y. X. Jing, G. S. Lin, X. Q. Shan, and J. G. Li, *Nitrogen Fixation: Achievements and Objectives* (P. M. Gresshoff, L. E. Roth, G. Stacey, and W. E. Newton, eds.), Chapman and Hall, New York, p. 829 (1990).
231. S. E. Nielson and G. M. Soerensen, *Chem. Abstr., 107*: 235523q (1987).
232. C. Elmerich, *Biology and Biochemistry of Nitrogen Fixation* (M. J. Dilworth and A. R. Glenn, eds.), Elsevier, Amsterdam, p. 103 (1991).
233. G. Ditta, E. Virts, A. Palomares, and C-H. Kim, *J. Bacteriol., 169*: 3217 (1987).
234. M. David, M-L. Daveran, J. Batut, A. Dedieu, O. Domergue, J. Ghai, C. Hertig, P. Bolstard, and D. Kahn, *Cell, 54*: 671 (1988).
235. H. Hennecke, H-M. Fischer, M. Gubler, B. Thöny, D. Anthamatten, I. Kullik, S. Ebeling, S. Fritsche, and T. Zürcher, *Nitrogen Fixation: Hundred Years After* (H. Bothe, F. J. de Bruijn, and W. E. Newton, eds.), Gustav Fischer, Stuttgart, New York, p. 339 (1988).
236. M. G. Fortin, N. A. Morrison, and D. P. S. Verma, *Nucl. Acids. Res.., 15*: 813 (1987).
237. P. Katnakis and D. P. S. Verma, Proceedings National Academy of Sciences (U.S.), *82*: 4157 (1985).
238. R. B. Mellor, C. Garbers, and D. Werner, *Plant Mol. Biol., 12*: 307 (1989).
239. N. J. Brewin, J. G. Robertson, E. A. Wood, B. Wells, A. P. Larkins, G. Galfre, and G. W. Butcher, *EMBO. J., 4*: 605 (1985).

240. J. G. Robertson, P. Lyttleton, S. Bullivant, and G. F. Grayston, *J. Cell Sci.,* *30*: 129 (1978).

241. P. Putnoky, E. Grosskopf, D. T. Cam Ha, G. B. Kiss, and A. Kondorosi, *J. Cell Biol., 106*: 597 (1988).

242. S. A. Lancelle and J. G. Torrey, *Can. J. Bot., 63*: 25 (1984).

243. M. J. Trinick and T. A. Hadobas, *Plant and Soil, 110*: 177 (1988).

244. S. M. de Faria, J. M. Sutherland, and J. I. Sprent, *Plant Sci., 45*: 143 (1986).

245. S. M. de Faria, S. McInroy, and J. I. Sprent, *Can. J. Bot., 65*: 533 (1987).

246. J. P. Nap and T. Bisseling, *Physiol. Plant., 79*: 407 (1990).

247. F. J. Bergersen, *Root Nodules of Legumes. Structure and Functions*, Research Studies Press, Toronto, p. 97 (1982).

248. M. R. O. Brian, P. M. Kirschbaum, and R. J. Maier, *Proceedings National Academy of Sciences (U.S.), 84*: 8390 (1987).

249. D. P. S. Verma, R. Haugland, N. Brisson, R. P. Legocki, and L. Lacroix, *Biochim. Biophys. Acta, 653*: 98 (1981).

250. V. P. Mauro, T. Nguyen, P. Katinakis, and D. P. S. Verma, *Nucl. Acids Res., 13*: 239 (1985).

251. D. P. S. Verma, M. G. Fortin, J. Stanley, V. P. Mauro, S. Purohit, and N. Morrison, *Plant Mol. Biol., 7*: 51 (1986).

252. D. P. S. Verma, *Plant Nitrogen Metabolism* (J. E. Poulton, J. T. Romeo, and E. E. Conn, eds.), Plenum Publishers, New York, p. 43 (1989).

253. J. V. Cullimore, D. A. Lightfoot, M. J. Bennett, F. L. Chem, C. Gebhardt, J. Oliver, and B. G. Forde, *Molecular Genetics of Plant-Microbe Interactions* (D. P. S. Verma and N. Brisson, eds.), p. 340 (1987).

254. C. Sengupta-Gopalan and J. W. Pitas, *Plant Mol. Biol., 7*: 189 (1986).

255. E. F. Henzell, *Plant and Soil, 108*: 15 (1988).

4

Molecular and Applied Aspects of Nitrate Assimilation

John L. Wray

University of St. Andrews
St. Andrews, Scotland

Michael T. Abberton

AFRC Institute for Grasslands
and Environmental Research
Aberystywth, Dyfed, Wales

INTRODUCTION

Nitrate is the major nitrogen source for most crop plants growing in well-aerated soils (reviewed in [1]). Nitrate is assimilated by a pathway involving a nitrate uptake step, and two reductive steps catalyzed by the enzymes nitrate reductase (NR) and nitrite reductase (NiR). Several recent reviews covering various aspects of this pathway in higher plants are available [2–5]. The ammonium ion product of the pathway is then assimilated by the glutamine synthetase/glutamate synthase cycle [6] and thus becomes available for the synthesis of organic nitrogen compounds within the plant, quantitatively the most important of which is protein.

The three steps of the nitrate assimilation pathway are localized in different subcellular compartments; namely, the plasma membrane and probably the tonoplast (nitrate transport), the cytoplasm (nitrate reductase), and the chloroplast/plastid (nitrite reductase). Nitrate assimilation has been reported to occur both in the leaves and roots of plants. The extent of nitrate assimilation in these two organs depends on the plant species and on the concentration and duration of nitrate supply.

The pathway is tightly regulated. The component steps are absent, or present at low level, in most plants grown in the absence of nitrate. Increase in the capacity of the steps of the pathway occurs on nitrate treatment and, in the green parts of the plant, light together with nitrate is required for the maximal development of the pathway. The effects of light are complex and a number of ways in which light can influence the development and the activity of the nitrate assimilation pathway have been reported [7, reviewed in 8].

Other environmental factors such as oxygen and carbon dioxide tension, temperature and water potential, provision of ammonium as opposed to nitrate as a sole nitrogen source, as well as exogenous application of cytokinins and other chemicals have been reported to influence nitrate assimilation [9]. In some plant species, exogenous application of pathway end products such as glutamine has been suggested to down-regulate the nitrate reduction step of the pathway [10].

Drawing on the observations that NR catalyses the first intracellular step of the pathway, that its activity is tightly regulated and generally much lower than that of NiR, and that nitrite rarely accumulates within the cell, Beevers and Hageman [11] proposed in 1969 that NR catalyzes the rate-limiting step of the nitrate assimilation pathway. This view has been reiterated by most subsequent commentators.

There have been many attempts to relate the activity level of the enzyme to yield and, indeed, to use NR activity levels as a yield indicator. These topics are the subject of a number of recent reviews which should be referred to for a detailed analysis [12–21].

Nitrogen supply does of course play a major role in determining crop productivity. Indeed, in cereals, nitrogen fertilizer generates more yield response than any other factor, particularly under intensive cultivation [22]. Numerous other factors, however, not all related to nitrogen metabolism, also affect productivity and these act with variable intensity throughout the life of the plant. Studies do suggest, however, that not only NR activity, but also nitrate uptake and nitrate flux through the plant, correlate to some extent with productivity [23–27]. That nitrate uptake and nitrate flux also correlate with productivity is not surprising since the ability of the plant to acquire nitrate from the environment influences nitrate flux through the plant and hence the development of the component steps of the pathway. It has been argued recently that the increased activity of the steps of the nitrate assimilation pathway seen in the presence of nitrate may represent primary responses to an activated constitutive nitrate–sensor–protein system, which regulates transcription of genes necessary for the expression of these processes in higher plants [28].

Because of the significance of nitrate uptake and reduction in the nitrogen economy of the plant, and indeed also, the deleterious effects of nitrate on the environment [29], the nitrate assimilation pathway has become the target of investigation not only at the physiological and biochemical level but also at the genetic and molecular level. These latter studies aim to identify and clone genes involved in the operation of the pathway, to characterize the structure of these genes and the proteins they encode, and to understand how expression of these genes is regulated by environmental and endogenous factors.

Here we review our present understanding of the nitrate assimilation pathway at the molecular and genetic level and consider ways in which this knowledge might be exploited. Various aspects of the molecular genetics of nitrate assimilation in higher plants have been the subject of a number of recent reviews [30–44].

MOLECULAR AND GENETIC ASPECTS OF NITRATE UPTAKE AND TRANSLOCATION

Some 5 years ago one of us wrote that, despite being major targets for molecular analysis, nitrate uptake, its subsequent allocation to storage, transport, or reduction, and its remobilization within the plant are the most poorly characterized steps of the higher plant nitrate assimilation pathway in biochemical, let alone, molecular terms [31]. This remains largely true at the time of writing.

Our understanding of nitrate uptake by the root from the ambient solution rests mainly on a large body of physiological data, gleaned from a relatively small number of crop plants. Nitrate uptake is mediated by an active, multiphasic transport system. At least two transport systems are present which can be distinguished on the basis of their kinetic properties [45,46]. Both the low affinity system (K_m greater than 0.5 mM) and the high affinity system (K_m between 10 and 300 μM) rapidly depolarize the plasma membrane in response to nitrate, indicative perhaps of an electrogenic proton-nitrate symporter [47,48]. A low-level, "constitutive," uptake system is present in plants which have not been exposed to nitrate, whilst an inducible system develops after a lag of one to a few hours following exposure of roots to ambient nitrate [49–51]. Protein and RNA synthesis inhibitor studies suggest that a protein component of the induced nitrate uptake system is synthesized de novo in response to the presence of nitrate [52,53]. As nitrate accumulates in the cell, uptake activity declines as the system is repressed [34].

Attempts have been made to identify nitrate-inducible polypeptides which mediate nitrate transport. ^{35}S-methionine labeling studies in maize roots followed by membrane isolation, protein solubilisation, two-dimensional gel electrophoresis and autoradiography identified a tonoplast/endoplasmic reticulum polypeptide of about 31 kD whose synthesis was enhanced by nitrate [54]. A separate study reported that at least four integral plasma membrane polypeptides were induced by nitrate but synthesis of some also increased when roots were incubated with chloride [55]. Another approach [56] has been to use phenylglyoxal, a diketone which covalently binds the guanidinium group of arginine [57]. Phenylglyoxal strongly inhibits chloride transport in mammalian red blood cells and ^{14}C-phenylglyoxal has been used to label specifically, and to identify, the red blood cell anion transporter (the Band 3 protein) [58]. Since phenylglyoxal also inhibits nitrate uptake into excised maize roots, presumably by interacting with arginyl residues of a protein component of the nitrate transport system which is accessible on the outer surface of the plasma membrane [56], it may be possible to use ^{14}C-phenylglyoxal to label the nitrate transport protein (or a component thereof) and identify it by gel electrophoresis. To date, however, no protein has been isolated or shown to be unequivocally involved in nitrate transport. Although, in the studies described above, proteins labeled with ^{35}S-methionion were induced in parallel with increases in nitrate uptake rate [54,55] this does not prove that they are components of the uptake system. Unequivocal confirmation that a protein identified by labeling is indeed a component of a nitrate transport system will not be a trivial task and will depend on the demonstration of a functional role in transport [31].

More recently some success toward cloning genes which encode nitrate transporters has been reported [59]. This approach made use of the T-DNA from *Agrobacterium tumefaciens* as an insertional mutagen for gene tagging [60,61] and has led to the identification of a mutant which is allelic to a previously reported chlorate-resistant mutant of *Arabidopsis* (chl 1-1) defective at the CHL1 gene (see below). The new mutant (*chl1*-2::T-DNA) was used to clone the CHL1 gene which was shown to encode a protein with 12 putative membrane-spanning segments. As expected, CHL1 mRNA levels were increased in roots of plants treated with nitrate, reaching a maximum after 2h and decreasing after 4h. Little CHL1 mRNA was found in leaves. The CHL1 gene was shown to encode a nitrate transporter using *Xenopus* oocytes as an expression system to assay transport activity of the CHL1 protein. Nitrate, but not malate, induced a depolarization of the plasma membrane of CHL1 mRNA-injected oocytes. A significant increase in retention of nitrate in injected oocytes could be detected compared to uninjected oocytes. Since *Arabidopsis* retains the ability to grow on nitrate when the CHL1 gene is deleted, however, at least one other transporter must be functional [59].

Mutants defective in nitrate transport have been isolated in *Arabidopsis* and barley. In *Arabidopsis*, early studies led to the isolation of a *chl1* mutant (B1) by selecting within M_2 populations, mutagenized with EMS in the M_1, for chlorate resistance [45,62,63]. This B1 mutant was shown to be defective in the low-affinity nitrate transporter [45,62] as well as in the transport of chloride and potassium [64]. Since the *chl1*-2::TDNA mutant isolated by insertion mutagenesis (described above) is allelic to this B1 mutant it would seem that the CHL1 gene encodes a low-affinity nitrate transporter which transports ions other than nitrate. In barley, uptake mutants have been isolated by screening M_2 populations, mutagenized with azide in the M_1, for individuals which exhibit "nitrogen stress" during growth on low nitrate or by direct tests for the inability of individual M_2 seedlings to take up nitrate over a 24 h period (as assessed with a nitrate-specific electrode) [35].

Identification of a nitrate transport protein and/or gene allows access to the characterization of the uptake process per se but may also allow studies on other nitrate transport systems present in plant cells such as those presumably involved in the movement of nitrate to the xylem through symplastic transport via the cortex and endodermis [65], its subsequent unloading [66], and eventual uptake by cells of the aerial part of the plant, and in movement between the vacuole and cytoplasm [67].

MOLECULAR AND GENETIC ASPECTS OF NITRATE REDUCTION

The Nitrate Reductase Enzymes

Nitrate reduction is performed by the NR enzymes which catalyze the 2-electron reduction of nitrate to nitrite. The most widely distributed NR is the NADH-linked enzyme (EC 1.6.6.1) which has a pH optimum of 7.5 and has been identified in almost all plant species examined. Some plant species possess a second NR (EC 1.6.6.2) which uses NADH or NADPH as electron donor [68–70]. A third type of NR has been identified in soybean. This is a constitutive NADH-NR with a pH optimum of 6.5, rather than 7.5 [71]. Constitutive NR species have also been reported for several other leguminous species [72].

NADH-NR has been purified to homogeneity and biochemically characterised [73], whilst a number of genes which encode the NR apoprotein have been cloned and sequenced (see p. 115). The NR enzymes are homodimeric, multicenter redox proteins. Each monomer subunit of ca. 100 kDa contains three functional, catalytic domains which are involved in binding the three prosthetic groups FAD, heme and molybdenum-cofactor (MoCo) in a 1:1:1 ratio. These catalytic domains, linked together by protease sensitive hinge regions [74], have been identified in the primary amino acid sequence of the enzyme, deduced from the nucleotide sequence of the cloned gene or cDNA, by comparison of amino acid sequence similarity with cytochrome b_5 reductase (FAD-domain), cytochrome b_5 (heme domain) and sulphite oxidase (MoCo-binding domain) [39,75,76]. These sequence similarity comparisons, together with limited proteolysis studies [77,78], show that the MoCo binding domain is located toward the N-terminus of the NR subunit and carries the nitrate reducing active site. Electron flow from NAD(P)H to this nitrate-reducing site is via the FAD-domain (located toward the C-terminus of the NR subunit) and the heme domain. Conserved cysteine residues involved in binding of NADH and MoCo, and in dimerization of the monomer subunits via the MoCo domain, have been identified [39,76,79].

In addition to the overall physiological reaction (NAD(P)H dependent nitrate reduction) the NR enzymes have partial activities which can be assayed in vitro and which require the participation of one or more of the three functional domains. Inhibitor studies, characterization of the products of limited proteolysis, and immunological studies of mutants altered in the apoprotein gene have allowed these domains to be identified [38]. Structure–function relationships between nitrate reductases and other multicentre redox enzymes have recently been reviewed [39,80].

Genetics of Nitrate Reduction

Mutants defective in nitrate reduction have been isolated, both at the cell/protoplast and whole plant level, in order to identify the number and types of genetic loci which encode a functional nitrate reduction step. Mutants have been isolated either by used of a quantitative test for the presence of in vivo leaf nitrate reductase activity [81], by selection for resistance to the toxic nitrate analog, chlorate (for example [82]) or by screening for the loss of the ability to grow with nitrate as a sole nitrogen source [83,84]. Selection at the whole plant level is preferred since it bypasses problems which may arise in the regeneration of plants from mutant cell lines.

These studies have provided information on two types of genes: those encoding the nitrate reductase apoprotein and those encoding steps in the biosynthetic pathway of the molybdenum cofactor.

Nitrate Reductase Apoprotein Gene Mutants

Apoprotein gene mutants have been isolated in a number of crop species as well as in plants which might more properly be considered model species (reviewed in [10,30,32,36–38,41]). In *Nicotiana plumbaginifolia* a single apoprotein gene locus has been identified [85]. The two unlinked apoprotein gene loci *nia1* and *nia2* identified in *N. tabacum* are probably derived from its progenitor species *N. tomentosiformis* and *N. sylvestris* [86]. However, in barley two unlinked loci, *nar1* and *nar7*, have been identified which encode the NADH-NR and NADPH-NR species respectively [30,87].

Two unlinked loci, *ch12* and *ch13*, mutations at which lead to chlorate resistance, have been described in *Arabidopsis thaliana* [88]. Since both mutants had reduced levels of NR activity compared to the wild-type, and apparently were not defective in either nitrate uptake or in molybdenum cofactor biosynthesis, the loci were suggested to encode NR apoprotein. This has been confirmed for the *ch13* locus which has been shown to encode the NIA2 apoprotein gene [89,90]. However, the other apoprotein gene identified, NIA1, maps to a site distinct from *ch12* [89].

Molybdenum Cofactor Biosynthetic Gene Mutants

Using the selection approaches outlined above, mutants altered in one or other steps of the biosynthetic pathway leading to the molybdenum cofactor have been isolated from several species [reviewed in 30–32,36,41]. Identification has been based on the simultaneous loss of nitrate reductase activity and the activity of another molybdenum cofactor-containing enzyme, xanthine dehydrogenase. The most comprehensive study of the pathway has been carried out in *Nicotiana plumbaginifolia* where genetic analysis, performed either through complementation via protoplast fusion, or by conventional sexual analysis of plants regenerated from chlorate-resistant cell lines, shows that the mutants define one or other of six complementation groups (*cnxA*, *cnxB*, *cnxC*, *cnxD*, *cnxE*, and *cnxF*) [91–93]. The *cnxA* gene locus is believed to encode a function required for insertion of Mo into the molybdopterin moiety of the molybdenum cofactor, whilst the other loci probably encode functions required for molybdopterin biosynthesis, related to synthesis of the 4-carbon side-chain, its phosphorylation, and the attachment of the sulphur ligands [36]. Recent studies in *Arabidopsis thaliana* show that the *ch12* mutant is defective in the molybdenum cofactor, describe a new moybdenum cofactor mutant, *ch17*, and provide a new diagnostic test for molydenum cofactor mutants, sensitivity to tungstate [94].

Molybdenum cofactor genes which are probably the equivalent of those identified in *N. plumbaginifolia* have also been identified in the crop plants *Hordeum vulgare* [30,95–99] and *Pisum sativum* [100].

Molecular Cloning of Genes Involved in Nitrate Reduction

The molecular cloning of genes involved in nitrate reduction is significant for a number of reasons. Sequencing of the gene or cDNA allows the subsequent deduction of the primary amino acid sequence of the apoprotein, and by comparison of sequence similarity with homologous enzymes has contributed substantially to our understanding of the organization of the NR apoprotein and of structure–function relationships within it (outlined p. 113 and discussed further in [39]). Second, the cloned DNA sequences can be used as hybridization probes to study gene expression, its regulation at the mRNA level, and the role of environmental factors, e.g., nitrogen source, light, temperature, water potential, CO_2 and O_2 tension, exogenous applications of cytokinins and other chemicals, which are known to influence the activity of one or more of the component steps of the nitrate assimilation pathway [9].

Third, the availability of genomic sequences should allow the identification of *cis*-acting regulatory sequences in the 5′ upstream region of the gene, and *trans*-acting proteins, which might mediate organ, tissue and cell-specific expression, and the effects of environmental factors [101]. Finally, the availability of cloned genes, mutants, and gene transfer systems will allow the manipulation of nitrate assimilation in a number of ways (see pp. 125–127).

Nitrate Reductase Apoprotein Genes

The complete or partial nucleotide sequence of a number of NR apoprotein genes or cDNAs has been determined (Table 1). As indicated above (p. 113) the three catalytic domains of the NR protein, as well as conserved sites involved in catalysis, have been identified (discussed in detail in [41]) and a

Table 1 Molecular Cloning of Higher Plant Nitrate Reductase Apoprotein Genes

Plant species	Clone	Ref.
Barley	Nar1	
	cDNA[a]	102
	Nar7	
	cDNA	103
Squash	cDNA	104,105
Arabidopsis thaliana	NIA1	
	cDNA	89
	genomic	89
	NIA2	
	cDNA	89
	genomic	90
Nicotiana tabacum	nia-1	
	cDNA[a]	106
	genomic	107,108
	nia-2	
	genomic	107,108
Tomato	genomic	109
Rice	genomic	110,111
Maize	cNDA[a]	112
Spinach	cDNA	113
Betula pendula	cDNA	114
Phaseolus vulgaris	genomic	115

[a] Denotes partial sequence.

site-directed mutagenesis approach to the study of structure–function relationships within the NR apoprotein is now possible. The FAD domain [116] and the linked FAD-heme domains [117] of maize NR have been expressed in *E. coli* and preliminary crystallographic studies of the FAD domain have been reported [118]. Tobacco NR expressed in *Saccharomyces cerevisiae* cells lacked NR activity but retained cytochrome c reductase activity. Whilst the enzyme contained functional FAD and heme domains the molybdenum domain was inactive, presumably since yeast is unable to synthesize this cofactor. The enzyme was present as a dimer thus disproving the assumption that the molybdenum cofactor is involved in dimerization [119].

Molybdenum Cofactor Biosynthetic Genes

The molecular cloning of higher plant MoCo genes has not yet been reported. The procedure used for the isolation of *nia* genes (immunological screening of a cDNA expression library with antibody to the gene product [120]) is not available since the protein products of the *cnx* genes are unknown. Nor for the same reason is it possible to use oligonucleotide probes derived from a knowledge of the amino acid sequence of the *cnx* gene product.

An attempt to clone higher plant MoCo genes by functional complementation of *E. coli chl* mutants, defective in genes equivalent to *cnx*, has been reported [121]. An *N. tabacum* cDNA of 0.8 Kb reproducibly complemented an *E. coli chlG* mutant, but no other. However the clone was also shown to complement functionally the *N. plumbaginifolia* mutants *cnxB*, *cnxC*, *cnxD*, *cnxE*, and *cnxF* but not *cnxA*, when electroporated into their protoplasts and assayed 3–5 days later for the restoration of NR activity. Proof that the selected species is *cnx*-related will come from the stable genetic complementation of a *cnx* locus by its corresponding genomic sequence. Another possible route to *cnx* genes is by gene-tagging (see p. 112).

Molecular Aspects of the Regulation of Nitrate Reduction

Nitrate and light act to up-regulate NR levels in the leaves of green, white-light-grown plants. Induction of NR activity by nitrate was reported first by Tang and Wu [122] in 1957 and since has been shown to occur in a wide range of plant species [123]. Evidence that these nitrate-induced increases in NR activity are due to de novo synthesis of enzyme protein has been provided by both immunological [124,125] and density-labeling [126] techniques. More recently the use of NR apoprotein cDNA species as hybridization probes in northern blot analysis of mRNA has shown that in a number of plants (*Arabidopsis thaliana* [89], squash [104], tobacco [106], *Betula pendula* [114], *Phaseolus vulgaris* [115], tomato [109], and barley [102, 128]), the steady-state level of NR mRNA also increases after nitrate treatment. In soybean [129] and maize [130] nuclear run-off transcription assays show that this is due to increased transcriptional activity. In tobacco and tomato, NR mRNA levels were not significantly affected over a 12-day period when nitrate-grown plants were deprived of nitrate, although NR activity and NR protein levels decreased rapidly, suggesting that nitrate availability regulates the level of NR activity at the posttranscriptional, as well as the transcriptional, level [127].

The role of light in regulation of NR activity in green, light-grown plants is complex but does not seem to involve phytochrome [112,128]. Transfer of nitrate-treated, light-grown plants to the dark causes a decrease in the level of NR activity and mRNA levels [131,132], whilst subsequent addition of light causes an increase in NR mRNA levels followed by an increase in activity [131,133], suggesting that light acts at the transcriptional level. However, light also acts posttranscriptionally since transgenic *N. plumbaginifolia* plants expressing the NR mRNA constitutively from an NR gene under the control of a 35S CaMV promoter show a decrease in NR protein and activity after transfer to the dark [134].

Light is unlikely to act directly in transcriptional activation of the NR apoprotein genes. Rather, the light effect is mediated by sugars produced in the light via photosynthesis. In both *Arabidopsis thaliana* [135] and *Nicotiana plumbaginifolia* [136] exogenous sugars supplied in the dark are able to replace the light requirement for the increase in NR mRNA levels. Sugars probably act at the transcriptional level since expression in *N. plumbaginifolia* of a chimaeric NR gene under the control of the 35S CaMV promoter is not influenced by dark or sugar treatment whilst GUS reporter gene expression driven by the promoter of the tobacco *nia1* gene is repressed in darkness and induced by sugars in the dark [136].

A mechanism by which light may control the activity of NR posttranscriptionally, albeit by indirect means, became apparent very recently when it was shown that NR from spinach leaves is reversibly modulated, being active in the light and inactive in the dark. Exposure of spinach leaves to low CO_2 levels also resulted in inactivation of NR. Changes in activity appear to be the result of a covalent modification of the enzyme protein via phosphorylation at a specific serine residue, with the dephosphorylated form of the enzyme being active in the light [7,137–141].

During plastidogenesis in etiolated tissue light-induced increases in NR activity and mRNA levels are mediated by phytochrome [112,128,142–144]. In the absence of nitrate light is ineffective in bringing about increases in NR activity and mRNA steady-state levels in etiolated mustard seedling cotyledons [145,147]. Nitrate/light regulation of NR also requires a factor produced by functional chloroplasts [146]. Photooxidative damage to chloroplasts blocks expression of a number of nuclear genes whose protein products function in the chloroplast and it has been suggested that a plastidic factor, produced by functional chloroplasts, is normally involved as a positive regulator of the transcription of these nuclear genes (reviewed in [148]). The identity of this factor is unknown but it is unlikely to be a protein [149]. With the exception of NR cytosolic enzyme levels are not impaired.

Down-regulation of NR apoprotein gene transcription is probably exerted by nitrogenous end-products of nitrate assimilation. In vivo inhibition of glutamine synthetase by phosphinothricin decreases the tissue glutamine level and abolishes the decrease in NR mRNA level seen as the day progresses [150] suggesting that glutamine and/or other nitrogen metabolites may exert a negative control on NR expression. Further evidence that nitrate reduction may be regulated by nitrate-derived metabolites comes from the observation that when a functional *N. plumbaginifolia nia* locus is introduced by a sexual cross into an NR-deficient mutant of *N. plumbaginifolia* which has been transformed with a functional tomato *nia* gene the amount of NR mRNA synthesized from the tomato *nia* gene is reduced [151]. This can be explained by assuming that the high NR activity expressed by the *N. plumbaginifolia nia* locus results in the accumulation of nitrogen metabolites which repress the expression of the tomato *nia* gene. The supply of exogenous glutamine or glutamate to detached *N. plumbaginifolia* leaves caused a decrease in the steady-state level of NR mRNA levels [136]. In barley, a mutant, *nir1*, defective in synthesis of NiR overexpressed NADH-NR activity suggesting a regulatory perturbation of the *nar1* gene [152] whilst in *N. plumbaginifolia* down-regulation of NiR gene expression via expression of NiR antisense RNA led to the elevation of NR gene expression and a decrease in the leaf glutamine pool size [153].

Down-regulation by nitrogenous end-products of nitrate assimilation is believed also to be involved in the circadian rhythm seen in NR activity and steady-state mRNA levels of plants grown in a continuous light-dark cycle. When tomato plants are grown in this way with a constant supply of nitrate the steady-state level of NR mRNA increases rapidly during the dark period and reaches a maximum at the start of the day. At the end of the day the amount of NR mRNA is decreased by at least a factor of 100 compared to sunrise. The peak of NR activity occurs some 2–4 h after sunrise [127]. The circadian rhythm in NR mRNA levels is maintained in continuous light but fades on transfer of plants to the dark. Disruption of NR activity, either by mutation in the apoprotein gene or in a MoCo biosynthetic gene [154], or by growth in the presence of tungsten [155], which substitutes for molybdenum in the MoCo [156], abolishes the light-dark fluctuation in NR mRNA level, and the steady-state level of NR mRNA becomes high and constant (although still under nitrate induction).

MOLECULAR AND GENETIC ASPECTS OF NITRITE REDUCTION

The Nitrite Reductase Enzymes

The nitrite reductase (NiR) enzymes catalyze the 6-electron reduction of nitrite to ammonium in the third step of the nitrate assimilation pathway (reviewed in [33]). The enzymes are isolated usually as monomeric polypeptides of ca. 63kDa with very similar amino acid composition and contain 1 mole of siroheme and 1 mole of a 4Fe/4S center per mole of enzyme [157]. Evidence for higher molecular weight forms has been presented [158,159] but their relationship to the 63 kDa form is unclear. The enzymes are located in the chloroplasts of leaves and the plastids of roots and other non-green tissue. Isoforms of NiR have been identified in a number of plant species (see, for example, [160]).

The nitrite reductase (EC 1.7.7.1) present in the leaf uses reduced ferredoxin as electron donor [161] whilst the enzyme from nongreen tissue, such as root, probably receives electrons from NADPH generated in the oxidative pentose phosphate pathway [162] via a pyridine nucleotide reductase and a nonheme iron-containing protein with antigenic similarities to spinach leaf ferredoxin [163]. However, in vitro, both the leaf and root enzyme can use dithionite-reduced methylviologen dye as an electron donor.

Molecular Cloning of Nitrite Reductase Apoprotein Genes

The complete or partial nucleotide sequence of a number of NiR apoprotein genes or cDNA species has been determined (Table 2). Deduced amino acid sequence similarity ranges between 76 and 79% for the spinach, maize, and birch proteins [169] and the tobacco *nir1* protein [153]. Comparison of the amino acid sequences of the higher plant enzyme revealed nine well-conserved regions with a minimum of 10 amino acids that are separated by less conserved regions [169]. When compared to another siroheme-containing enzyme, bacterial NADPH sulfite reductase (EC 1.8.1.2), a highly homologous region was found (residues 437–524 of the alignment) in the C-terminal part of the enzyme which contained four conserved cysteine residues at positions 443, 449, 484, and 488 of the alignment (positions 473, 479, 514, and 518 of spinach NiR [164]). These residues are involved in binding the siroheme/4Fe4S center at the reducing site of the enzyme with the bridging ligand between the siroheme and the 4Fe4S center being assigned to either cys443 or cys449 of the alignment (cys473 or cys479 of spinach NiR) [164,169,171]. Sequence comparisons with ferredoxin-NADP reductase (EC 1.18.1.2) reveal the presence of a short conserved sequence at the N-terminus containing a cluster of 5–6 positively charged amino acids, which probably participate in binding the electron donor, reduced ferredoxin [169].

Table 2   Molecular Cloning of Higher Plant Nitrite
Reductase Apoprotein Genes

Plant species	Clone	References
Spinach	cDNA	164
	genomic	165
Maize	cDNA	166
Rice	cDNA[a]	167
Arabidopsis thaliana	cDNA[a]	168
	genomic[a]	168
Birch	cDNA	169
Tobacco	cDNA	
	nir1[a]	153
	nir2[a]	170
	nir3[a]	170
	genomic	153

[a] Denotes partial sequence.

Enzyme polymorphism [172] and mutational [152] studies indicate that
the NiR apoprotein gene is encoded by nuclear DNA. In common with other
proteins which function in the chloroplast but are nuclear-encoded, NiR is
synthesized as a precursor carrying an N-terminal extension, the transit peptide, which acts to target the protein to, and within, the chloroplast and which
is cleaved during import (reviewed in [173]). The transit peptide of the spinach enzyme [164] is 32-amino acids in length whilst that of the birch enzyme
[169] is shorter. Both are rich in the hydroxylated amino acids threonine and
serine and, in common with other transit peptides, have a net positive charge.
    The number of NiR apoprotein genes differs between plant species.
Whereas spinach [165] and barley (A. Sherman and J. L. Wray, unpublished)
possess a single NiR apoprotein gene, copy number determination in maize
suggests at least two genes per haploid genome although only a single cDNA
species has been cloned [166]. In *Nicotiana tabacum*, a combination of cDNA
cloning and Southern hybridization reveals the presence of four genes, all
of which are expressed [170]. Two of these genes (*nir1* and *nir2*) are believed
to be derived from *N. tomentosiformis* and two (*nir3* and *nir4*) from *N. sylvestris*, the two ancestral progenitors of *Nicotiana tabacum*.
    Several investigators have addressed the question of whether the assay-
able NiR activity detected in green, photosynthetic leaf tissue and in nonphotosynthetic, heterotrophic root tissue of higher plants is the property of
the same protein species (e.g., [174,175]). Clearly in barley and spinach,
which possess a single NiR apoprotein gene, they must be. In barley, loss
of NiR in both leaf and root of a *nir1* mutant, which is probably defective

in the NiR apoprotein gene, confirms this [152]. However, in tobacco, northern hybridisation showed that the four nitrite reductase apoprotein genes are expressed differentially; the *nir1* and *nir3* genes being expressed predominantly in the leaf and the *nir2* and *nir4* genes mostly in the root [170].

Molecular Aspects of the Regulation of Nitrite Reduction

During plastidogenesis nitrate and light, as well as an unidentified factor — plastidic factor — produced by functional plastids, is required for the formation of NiR activity (reviewed in [42]). The light requirement operates via phytochrome, but the nature of the interaction between these effectors, and the way in which they exert their regulation, appears to differ between plant species. In a comparative study between mustard, spinach, tobacco, and barley it was shown that in etiolated mustard seedling cotyledons a strong synergism existed between nitrate and light with respect to enzyme synthesis, but that the NiR transcript level was determined by light [176]. In etiolated spinach seedling cotyledons the action of light on NiR activity level was superimposed multiplicatively on the action of nitrate, indicating that each factor acted independently, but transcript level was determined by nitrate [177]. In contrast, in etiolated tobacco seedling cotyledons [178] and in etiolated barley shoots [179] no light effect on enzyme activity was seen in the absence of nitrate, and a coaction of nitrate and light was required to bring about a high transcript level. However, regardless of the species, the plastidic factor is a prerequisite for nitrate and/or light action.

In leaves of green, white-light-grown plants, up-regulation of NiR transcript level is controlled by nitrate and light (spinach [164]; maize [132, 166]; *Nicotiana plumbaginifolia* [133,134]; *Betula pendula* [169]). How light operates to control transcript level and NiR activity is unclear but it is unlikely to act directly and attempts to demonstrate a role for phytochrome have been inconclusive [180]. The observation that the decrease in NR activity which results when green, light-grown plants are transferred to the dark can be reversed by an exogenous supply of glucose [181] suggested that light might indirectly affect nitrate assimilation via carbohydrates synthesised through photosynthesis. However, although exogenous sugars could replace the light requirement for NR gene expression in *Arabidopsis thaliana* [135] and *Nicotiana plumbaginifolia* [136] neither NiR or *rbcS* transcript levels were increased. The observations that whilst NR transcript levels were similar in leaves of plants grown at high or low light intensity NiR and *rbcS* transcript levels were lower in leaves of plants grown at low light intensity have prompted the conclusion that light regulation of NiR genes is related more closely to that of photosynthetic genes than to the NR gene [136].

Analysis of the DNA elements which regulate expression of the NiR apoprotein gene has been studied in spinach by fusing a 3.1 kb fragment of the gene, upstream from the transcription initiation site, to the GUS reporter gene and examing GUS expression in transgenic tobacco plants [165]. These studies show that the regulatory elements required for nitrate regulation are present within this 3.1 kb fragment and that expression of the spinach gene is regulated by nitrate at the level of transcription. Analysis of promoter-deletion constructs fused to the GUS gene showed that a 130 bp region, located between -330 and -200 with respect to the transcription initiation site, is sufficient to achieve full nitrate-inducible, tissue-specific expression in leaf mesophyll cells as well as in the vascular tissue of stem and root [182].

Down-regulation of NiR gene expression, like that of NR, is controlled by nitrogenous end-products of nitrate assimilation. In NR-deficient mutants of *Nicotiana plumbaginifolia* both NR and NiR transcript levels are elevated [133,154], whereas feeding glutamine, glutamate, or asparagine to detached leaves decreased NiR transcript level, although not to the same extent as for NR [136]. Expression of NiR antisense RNA in *N. tabacum* inhibited NiR activity and led to a reduction in the size of the glutamine pool and overexpression of NR activity and transcript level [153]. Circadian fluctuations observed in NR and NiR transcript levels, being high at the end of the night period and decreasing during the succeeding light period [132,133], correlate inversely with leaf glutamine pool size [150].

In roots of green, white-light-grown maize plants NiR transcript levels increase on nitrate addition and decline on removal of nitrate with a half-time of less than 30 min [166,183] but a role for light was not considered. In barley, synthesis of NiR is regulated differently in leaf and root: both nitrate and light are required in the leaf but in the root nitrate alone is able to bring about synthesis of NiR molecules [184]. Since the available evidence suggests that barley possesses a single NiR apoprotein gene [152] the regulatory differences observed between leaf and root are most probably located in the pathway that transduces the environmental signals, nitrate and light, and are not due to differences that might exist between the *cis*-acting DNA regulatory elements of a leaf-active NiR apoprotein gene and a different, root-active NiR apoprotein gene.

Genetics of Nitrite Reduction

Mutants defective in nitrite reduction have been isolated only in barley (*Hordeum vulgare L.*) [152,185]. Mutants were isolated in M_2 populations of barley (mutagenized with azide in the M_1) by virtue of the fact that, unlike wild-type plants, they accumulate nitrite in their leaf tissue when treated with nitrate in the light. A priori a number of different mutations might be expected to lead to loss of function of the nitrite reduction step and thus be potentially capable of detection by such a screen, viz., within the NiR structural gene affecting catalytic activity or the function of the transit sequence [164]; in the chloroplast envelope affecting recognition of the transit sequence and thus not allowing import of NiR into the chloroplast (a protein receptor mediating import of the *rbcS* gene product has been identified [186]); in prosthetic group synthesis (mutation in the *Escherichia coli cysG* locus affects siroheme synthesis [187]); in the transport of nitrite into the chloroplast, if protein mediated [188]; and in regulation. However, mutation in some of these functions may be lethal to plant survival.

The mutant STA3999, isolated from the barley cultivar Tweed, was shown to carry a recessive mutation in a single nuclear gene which has been designated *Nir1* [152]. The homozygous *nir1* mutant lacked the low basal level of NiR cross-reacting material present in the leaf and root of wild-type plants not treated with exogenous nitrate [184] as well as the higher levels of cross-reacting material synthesized de novo in leaf and root tissue in response to nitrate [152]. Two main possibilities present themselves with respect to the identity of the *Nir1* locus, the most likely being that the locus encodes the NiR apoprotein gene. If this is the case then it would seem that in barley a single, nuclear NiR apoprotein gene encodes the "constitutive" and the "inducible" NiR protein seen in leaf and root. Less likely is the possibility that the *Nir1* locus encodes a regulatory gene.

REGULATORY GENES FOR NITRATE ASSIMILATION

Two types of regulatory gene associated with the nitrate assimilation pathway have been identified in the lower eukaryotes *Aspergillus nidulans* [189] and *Neurospora crassa* [190] by mutational analysis.

nirA is a pathway-specific regulatory gene of *A. nidulans* which acts in a positive regulatory fashion such that its product is required for the expression of the apoprotein genes (*niaD* and *niiA*) of NR and NiR. *nirA*-mutations lead to loss of ability to produce NR and NiR protein and activity above the low levels seen in wild-type strains in the absence of nitrate. *nirA*^{c} mutations lead to the constitutive expression of both activities, irrespective of the presence of nitrate [189]. An equivalent gene, *nit4*, is found in *N. crassa* [190]. *nirA* [191], and *nit4* [192] have been cloned and characterized and shown to encode DNA-binding proteins.

The other regulatory gene identified in these organisms, *areA* in *A. nidulans* and *nir2* in *N. crassa*, is a "global" control gene and mediates the nitrogen catabolite repression of the nitrate assimilation pathway genes as well as the genes of other ammonium-repressible pathways [189,190]. *areA*

[193] and *nit2* [194,195] have also been cloned and characterized and shown to encode DNA-binding proteins. In vitro studies show that NIT2, the regulatory protein product of the *nit2* gene, binds to DNA sequences upstream of *nia*, the tomato NR apoprotein gene, suggesting the existence of a NIT2-like homologue in higher plants [196].

Despite extensive screening, however, either by use of a qualitative test for in vivo NR activity [30], by chlorate-resistance [36,93], or ability to grow on nitrate as sole nitrogen source [84], all mutants so far identified in higher plants are defective either in an apoprotein gene or a MoCo gene and no regulatory mutants have been identified. Attempts to identify *nirA*-type mutations may require further chlorate-screening, whilst *nirA*[c]-type mutants might be identified, through whole-plant screening, as individuals which have "induced" levels of NR and NiR activity, even in the absence of nitrate [31].

In *Arabidopsis thaliana* a locus, *cop1* (constitutively photomorphogenic), has been identified by screening for mutants that develop in darkness as wild-type seedlings do in the light [197]. Recessive mutations in this locus lead to constitutive expression of an array of normally light-regulated genes (including that for NR) in dark-grown seedlings, suggesting that the wild-type *cop1* product normally acts in darkness to repress the expression of these genes. This screening approach may lead to the identification of other loci which encode components of the light signal transduction pathway.

FUTURE PROSPECTS FOR THE GENETIC MANIPULATION OF NITRATE ASSIMILATION

As discussed in the introduction, a number of investigators have proposed that crop yield may be correlated positively with the efficiency of nitrate uptake and/or nitrate reduction within the plant, making these two steps of nitrate assimilation likely targets for manipulation. The availability of cDNA and genomic sequences for a number of NR species makes it likely that initial attempts will be directed at the nitrate reduction step of nitrate assimilation. Already transformation of a model plant species, *Nicotiana tabacum*, with an NR cDNA under the control of the "constitutive" 35S promoter from cauliflower mosaic virus has been reported [134]. The recipient was a *nia* mutant not expressing endogenous NR activity. Of nine independent transgenic plants analyzed one grew poorly on nitrate and expressed only around 20% of the wild-type activity and in others NR activity was at most three times that of the wild-type level. Amongst other things these studies showed that stringent regulation of NR gene expression is not essential for vegetative development and seed production [134].

The ability to perturb nitrate assimilation by down-regulation of NR activity through antisense RNA approaches (reviewed in [198]) is also technically feasible. The availability of organ-specific promoters, such as for shoot and root, allows not only down-regulation throughout the plant but also the targeting of down-regulation to specific organs, making it possible to determine the role these organs play in the overall nitrogen economy of the plant and perhaps allowing insights into interactions between nitrogen and carbon metabolism. Further, if sufficient sequence dissimilarity exists between different NR apoprotein genes within a species then it should be possible to down-regulate individual types of NR gene and delineate the role that individual NR isoforms play. Of course, this latter approach may not be necessary if it is possible to isolate mutants in different NR apoprotein genes, as accomplished in *Arabidopsis* and barley (see p. 114).

An alternative approach to modifying the rate of the nitrate reduction step might be via improvement in the catalytic efficiency of NR enzymes by site-directed mutagenesis. This approach has already been used in *Neurospora crassa* to examine the role of individual amino acids within the heme domain of the enzyme [199].

Equivalent approaches to the above will become possible as genes encoding nitrate transporters become available.

However, meaningful attempts at improving the efficiency of nitrate assimilation, that is, the efficiency with which nitrate is recovered from the soil, and the efficiency with which internal plant nitrate is translated into yield, is still some way in the future since it is clear that there are many fundamental aspects of nitrate assimilation and related aspects of nitrogen metabolism that we know little or nothing about. These fundamental aspects include: (1) the molecular basis of nitrate uptake and its regulation which we discussed above; (2) the significance of nitrate efflux from the root, which can be considerable [200] and its interaction, if any, with nitrate uptake; (3) the mechanisms underlying partitioning of nitrate inside the root to reduction, to storage, and to transport in the xylem to the aerial part of the plant, and its partitioning there; (4) the nature of the processes involved in xylem loading and unloading, in the transport of nitrate into and out of the plant cell and the vacuole, and its regulation; (5) the significance of root versus shoot nitrate assimilation; (6) the role of different NR enzymes within a single species; (7) the extent to which temporal changes in the level of nitrate nitrate uptake, nitrate reduction and nitrate translocation affect yield; (8) the energy requirements for nitrate uptake, translocation, and reduction and the generation of grain protein [201,202]; (9) the nature of the regulatory mechanisms which integrate nitrogen and carbon metabolism [203]; (10) whether there are pathway-specific regulatory genes controlling nitrate assimilation; (11) control by end products; (12) the nature of the interactions between light, acting through photosynthesis and through the phytochrome system, on nitrate assimilation; (13) the role of plant growth substances and (14) the metabolic and physiological processes underlying grain filling and plant and fruit senescence. The reader will undoubtedly be able to add to this list.

Even if we accept that there is a link between yield and nitrate assimilation it does not necessarily follow that manipulation of the nitrate assimilation pathway will lead to an improvement in yield since another step in the complex, interlocked processes of metabolism and development may become rate-limiting. Identification of such a new, rate-limiting step may prove difficult but, if it lies within a biochemical pathway, it may be possible to identify it by the application of control analysis [204–206]. The application of control analysis to the study of photosynthesis in a series of *N. tabacum* plants transformed with an antisense *rbcS* gene and expressing a range of ribulose-1,5-bisphosphate carboxylase/oxygenase activities has recently been described [207–211]. Studies performed at different N supplies show that N availability and the availability of photosynthate both regulate storage and allocation of biomass to optimize resource allocation, but achieve this via different mechanisms [210]. Manipulation of nitrate assimilation may also adversely affect energy requirements [202, 203] and the interaction of nitrogen metabolism with carbon and sulphur metabolism.

It is clear that the problems associated with the genetic manipulation of nitrate assimilation stems not from our inability to carry out such manipulations but from our poor understanding of the physiology and biochemistry of the process, and our inability to predict the physiological and biochemical consequences of the molecular changes brought about. These problems are, however, likely to be a spur, rather than a deterrent to their resolution.

ACKNOWLEDGMENTS

This work was supported by SERC grant GR/D/81442 to J.L.W. We thank Michael P. Ward and Michael A. Roberts for critically reading the manuscript.

REFERENCES

1. R. J. Haynes, *Mineral Nitrogen in the Plant-Soil System* (R. J. Haynes, ed.), Academic Press, New York (1986).
2. W. H. Campbell and J. Smarrelli, *Biochemical Basis of Plant Breeding, Vol 1. Nitrogen Metabolism* (C. A. Neyra, ed.), CRC Press, Boca Raton, FL, p. 1 (1986).

3. H. Lambers, J. J. Neeteson, and I. Stulen, *Functional, Ecological and Agricultural Aspects of Nitrogen Metabolism in Higher Plants* (H. Lambers, J. J. Neeteson, and I. Stulen, ed.), Martinus Nijhoff, Dordrecht, The Netherlands (1986).
4. J. L. Wray and J. R. Kinghorn, *Molecular and Genetic Aspects of* Nitrate *Assimilation* (J. L. Wray and J. R. Kinghorn, ed.), Oxford Science Publications, Oxford (1989).
5. Y. P. Abrol, *Nitrogen in Higher Plants* (Y. P. Abrol, ed.), John Wiley, New York (1990).
6. P. J. Lea, S. A. Robinson, and G. R. Stewart. *The Biochemistry of Plants* (B. J. Miflin and P. J. Lea, ed.) Vol. 16, Academic Press, New York, p. 121 (1990).
7. J. L. Huber, S. C. Huber, W. H. Campbell, and M. G. Redinbaugh, *Arch. Biochem. Biophys.*, *296*: 58 (1992).
8. S. K. Sawhney and M. S. Naik, *Nitrogen in Higher Plants* (Y. P. Abrol, ed.), John Wiley, New York, p. 93 (1990).
9. M. G. Guerrero, J. M. Vega, and M. Losada, *Ann. Rev. Plant Physiol.*, *32*: 169 (1981).
10. R. L. Langendorfer, M. T. Watters, and J. Smarrelli, *Plant Sci., 57*: 119 (1988).
11. L. H. Beevers and R. H. Hageman, *Ann. Rev. Plant Physiol.*, *20*: 495 (1969).
12. Y. P. Abrol and T. V. R. Nair, *Nitrogen Assimilation and Crop Productivity* (S. P. Sen, Y. P. Abrol, and S. K. Sinha, ed.), Associated Publishing Company, New Delhi, India, p. 113 (1978).
13. R. H. Hageman, *Nitrogen Assimilation of Plants* (E. J. Hewitt and C. V. Cutting, ed.), Academic Press, London, p. 591 (1979).
14. M. S. Naik, Y. P. Abrol, and T. V. R. Nair, *Phytochem.*, *21*: 495 (1982).
15. Y. P. Abrol, P. A. Kumar, and T. V. R. Nair, *Adv. Cereal Sci. Tech.*, *6*: 1 (1984).
16. P. B. Cregan and P. van Berkum, *Theoret. App. Genet.*, *67*: 97 (1984).
17. J. H. J. Spiertz, N. M. de Vos, and L. ten Holte, *Cereal Production* (E. J. Gallagher, ed.), Butterworths, London, p. 249 (1984).
18. J. H. Sherrard, R. J. Lambert, F. E. Below, R. T. Dunand, M. J. Messmer, M. R. Wilson, C. S. Winkles, and R. H. Hageman, *Biochemical Basis of Plant Breeding. Vol. 2. Nitrogen Metabolism* (C. A. Neyra, ed.), CRC Press, Boca Raton, Florida, p. 109 (1986).
19. T. V. R. Nair and S. R. Chatterjee, *Nitrogen in Higher Plants* (Y. P. Abrol, ed.) John Wiley, New York, p. 367 (1990).
20. M. L. Reilly, *Nitrogen in Higher Plants* (Y. P. Abrol, ed.), John Wiley, New York, p. 335 (1990).
21. R. H. Hageman and F. E. Below, *Nitrogen in Higher Plants* (Y. P. Abrol, ed.), John Wiley, New York, p. 103 (1990).
22. A. Falisse and B. Bodson, *Cereal Production* (E. J. Gallagher, ed.), Butterworths, London, England, p. 273 (1984).
23. G. A. Meeker, A. C. Purvis, C. A. Neyra, and R. H. Hageman, *Mechanism and Regulation of Plant Growth* (R. L. Bielski, A. R. Ferguson and M. M. Cresswell, ed.), Royal Society of New Zealand, Wellington, p. 49 (1974).
24. D. L. Shaner and J. S. Boyer, *Plant Physiol.*, *58*: 499 (1976).

25. A. J. Reed and R. H. Hageman, *Plant Physiol.*, *66*: 1179 (1980).
26. A. J. Reed and R. H. Hageman, *Plant Physiol.*, *66*: 1184 (1980).
27. R. H. Hageman and F. E. Below, *Nitrogen in Higher Plants* (Y. P. Abrol ed.), John Wiley & Sons Ltd., New York, p. 313 (1990).
28. M. G. Redinbaugh and W. H. Campbell, *Physiol. Plant.*, *82*: 640 (1991).
29. W. B. Wilkinson and L. A. Greene, Philosophical Trans. Royal Society (London). *Ser. B.*, *296*: 459 (1982).
30. A. Kleinhofs, R. L. Warner, J. M. Lawrence, J. M. Melzer, J. M. Jeter, and D. A. Kudrna, *Molecular and Genetic Aspects of Nitrate Assimilation* (J. L. Wray and J. R. Kinghorn ed.), Oxford Science Publications, Oxford, England, p. 197 (1989).
31. J. L. Wray, *Plant Cell Environ.*, *22*: 369 (1988).
32. J. L. Wray, *A Genetic Approach to Plant Biochemistry* (A. D. Blonstein & P. J. King ed.), Springer Verlag, Vienna, Austra, p. 101 (1986).
33. J. L. Wray, *Molecular and Genetic Aspects of Nitrate Assimilation* (J. L. Wray & J. R. Kinghorn, ed.), Oxford Science Publications, Oxford, England, p. 244 (1989).
34. C.-M. Larsson and B. Ingemarsson, *Molecular and Genetic Aspects of Nitrate Assimilation* (J. L. Wray and J. R. Kinghorn, ed.), Oxford Science Publications, Oxford, England, p. 3 (1989).
35. R. M. Wallsgrove, H. Hasegawa, A. C. Kendall, and J. C. Turner, *Molecular and Genetic Aspects of Nitrate Assimilation* (J. L. Wray and J. R. Kinghorn, ed.), Oxford Science Publications, Oxford, England, p. 15 (1989).
36. A. J. Müller and R. R. Mendel, *Molecular and Genetic Aspects of Nitrate Assimilation* (J. L. Wray and J. R. Kinghorn, ed.), Oxford Science Publications, Oxford, England, p. 166 (1989).
37. M. Caboche, I. Cherel, F. Galangau, M.-A. Grandbastien, C. Meyer, T. Moureaux, F. Pelsy, P. Rouzé, H. Vaucheret, F. Vedele, and M. Vincentz, *Molecular and Genetic Aspects of Nitrate Assimilation* (J. L. Wray and J. R. Kinghorn, ed.), Oxford Science Publications, Oxford, England, p. 186 (1989).
38. M. Caboche and P. Rouzé, *Trends Genet.*, *6*: 187 (1990).
39. W. H. Campbell and J. R. Kinghorn, *Trends Biochem. Sci.*, *15*: 315 (1990).
40. N. M. Crawford and W. H. Campbell. *The Plant Cell*, *2*: 829 (1990).
41. P. Rouzé and M. Caboche, Inducible Plant Proteins: Their Biochemistry and Molecular Biology (J. L. Wray, ed.), Society for Experimental Biology Seminar Series: 49, Cambridge University Press, Cambridge, England, p. 45 (1992).
42. H. Mohr, A. Neininger and B. Seith, *Bot. Acta*, *105*: 81 (1992).
43. J. L. Wray, *Physiol. Plant.*, *89*: 607 (1993).
44. R. L. Warner and A. Kleinhofs, *Physiol. Plant.*, *85*: 245 (1992).
45. H. Doddema and G. P. Telkamp, *Physiol. Plant*, *45*: 332 (1979).
46. M. Y. Siddiqui, A. D. M. Glass, T. J. Ruth, and T. W. Rufty Jr., *Plant Physiol.*, *90*: 1426 (1990).
47. W. R. Ullrich and A. Novacky, *Plant Sci. Lett.*, *22*: 211 (1981).
48. A. D. M. Glass, J. E. Shaff, and L. V. Kochian, *Plant Physiol.*, *99*: 456 (1992).

49. W. A. Jackson, W. L. Pan, R. H. Moll, and E. J. Kamprath, *Biochemical Basis of Plant Breeding, Vol. 2. Nitrogen Metabolism* (C. A. Neyra, ed.), CRC Press, Boca Raton, Florida, p. 73 (1986).

50. R. B. Lee and M. C. Drew, *J. Exptl. Bot., 37*: 1768 (1986).

51. R. Behl, R. Tischner, and K. Raschke, *Planta, 176*: 235 (1988).

52. G. A. Tompkins, W. A. Jackson, and R. J. Volk, *Physiol. Plant., 43*: 166 (1978).

53. K. P. Rao and D. W. Rains, *Plant Physiol., 57*: 55 (1976).

54. P. R. McClure, T. E. Omholt, G. M. Pace, and P.-Y. Bouthyette, *Plant Physiol., 84*: 52 (1987).

55. K. S. Dhugga, J. G. Waines, and R. T. Leonard, *Plant Physiol., 87*: 120 (1988).

56. K. S. Dhugga, J. G. Waines, and R. T. Leonard, *Plant Physiol., 86*: 759 (1988).

57. K. Takahashi, *J. Biol. Chem., 243*: 6171 (1968).

58. P. J. Bjerrum, J. Weith, and C. L. Borders, Jr., *J. Gen. Physiol., 81*: 453 (1983).

59. Y.-F. Tsay, J. I. Schroeder, K. A. Feldmann, and N. M. Crawford, *Cell, 72*: 705 (1993).

60. K. A. Feldmann, M. D. Marks, M. L. Christanson, and R. L. Quatrano, *Sci., 243*: 1351 (1989).

61. K. A. Feldmann, *Plant J., 1*: 71 (1991).

62. H. Doddema, J. Hofstra, and W. Feenstra, *Physiol. Plant., 43*: 343 (1978).

63. F. J. Braaksma and W. J. Feenstra, *Theor. Appl. Genet., 64*: 83 (1982).

64. H. J. Scholten and W. J. Feenstra, *Physiol. Plant., 66*: 265 (1986).

65. T. W. Rufty, J. F. Thomas, J. L. Remmler, W. H. Campbell, and R. J. Volk, *Plant Physiol., 82*: 675 (1986).

66. F. N. Ezeta and W. A. Jackson, *Plant Physiol., 56*: 148 (1975).

67. M. Aslam, A. Oaks, and R. C. Huffaker, *Plant Physiol., 58*: 588 (1976).

68. T. C. Shen, E. A. Funkhauser, and M. G. Guerrero, *Plant Physiol., 58*: 292 (1976).

69. M. G. Redinbaugh and W. H. Campbell, *Plant Physiol., 68*: 115 (1981).

70. R. S. Nelson, L. Streit, and J. E. Harper, *Physiol. Plant., 61*: 384 (1984).

71. L. Streit, R. S. Nelson, and J. E. Harper, *Plant Physiol., 84*: 654 (1987).

72. M. Andrews, S. M. De Faria, S. G. McInroy, and J. I. Sprent, *Phytochem., 29*: 49 (1990).

73. M. G. Redinbaugh and W. H. Campbell, *J. Biol. Chem., 260*: 3380 (1985).

74. J. Brown, I. S. Small, and J. L. Wray, *Phytochem., 20*: 389 (1981).

75. N. M. Crawford, M. Smith, D. Bellissimo, and R. W. Davis, Proceedings of the National Academy of Sciences (U.S.), *85*: 5006 (1988).

76. P. J. Neame and M. J. Barber, *J. Biol. Chem., 264*: 20894 (1989).

77. L. P. Solomonson and M. J. Barber, *Molecular and Genetic Aspects of Nitrate Assimilation* (J. L. Wray and J. R. Kinghorn, ed.), Oxford Science Publications, Oxford, England, p. 88 (1989).

78. Y. Kubo, N. Ogura, and H. Nakagawa, *J. Biol. Chem., 263*: 19684 (1989).

79. G. E. Hyde, J. A. Wilberding, A. L. Meyer, E. R. Campbell, and W. H. Campbell, *Plant Mol. Biol., 13*: 233 (1989).

80. J. R. Kinghorn and E. I. Campbell, *Molecular and Genetic Aspects of Nitrate Assimilation* (J. L. Wray and J. R. Kinghorn, ed.), Oxford Science Publications, Oxford, England, p. 385 (1989).

81. R. L. Warner, C. J. Lin, and A. Kleinhofs, *Nature, 269*: 406 (1977).
82. F. J. Oostindier-Braaksma and W. J. Feenstra, *Arabid. Inf. Serv., 9*: 9 (1972).
83. A. Strauss, F. Bucher, and P. J. King, *Planta, 153*: 75 (1981).
84. F. Pelsey, J. Kronenberger, J. M. Pollien, and M. Caboche, *Plant Sci., 76*: 109 (1991).
85. I. Negrutiu, R. Dirks, and M. Jacobs, *Theoret. App. Genet., 66*: 341 (1983).
86. A. J. Müller, *Mol. Gen. Genet., 192*: 275 (1983).
87. R. L. Warner, K. Narayanan, and A. Kleinhofs, *Theoret. App. Genet., 74*: 714 (1987).
88. F. J. Braaksma and W. J. Feenstra, *Theoret. App. Genet., 64*: 83 (1982).
89. C. L. Cheng, J. Dewdney, H.-G. Nam, B. G. W. den Boer, and H. M. Goodman, *EMBO J., 7*: 3309 (1988).
90. J. Q. Wilkinson and N. M. Crawford, *Plant Cell, 3*: 461 (1991).
91. R. Dirks, V. Sidorov, and M. Jacobs, *Mol. Gen. Genet., 201*: 339 (1985).
92. R. Grafe, A. Marion-Poll, and M. Caboche, *Theoret. Appl. Genet., 73*: 299 (1986).
93. J. Gabard, A. Marion-Poll, I. Cherel, C. Meyer, A. Müller, and M. Caboche, *Mol. Gen. Genet., 209*: 596 (1987).
94. S. T. LaBrie, J. Q. Wilkinson, Y.-F. Tsay, K. A. Feldmann, and N. M. Crawford, *Mol. Gen. Genet., 233* (1992).
95. A. Kleinhofs, R. L. Warner, and K. R. Narayanan, *Oxford Surveys Plant Mol. Cell Biol. Vol. 2* (B. J. Miflin, ed.), Springer Verlag, Vienna, p. 190 (1985).
96. B. I. Tokarev and V. K. Shumny, *Genetika, 13*: 2097 (1977).
97. S. W. J. Bright, P. B. Norbury, J. Franklin, D. W. Kirk, and J. L. Wray, *Mol. Gen. Genet., 189*: 240 (1983).
98. J. L. Wray, B. Steven, D. W. Kirk, and S. W. J. Bright, *Mol. Gen. Genet., 201*: 462 (1985).
99. B. J. Steven, D. W. Kirk, S. W. J. Bright, and J. L. Wray, *Mol. Gen. Genet., 219*: 421 (1989).
100. W. J. Feenstra and E. Jacobsen, *Theoret. App. Genet., 58*: 39 (1980).
101. C. Kuhlemeier, P. J. Green, and N.-H. Chua, *Ann. Rev. Plant Physiol., 38*: 221 (1987).
102. C.-L. Cheng, J. Dewdney, A. Kleinhofs, and H. M. Goodman, Proceedings of the National Academy of Sciences (U.S.), *83*: 6825 (1986).
103. J. Miyazaki, M. Juricek, K. Angelis, K. M. Schnorr, A. Kleinhofs, and R. L. Warner, *Mol. Gen. Genet., 228*: 329 (1991).
104. N. M. Crawford, W. H. Campbell, and R. W. Davis, Proceedings of the National Academy of Sciences (U.S.), *83*: 8073 (1986).
105. G. E. Hyde and N. M. Crawford, *J. Biol. Chem., 266*: 23542 (1992).
106. R. Calza, E. Huttner, M. Vincentz, P. Rouzé, F. Galangau, H. Vaucheret, I. Cherel, C. Meyer, J. Kronenberger, and M. Caboche, *Mol. Gen. Genet., 209*: 552 (1987).
107. H. Vaucheret, M. Vincentz, J. Kronenberger, M. Caboche, and P. Rouzé, *Mol. Gen. Genet., 216*: 10 (1989).
108. H. Vaucheret, J. Kronenberger, P. Rouzé, and M. Caboche, *Plant Mol. Biol., 12*: 597 (1989).

109. F. Daniel-Vedele, M.-F. Dorbe, M. Caboche, and P. Rouzé, *Gene, 85*: 371 (1989).
110. H. M. Hamat, A. Kleinhofs, and R. L. Warner, *Mol. Gen. Genet., 218*: 93 (1989).
111. H. K. Choi, A. Kleinhofs, and G. An, *Plant Mol. Biol., 13*: 731 (1989).
112. G. Gowri and W. H. Campbell, *Plant Physiol., 90*: 792 (1989).
113. I. M. Prosser and C. M. Lazarus, *Plant Mol. Biol., 15*: 187 (1990).
114. A. Friemann, K. Brinkman, and W. Hachtel, *Mol. Gen. Genet., 227*: 97 (1991).
115. T. Hoff, B. M. Stummann, and K. W. Henningsen, *Physiol. Plant, 82*: 197 (1991).
116. G. E. Hyde and W. H. Campbell, *Biochem. Biophys. Res. Commun., 168*: 1285 (1990).
117. W. H. Campbell, *Plant Physiol., 99*: 693 (1992).
118. G. Lu, W. Campbell, Y. Lindqvist, and G. Schneider, *J. Mol. Biol., 224*: 277 (1992).
119. H.-N. Truong, C. Meyer, and F. Daniel-Vedele, *Biochem. J., 278*: 393 (1991).
120. R. A. Young and R. W. Davis, Proceedings of the National Academy of Sciences (U.S.), *80*: 1194 (1983).
121. J. Schiemann, D. Inzé, and R.-R. Mendel, Cloning of MoCo Genes (*cnx*) from Tobacco, Proceedings Third International Symposium Nitrate Assimilation—Molecular and Genetic Aspects, Bombannes, France, pp. 141–142 (1990).
122. P.-S. Tang and H.-Y. Wu, *Nature, 179*: 1355 (1957).
123. H. S. Srivastava, *Phytochem., 19*: 725 (1980).
124. D. A. Somers, T. M. Kuo, A. Kleinhofs, R. L. Warner, and A. Oaks, *Plant Physiol., 72*: 949 (1983).
125. J. L. Remmler and W. H. Campbell, *Plant Physiol., 80*: 442 (1986).
126. H. R. Zielke and P. Filner, *J. Biol. Chem., 246*: 1772 (1971).
127. F. Galangau, F. Daniel-Vedele, T. Moureaux, M.-F. Dorbe, M.-F. Leydecker, and M. Caboche, *Plant Physiol., 88*: 383 (1988).
128. J. M. Melzer, A. Kleinhofs, and R. L. Warner, *Mol. Gen. Genet., 217*: 341 (1989).
129. J. J. Callaci and J. Smarrelli, Jr., *Biochim. Biophys. Acta, 1008*: 127 (1991).
130. C. Lillo, *Plant Sci., 73*: 149 (1991).
131. M.-D. Deng, T. Moureaux, M.-T. Leydecker, and M. Caboche, *Planta, 180*: 257 (1990).
132. C. J. Bowsher, D. M. Long, A. Oaks, and S. J. Rothstein, *Plant Physiol., 95*: 281 (1991).
133. J. D. Faure, M. Vincentz, J. Kronenberger, and M. Caboche, *Plant J., 1*: 107 (1991).
134. M. Vincentz and M. Caboche, *EMBO J., 10*: 1027 (1991).
135. C.-L. Cheng, G. Acedo, M. Cristinsin, and M. A. Conkling, Proceedings of the National Academy of Sciences (U.S.), *89*: 1861 (1992).
136. M. Vincentz, T. Moureaux, M.-T. Leydecker, H. Vaucheret, and M. Caboche, *Plant J., 3*: 315 (1993).
137. W. M. Kaiser and E. Brendle-Behnisch, *Plant Physiol., 96*: 363 (1991).

138. W. M. Kaiser and D. Spill, *Plant Physiol., 96*: 368 (1991).
139. W. M. Kaiser, D. Spill, and E. Brendle-Behnisch, *Planta, 186*: 236 (1992).
140. B. Riens and H. W. Heldt, *Plant Physiol., 98*: 573 (1992).
141. C. Mackintosh, *Biochim. Biophys. Acta, 1137*: 121 (1992).
142. A. K. Sharma and S. K. Sopory, *Phytochem. Photobiol., 39*: 491 (1984).
143. V. K. Rajasekhar, G. Gowri, and W. H. Campbell, *Plant Physiol., 88*: 242 (1988).
144. C. Schuster and H. Mohr, *Planta, 184*: 74 (1990).
145. C. Schuster, R. Oelmuller, and H. Mohr, *Planta, 171*: 136 (1987).
146. C. Schuster, C. Schmidt, and H. Mohr, *Planta, 177*: 74 (1989).
147. C. Schuster and H. Mohr, *Planta, 181*: 125 (1990).
148. R. Oelmuller, *Photochem. Photobiol., 49*: 229 (1989).
149. R. Oelmuller, I. Levitan, R. Bergfeld, V. K. Rajasekhar, and H. Mohr, *Planta, 168*: 482 (1986).
150. M.-D. Deng, T. Moureaux, I. Cherel, J.-P. Boutin, and M. Caboche, *Plant Physiol. Biochem., 29*: 239 (1991).
151. M.-F. Dorbe, M. Caboche, and F. Daniel-Vedele, *Plant Mol. Biol., 18*: 363 (1992).
152. E. Duncanson, A. F. Gilkes, D. W. Kirk, A. Sherman, and J. L. Wray, *Mol. Gen. Genet., 236*: 275 (1993).
153. H. Vaucheret, J. Kronenberger, A. Lepingle, F. Vilaine, J.-P. Boutin, and M. Caboche, *Plant J., 2*: 559 (1992).
154. S. Pouteau, I. Cherel, H. Vaucheret, and M. Caboche, *Plant Cell, 1*: 1111 (1989).
155. M. Deng, T. Moureaux, and M. Caboche, *Plant Physiol., 91*: 304 (1989).
156. J. L. Wray and P. Filner, *Biochem. J., 119*: 715 (1970).
157. J. M. Vega and H. Kamin, *J. Biol. Chem., 252*: 896 (1977).
158. Y. Ishiyama and G. Tamura, *Plant Sci. Lett., 37*: 251 (1985).
159. M. Hirasawa, R. W. Shaw, G. Palmer, and D. B. Knaff, *J. Biol. Chem., 262*: 12428 (1987).
160. M. Kutscherra, W. Jost, and D. Schlee, *J. Plant Physiol., 129*: 383 (1987).
161. K. W. Joy and R. W. Hageman, *Biochem. J., 100*: 263 (1966).
162. C. G. Bowsher, D. P. Hucklesby, and M. J. Emes, *Planta, 177*: 359 (1989).
163. A. Suzuki, A. Oaks, J.-P. Jacquot, J. Vidal, and P. Gadal, *Plant Physiol., 78*: 374 (1985).
164. E. Back, W. Burkhart, M. Moyer, L. Privalle, and S. Rothstein, *Mol. Gen. Genet., 212*: 20 (1988).
165. E. Back, W. Dunne, A. Schneiderbauer, A. de Framond, R. Rastogi, and S. J. Rothstein, *Plant Mol. Biol., 17*: 9 (1991).
166. K. Lahners, V. Kramer, E. Back, L. Privalle, and S. Rothstein, *Plant Physiol., 88*: 741 (1988).
167. J. Matsui, G. Takeba, and S. Ida, *Agric. Biol. Chem., 54*: 3069 (1990).
168. S. T. LaBrie, J. Q. Wilkinson, and N. M. Crawford, *Plant Physiol., 97*: 873 (1991).
169. A. Friemann, K. Brinkmann, and W. Hachtel, *Mol. Gen. Genet., 231*: 411 (1992).

170. J. Kronenberger, A. Lepingle, M. Caboche, and H. Vaucheret, *Mol. Gen. Genet., 236*: 203 (1993).
171. L. M. Siegel and J. O. Wilkerson, *Molecular and Genetic Aspects of Nitrate Assimilation* (J. L. Wray and J. R. Kinghorn, eds.), Oxford Science Publications, Oxford, England, p. 263 (1989).
172. S. Heath-Paglioso, R. C. Huffaker, and R. W. Allard, *Plant Physiol., 76*: 353 (1984).
173. E. K. Archer and K. Keegstra, *J. Bioenerg. Biomembr., 22*: 789 (1990).
174. M. Hirasawa, K. Fakushima, G. Tamura, and D. M. Knaff, *Biochem. Biophys. Acta, 791*: 145 (1984).
175. C. E. Bowsher, M. J. Emes, R. Cammack, and D. P. Hucklesby, *Planta, 175*, 334 (1988).
176. C. Schuster and H. Mohr, *Planta, 181*: 327 (1990).
177. B. Seith, C. Schuster, and H. Mohr, *Planta, 184*: 74 (1991).
178. A. Neininger, J. Kronenberger, and H. Mohr, *Planta, 187*: 381 (1992).
179. B. Seith, A. Sherman, J. L. Wray, and H. Mohr, *Planta, 192*: 110 (1993).
180. T. W. Becker, C. Foyer, and M. Caboche, *Planta, 188*: 39 (1992).
181. J. C. Nicholas, J. E. Harper, and R. H. Hageman, *Plant Physiol., 58*: 731 (1976).
182. R. Rastogi, E. Back, A. Schneiderbauer, C. G. Bowsher, B. Moffatt, and S. J. Rothstein, *Plant J., 4*: 317 (1993).
183. V. Kramer, K. Lahners, E. Back, L. S. Privalle, and S. Rothstein, *Plant Physiol., 90*: 1214 (1989).
184. E. Duncanson, S.-M. Ip, A. Sherman, D. W. Kirk, and J. L. Wray, *Plant Sci., 87*: 151 (1992).
185. J. L. Wray, E. Duncanson, A. F. Gilkes, and D. W. Kirk, *Barley Genetics VI. Vol 1* (L. Munck, ed.), 6th International Barley Genetics Symposium, Helsingborg, Sweden, p. 104 (1991).
186. D. Pain, Y. S. Kanawar, and G. Blobel, *Nature, 331*: 232 (1988).
187. H. MacDonald and J. A. Cole, *Mol. Gen. Genet., 200*: 328 (1985).
188. E. Krämer, R. Tischner, and A. Schmidt, *Planta, 176*: 28 (1988).
189. C. Scazzocchio and H. N. Arst, Jr., *Molecular and Genetic Aspects of Nitrate Assimilation* (J. L. Wray and J. R. Kinghorn, ed.), Oxford Science Publications, Oxford, England, p. 299 (1989).
190. G. Marzluf and Y.-H. Fu, *Molecular and Genetic Aspects of Nitrate Assimilation* (J. L. Wray and J. R. Kinghorn, eds.), Oxford Science Publications, Oxford, England, p. 314 (1989).
191. G. Burger, J. Strauss, C. Scazzocchio, and B. F. Lang, *Mol. Cell. Biol., 11*: 5746 (1991).
192. G. F. Yuan, Y. H. Fu, and G. Marzluf, *Mol. Cell. Biol., 11*: 5735 (1991).
193. B. Kudla, M. X. Caddick, T. Langdon, N. M. Martinez Rossi, C. F. Bennett, S. Sibley, R. W. Davies, and H. N. Arst, *EMBO J., 9*: 1355 (1990).
194. Y.-H. Fu and G. Marzluf, *Mol. Cell. Biol., 10*: 1056 (1990).
195. Y.-H. Fu and G. Marzluf, Proceedings of the National Academy of Sciences (U.S.), *87*: 5331 (1990).

196. G. Jarai, H.-N. Truong, F. Daniel-Vedele, and G. Marzluf, *Curr. Genet., 21*: 37 (1992).

197. X.-W. Deng, T. Caspar, and P. H. Quail, *Genes Devel., 5*: 1172 (1991).

198. A. Coleman, *J. Cell Sci., 97*: 399 (1990).

199. P. M. Okamoto and G. A. Marzluf, *Mol. Gen. Genet., 240*: 221 (1993).

200. C. E. Deane-Drummond, *Nitrogen in Higher Plants* (Y. P. Abrol, ed.), John Wiley, New York, p. 1 (1990).

201. D. J. Pilbeam and E. A. Kirby, *Nitrogen in Higher Plants* (Y. P. Abrol, ed.), John Wiley, New York, p. 39 (1990).

202. C. R. Bhatia and R. Mitra, *Nitrogen in Higher Plants* (Y. P. Abrol, ed.), John Wiley, New York, p. 427 (1990).

203. I. Stulen, *Nitrogen in Higher Plants* (Y. P. Abrol, ed.), John Wiley, New York, p. 297 (1990).

204. H. Kacser and J. A. Burns, *Symp. Soc. Exper. Biol., 27*: 65 (1973).

205. R. Heinrich and T. A. Rapoport, *Eur. J. Biochem., 42*: 107 (1974).

206. H. Kacser and J. W. Porteus, *Trends Biochem. Sci., 12*: 5 (1987).

207. W. P. Quick, U. Schurr, R. Scheibe, E.-D. Schulze, S. R. Rodermel, L. Bogorad, and M. Stitt, *Planta, 183*: 542 (1991).

208. M. Stitt, W. P. Quick, U. Schurr, E.-D. Schulze, S. R. Rodermel, and L. Bogorad, *Planta, 183*: 555 (1991).

209. W. P. Quick, K. Fichtner, E.-D. Schulze, R. Wendler, R. C. Leegood, H. Mooney, S. R. Rodermel, L. Bogorad, and M. Stitt, *Planta, 188*: 522 (1992).

210. K. Fichtner, W. P. Quick, E.-D. Schulze, H. A. Mooney, S. R. Rodermel, L. Bogorad, and M. Stitt, *Planta, 190*: 1 (1993).

211. M. Lauerer, D. Saftic, W. P. Quick, C. Labata, K. Fichtner, E.-D. Schulze, S. R. Rodermel, L. Bogorad, and M. Stitt, *Planta, 190*: 332 (1993).

5

Nutrient Deficiencies and Vegetative Growth

Nancy Longnecker

The University of Western Australia
Nedlands, Western Australia, Australia

INTRODUCTION

Plants must get the nutrients they need from the environment in which they grow and exclude or tolerate substances that are present in excess. The flexibility which allows plants to live in difficult situations is amazing. The ways in which nutrient status affects plant growth and development is the subject of this chapter. Vegetative growth consists of growth and development of the primary axis (cell division and cell expansion) as well as initiation and growth of lateral meristematic tissue. Cell division and expansion are discussed in the section on cell growth, differentiation in the section on organ growth, and coordination between tissues in the section on plant growth. Much research has been reported since Moorby and Besford's review on mineral nutrition and growth [1] and Marschner's excellent book on plant nutrition [2]. This review will focus on more recent work, with particular emphasis on nitrogen (N) and phosphorus (P) deficiency; examples of other stresses are mainly used to illustrate comparisons and contrasts with these two readily remobilized elements.

For many years, agricultural research has focused on maximum production. There are major reasons for improving efficiency of agricultural production. In some instances, there is a lack of capital for inputs. In others, serious environmental problems are caused by excessive inputs. Improved efficiency of fertilizer use in crop production will benefit both intensive and extensive agricultural systems. Two ways to achieve this are: (1) improved diagnosis of nutrient deficiency and appropriate timing of fertilizer application in order to take maximum advantage of the added nutrient, and (2) use of plant species or cultivars which are more efficient at using a given nutrient supply. A better understanding of the mechanisms by which nutrient deficiency affects plant growth and development will improve both of these approaches.

Effects on Source or Sink?

An obvious effect of nutrient deficiency is decreased plant size. Another effect is decreased net photosynthesis. These two observations lead to a chicken and egg type of question: Is the primary effect of nutrient stress decreased growth, with secondary effects such as decreased sink strength, decreased leaf area, and decreased net photosynthesis? Or is the primary effect of nutrient stress a decreased rate of photosynthesis, with the secondary effect of decreased growth because of limited available assimilate? The answers to these questions probably depend on the nutrient in question and the severity of the stress. Manganese (Mn) deficiency is an example where the primary effect of deficiency may be on decreased photosynthesis (see p. 165). Calcium (Ca) is an example of a nutrient whose deficiency directly decreases growth rather than decreasing growth through an effect on photoassimilate availability (see the section on "Cell Growth"). Whether the primary effect of other deficiencies such as N or P is on photosynthesis in source tissue (which exports assimilates for growth of other tissue), or on the growth of sink tissue (which must import assimilates for growth), has been widely debated. Current evidence indicates that the primary effect of N and P deficiency is the limitation of growth, with feedback effects on photosynthesis. Another possibility is that neither CO_2 fixation per se nor growth is the primary limiting metabolic process; the first limitation by some nutrient stresses may be a change in partitioning of reduced carbon controlled by plant growth substances and subsequent decreased transport of assimilate to particular organs.

In evaluating the primary effect of a nutrient stress, it is useful to have information on rates of photosynthesis, levels of soluble carbohydrate, and a measure of growth (e.g., ethanol-insoluble dry weight) in the same experimental system. A useful experimental approach is the manipulation of source and sink [3] in order to examine their relationships. If decreased photosynthesis is the primary effect of a nutrient deficiency, this could be due to: (1) substrate limitation (e.g., decreased stomatal opening, increased mesophyll resistance, decreased energy capture, etc.) or (2) decreased biochemical activity (e.g., decreased enzyme synthesis or activity). With a limited rate of photosynthesis, carbohydrate levels might decrease in the source tissue. If decreased photosynthesis is a secondary effect of nutrient stress, however, this could be due to feedback inhibition of photosynthesis because of decreased requirement for assimilate. In this case, carbohydrates may accumulate in the source and even in the growing tissue.

A cautionary note is important. There is danger in making assumptions that the levels of carbohydrates in tissues indicate whether source or sink activity is most limited. Because carbohydrate content is usually measured at a single point in time, one measures differences in pool size. The same throughput can be achieved, however, with either a small pool cycling quickly or a large pool cycling slowly.

Is the Primary Effect of Nutrient Stress Decreased Growth?

It is frequently said that decreased nutrient supply decreases the leaf area rather than the efficiency of photosynthesis (e.g., [4–6]). This depends on the particular nutrient being examined. There are unequivocal examples of nutrient deficiencies such as calcium and boron (B) which cause decreased growth as their primary effect (see pp. 247–248). Nutrients such as nitrogen and phosphorus, which have direct roles in photosynthesis, but also function in most other metabolic process, are more likely to limit growth with subsequent feedback effects on photosynthesis. The evidence for decreased growth as the primary effect of N and P deficiencies, however, is less clear-cut and is discussed below. Because data on growth and carbohydrate levels in the sink tissue are available from the same experiments, evidence from studies on salinity stress are presented for comparison.

Nitrogen

The bulk of evidence suggests that decreased growth is the primary effect of N deficiency. In field studies, Wullschleger and Oosterhuis [7] found that within 5 days of the leaf unfolding, growth rates of slightly N-deficient cotton leaves were reduced by 40% compared to controls. Leaf area index (m^2 leaf per m^2 land area) was decreased by N deficiency within 76 days after sowing. Yet net photosynthesis of single leaves in the upper canopy was not affected by N treatment. Vessey and Layzell [8] also observed effects of N treatment on C and N partitioning in plants when there were no effects of N treatment on the rate of photosynthesis of the entire plant.

If light is not limiting, available N apparently determines shoot growth and demand for carbohydrates. Within the first day of decreased N supply, starch and sucrose accumulated in both the source and sink leaves of N-deficient plants [11] (see p. 158). Leaf growth decreased significantly by the second day of removal of N supply [11]. Expansion rate of new leaves was lower and final leaf area was smaller in N-deficient plants than in those resupplied with N [9,10].

Nitrogen deficiency decreases expansion of individual leaves as well as the growth of tillers or lateral branches with a resulting amplified effect on leaf number. Although N deficiency also decreases photosynthesis (see p. 163), the immediate, direct effect of N deficiency appears to be on sink growth; and the major, cumulative effect of low N on net C assimilation is due to decreased photosynthetic area rather than decreased rate of photosynthesis per unit leaf area [6,12].

Phosphorus

Phosphorus deficiency resulted in smaller soybean leaves even when there was no effect on rate of photosynthesis [13]. Fredeen and Terry [13] suggested that P deficiency decreases growth before any decrease in photosynthesis, perhaps because there is insufficient ATP for protein synthesis. In another study, shoot growth was reduced within 7 days of low-P treatment while rate of photosynthesis did not decrease until 19 days after treatment [14].

Salinity

For salt-stressed plants there is similar evidence that the decreased growth is due to effects in the growing tissue rather than effects on photosynthesis. Rate of leaf elongation decreased within 2 h of transfer of plants from 0.5 mM NaCl to 20 or 60 mM NaCl, but then recovered within 10 h of continued treatment [15]. The rate of leaf elongation of the less severely NaCl-stressed plants recovered to 100% of the control value while the more severely NaCl-stressed plants recovered only to 60–70% of the control value. Because of the speed at which leaf elongation decreased, the effect was interpreted as being due to direct effects on cell expansion rather than to reduced supply of carbohydrates. Also, the concentration of ethanol-soluble carbohydrates was greater in the elongating zone of barley leaves exposed to high NaCl than in control leaves, indicating that decreased growth in this zone was not due to limited supply of carbohydrates [16]. Levels of starch were similar in the growing zones of leaves receiving different treatments [16]. The decreased growth was postulated to be due to water deficit in the growing tissue and the inability to maintain turgor [15]. Flowers et al. [17] have provided evidence that salts accumulate in the apoplast of rice leaves treated with NaCl. They suggest that low osmotic potential in the apoplast could result in cell dehydration and be the initial cause of decreased growth. However, applying pressure to wheat or barley roots which presumably increased the turgor pressure in the leaves by the same amount did not compensate for the long-term depression of leaf-growth rate of NaCl-stressed plants [18].

CELL GROWTH

Growth is the irreversible, quantitative increase in plant size. It is a coordinated process which depends on the division and enlargement of individual cells and is restricted to certain zones of tissue. Thus, decreased growth can be caused by decreased cell division, cell elongation, or a combination of these two. The decreased growth which results from changes in differentiation (i.e., decreased initiation of leaves or growth of tillers, lateral branches, or root primordia) will be discussed in the section on "Organ Growth." Although there are instances of nutrient deficiencies causing decreased grain yield with no visible effects on vegetative growth (e.g., specific effects on pollen fertility of some trace elements such as copper (Cu) [19] and zinc (Zn) [20]), nutrient deficiencies usually cause decreased vegetative growth. There are numerous specific, rapid effects of nutrient stresses on cell division, expansion, and senescence; some examples are shown in Table 1.

Cell Division

In most instances, it is not clear whether nutrient stress primarily decreases growth through an effect on cell division or on cell expansion or a combination of the two. In a study examining why Bonsai plants are small, Körner et al. [22] concluded that the main difference between leaves from Bonsai plants and control leaves of the same species was in cell number. They showed that cell size was the same (or even bigger) in the Bonsai plants. They concluded that differences in cell size cannot explain differences in size of plants, and that genetic control of cell size is stronger than "the most severe environmental impact" [22]. There are instances, however, in which nutrient stresses undoubtedly result in decreased cell size. It is possible that the timing of stress is critical in determining whether the primary growth depression occurs via decreased cell number of decreased cell size. For example, there is evidence that drought stress can decrease cell numbers in tissue which is still dividing, but will decrease cell size in elongating tissue [23].

Nitrogen

Of the mineral nutrients, N has the most widespread and commonly observed effect on growth [24]. Nitrogen deficiency causes substantial decrease in leaf size (Figure 1). This is mainly due to decreased rate of leaf expansion rather than a change in duration of leaf expansion [6,25,26]. The number of cells per leaf is lower with N deficiency [26,27] and cell size is smaller [26–28]. Because they observed no effect of N treatment on cell numbers in the upper epidermis, Radin and Boyer [6] suggested that cell expansion was more sensitive to N deficiency in sunflower leaves than was cell division. In contrast, Terry [26] found larger decreases of cell number than of cell size in N-deficient sugar beet leaves.

Table 1 Examples of Disrupted Cell Division or Expansion Caused by Nutrient Stress

Process	Tissue	Ref.
Low N		
Decreased cell number	Root	26
Decreased cell expansion	Leaf	6
Decreased cell expansion	Root	26
Increased denucleation	Root cortex	186
Increased aerenchyma formation	Root cortex	21
Low P		
Decreased cell expansion	Leaf	39
Increased aerenchyma formation	Root	21
Low K		
Decreased cell size	Leaf	187
Low Ca		
Decreased cell division	Various	30
Decreased cell expansion	Pollen tube	48
Increased denucleation	Root cortex	186
Low Fe		
Development of transfer cells	Root	188
Increased cell division	Root	115
Low Mn		
Decreased cell expansion	Root	189
Decreased cell expansion	Root	190
Low B		
Decreased cell division	Root	191
Decreased cell division	Root	192
Decreased cell elongation	Pollen tube	164
Decreased cell elongation	Root	192
High Na		
Decreased cell division	Root	193
Decreased cell expansion	Root	193

Calcium

Because of the low phloem mobility of Ca (Table 2) and low transpiration of expanding tissue (hence low xylem stream delivery), Ca deficiency can result in dramatic stunting of new tissue. When Ca was removed from nutrient solution, elongation of bean roots decreased within 12 h [29]. Low external Ca supply (<0.1 mM) resulted in a decreased rate of root-cell division [30]. Levels of free cytoplasmic Ca may play a regulatory role in mitosis through control of the polymerization of microtubules [31].

Figure 1   Wheat plants (cv. Gamenya) grown for 31 days in a controlled-environment chamber. Seeds were sown in white sand and supplied 3 times weekly with complete nutrient solution which contained 50 μM N (Low N) or 4000 μM N (High N). Low N plants have smaller and fewer leaves.

Boron

Response to B deficiency is similar to Ca in that phloem mobility is limited and there is a rapid effect of low B supply on cell division in meristematic tissue [2]. Within 6 h of removal of B supply, mitosis in sunflower root tips decreased [32]. Shelp [33] has summarized data demonstrating effects of B deficiency on DNA and RNA synthesis. A working hypothesis emerging from his review is that B is required for formation of pyrimidine nucleotides and

Table 2   Ease of Retranslocation of Different Nutrients via the Phloem

Mobile	Variable mobility	Immobile
N, P, K	Mg, S, Cu, Zn, Mo	Ca, Mn, Fe, B

(Taken from [194])

that this results in the decline in RNA and DNA with consequent decreased cell division.

Cell Expansion

In order for a cell to expand, the rigid wall must be loosened and the turgor pressure of the protoplast must be maintained [34]. Nutrient stresses can affect both cell wall extensibility and cell turgor. Some examples given below, however, show that the role of turgor in determining the rate of cell extension is unclear. A change in turgor pressure had an immediate effect on rate of leaf elongation, but the rate soon returned to the equilibrium rate established before the change [18]. Also, longitudinal growth rates of cells in roots grown at low water potentials were not related to the turgor pressure (measured with a pressure probe) of those cells [35]. These results indicate that changes in the rate of cell expansion result from changes in cell wall properties. There is other evidence given in the next sections that nutrient deficiencies decrease turgor which plays a role in the resultant decreased growth.

Nitrogen

Indirect evidence from the relationship between pressure and cell volume suggested that cell walls of N-deficient cotton leaves were more rigid than those of control plants [28]. This may partially explain the increased drought resistance of N-deficient plants [28]. Almost all of the decreased growth of N-deficient sunflower leaves could be accounted for by a decrease in the daytime growth rate [6]. During the day, when transpiration was occurring, hydraulic conductivity of the roots was lower and water potential of N-deficient leaves was more negative; these results were interpreted to show that turgor was not maintained at a level sufficient for growth [6]. Nitrogen treatment had no apparent effect on extensibility or the threshold turgor for growth, but leaves from low N plants had lower turgor and lower growth rates than leaves from high N plants [6]. Leaf growth of high NaCl-stressed plants was also reduced more during the daytime than at night [15].

Phosphorus

Phosphorus deficiency also causes marked growth reduction. Phosphorus-deficient shoots are frequently darker than those of healthy plants [36] because cell growth is inhibited more than chlorophyll synthesis and the chlorophyll is concentrated in smaller cells [37]. Isoflavone [38] and anthocyanin [37] contents can also increase in P-deficient plants. Leaf area is decreased by P deficiency [5,14,39]. As in N-deficient plants, the decreased leaf growth was associated with a lower root hydraulic conductance and a lower leaf water potential [39].

Calcium

The need for Ca in cell expansion is not intuitively obvious since Ca plays a role in stabilizing the cell wall (Ca cross links of pectins [40]). However, Ca plays a role as a second messenger in the detection of many external signals which affect plant growth [41]. Auxin and Ca both stimulate cell elongation and their transport within the plant is interrelated [42]. Calcium redistribution and gravitropic response can both be inhibited by treatments of roots with auxin inhibitors such as triiodobenzoic acid (TIBA [43]). However, the growth of cells caused by exogenous Ca and by auxin appears to be mediated by different mechanisms [44]. According to the acid growth theory, auxin stimulates H^+ efflux which results in cell wall loosening [45]. There is conflicting evidence in the grasses that auxin stimulation of cell wall loosening may [46] or may not [47] occur via breakdown of xyloglucans. It is possible that auxin-stimulated cell growth is mediated by cell wall glucanases, while Ca-stimulated cell growth affects synthesis of new cell wall material [48].

Potassium

Cell extension decreases in plants receiving a low potassium (K) supply [2]. Stem internodes are much shorter in K-deficient plants [49]. Although one of the major roles of K is the activation of many enzymes, K also plays an important role in maintaining turgor [50]. Potassium salts in the vacuole are frequently the main contributor to the turgor pressure necessary for cell extension [51]. The K in the vacuole can be replaced in turgor maintenance by Na, Mg, or reducing sugars [51,52]. Potassium may also help maintain cell-wall loosening; its influx can balance the H^+-efflux which stimulates cell-wall loosening. The auxin-stimulated elongation of oat coleoptiles stopped after 6 to 12 hours when K was not supplied [53].

ORGAN GROWTH

Environmental stresses have been used to understand the processes regulating plant development. For example, varying light and temperature treatments have illustrated the flexibility of plant development in adapting to particular environmental constraints. Similarly, nutrient stresses are important, frequently limiting environmental stresses which can regulate plant development. The effect of nutrient stress on vegetative growth varies between tissues. Most nutrient stresses eventually decrease both shoot and root growth. However, most nutrient deficiencies initially decrease shoot growth more than root growth or decrease shoot growth and stimulate root growth. Either result in a higher root:shoot ratio (RSR) of dry matter (as discussed p. 157). Nutrient stress can also result in decreased lateral bud growth. The growth and developmental responses of different organs to nutrient stresses are discussed here.

Shoot Meristems

One of the more fascinating aspects of plant growth is that the shoot apical meristem gives rise to forms of such diversity and order (Figure 2). Growth of the shoot ultimately depends on the health of the meristem. Depending on the function and mobility of a particular nutrient, growth may depend on the supply of nutrient as well as photosynthate to the meristem. It is generally believed that transport to meristems of nutrients which are readily retranslocated via the phloem (Table 2) such as N is maintained under deficient conditions; however, there are not many direct measurements of nutrients in meristems in the literature. Older leaves senesce and mobile elements are thought to be remobilized to maintain their concentration in the meristematic tissue. In contrast, symptoms of deficiencies of immobile elements are first visible in young leaves and meristematic tissue [36]. For more mobile nutrients, stresses have large effects on leaf expansion (see pp. 144, 148), but have relatively less effect on the initiation of leaf primordia. Table 3 gives examples where nutrient stresses decrease the number of leaves on the main stem.

Table 3   Effects of Nutrient Stresses on the Number of Leaves on the Plant's Main Stem

Leaves on stressed plant (Leaves on control plant)	Plant	Ref.
Low N		
8 (9)	Barley	55
6 (8)	Wheat (cv. Gabo)	57
7 (8)	Wheat (cv. Aroona)	58
No effect	Wheat	54
32 (38)	Sugarbeet	27
30 (50) at 15°C	Sugarbeet	26
39 (48) at 25°C		
Low P		
9 (10)	Wheat (cv. Thatcher)	59
7 (8)	Wheat (cv. Falcon)	59
Low Cu		
6.7 (8)	Wheat (cv. Gamenya)	202
High NaCl		
10–11 (12–13)	Triticum aestivum	63
8–9 (11–12)	T. turgidum	63

Primordia Initiation

Some researchers have found no effect of N treatment on the number of leaves on the main stem of wheat [54] while others have observed fewer leaves on the main stem of N-deficient cereals [55–58] and dicots [9,27]. There were probably differences in severity of N deficiency in those studies. Similarly, phosphorus deficiency has been observed to decrease the number of leaves on the main stem ([59]; N. E. Longnecker and A. D. Robson, unpublished data) or to have no effect on leaf number [60]. In our laboratory, we have also observed an increase in total number of leaves on the main stem of Cu-deficient wheat after a foliar spray of $CuSO_4$ (J. Slater, N. E. Longnecker, and A. D. Robson, unpublished data). These observations mean that nutrient status of cereals can change the destiny of a specific primordia from becoming a leaf to becoming a spikelet. The mechanisms controlling differentiation of primordia are not known but are of obvious interest.

Many stresses are known to decrease the number of spikelet primordia initiated on cereal apices. Terminal spikelet production occurred earlier in drought stress [61,62] and salt stress [63], and there was less time for spikelet primordia initiation. The effect of nutrient deficiency depends on both the nutrient and the severity of stress. There were fewer days of spikelet development in N-deficient wheat, especially at higher temperatures [64]. However, the duration of the spikelet initiation phase of development of N-deficient wheat was not affected, even in severely N-deficient wheat [58]. In that study, the main cause of decreased primordia initiation in N-deficient wheat was a slower rate of initiation [58].

Rate of Development

Delayed maturity is frequently observed in nutrient-deficient plants (e.g., N [65]; P [65]; Cu [66]; Mn [67]). However, substantial effects of nutrient deficiency on rates of vegetative apical development are less common. Delays in apical development have been observed in N-deficient wheat [64,65,68], but days to double ridge are frequently not affected by N status unless N deficiency is very severe (e.g., <30% of maximum dry matter production [65]).

In the case of Mn deficiency, significant delays in apex development have been observed: main stem apices were at the double ridge stage in Mn-deficient barley when the apices of healthy barley plants were more advanced, at late double ridge to triple mound [58]. Phosphorus deficiency has been observed to delay floral initiation if severe [69,70] or to have no effect on floral initiation [60,69].

Transport to Apex

Relatively little is known about the effects of nutrient stress on differentiation in the shoot apex and transport of nutrients to and within the apex.

Nutrient transport to the apical meristem differs from long-distance transport to other plant organs because of the lack of developed vascular tissue and low rates of transpiration. There is also a large demand for some nutrients such as N and P in the meristem because of high rates of metabolic activities.

The little evidence available suggests that the N content of shoot apices is unaffected by change in N supply. The absolute amounts of N transported to the apices were similar when nitrate-fed white lupin plants went from adequate to deficient N supply [71]. However, the withdrawal of N resulted in proportionately less dry matter accumulation by the apices [71]. It would be interesting to have more precise measurements of growth and nutrient content of shoot apices after changes in nutrient supply in order to compare the effects of mobile and immobile nutrients on apical growth.

Smith and Wareing [72] investigated the transport of ^{32}P to dominant meristems. They trained birch (*Betula pubescens*) seedlings so that lateral meristems occupied the highest position. This looping caused the lateral highest on the loop to assume dominance [73] and resulted in the greatest accumulation of ^{32}P in that meristem [72]. The greatest accumulation of ^{32}P did not always occur in the dominant apex; in the control plants (with straight stems), there were greater amounts of ^{32}P in the lateral buds immediately below the dominant apex. Smith and Wareing [72] suggested that nutrients were diverted toward the dominant apex and that the level of apical dominance affected the level of accumulation. Auxin may be the signal involved; auxin applied to the stem of an excised pea plant increased accumulation of ^{32}P in that region of auxin application [74].

Lazof and Läuchli [75] used electron-probe microanalysis to show that salinity stress resulted in increased Na and Cl and decreased Ca, P, and K concentrations in basal regions (especially 200–500 μm below the apex) of the shoot apex, but that nutrient concentrations in the apical 10–50 μm were less affected by the increase in Na and Cl. These results indicate that the apical 10–50 μm were buffered against the increased Na and Cl to a greater extent than the basal regions.

Leaves

Both size and number of leaves determine total leaf area per plant (Figure 1). The effects of nutrient status on leaf size and number are discussed here. The effects of nutrient deficiencies on cellular growth which determine cell number and size were discussed in the section "Cell Growth."

Leaf Size

There are many examples of decreased leaf area or mass per leaf on nutrient-deficient plants, including: N [6,25,26], P [5,14,39,76], Mg [77], Cu [67], and

Figure 2 Apex of a main stem of wheat (cv. Aroona) at the glume primordium stage of development. The large lobes differentiate to become spikelets on the head of wheat. Leaf primordia were formed previously from this apex. Seeds were sown in white sand and plants were grown for 29 days in a controlled-environment chamber and received complete nutrient solution 3 times weekly.

B [78]. In the case of B deficiency, there was a decreased rate of leaf expansion, so that leaves grew more slowly but eventually reached a similar final size [78]. This illustrates the importance of harvests over time in some cases. Improved P status as a result of mycorrhizal infection resulted in a greater leaf area per g leaf tissue [13]. In this example, differences in starch content of the leaves could not explain the increased leaf mass.

Leaf Number

In addition to decreasing the size of leaves, nutrient stresses can decrease the total number of leaves per plant through effects on leaf numbers on the main stem (p. 147) and on lateral shoot emergence (p. 150). Nutrient stresses can also delay leaf emergence as illustrated for P-deficient barley in Figure 3. Leaf emergence is also slower in N-deficient wheat [55,58], Cu-deficient wheat [66,202], Zn-deficient wheat [79], and wheat exposed to high NaCl [63].

Leaf Senescence

Persistence of the leaf is also an important determinant of its ultimate contribution of photosynthate. Timing of leaf senescence is markedly affected by plant nutrient status. Once again, the effect of deficiency on senescence depends on the nutrient, a useful fact for the visual diagnosis of nutrient deficiencies. Deficiency of readily retranslocated nutrients such as N, P, and K (Table 2) cause earlier and faster senescence of older leaves. For example, Thorne and Watson [80] reported leaf area duration of 5.6 weeks for N-deficient wheat compared to 7.1 weeks after early N application and 6.8 weeks after late N application. Early senescence releases some nutrients for remobilization to the growing leaves and apices [e.g., 76,81]. In contrast, there can be delayed leaf senescence in plants deficient in other elements, for example in Cu-deficient wheat [82]. There are good reviews of the remobilization of nutrients and leaf senescence [83,84].

Lateral Shoot Buds

Lateral shoots are frequently an important contributor to yield potential. One commonly reported effect of nutrient deficiency is decreased growth of lateral shoots or tillers. Deficiencies of nutrients such as P, Cu, and Mn decrease tiller or lateral numbers, as shown in Table 4. Potassium has been reported to have no effect on tillering [85].

In particular, N deficiency greatly decreases the number of tillers in grasses [58,86–89]. This well-known response is illustrated in Figure 4. The main reason for decreased tiller number on N-deficient plants is that the tiller buds which are initiated do not grow [58,90]. There is a decreased initiation

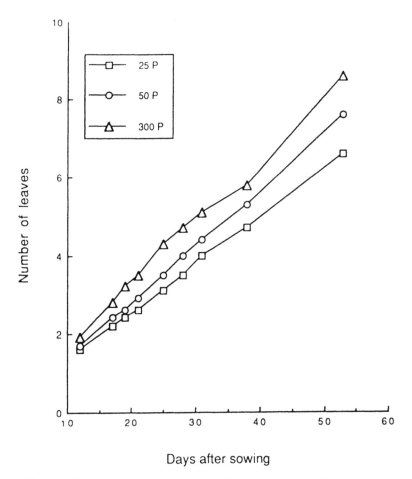

Figure 3 Phosphorus deficiency delayed emergence of leaves on the main stem of barley (cv. O'Connor). Plants were grown with 25, 50, and 300 mg P per kg of Yalanbee soil. (N. E. Longnecker and A. D. Robson, unpublished data)

of tiller buds in N-deficient wheat, but this is because there are fewer emerged leaves and therefore fewer axils in which tiller buds can form, not because of a specific effect of the N treatment on bud initiation [58].

The type of N supply can also affect tillering. More tillers emerged on wheat grown with mixed N (NH_4^+ and NO_3^-) than with either NH_4^+ or NO_3^- alone [85,91].

Table 4   Effects of Nutrient Stresses on Numbers of Lateral Shoots Which Emerge

Laterals on stressed plant (Laterals on control plant)	DAS	Plant	Ref.
Low N			
3 (5.8)	35	Barley (cv. Proctor)	88
4 (23)	45	Wheat (cv. Aroona)	58
Low P			
0–2 (4–5)	47	Subterranean clover	69
0–1 (8–10)	84	Subterranean clover	69
1–2 (10)	53	Barley (cv. Forrest)	N. Longnecker and A. D. Robson, unpublished data
Low Mn			
7.5 (12)	30	Barley (cv. Galleon)	65
25–30 (16)	100	Barley (cv. Galleon)	65
2.7 (11.3)	26	Wheat (cv. Stacy)	194a
1 (3)	60	Wheat (cv. Neepawa)	195
Low Cu			
0.3 (1)	14	Wheat (cv. Gamenya)	67
1 (3)	60	Wheat (cv. Neepawa)	195
0 (1.5)	26	Wheat (cv. Gamenya)	202
3 (2.5)	80	Wheat (cv. Gamenya)	
High NaCl			
Values not reported	—	Wheat (cv. Probred)	63
3–4 (5–6)	44	Wheat	85

A less common observation is an increase in tiller numbers of nutrient-deficient plants. For example, in experiments on Mn-deficient barley, tiller numbers were reduced during early tillering (Figure 5). But severely deficient plants continued to produce tillers until they died, resulting in 25–30 tillers per plant, compared to 16 tillers on control plants [67]. This was apparently because the main stem apex died in severely Mn-deficient plants and released the tiller buds from apical dominance. With less severe deficiency, rate of tillering was decreased but the same maximum number was produced and the main stem apex advanced from vegetative to reproductive development and produced grain. This is another example of the need to take into account the severity of the deficiency when describing the effects of nutrient deficiency. Other examples of nutrient deficiencies causing increased branching are in Mn-deficient lupins [92], Zn-deficient citrus [93], and B-deficient rapeseed [93].

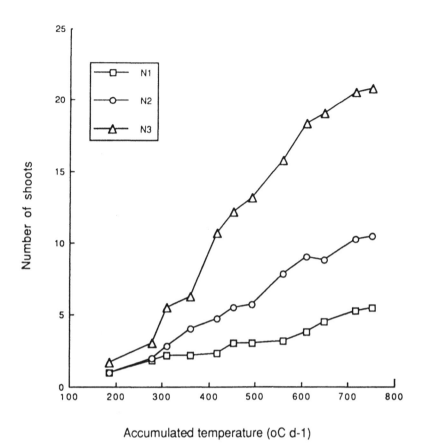

Figure 4   Total number of shoots produced per plant on wheat (cv. Gamenya) grown with 50 μM N (N1), 200 μM N (N2), and 800 μM N (N4). (N. E. Long-necker, E. J. M. Kirby, and A. D. Robson, unpublished data)

Roots

Root development is not as well documented as shoot development; how-ever, basic root morphology has been described. Dicotyledons are character-ized by a dominant taproot while monocotyledonous roots can be classified as seminal (produced from primordia present in the seed) or nodal (produced from primordia at the nodes after germination) [94]. Wheat roots grow in an orderly sequence which is related to shoot development [95-97].

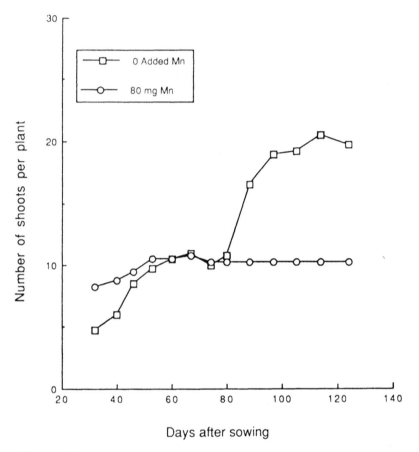

Figure 5 Severe Mn deficiency can result in increased number of tillers produced in barley grown with 0 added or 80 mg Mn kg^{-1} soil (soil described in Longnecker et al., 1990). (N. E. Longnecker and R. D. Graham, unpublished data).

The result of nutrient deficiency is usually an increased root:shoot ratio (RSR). This can occur because of increased root growth or because root growth is decreased less than shoot growth. Stimulation of root growth can occur in nutrient-deficient conditions. Within 2 days of changing from adequate to deficient N supply, roots of N-deficient tobacco plants had greater dry weights while shoots of N-deficient plants had lower weights than control plants [98]. Phosphorus deficiency results in increased root length [99] and decreased root diameter [100]. Zinc deficiency can result in an initial increase in root weight [79,101].

The increased RSR with nutrient deficiency is not surprising considering the direct role of roots in nutrient uptake; there is an apparent ecological advantage in maintaining root mass to "search for" nutrients since the size and distribution of roots in soils has a large impact on nutrient uptake [102]. The uptake capacity for a given nutrient can also be specifically increased by deficiency of that nutrient; this has been well documented (e.g., N [98,103–106]; K [107–109]; P [110–112]; Fe [113–117]).

The growth of roots in response to nutrient supply is examined here. Nutrient deficiencies can decrease the size and number of roots (Table 5) in analogy to the decreased size and number of shoots and leaves. Hackett [118] showed that root branching increased with localized nitrate supply, but only if the nitrate was supplied to segments of the root within 10 mm of the tip. In that study, localized nitrate stimulated lateral root elongation. Drew and Saker observed localized growth of roots supplied with N [119, 120] and P [121]. Brouwer [122] showed that increased growth of roots in half of a split-root system receiving O_2 could compensate for decreased roots in an unaerated half. Thus, total dry matter of roots remained the same. In general, however, increased root growth in a well-supplied region does not compensate for poor root growth in regions with low nutrient supply (e.g., [122a]). When root growth is stimulated in regions with high fertility, the signal which causes this increased growth is unknown.

Localized nutrient supply does not always stimulate root growth. The response of root growth apparently depends on the nutrient which is deficient. Four examples from split-root experiments are given here: (1) Localized nitrate stimulated lateral root growth only when the rest of the root system was lacking in N supply [121]. (2) In a split-root system, lupin roots grew well only when Fe was supplied in the immediate environment [123]. Growth of roots in the + Fe chamber did not compensate for the poor root growth in the − Fe chamber, even though Fe concentration in shoots was adequate [123]. (3) The dry weights of wheat roots were similar, whether they were grown in the + Zn or − Zn chambers of a split-root system [124]. (4) In a clever system used to ensure that both parts of the root system were necessary for plant growth, Jeschke and Wolf [51] grew part of the root system supplied with K^+ and without NO_3^- and part without K^+ and with NO_3^-. They found that an external supply of K^+ was not necessary for root growth since root growth was greatest in the compartment without K^+ and with NO_3^- [51].

In addition to their function in nutrient and water uptake, roots are the source of numerous plant hormones and as such, may be the source of signals controlling carbon partitioning for root and shoot growth. Roots are necessary for hormonal production as well as nutrient and water uptake. Thus, when wheat roots were restricted by a small soil volume, plant growth

Table 5 Effects of Nutrient Supply on Root Growth

Process	Tissue	Ref.
Delayed N		
Decreased number of primary and secondary laterals	Barley	196
Added N		
Increased number of laterals	Wheat	118
	Barley	119,121
Increased cell number in young roots	Sugar beet	26
Low P		
Increased root length per mass	*Carex coriacea*	100
Decreased number of nodals	Barley	197
Decreased length of primary laterals	Barley	197
Decreased total root length	Cotton	39
Added P		
Increased number of laterals	Barley	120
Low K		
Decreased number of nodals	Barley	197
Decreased length of primary laterals	Barley	197
Inhibited formation of secondary laterals	Barley	197
Added S		
Increased total root length	Ryegrass	137
Decreased total root length	Subclover	137
Increased root diameter	Ryegrass and subclover	137
Low Ca		
Decreased primary root extension	Bean	29
Low Fe		
Enhanced root hair formation	Sunflower	115
Low B		
Decreased elongation of primary laterals	Soybean	78
Decreased elongation of primary laterals	Squash	167
Decreased root tip elongation	Tomato	163
High Al		
Decreased root hair development	Soybean	198
Decreased tap root elongation	Many legumes	199
High Na		
Decreased root elongation	Cotton	200
	Maize	201

was reduced, even though nutrients and water were always adequately supplied [125]. The interaction of nutrient status and plant hormones in the control of plant growth and development is a complex topic which deserves a review in its own right.

PLANT GROWTH

Nutrient deficiencies can cause immediate decreases in growth at the cellular level (p. 141). The response of cell growth to nutrient deficiency depends on the tissue (p. 145). Differences in growth of roots and shoots of nutrient-deficient plants frequently result in an increased RSR [126,127]. This appears to be mediated by a change in the partitioning of assimilate in the plant, so that available photosynthate is preferentially transported to the root. The mechanisms by which whole plant growth is coordinated to respond to nutrient and other environmental stresses are not clearly understood. For example, the signal which stimulates change in partitioning has not been identified. In this section, the control of the RSR and the effects of nutrient deficiencies on the metabolic processes of carbon fixation and partitioning are discussed.

Control of Root:Shoot Ratio (RSR)

First Come, First Served,
or Feedback Through the Shoot?

A common view of root and shoot growth during periods of nutrient deficiency is that RSR increases because a greater proportion of the limiting nutrient is used in the root which is nearer to the initial entry of the nutrient into the plant [1,128]. According to this view, nutrient deficiency results in proportionately greater root growth than shoot growth and a greater partitioning of the carbon to the roots. However, the ready cycling of nutrients within the plant system [8,71,129–132] does not support this model of immediate nutrient use in the absorbing tissue. Even more conclusive evidence against this model is the report by Smith et al. [140] who showed that when *Stylosanthes* plants were transferred from solution with adequate P to solution with no P, root growth was initially stimulated in response to P deficiency.

The rate of soybean shoot growth decreased within 7 days after low P treatment was imposed while root weights were not affected until 17 days after treatment [14]. Thus, the increased RSR observed within the first 7 days after treatment was due to decreased shoot growth. In that experiment, a higher proportion of the total plant P was in the root system of P-deficient plants compared to control plants (a RSR of P of 1.6 in P-deficient plants and of 0.5 in control plants [14]). Was the greater root growth compared to shoot growth due to higher root P or was higher root P due to greater root growth? The subsequent decrease in root growth was apparently not due to lower transport of photosynthate to the roots; there was an accumulation of starch and sucrose in fibrous roots [14].

A possible mechanism by which N status controls RSR was summarized by Rufty et al. [98] as follows: With decreased N supply, there could be decreased transport of reduced-N to shoot sinks. This would cause decreased growth and carbohydrate metabolism of meristems and other existing sink tissue (e.g., expanding leaves). More photosynthate could then be transported to the roots, stimulating root growth. Within 2 days of changing from adequate to deficient N supply, starch and sucrose concentrations were higher in roots of N-deficient tobacco plants [98]. Other researchers have also found greater amounts of carbohydrates in the roots of N-deficient plants [133, 134].

The current evidence supports the hypothesis that the exchange of C and N between leaves and roots is involved in the whole-plant regulation of growth [98,135]. Decreased N uptake can limit leaf initiation and expansion which decreases the carbohydrate demand in the shoot sinks. This allows proportionately greater transport of soluble carbohydrates to roots with a resulting increase in RSR. The concept of a pool of amino-N cycling within the plant may help explain the coupling of root and shoot growth with nutrient supply [131].

One problem with this scenario is the proposed decreased transport of assimilated-N to shoot meristems when plants are faced with a decreased N supply. There is evidence that the N transported to shoot apices of white lupins was similar in control plants and in plants transferred to N-deficient supply [71]. Kuiper [136] suggested that decreased shoot growth after transfer to low nutrient supply was due to an effect of nutrient supply on cytokinin synthesis rather than a direct effect of the nutrient content. However, the nutrient content of the growing tissue was not measured separately in that study [136] and changes may have been masked in the measurement of shoot concentrations.

Nutrient Cycling Within the Plant

Xylem sap is traditionally thought to include predominantly recently absorbed nutrients which are being transported to the shoot. However, over 50% of the N in xylem is N cycling between shoots and roots rather than N being taken up and transported directly from the growing medium [129, 131]. Other nutrients cycle similarly. For example, if the main site of SO_4^{2-} reduction is the chloroplast [138], reduced S used in root metabolism must be recycled from the shoots [137]. Similarly, the main site of NO_3^- reduction and assimilation in many plants is the shoot. Therefore, the reduced

N used in root growth must be recycled from the shoot unless nitrate reductase activity in the root is high enough to account for the root's requirements. The amount of K^+ cycling from shoots to roots has been proposed as the signal by which shoot K status influences K^+ uptake by the roots [107, 109]; however, there is also evidence that shoot effects on K^+ uptake by roots are indirect [139].

Importance of Previous Nutrient Supply

Response to nutrient stress depends on previous supply. Many studies have shown that internal nutrient pools can be shifted during nutrient stress. Changing the NO_3^- supply from 5 mM to 0 mM resulted in an increased proportion of total plant N being in the roots of white lupins [71]. In addition to the proportion increasing after changing to N-deficient supply, the total amount of reduced-N was greater in N-deficient soybean roots than in control roots [141]. Similar results were observed in tobacco plants; within 2 days of changing to N-deficient supply, the amount of reduced-N was greater in roots of N-deficient tobacco plants than in roots of control plants [98].

Changing from adequate P supply to deficient P supply resulted in an increased P content in the roots of *Stylosanthes hamata* [140]. In contrast, changing from luxury P supply to deficient P supply resulted in a decreased P content in the roots [140]; apparently mobility of P in these plants is very high and amount and direction of transport is dependent on previous as well as current supply.

Phosphorus deficiency resulted in a RSR of 0.65 for plants grown with low P from sowing, compared to a RSR of 0.36 in control, high P plants [142]. However, for plants given an initial high P supply and then changed to low P solutions, the RSR did not change within 14 days of low P treatment [142]. Khamis et al. [143] also showed that growth rate of plants supplied low P could be maintained temporarily, depending on previous supply. This was probably due to the use of vacuolar inorganic P reserves [144].

Nutrient Deficiency and Changes in Carbon Partitioning

Nitrogen

The reducing power generated by the oxidation of water to O_2 in photosynthetic cells can be used for the reduction of NO_3^- as well as CO_2. Thus, it is not surprising that the flow of carbon to starch, sucrose, and amino acids is affected by the N status of the plant (Figure 6). Reserve carbohydrates accumulate in N-deficient leaves [145,146]. Accumulation of starch or other reserve carbohydrates in nutrient-deficient plants has been used as evidence that growth is limited to a greater extent than photosynthesis [147]. Another

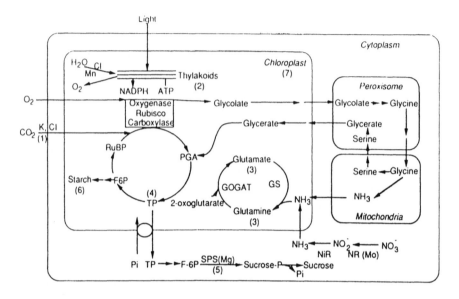

Figure 6 A simplified scheme of the role of various nutrients in carbon assimilation by green leaves of C_3 plants. (1) K and Cl function in stomatal regulation.
organic anions such as malate can replace Cl in some plant species as the anion which
compensates for K flux. (2) Many nutrients function in energy capture and transfer
in the thylakoids, either structurally (N, S, Mg, Ca) or in electron transfer (Fe, Cu,
Mn, Zn, P). (3) In leaves with low N status, glutamate is the major form of assimilated N; in leaves with high N status, glutamine becomes the major form. (4) Nitrogen and P deficiency result in decreased sucrose synthesis and starch accumulation.
The transport of triose phosphate from the chloroplast via the Pi translocator is
thought to be a key step controlling carbon partitioning in photosynthetic cells. The
rate of sucrose synthesis is also a key regulatory step and is coordinated with the
rate of export to sinks and to the rate of photosynthesis. (5) The activity of SPS increased with leaf N concentration up to 5 g kg^{-1} [149], but decreased with excess
applied NO_3^- [155]. (6) Starch degradation in leaf cells is mainly a phosphorolytic
process, controlled by phosphorylase and phosphoglucomutase. (7) For CO_2 fixation
to proceed, CA is maintained at very low concentrations (~ 1 μM) in the stroma of
the chloroplast. Abbreviations: TP: triose phosphate; Rubisco: ribulose bisphosphate
carboxylase oxygenase; RuBP: ribulose bisphosphate; PGA: phosphoglycerate; F6P:
fructose 6-phosphate; Pi: inorganic phosphate; SPS: sucrose phosphate synthase;
GS: glutamine synthase; GOGAT: glutamate synthase; NR: nitrate reductase; NiR:
nitrite reductase. (Information from [2,147,149,155,158].)

interpretation is possible. Ariovich and Cresswell [145] suggested that the starch accumulation in low N leaves occurred because of decreased transport of triose phosphates from the chloroplast to the cytoplasm. According to this hypothesis, CO_2 fixation is not limited per se nor is starch accumulating because of limited growth. Instead, decreased transport of assimilates from photosynthesizing tissue may limit growth. However, there is evidence of accumulation of carbohydrates in nonsource tissues of N-deficient plants (e.g., fructans in stems of N-deficient wheat [148]).

What is the role of N status in control of sucrose synthesis? Nitrogen concentration in soybean leaves (up to 5 g kg^{-1}) was positively correlated with sucrose phosphate synthase activity [149]. Rufty and Huber [150] suggested that control of sucrose synthesis may occur through regulation by the plant growth substance, abscissic acid. Since sucrose synthesis may regulate the rate of sucrose transport out of photosynthetic cells, a depression of sucrose phosphate synthase activity may explain why interorgan assimilate transport is decreased in low N plants [149,152,153,154].

Excess NO_3^- uptake (40 mM NO_3^- fed to detached leaves) can also depress sucrose synthesis [155]. Excess NO_3^- in wheat leaves decreased the rate of CO_2 fixation by 10% and decreased incorporation of ^{14}C into sucrose by 40% [155]. The mechanism by which excess NO_3^- causes the decrease in sucrose synthesis is unclear, but evidence using high levels of other anions (Cl^-) and addition of tungstate (to inhibit NO_3^- reduction), indicate that the decrease is due to N assimilation, and not to the presence of the NO_3^- [155]. The depression of C flow into sucrose in high N leaves is thought to occur through competition for photosynthetic energy required for reduction and carbon skeletons required for assimilation of CO_2 into triose phosphates and NO_3^- into amino acids.

Phosphorus

In addition to the evidence for starch accumulation in N-deficient plants, there is a large body of literature on the effects of plant P status on carbon partitioning (Figure 6). Starch accumulates in P-deficient plants [14,144, 156,157]. In P-deficient leaves, there is less transport of triose phosphates from the chloroplast to the cytoplasm, where sucrose is synthesized (for review, see [158]). The build-up of triose phosphates in the chloroplast is associated with increased starch formation [159].

Phosphorus deficiency affects more than the export of triose-P from the chloroplasts. A primary effect appears to be the control of starch accumulation and degradation [157,159]. Fredeen et al. [14] observed 30-fold starch increases in young expanding leaves (sink tissue) which were stunted by P deficiency, indicating that neither photosynthesis nor transport to growing tissue were limiting growth. Starch can also accumulate in stems of nutrient-deficient plants [147,157]. Starch accumulated while sucrose levels were lower in the stems of P-deficient soybeans [157].

Potassium

Potassium (K) is known to affect assimilate transport [2]. Potassium differs from N and P in that it functions primarily as an activator of enzymes and an osmoticum, while N and P are structural components. Potassium plays a role in phloem loading of sucrose, although evidence indicates that the role is an enhancement of sucrose loading and transport through an effect on the membrane potential (driving force), rather than an obligatory function (e.g., co-transport). Exchange of K^+ and soluble sugars has been observed in roots of sugar beet [160] and *Ricinus communis* [161] and interpreted as vacuolar solute exchange [52]. As such, K^+ is also important in regulation of cell turgor and thus root cell expansion is discussed p. 145.

Boron

There have been suggestions that the formation of a complex of sucrose and B is involved in sucrose transport across membranes or in long-distance sucrose transport (for review, see [162]). The evidence for this model is not strong. Root elongation in B-deficient tomatoes was lower than controls even though contents of total soluble and hydrolyzable carbohydrates (per 100 mm segment of root tip) were higher in low B root tips [163]. Increased sugar leakage has been observed in low B lily pollen [164], but this might be related to a more general effect of low B on increasing membrane permeability [165].

An early effect of B deficiency is decreased elongation rate of young leaves [78,166] and roots [167]. The correlation between sucrose transport and B status may be due to negative feedback on sucrose transport caused by poor sink growth.

Nutrient Deficiency and Decreased Photosynthesis

Net photosynthesis depends on total photosynthetic area, the duration of photosynthetic activity in that tissue, and the capacity of the photosynthetic apparatus (the amount and activity of soluble enzymes as well as proteins associated with thylakoid membranes). While all nutrient deficiencies can result in a decrease in net photosynthesis through effects on one or more of the above components [168], the effect of a nutrient stress on photosynthesis is not usually the primary effect of that stress. However, this depends on the nutrient in question and the metabolic function of that nutrient (e.g., Figure 6). While most effects of nutrient stress on photosynthesis appear to be secondary (see p. 139), there is evidence for at least two nutrients that the primary effect of their deficiency is a decreased rate of photosynthesis. The primary effect of Mn deficiency appears to be decreased photosynthetic oxygen evolution and Mg deficiency apparently causes decreased photosynthesis before it causes measurable decreases in growth.

Nitrogen

Nitrogen treatment had no measurable effect on net assimilation rate (NAR; the rate of increase of plant dry weight per unit leaf area) in some early studies [169,170]. This has been interpreted as showing that low N status decreases yield by decreasing the total amount of photosynthetic tissue rather than the rate of photosynthesis in that tissue (e.g., [6]; see p. 139 for further discussion). Lower net photosynthesis of low N plants cannot always be explained by differences in light interception (i.e., lower leaf area). Robson and Parsons [171], imposed N treatments after leaf area was large enough for complete light interception (Figure 7) and still observed decreased photosynthesis in low N treated plots. Bouma [172] observed decreases of NAR in low N plants and suggested that the reason that NAR was not affected by N status in some earlier work was because of the low light intensities used in those studies. There is a strong interaction between light intensity and N status on photosynthesis, with the decrease in photosynthesis caused by N deficiency being greatest at high light intensity (Figure 7), though some early results were observed in the field [170]. Differences in severity of stress may be a better explanation for the differences in results. Williams [169] suggested that he might have observed effects of N treatment on NAR if the N deficiency in his study had been more severe. Subsequent research has shown a strong positive correlation between CO_2 fixation and leaf nitrogen content in a wide range of plant species (reviews [173,174]) with the possible exception of evergreen conifers [175]. This correlation depends on the developmental age of the leaves being measured and holds as long as fully developed, nonsenescing leaves are measured and light is nonlimiting [175].

How does N deficiency limit CO_2 fixation? The observation that N-deficient plants are less susceptible to drought stress than control plants has led to many studies on the interaction of water and N stress [28]. Diffusion is the main process by which CO_2 moves into the leaves. The conductance of CO_2 decreases when leaf water potential drops. The water potential of low N leaves drops earlier than water potential of high N leaves [176,177]. The mesophyll conductance of CO_2 decreases more than stomatal conductance in N-deficient plants (with low mineral N supply: [177], or with low rates of N_2 fixation: [178]). Longstreth and Nobel [179] showed that the primary component of this decreased conductance was the cellular CO_2 conductance per mesophyll surface area. This measurement includes diffusion into the

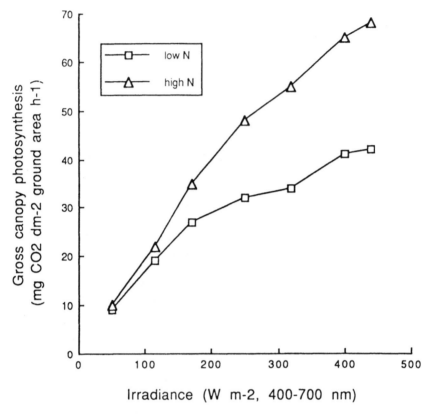

Figure 7   Interaction between light intensity and N status on net photosynthesis of ryegrass. (Modified from [171].)

mesophyll cells and the initial biochemical step of CO_2 fixation. Within 2 days of rewatering after drought stress, net CO_2 fixation by low N plants recovered to predrought levels [177]. Net CO_2 fixation by high N plants partially recovered, but not to pre-drought levels. Poor stomatal reopening could not explain the incomplete recovery in high N plants. These results using drought and N stress are compatible with the view that N deficiency decreases photosynthesis by negative feedback or by decreased content and/or activity of photosynthetic enzymes.

A large proportion of total plant nitrogen is present as carboxylating enzymes. In plants with the C_3 photosynthetic pathway, Evans [173] reviewed results showing that ribulose 1,5 bisphosphate carboxylase accounts for 10 to 30% of the total leaf N. The amount of various photosynthetic enzymes decreases in plants with low N status [173,180]. After addition of N, the accumulation of photosynthetic enzymes was due to an accumulation of corresponding messenger RNA during the recovery from N deficiency, not to a change in protein degradation [180]. This indicates that the regulation of enzyme synthesis by N status of the plant occurs at the level of gene expression rather than the availability of substrate for protein synthesis. Another effect of N deficiency is decreased respiration rates: Pearman et al. [181] observed lower respiration rates in low N plants which were not totally accounted for by decreased dry weights.

Magnesium

Bouma [182] found that Mg deficiency decreased net photosynthesis before decreased leaf area or leaf or plant dry weight were detected. When photosynthesis in mature leaves decreases as a result of nutrient stress, higher rates of photosynthesis in other plant parts might compensate. For example, Bouma et al. [183] found a decrease in net photosynthesis of old leaves of subterranean clover (*Trifolium subterraneum*) within 4 days of transfer to the 0 Mg treatment. There was no change in plant dry weight until 10 days after the transfer. Bouma et al. [183] suggested that the photosynthesis of young leaves compensated for the decreased photosynthesis of the older leaves. Other interpretations of the results are: (1) photosynthesis was not limiting growth; (2) respiration increased and stored carbohydrates were used to maintain growth; or (3) rates of maintenance respiration decreased in the short-term so that net carbon fixation was maintained.

Manganese

Manganese deficiency may decrease the rate of photosynthesis before there are measurable effects on growth. Nable et al. [184] conducted experiments where healthy plants were transferred to nutrient solutions without added Mn. Both Mn concentration in the youngest open leaf and photosynthetic oxygen evolution (POE) decreased before there were noticeable effects on chlorophyll concentrations or plant dry matter. It would be useful to have measures of growth such as leaf area or ethanol-insoluble dry weight in addition to dry weight. There was a full recovery of POE after incubation of detached leaves with 10 mM $MnSO_4$ for 24 hours. It would be worthwhile to reevaluate the primary effects of both Mg and Mn deficiency with more precise measurements of plant growth after changing from adequate to deficient conditions.

Another factor in the effect of nutrient stress on photosynthesis is the severity of the stress. For example, Simpson and Robinson [185] found that Mn deficiency caused disruption of the thylakoid system in more severely deficient plants, but that this was a secondary effect which occurred after loss of particles on the thylakoid membranes. They noted that different ultra-structural changes reported by other authors could probably be accounted for by different severity of Mn deficiency in the test plants. It would be extremely useful to have an index of stress severity in plant nutrition. The growth of plants receiving nutrient treatments should be reported (as % of maximum growth) since "low" supply can correspond to deficient or control plants, and "high" supply can correspond to control or plants with excess nutrient.

CONCLUSIONS

To minimize decreases in crop yield caused by nutrient deficiencies, it is important to understand the specific effects of nutrient deficiencies on individual yield components. To do that, we must improve our understanding of how nutrient deficiencies affect plant growth and development. In particular, how nutrient deficiencies alter the control of growth of different tissues is a tantalizing question. Perhaps nutrient status affects the sensitivity of different tissues to plant growth substances. A better understanding of this control will improve our efficiency of diagnosis of nutrient deficiencies and of application of fertilizers.

A final word of caution to experimentalists: the interpretation of experimental results can be difficult or impossible when control plants are grown in nutritionally inadequate solutions. Yet plants are still frequently grown with only $CaSO_4$ for relatively long periods of time (e.g., 2–3 weeks), especially for physiological studies. This practice continues, despite evidence that poor nutrition can result in large alterations in plant growth and development at every level of organization. For example, nitrogen starvation resulted in complete lysis of cells in over 30% of the area of root cortex in maize root tips within 4 days [21]. Whenever possible, particularly in studies where structural integrity is paramount (i.e., experiments on efflux), plants must be grown in complete nutrient solution.

ACKNOWLEDGMENTS

I am grateful to the following colleagues for useful discussions and criticisms of this manuscript: Rob Bramley, Steve Carlin, Jane Gibbs, Hank Greenway, Lorraine Osbourne, Paul Pheloung, Alan Robson and Caixan Tang. Thanks to Caixan Tang for Chinese translation, Cordula Schulze for German translation, Julie Crosbie and Lesley Parriss for photographs and many odd jobs, and Caixan Tang for technical assistance with figures.

REFERENCES

1. J. Moorby and R. T. Besford, *Enc. Plant Physiol; New Series*, 15B: 481 (1983).
2. H. Marschner, *Mineral Nutrition in Higher Plants*, Academic Press, New York (1986).

3. A. J. Hall and F. L. Milthorpe, *Austral. J. Plant Physiol., 5*: 1 (1978).
4. D. J. Watson, *Adv. Agron., 4*: 101 (1952).
5. I. M. Rao and N. Terry, *Plant Physiol., 90*: 814 (1989).
6. J. W. Radin and J. S. Boyer, *Plant Physiol., 69*: 771 (1982).
7. S. D. Wullschleger and D. M. Oosterhuis, *J. Plant Nutr., 13*: 1141 (1990).
8. J. K. Vessey and D. B. Layzell, *Plant Physiol., 83*: 341 (1987).
9. B. Steer and P. Hocking, *Ann. Bot., 52*: 267 (1983).
10. L. C. Tolley-Henry and C. D. Raper, *Bot. Gaz., 147*: 400 (1986).
11. T. W. Rufty, S. C. Huber, and R. J. Volk, *Plant Physiol., 88*: 725 (1988).
12. R. F. Sage and R. W. Pearcy, *Plant Physiol., 84*: 954 (1987).
13. A. L. Fredeen and N. Terry, *Can. J. Bot., 66*: 2311 (1988).
14. A. L. Fredeen, I. M. Rao, and N. Terry, *Plant Physiol., 89*: 225 (1989).
15. R. Delane, H. Greenway, R. Munns, and J. Gibbs, *J. Exp. Bot., 33*: 557 (1982).
16. R. Munns, H. Greenway, R. Delane, and J. Gibbs, *J. Exp. Bot., 33*: 574 (1982).
17. T. J. Flowers, M. A. Hajubagheri, and A. R. Yeo, *Plant, Cell and Environ., 14*: 319 (1991).
18. A. Termaat, J. B. Passioura, and R. Munns, *Plant Physiol., 77*: 869 (1985).
19. R. D. Graham, *Nature, 254*: 514 (1975).
20. P. N. Sharma, C. Chatterjee, S. C. Agarwala, and C. P. Sharma, *Plant Soil, 124*: 221 (1990).
21. M. C. Drew, C. J. He, and P. W. Morgan, *Plant Physiol., 91*: 266 (1989).
22. Ch. Körner, S. P. Menendez-Riedl, and P. C. L. John, *Austral. J. Plant Physiol., 16*: 443 (1989).
23. H. C. Randall and T. R. Sinclair, *Plant, Cell, Environ., 11*: 835 (1988).
24. N. Terry, L. J. Waldron, and S. E. Taylor, *The Growth and Functioning of Leaves* (J. E. Dale and F. L. Milthorpe, eds.), Cambridge University Press, Cambridge, Massachusetts, p. 179 (1983).
25. M. J. Robson and M. J. Deacon, *Ann. Bot., 42*: 1199 (1978).
26. N. Terry, *J. Exp. Bot., 21*: 477 (1970).
27. A. G. Morton and D. J. Watson, *Ann. Bot., 12*: 281 (1948).
28. J. W. Radin and L. L. Parker, *Plant Physiol., 64*: 495 (1979).
29. H. Marschner and C. Richter, *Plant Soil, 40*: 193 (1974).
30. R. G. Wyn Jones and O. R. Lunt, *Bot. Rev., 33*: 407 (1967).
31. P. K. Hepler and R. O. Wayne, *Ann. Rev. Plant Physiol., 36*: 397 (1985).
32. H. M. Moore and A. M. Hirsch, *Amer. J. Bot., 70*: 165 (1983).
33. B. J. Shelp, *Boron and its Role in Crop Production* (U. C. Gupta, ed.), CRC Press, Boca Raton, Florida (in press).
34. L. Taiz, *Ann. Rev. Plant Physiol., 35*: 585 (1984).
35. W. G. Spollen and R. E. Sharp, *Plant Physiol., 96*: 438 (1991).
36. K. Snowball and A. D. Robson, *Symptoms of Nutrient Deficiencies: Subterranean Clover and Wheat*, University of Western Australia Press, Nedlands, Western Australia (1983).
37. Ch. Hecht-Buchholz, *Z. Pflanzenernahr Bodenk., 118*: 12 (1967).
38. R. C. Rossiter and A. B. Beck, *Austral. J. Agric. Res., 17*: 447 (1966).
39. J. W. Radin and M. D. Eidenbock, *Plant Physiol., 75*: 372 (1984).
40. T. Tagawa and J. Bonner, *Plant Physiol., 32*: 207 (1957).
41. B. W. Poovaiah and A. S. N. Reddy, *CRC Crit. Rev. Plant Sci., 6*: 47 (1987).

42. F. Bangerth, *Ann. Rev. Phytopathol., 17*: 97 (1979).
43. J. S. Lee, T. J. Mulkey, and M. L. Evans, *Planta, 160*: 536 (1984).
44. S. Virk and R. E. Cleland, *Planta, 182*: 559 (1990).
45. R. E. Cleland, *Enc. Plant Physiol., New Series* (W. Tanner and F. A. Loemus, eds.), Vol. 13B, p. 255 (1981).
46. M. Inouhe and D. J. Nevins, *Plant Physiol., 96*: 426 (1991).
47. T. Hoson, Y. Masuda, Y. Sone, and A. Misaki, *Plant Physiol., 96*: 551 (1991).
48. H. D. Reiss and W. Herth, *Planta, 145*: 225 (1979).
49. M. D. de la Guardia and M. Benlloch, *Physiol. Plant., 49*: 443 (1980).
50. R. A. Leigh and R. G. Wyn Jones, *New Phytol., 97*: 1 (1984).
51. W. D. Jeschke and O. Wolf, *J. Exp. Bot., 39*: 1149 (1988).
52. C. A. Perry, R. A. Leigh, A. D. Tomos, R. E. Wyse, and J. L. Hall, *Planta, 170*: 353 (1987).
53. H. P. Haschke and U. Lüttge, *Z. Pflanzenphysiol. Bd., 76*: 450 (1975).
54. R. H. M. Langer and F. K. Y. Liew, *Austral. J. Agric. Res., 24*: 647 (1973).
55. J. E. Dale and R. G. Wilson, *J. Agric. Sci., 90*: 503 (1978).
56. J. E. Dale and R. G. Wilson, *Ann. Bot., 44*: 537 (1979).
57. W. V. Single, *Austral. J. Exp. Agric. An. Husb., 4*: 165 (1964).
58. N. E. Longnecker, E. J. M. Kirby, and A. D. Robson, *Crop Sci., 33*: 154 (1993).
59. A. N. Smith, *Physiol. Plant, 22*: 371 (1969).
60. M. S. Rahman and J. H. Wilson, *Austral. J. Agric. Res., 28*: 183 (1977).
61. D. M. Oosterhuis and P. M. Cartwright, *Crop Sci., 23*: 711 (1983).
62. A. B. Frank, A. Bauer, and A. L. Black, *Crop Sci., 27*: 113 (1987).
63. E. V. Maas and C. M. Grieve, *Crop Sci., 30*: 1309 (1990).
64. A. B. Frank and A. Bauer, *Agron. J., 74*: 504 (1982).
65. D. E. Peaslee, *Commun. Soil Sci. Plant Anal., 8*: 373 (1977).
66. J. F. Loneragan, K. Snowball, and A. D. Robson, *Ann. Bot., 45*: 621 (1980).
67. N. E. Longnecker, R. D. Graham, and G. Card, *Field Crops Res., 28*: 85 (1991).
68. N. J. Halse, E. A. N. Greenwood, P. Lapins, and C. A. P. Boundy, *Austral. J. Agric. Res., 20*: 987 (1969).
69. R. C. Rossiter, *Ann. Bot., 42*: 325 (1978).
70. K. D. Shepherd, P. J. M. Cooper, A. Y. Allan, D. S. H. Drennan, and J. D. H. Keatinge, *J. Agric. Sci. Camb., 108*: 365 (1987).
71. J. S. Pate, D. B. Layzell, and C. A. Atkins, *Plant Physiol., 64*: 1983 (1979).
72. H. Smith and P. F. Wareing, *Planta, 70*: 87 (1966).
73. P. F. Wareing and T. A. A. Nasr, *Ann. Bot., 25*: 321 (1961).
74. C. R. Davies and P. F. Wareing, *Planta, 65*: 139 (1965).
75. D. Lazof and A. Läuchli, *Planta, 184*: 334 (1991).
76. H. Greenway and A. Gunn, *Planta, 71*: 43 (1966).
77. B. J. Scott and A. D. Robson, *Austral. J. Agric. Res., 41*: 499 (1990).
78. G. J. Kirk and J. F. Loneragan, *Agron. J., 80*: 758 (1988).
79. M. J. Webb and J. F. Loneragan, *Soil Sci. Soc. Amer. J., 52*: 1676 (1988).
80. G. N. Thorne and D. J. Watson, *J. Agric. Sci., 46*: 449 (1955).
81. H. Greenway and M. G. Pitman, *Austral. J. Biol. Sci., 18*: 235 (1965).
82. J. Hill, A. D. Robson, and J. F. Loneragan, *Austral. J. Agric. Res., 29*: 925 (1978).

83. J. F. Loneragan, K. Snowball, and A. D. Robson, *Transport and Transport Processes* (I. F. Wardlaw and J. B. Passioura, eds.), Academic Press, London, England (1976).
84. J. Hill, *J. Plant Nutr., 2*: 407 (1980).
85. M. Silberbush and S. H. Lips, *J. Plant Nutr., 14*: 751 (1991).
86. C. J. Birch and K. E. Long, *Austral. J. Exp. Agric., 36*: 237 (1990).
87. P. M. Bremner, *J. Agric. Sci., Camb., 72*: 273 (1969).
88. G. N. Thorne, *Ann. Bot., 26*: 37 (1962).
89. D. J. Watson, *J. Agric. Sci., 26*: 391 (1936).
90. G. M. Fletcher and J. E. Dale, *Ann. Bot., 38*: 63 (1974).
91. F. E. Below and J. A. Heberer, *J. Plant Nutr., 13*: 667 (1990).
92. M. W. Perry and J. W. Gartrell, *J. West Austral. Dept. Agric., 17*: 20 (1976).
93. S. C. Qin, *Diagnoses and Corrections of Nutritional Disorders in Crops*. Zhegiang Science and Technology Press (1988) (in Chinese).
94. R. Scott Russell, *Plant Root Systems: Their Function and Interaction with the Soil*, McGraw-Hill, New York (1977).
95. R. K. Belford, B. Klepper, and R. W. Rickman, *Agron. J., 79*: 310 (1987).
96. B. Klepper, R. K. Belford, and R. W. Rickman, *Agron. J., 76*: 117 (1984).
97. R. W. Rickman, B. Klepper, and R. K. Belford, *Wheat Growth and Modelling* (W. Day and R. Atkin, eds.), Plenum Press, p. 9 (1985).
98. T. W. Rufty, C. T. MacKown, and R. J. Volk, *Physiol. Plant., 79*: 85 (1990).
99. W. Römer, J. Augustin, and G. Schilling, *Structural and Functional Aspects of Transport in Roots* (B. C. Loughman et al., eds.), Kluwer Academic Publishers, Dordrecht, The Netherlands, p. 123 (1989).
100. C. Ll. Powell, *Plant Soil, 41*: 661 (1974).
101. J. F. Loneragan, G. J. Kirk, and M. J. Webb, *J. Plant Nutr., 10*: 1247 (1987).
102. P. H. Nye and P. B. Tinker, *Solute Movement in the Soil: Root System*, Blackwell, Oxford, England (1977).
103. D. J. Hole, A. M. Emran, Y. Fares, and M. C. Drew, *Plant Physiol., 93*: 642 (1990).
104. R. B. Lee and M. C. Drew, *J. Exp. Bot., 37*: 1768 (1986).
105. R. B. Lee and K. A. Rudge, *Ann. Bot., 57*: 471 (1986).
106. R. H. Teyker, W. A. Jackson, R. J. Volk, and R. H. Moll, *Plant Physiol., 86*: 778 (1988).
107. M. G. Pitman, *Austral. J. Biol. Sci., 25*: 905 (1972).
108. A. D. M. Glass, *Austral. J. Plant Physiol., 4*: 313 (1977).
109. M. C. Drew and L. R. Saker, *Planta, 160*: 500 (1984).
110. D. H. Colgiatti and D. T. Clarkson, *Physiol. Plant, 58*: 287 (1983).
111. M. C. Drew, L. R. Saker, S. A. Barber, and W. Jenkins, *Planta, 160*: 490 (1984).
112. D. D. Lefebvre and A. D. M. Glass, *Physiol. Plant., 54*: 199 (1982).
113. R. L. Chaney, J. C. Brown, and L. O. Tiffin, *Plant Physiol., 50*: 208 (1972).
114. D. T. Clarkson and J. Sanderson, *Plant Physiol., 61*: 731 (1978).
115. V. Römheld and H. Marschner, *Physiol. Plant., 53*: 354 (1981).
116. N. E. Longnecker and R. M. Welch, *Plant Physiol., 92*: 17 (1990).
117. M. A. Grusak, R. M. Welch, and L. V. Kochian, *Plant Physiol., 93*: 976 (1990).

118. C. Hackett, *Austral. J. Biol. Sci., 25*: 1169 (1972).

119. M. C. Drew and L. R. Saker, *J. Exp. Bot., 26*: 79 (1975).

120. M. C. Drew and L. R. Saker, *J. Exp. Bot., 29*: 435 (1978).

121. M. C. Drew, L. R. Saker, and T. W. Ashley, *J. Exp. Bot., 24*: 1189 (1973).

122. R. Brouwer, *Plant Soil, 63*: 65 (1981).

122a. R. B. Stryker, J. W. Gilliam, and W. A. Jackson, *Soil Science Society America Proceedings, 38*: 334 (1974).

123. C. Tang, A. D. Robson, and M. J. Dilworth, *New Phytol., 115*: 61 (1990).

124. M. J. Webb and J. F. Loneragan, *J. Plant Nutr., 13*: 1499 (1990).

125. C. M. Peterson, B. Klepper, F. V. Pumphrey, and R. W. Rickman, *Agron. J., 76*: 861 (1984).

126. A. J. Bloom, F. S. Chapin, III, and H. A. Mooney, *Ann. Rev. Ecol. Syst., 16*: 363 (1985).

127. F. S. Chapin, *Ann. Rev. Ecol. Syst., 11*: 233 (1980).

128. R. Brouwer, *The Growth of Cereals and Grasses* (F. L. Milthorpe and D. J. Irvins, eds.), Butterworth, London, England, p. 153 (1966).

129. R. J. Simpson, H. Lambers, and M. J. Dalling, *Physiol. Plant., 56*: 11 (1982).

130. W. D. Jeschke, C. A. Atkins, and J. S. Pate, *J. Plant Physiol., 117*: 319 (1985).

131. H. D. Cooper and D. T. Clarkson, *J. Exp. Bot., 40*: 753 (1989).

132. C. M. Larsson, M. Larsson, J. V. Purves, and D. T. Clarkson, *Physiol. Plant., 82*: 345 (1991).

133. A. Talouizte, M. L. Champigny, E. Bismuth, and A. Moyse, *Physiol. Veg., 22*: 19 (1984).

134. D. C. Bowman and J. L. Paul, *Plant Physiol., 88*: 1303 (1988).

135. J. K. Vessey, C. D. Raper, and L. Tolley-Henry, *J. Exp. Bot., 41*: 1579 (1990).

136. D. Kuiper, *Plant Physiol., 87*: 555 (1988).

137. M. Gilbert and A. D. Robson, *Plant Soil, 77*: 377 (1984).

138. J. A. Schiff, *Enc. Plant Physiol., New Series* (A. Läuchli and R. L. Bieleski, eds.), Springer Verlag, New York, Vol. 15A, p. 401 (1983).

139. M. Y. Siddiqi and A. D. M. Glass, *J. Exp. Bot., 38*: 935 (1987).

140. F. W. Smith, W. A. Jackson, and P. J. Van den Berg, *Austral. J. Plant Physiol., 17*: 451 (1990).

141. T. W. Rufty, C. D. Raper, S. C. Huber, *Can. J. Bot., 62*: 501 (1984).

142. V. G. Breeze, A. D. Robson, and M. J. Hopper, *J. Exp. Bot., 36*: 725 (1985).

143. S. Khamis, T. Lamaze, Y. Lemoine, and C. Foyer, *Plant Physiol., 94*: 1436 (1990).

144. C. Foyer and C. Spencer, *Planta, 167*: 369 (1986).

145. D. Ariovich and C. F. Cresswell, *Plant Cell Environ., 6*: 657 (1983).

146. R. N. Gallaher and R. H. Brown, *Crop Sci., 17*: 85 (1977).

147. I. F. Wardlaw, *New Phytol., 116*: 341 (1990).

148. W. M. Blacklow, B. Darbyshire, and P. C. Pheloung, *Plant Sci. Letters, 36*: 213 (1984).

149. P. S. Kerr, D. W. Israel, S. C. Huber, and J. T. W. Rufty, *Can. J. Bot., 64*: 2020 (1986).

150. T. W. Rufty and S. C. Huber, *Plant Physiol., 72*: 474 (1983).
152. P. S. Kerr, S. C. Huber, and D. W. Israel, *J. Plant Physiol., 75*: 483 (1984).
153. J. G. C. Small and O. A. Leonard, *Am. J. Bot., 56*: 187 (1969).
154. D. J. Ursino, D. M. Hunter, R. D. Laing, and J. L. S. Keighley, *Can. J. Bot., 60*: 2665 (1982).
155. L. Van Quy, T. Lamaze, and M. L. Champigny, *Planta, 185*: 53 (1991).
156. B. D. Thomson, A. D. Robson, and L. K. Abbott, *New Phytol., 103*: 751 (1986).
157. J. Qiu and D. W. Israel, *Plant Physiol.* (in press).
158. M. L. Champigny, *Photosynth. Res., 6*: 273 (1985).
159. I. M. Rao, A. L. Fredeen, and N. Terry, *Plant Physiol., 92*: 29 (1990).
160. R. A. Saftner and R. E. Wyse, *Plant Physiol., 66*: 884 (1980).
161. S. Chapleo and J. L. Hall, *New Phytol., 111*: 381 (1989).
162. W. M. Dugger, *Enc. Plant Physiol; New Series* (A. Läuchli and R. L. Bieleski, eds.), Vol. 15B, p. 628 (1983).
163. R. Y. Yih and H. E. Clark, *Plant Physiol., 40*: 312 (1965).
164. D. B. Dickinson, *J. Amer. Soc. Hort. Sci., 103*: 413 (1978).
165. A. J. Pollard, A. J. Parr, and B. C. Loughman, *J. Exp. Bot., 28*: 831 (1977).
166. R. W. Bell, L. McLay, D. Plaskett, B. Dell, and J. F. Loneragan, *Plant Nutrition — Physiology and Applications* (M. L. van Beusichem, ed.), Kluwer Academic Publishers, Dordrecht, The Netherlands, p. 275 (1990).
167. C. W. Bohnsack and L. S. Albert, *Plant Physiol., 59*: 1047 (1977).
168. L. Natr, *Photosynth., 6*: 80 (1972).
169. R. F. Williams, *Ann. Bot., 10*: 41 (1946).
170. D. J. Watson, *Ann. Bot., 11*: 375 (1947).
171. M. J. Robson and A. J. Parsons, *Ann. Bot., 42*: 1185 (1978).
172. D. Bouma, *Ann. Bot., 34*: 1131 (1970).
173. J. R. Evans, *Oecologia, 78*: 9 (1989).
174. T. R. Sinclair and T. Horie, *Crop Sci., 29*: 90 (1989).
175. P. B. Reich, M. B. Walters, and D. S. Ellsworth, *Plant, Cell and Environ., 14*: 251 (1991).
176. J. W. Radin and R. C. Ackerson, *Plant Physiol., 67*: 115 (1981).
177. J. Ghashghaie and B. Saugier, *Plant Cell Environ., 12*: 261 (1989).
178. T. M. DeJong and D. A. Phillips, *Plant Physiol., 68*: 309 (1981).
179. D. J. Longstreth and P. S. Nobel, *Plant Physiol., 65*: 541 (1980).
180. B. Sugiharto, K. Miyata, M. Nakamoto, H. Sasakawa, and T. Sugiyama, *Plant Physiol., 92*: 963 (1990).
181. I. Pearman, S. M. Thomas, and G. N. Horne, *Ann. Bot., 47*: 535 (1981).
182. D. Bouma, Proceedings XI International Grassland Congress, University of Queensland Press, St. Lucia, Australia, p. 347 (1970).
183. D. Bouma, E. J. Dowling, and H. Wahjoedi, *Ann. Bot., 43*: 529 (1979).
184. R. O. Nable, A. Bar-Akiva, and J. F. Loneragan, *Ann. Bot., 54*: 39 (1984).
185. D. J. Simpson and S. P. Robinson, *Plant Physiol., 74*: 735 (1984).
186. D. Lascaris and J. W. Deacon, *New Phytol., 118*: 391 (1991).
187. W. W. Arneke, Ph.D. Thesis, Universität Gissen (1980), cited in [2].

188. D. Kramer, V. Römheld, E. Landsberg, and H. Marschner, *Planta, 147*: 335 (1980).
189. A. J. Abbott, *New Phytol., 66*: 419 (1967).
190. K. H. Neumann and F. C. Steward, *Planta, 81*: 333 (1968).
191. M. S. Cohen and R. Lepper, *Plant Physiol., 59*: 884 (1977).
192. H. Kouchi, *Soil Sci. Plant Nutr., 23*: 113 (1977).
193. E. Kurth, G. C. Cramer, A. Läuchli, and E. Epstein, *Plant Physiol., 82*: 1102 (1986).
194. A. D. Robson and K. Snowball, *Plant Analysis, an Interpretation Manual* (D. J. Reuter and J. B. Robinson, eds.), Inkata Press, Melbourne, Australia, p. 13 (1986).
194a. K. Ohki, *Agron. J., 76*: 213 (1984).
195. R. E. Karamanos, J. G. Fradette, and P. D. Gerwing, *Can. J. Soil Sci., 65*: 133 (1985).
196. C. Marriott and J. E. Dale, *Z. Pflanzenphysiol. Bd., 81*: 377 (1977).
197. C. Hackett, *New Phytol., 67*: 287 (1968).
198. Ch. Hecht-Buchholz, D. J. Brady, C. J. Asher, and D. G. Edwards, *Plant Nutrition — Physiology and Applications* (M. L. van Beusichem, ed.), Kluwer Academic Publishers, Dordrecht, The Netherlands, p. 335 (1990).
199. S. Suthipradit, D. G. Edwards, and C. J. Asher, *Plant Nutrition — Physiology and Applications* (M. L. van Beusichem, ed.), Kluwer Academic Publishers, Dordrecht, The Netherlands, p. 375 (1990).
200. G. R. Cramer, E. Epstein, and A. Läuchli, *Plant Physiol., 81*: 792 (1986).
201. I. Zidan, H. Azaizeh, and P. M. Neumann, *Plant Physiol., 93*: 7 (1990).
202. N. Longnecker, J. Slater, and A. D. Robson, *Plant Nutrition — From Genetic Engineering to Field Practice* (N. J. Barrow, ed.). Kluwer Academic Publishers, Dordrecht, The Netherlands, p. 673 (1993).

6

Growth Regulators
and Crop Productivity

Peter Hedden and Gordon Victor Hoad*

*University of Bristol and
AFRC Institute of Arable Crops Research
Long Ashton, Bristol, England*

INTRODUCTION

The historical developments that led to the commercial use of growth-regulating chemicals have been reviewed extensively by Tukey [1], Wittwer [2], and Weaver [3] and will be discussed only briefly here.

Although scientists had suggested already in the last century that plants may contain chemical substances that control their growth and development, isolation of the first natural plant bioregulator, indole-3-acetic acid (IAA), was not achieved until the 1940s [4]. Synthetic indoles and naphthalene derivatives, available from about 1904, were first applied to plants in the 1930s, after Went had demonstrated the existence of a natural growth regulator, termed *auxin* by Kögl and Haagen-Smit [5]. Synthetic auxins were quickly developed for horticultural application and were first used to delay fruit drop in apples in 1939 [6], and to promote thinning of certain top fruit in 1943 [7]. During the 1940s a number of new synthetic auxins, including the chlorophenoxy acetic acids, 2,4-D and 2,4,5-T, were produced and found commercial application, particularly as herbicides.

Discovery of the gibberellins and cytokinins followed in the next decade. The growth promoting activity of the former compounds were known in Japan as early as 1926, but this information was not available in the Western world until about 1950. An excellent review of the early history of the gib-

---

Current affiliation: University of Bahrain, Bahrain

berellins was written by Phinney [8] in 1983. Cytokinins, which promote cell division, were characterized in 1955 by Miller et al. [9] and synthetic compounds, with cytokininlike activity, including 6-benzyladenine, were developed in the 1960s.

The biological activity of unsaturated gases, such as ethylene and acetylene, in smoke was noted in the 1920s and burning vegetation to stimulate flowering in pineapple represents one of the earliest practical uses of a chemical to modify plant development [10]. Difficulty in applying ethylene to crops delayed its extensive use in practical situations. The availability of ethylene-releasing agents such as ethrel [11] overcame many of the application problems and the use of these compounds has gained widespread acceptance in agriculture and horticulture.

The latest major group of endogenous plant hormones to be recognized were the "abscisins," represented by abscisic acid. The name suggests it is an abscission-causing agent, but it was isolated as a growth inhibitor [12] and has since been recognized as being active in stomatal control and in the regulation of gene action.

Although the first synthetic growth regulators were tested on the basis of their structural similarity to indoleacetic acid, most subsequent growth regulators have arisen fortuitously from mass, multipurpose screening programs, often where the primary targets were herbicides or fungicides. It is, therefore, not surprising that a basic understanding of their mechanism has in many cases lagged some considerable time behind their discovery, or even their introduction into agronomic use. In fact, the precise mode of action of some commonly used growth regulators is still uncertain. Since the empirical approach for discovering new growth regulators is inefficient and will probably become decreasingly cost effective, it is now becoming more attractive to target particular molecular events involved in the physiological processes of interest. Detailed knowledge of the mechanism by which existing growth regulators function will not only aid the search for new products, but will be of value in predicting possible secondary effects of potentially marketable compounds, particularly with regard to their impact on the environment.

Growth regulators form a heterogeneous group of compounds, many of which function by modifying the hormonal status of plants, so that their modes of action are linked closely to those of the hormones themselves. In the following discussion the more important compounds in terms of agricultural use will be grouped according to the plant growth hormone whose action they mimic. In addition, many important growth regulators, including the growth retardants, interfere with hormone metabolism or transport. There are also some widely used growth regulators whose modes of action are unrelated to hormone metabolism or action. These include inhibitors of cell division and certain herbicides that are used to induce processes related to senescence.

GIBBERELLINS

Gibberellins (GAs) induce a variety of physiological responses in plants, including stem extension and leaf enlargement, fruit set and development (in the absence of fertilization), flower formation and seed germination. They also cause bolting in rosette plants by overcoming the cold or long-day requirement. The best understood mechanism of GA action is in the germinating cereal grain [13]. During imbibition, gibberellin, probably GA_1, is produced in the embryo and migrates to the aleurone layer where it stimulates the synthesis and release of hydrolytic enzymes (amylases, proteases, ribonucleases, etc.), which break down the macromolecular reserves in the endosperm for utilization by the growing embryo. There is clear evidence that GAs increase the rates of transcription of several genes, including that for α-amylase, the most studied enzyme [14,15]. It is proposed that GAs induce the formation of proteins that bind to upstream, presumably regulatory, regions of the gene [16,17]. Hooley et al. [18] have demonstrated that the sites of gibberellin perception on protoplasts from *Avena fatua* aleurone cells are located on the plasmalemma.

The most important GA in commercial application is GA_3, which is prepared from fermentations of the fungus *Gibberella fujikuroi* [19]. The structure and biological activity of GA_3 is very similar to that of GA_1, which is considered to be the physiologically active GA for vegetative growth in most species [20]. It is now clear, however, that GA_3 also occurs naturally in higher plants. Gibberellins A_4 and A_7 are also produced from *G. fujikuroi*, in which they are biosynthetic precursors of GA_3 as illustrated in Figure 1 [21]. The yields of GA_4 and GA_7, which are not readily separated chemically, can be increased relative to that of GA_3 by judicious choice of the fungal strain and fermentation conditions [21]. The fungus *Sphaceloma manihoticola* produces GA_4 free from GA_7 and GA_3 [22].

Despite their wide availability and relative cheapness of production, GAs are not used extensively in crop production. Indeed, their major usage is in the malting of barley where they enhance the production of hydrolytic enzymes, including α-amylase, leading to a more rapid production of sugars for alcoholic fermentation [23]. Within the horticultural sector, GAs are used predominantly in the production of seedless grapes where they increase the size and quality of the berries [24]. Sprays applied prebloom will induce the rachis to elongate, creating looser clusters which are less susceptible to fungal attack later in the season. Gibberellins will also decrease pollen viability and reduce ovule fertility when applied at anthesis, reducing the number of

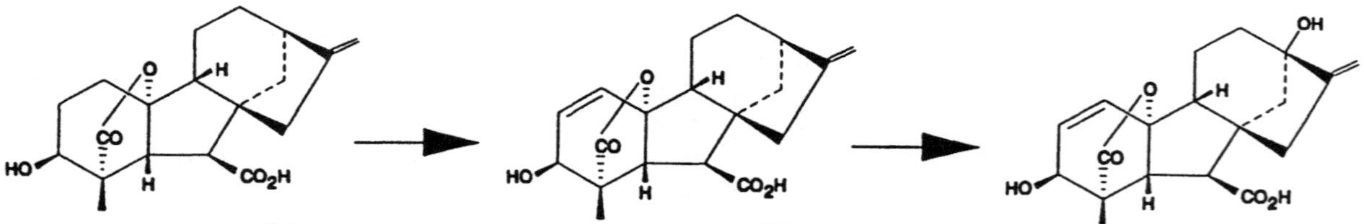

GA_4 → GA_7 → GA_3

Figure 1 The biosynthetic relationship of gibberellins produced from cultures of the fungus *Gibberella fujijuroi*.

berries per cluster and increasing the weight of the remaining fruit. Later applications, during the time of set, have been found to increase berry size further.

The importance of GAs in fruit set and growth is well documented [25]. In the absence of pollination a single application of GA_3 to flowers will induce parthenocarpic fruit set in many species. Gibberellin appears to act as a trigger, and subsequent fruit growth, due predominantly to cell expansion in the mesocarp [26,27], is comparable to that of seeded fruit. Stimulation of fruit set by GA application is of particular importance for pome and citrus fruits. In Europe frost during the flowering of apple and pear trees can reduce or even eliminate set. Applications of GA, either alone, or as part of a hormone mixture with an auxin and a cytokinin, immediately after frost damage has occurred to the flowers, will often induce parthenocarpic fruit formation and save a lost crop [28,29]. Gibberellins are also applied to citrus trees during full bloom to improve set, particularly in mandarin oranges which often set only lightly [30].

In some pome fruits and citrus an excess of set fruit can lead to poor flower initiation the following year, a phenomenon known as biennial bearing [31]. In extreme cases this can lead to a complete absence of fruit on the same tree every 2 years (in "off" years). Applications of GAs during the off year can reduce flower initiation, leading to less fruit in the following "on" year and a greater number of initiated flowers in that season, thus reducing or eliminating the problem of bienniality.

In Delicious-type apples, high temperatures during flowering can reduce the length to diameter ratio, resulting in a round fruit, rather than the elongated applies which are characteristic of these cultivars. Prolamin, a mixture of benzyladenine and GA_4/GA_7, applied at full bloom will elongate the apples, giving more normal looking fruit [32]. Prolamin has also been used to modify tree shape in pome fruits. Young trees often grow vigorously with a strong central leader and only a few upright branches. Wider "crotch angles" and an increase in branch number, which are desirable for fruit production, can be stimulated by the application of Prolamin.

Ripening of citrus fruit is easily manipulated by the application of GAs and ethylene-releasing agents. Gibberellins will delay ripening and are used to increase the availability of lemon fruit during the summer months in the northern hemisphere when demand is high [33]. Gibberellin is applied in November and December, delaying the harvest date and increasing the shelf life of the fruit. Fruit abscission in grapefruit and Navel oranges can be controlled by the application of 2,4-D, but the compound causes skin blemishes and enhanced senescence leading to a decrease in fruit quality. Gibberellins reduce the physiological disorders, such as rind staining, creasing, and water spot, by delaying senescence [34]. Some cultivars of apple are also susceptible to skin blemishes, e.g., russeting, and treatment of the fruit with GA_4 has been shown to improve fruit quality. Gibberellin A_7, although structurally similar to GA_4, is considerably less effective [35].

There have been many attempts to improve the production of field crops by the application of GAs, but with little success. In sugarcane, however, extension of internodes 3–5 may be enhanced by the application of GA, overcoming the reduced growth which occurs during the cooler winter season, and producing yield increases [36].

Juvenility in tree species is a major problem for breeders, as some species may take 20 to 30 years before they produce fruits or cones. Juvenile conifers can be induced to flower by application of GAs or GA/auxin mixtures, so dramatically shortening the breeding programs for these species [37]. It is of interest that the effectiveness of different GAs for flower formation varies according to family. Thus GA_3 is used on members of the Cupressaceae and Taxodiaceae, whereas GA_4/GA_7 mixtures or GA_9 are more effective on the Pinaceae [38].

A large number of compounds displaying GA-like biological activity have been investigated [39]. The N-substituted phthalimides hold out some promise as potential growth regulators [40]. One of the most active, 1-(3-chlorophthalimido)cyclohexane carboxamide (Figure 2), has been shown to stimulate α-amylase and germination in *Avena fatua* [41] and to induce male flowers in female plants of *Morus nigra* [42]. It does not appear to be related structurally to the GAs and its classification as a true gibberellin agonist must await studies with purified receptors.

ETHYLENE-RELEASING COMPOUNDS

As a gaseous hormone, ethylene is quite unusual and enables communication between plants as well as between organs or cells. Its physiological ac-

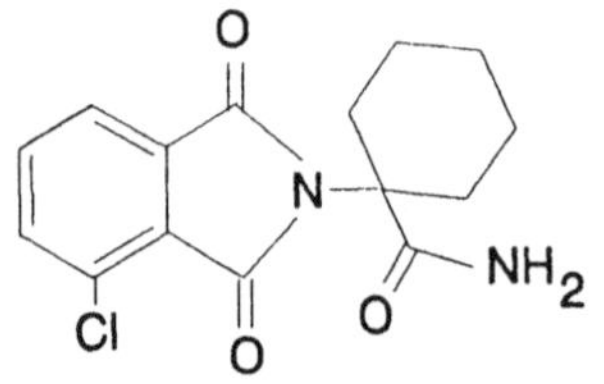

Figure 2   Structure of the gibberellin agonist 1-(3-chlorophthalimido)cyclohexane carboxamide (AC 94,377).

tivities include promotion of fruit ripening, leaf and fruit abscission in dicotyledonous species, flower senescence, stem extension of aquatic plants, gas space (aerenchyma) development in roots, leaf epinasty, stem and shoot swelling, femaleness in curcubits, apical hook closure in etiolated shoots, root hair formation, flowering in the Bromeliaceae and diageotropism of etiolated shoots. Since its synthesis is stimulated by auxin [43], some of the effects of auxin treatment are undoubtedly due to ethylene. The action of ethylene appears to be mediated by changes in the rate of expression of specific genes, such as those for cell-wall degrading enzymes during fruit ripening [44].

The discovery of compounds that release ethylene when applied to plants has rendered this hormone much more accessible to commercial exploitation. The first such compound to be developed was 2-chloroethylphosphonic acid (CEPA or ethephon; trade name "Ethrel"). This compound decomposes at pH >4 to produce ethylene, chloride, and phosphate [45,46], as shown in Figure 3. Other ethylene-releasing compounds, such as etacelasil and 2-chloroethyl-methyl-bis(phenylmethoxy)silane (CGA 15281) (Figure 3) have been introduced. These break down on contact with water and release ethylene more rapidly than does ethephon.

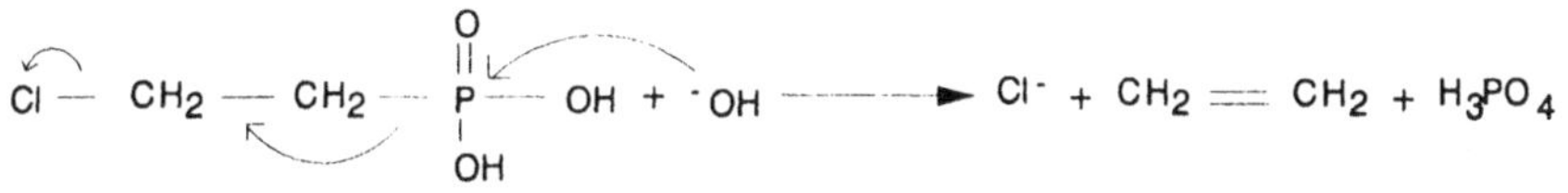

ethephon

$$(CH_3OCH_2CH_2O)_3SiCH_2CH_2Cl$$

etacelasil

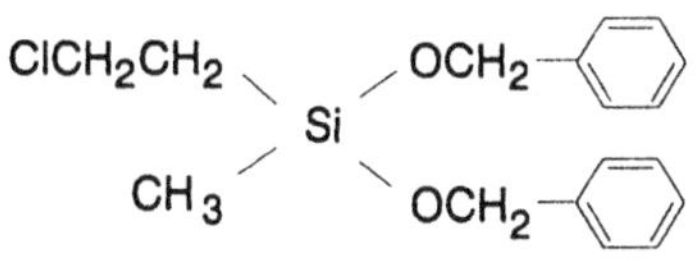

CGA 15281

Figure 3 Ethylene-releasing agents, showing the decomposition of ethephon at pH >4.

The suitability of ethylene-generating compounds for practical purposes depends on factors such as uptake and transport rates, speed of ethylene evolution, and mode of liberation. It is in the areas of fruit and nut ripening, senescence, and abscission that such compounds have found their major uses. Crops with which ethylene-releasing agents are used commercially include tomato, cherry, apple, table grape, raisin, blackberry, blueberry, cantaloupe, lemon, tangerine, pepper, filberts, and walnut. However, the major application currently is with cotton where ethylene releasers aid boll opening and harvesting.

Etacelasil is used in preference to ethephon prior to the mechanical harvesting of olives as it does not cause defoliation. Leaf abscission, which can take place after ethephon treatment, may require a more prolonged exposure to ethylene than occurs with etacelasil [47].

In addition to enhancing the ripening and abscission of fruit at the time of, or immediately prior to maturity, ethylene-releasing compounds may be applied earlier during the growing season to thin fruit and prevent an excessive number reaching maturity. In apple and peach, ethrel and CGA 15281 are used for this purpose; the latter compound has the advantage that it can be used to thin peaches without defoliation.

Ethephon will delay bud expansion and anthesis in cherry and peach if it is applied in the fall or early winter months. In this way, damage from spring frosts is avoided.

Latex flow from the tapping cut of rubber trees can be enhanced by applications of ethrel and increases in yields in the region of 50–100% may be obtained. The mechanism is not established, but it has been suggested that cell wall thickening of the latex vessels is stimulated, making them less likely to contract during tapping, thus stabilizing the lutoid bodies and allowing an increase in latex flow [48].

In Western Europe Terpal, a mixture of ethephon and the growth retardant mepiquat chloride, is applied to barley plants to shorten the straw and increase lodging resistance [49]. For reasons which are, as yet, unclear barley is much more responsive than wheat to this mixture.

Despite extensive studies the greatest use of ethrel and other ethylene releasing compounds is with horticultural and fruit crops. More widespread practical applications may be established once methods are found for overcoming some of the undesirable side effects of ethylene.

AUXINS

Auxins stimulate cell elongation, cell division, and differentiation. There is, however, some controversy over the action of this class of hormone at the molecular level. According to the acid growth theory they stimulate H^+-

ATPases, resulting in hyperpolarization of the plasma membrane and acidification of the apoplast [50]. It is proposed that the decrease in apoplastic pH causes cell-wall loosening, possibly by activating hydrolytic enzymes [51]. Auxins are also known to induce the synthesis of a number of proteins, although their function is unknown [52]. They may be involved in cell-wall loosening and/or in the maintenance of cell turgor. Auxin-binding proteins have been isolated and cloned [53]. Although most of these putative receptors are localized in the endoplasmic reticulum, there is evidence from physiological and immunological studies that they are present in active form on the outside of the plasma membrane [54].

The most prevalent natural auxin is indole-3-acetic acid (IAA), which, because it is chemically unstable and rapidly metabolized in plant tissues, is little used as a growth regulator. As a result of extensive screening programmes there are a large number of synthetic auxins (IAA agonists), many of which are commercially important (Figure 4).

Synthetic auxins have a range of applications, especially in horticulture. A recent review by Miller [55] deals with their use in apple and pear production. Auxins are used to induce parthenocarpic fruit set, particularly in adverse weather conditions, and are probably substituting for endogenous IAA normally supplied by the zygote or pollen [56]. For example, in California, treatment of the early spring tomato crop with 4-CPA at 25–50 ppm will stimulate fruit set at the time of the year when cool night temperatures normally inhibit fruit initiation. Auxins are also used to prevent the inappropriate abscission of certain top fruit during midfruit development, or of mature fruit before harvest. If auxins such as 2,4,5-T, NAA, and 2,4-D are applied to apple, pear, and some citrus fruits from the midstages of fruit growth up to the beginning of fruit drop, abscission may be delayed at the time of commercial harvest. Luckwill [57] has shown that the tendency of apple fruit to abscind is inversely related to the concentration of IAA in the fruit. Auxin application would then supplement the endogenous hormone in the fruit. In fact, it may be the amount of auxin that diffuses from the fruit that is important. Bangerth [58] has argued that the ability of organs such as fruit to compete for assimilates is related to the magnitude of diffused auxin gradients. Since auxin is known to stimulate the differentiation of phloem and xylem [59], fruits with high rates of auxin diffusion would develop most rapidly and better maintain vascular connections [58].

Fruit set in apple and pear may be reduced by the application of NAA and NAD 7–20 days after full bloom [60]. In this case auxins may act by stimulating the production of ethylene [61]. The action of these compounds appears to be more carefully controlled than that of ethrel, whose activity may be affected by the weather.

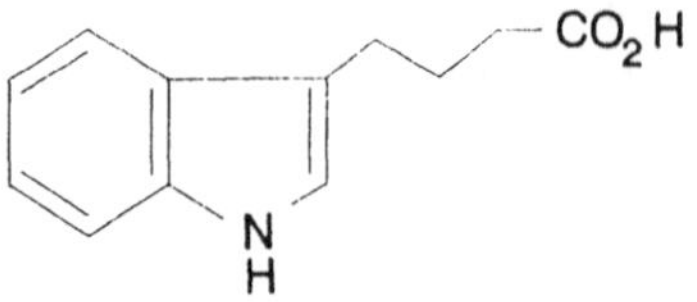

4-(Indol-3-yl)butyric acid
(IBA)

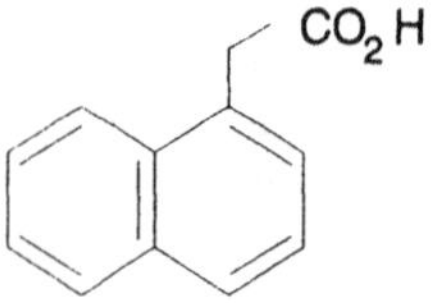

1-naphthylacetic acid
(NAA)

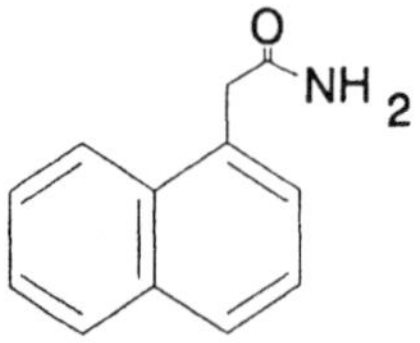

2-(1-naphthyl)acetamide
(NAD)

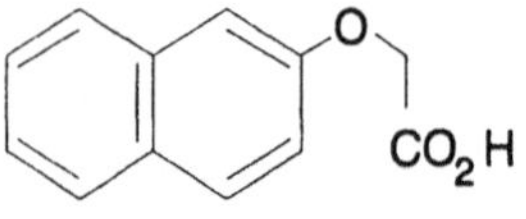

2-naphthyloxyacetic acid
(BNOC)

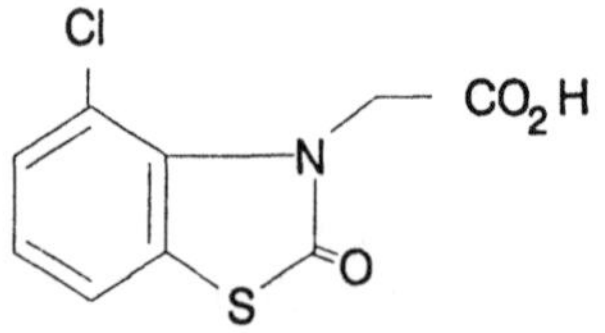

4-chloro-2-oxobenzthiazolin-3-
ylacetic acid (benazolin)

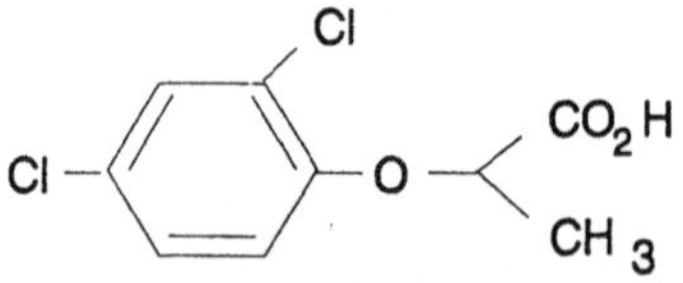

2-(2',4'-dichlorophenoxy)-
propionic acid (dichlorprop)

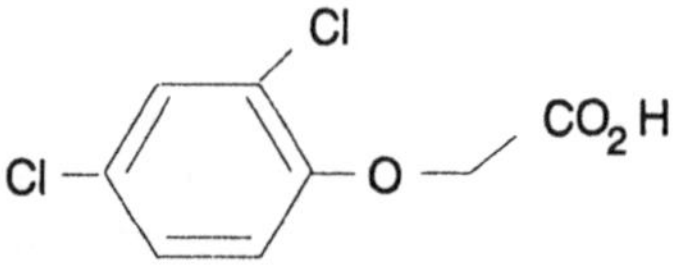

2,4,-dichlorophenoxyacetic
acid (2,4-D)

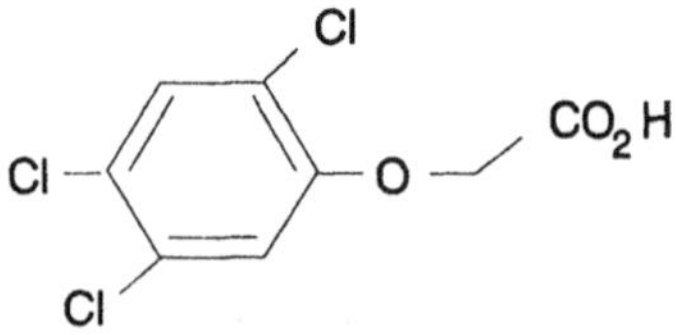

2,4,5-trichlorophenoxyacetic
acid   (2,4,5-T)

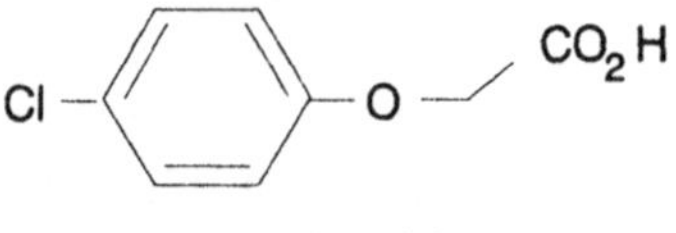

4-chlorophenoxyacetic acid
(4-CPA)

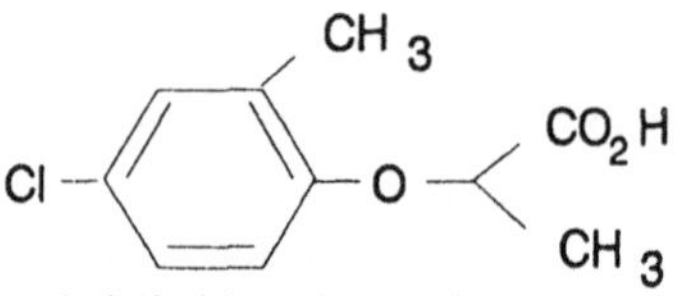

2-(2'-methyl-4'-chlorophenoxy)-
propionic acid (mecoprop)

Figure 4   Synthetic auxins of agricultural importance.

Stimulation of cell division and differentiation by auxin is utilized in micropropagation. Combinations of auxin and cytokinins are used to initiate and maintain callus; auxin alone, or as the major growth regulator in auxin/cytokinin mixtures, will induce the differentiation of root initials from callus tissue or shoot cuttings. In fact, the earliest use of synthetic auxins was to stimulate the rooting of cuttings which, in many species, is difficult without auxin treatment. The weak auxin, IBA, has been shown to be very effective for this purpose. Other synthetic auxins, such as 2,4-D and NAA will promote root initiation but their translocation to other parts of the cuttings, where they may exert toxic effects, limits their use [62].

The herbicidal action of certain synthetic auxins, such as 2,4-D, when applied at high concentrations, accounts for the major use of auxins in agriculture. Typical effects are inhibition of elongation growth, epinastic bending of leaves, radial expansion, and disorganized cell proliferation. Some symptoms, such as epinastic bending, may be accounted for by auxin-induced ethylene production, although most effects appear unrelated to ethylene. Treatment of soybean seedlings with high (1–2 mM) levels of 2,4-D results in cessation of cell division in the apical meristem and of cell expansion in the subapical region [63]. There is then radial enlargement of the mature, basal region of the hypocotyl due to cell proliferation, accompanied by increases in DNA, RNA and protein content. The higher RNA levels result from an increase in ribosomal RNA synthesis due to enhanced activity of RNA polymerase I [63,64]. Elevated RNA levels seem to be a general effect of auxin-herbicide treatment [65]. Furthermore, higher levels of RNAase are found in 2,4-D-resistant plants than in sensitive ones.

CYTOKININS

Although cytokinins are known to induce cell division, cell expansion, and differentiation there is, as yet, little indication as to how they produce these effects. There are no large-scale applications of cytokinins in agriculture. As cell-division factors they are used in tissue culture and micropropagation in combination with auxin to stimulate callus growth, and, when present in excess with respect to auxin, to induce shoot formation. The most common synthetic cytokinin is 6-benzylaminopurine, which is known to occur naturally in crown gall tumours of tomato [66]. Its major use is in tissue culture, but it also finds applications in the shaping of fruit trees, fruit thinning, and, in combination with gibberellins, as a fruit setting agent.

One of the physiological effects of cytokinins is to delay senescence, and this attribute has potential application in enhancing the shelf life of fruit and vegetables. There is, however, some resistance to their use for certain applications because they stimulate cell division.

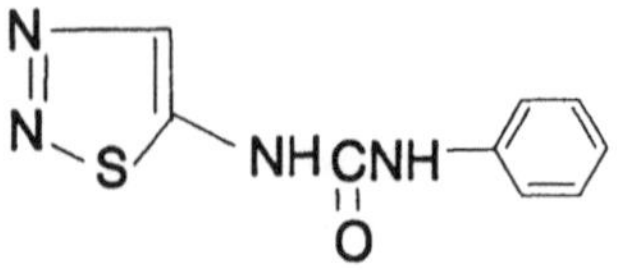

Figure 5   Structure of thidiazuron.

A number of compounds have been synthesized which exhibit cyto-kininlike activity including N^6-substituted adenine analogues, diphenylureas, pyridylureas, thiadiazolylureas, benzimidazole, and pyrimidines [67]. The highly active thiadiazolyl urea, thidiazuron (TDZ) (Figure 5) defoliates cotton very effectively, even when the leaves are still green, and is used as an harvesting aid in this crop. The phenylureas may function by inhibiting cytokinin oxidase, which catalyzes the cleavage of the cytokinin side chain [68], and thereby enhance endogenous cytokinin levels. This has been confirmed for thidiazuron, which caused the accumulation of cytokininlike activity in soybean callus [69] and inhibited partially purified cytokinin oxidase from *Phaseolus vulgaris* callus tissue [70].

ABSCISIC ACID

In recent years there have been considerable advances in understanding the mode of action of abscisic acid (ABA). By regulating the expression of specific genes it plays a major role in the development and germination of seeds, during which it induces the synthesis of storage proteins, confers desiccation tolerance and maintains dormancy [71]. For example, in the germinating cereal grain it suppresses transcription of genes such as α-amylase that are up-regulated by GA, as well as increasing the rate of transcription of other genes [13]. In the plant as a whole ABA serves as a stress hormone, its synthesis increasing dramatically, for example, under conditions of water deficit when it acts to close the stomata.

The involvement of ABA in physiological processes of agronomic importance, as typified by its ability to reduce the rate of transcription, has led to the search for analogs which might be used to increase water-use efficiency in crop plants. An example is the acetyleneacetate LAB 144 143 (Figure 6), which was produced by BASF and found to reduce transpiration in a number of cereals [72]. ABA itself, although available from cultures of the fungus *Cercospera rosicola* [73], is rapidly metabolized by plants as well as being highly sensitive to isomerization by UV light.

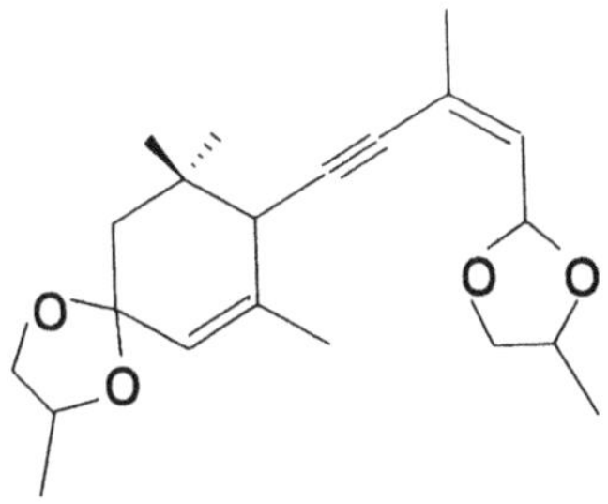

Figure 6 Structure of the synthetic abscisic acid-analogue LAB 144 143.

GROWTH RETARDANTS: INHIBITORS OF GIBBERELLIN BIOSYNTHESIS

By acting primarily to inhibit cell expansion in the subapical meristems of dicotyledons or basal intercalary meristems of monocotyledons, growth retardants reduce plant height without affecting the pattern of development. The earliest group of retardants to be developed were the onium compounds such as AMO-1618 or chlorphonium chloride. These form a large group of quarternary ammonium, phosphonium, and sulphonium compounds [74], which are characterized by the presence of a permanent positive charge, or in some cases by a tertiary N-atom which would be positively charged at physiological pH. Some examples are presented in Figure 7. Dennis et al. [75] showed that AMO-1618, and other retardants, inhibited the conversion of geranylgeranyl pyrophosphate to ent-kaurene in the biosynthetic pathway to gibberellins. It has been proposed that sterol biosynthesis is an alternative site of action for the onium retardants [76], but this is likely only at very high doses. Duriatti et al. [77] reported no effect of 10^{-4} M AMO-1618 on oxidosqualene cyclase isolated from pea seedlings.

The most important onium retardants in agricultural terms are chlormequat chloride (chlorocholine chloride or CCC), and mepiquat chloride. Although the precise mode of action of CCC has been in some doubt [74], it has been shown recently that treatment with this inhibitor reduced the GA_1 content of wheat seedlings and, except at very high concentrations of CCC, the growth of treated seedlings could be fully restored by simultaneous application of GA_3 (Temple-Smith and Lenton, unpublished data reported in [78]). Furthermore, Graebe et al. [79] have shown conclusively that CCC inhibits ent-kaurene synthesis in germinating wheat caryopses. Thus, although CCC is a relatively inefficient inhibitor of GA biosynthesis, as is also evidenced by the high rates that must be applied in the field, the GA pathway is its most likely primary site of action in wheat.

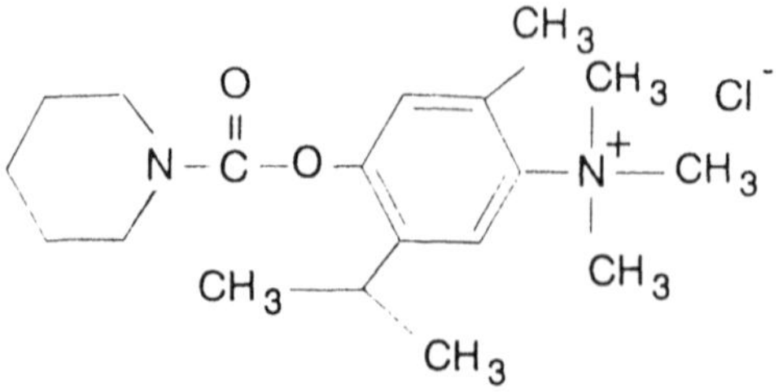

AMO 1618

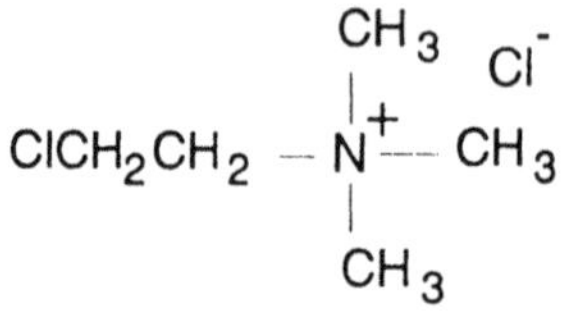

chlormequat chloride

(CCC)

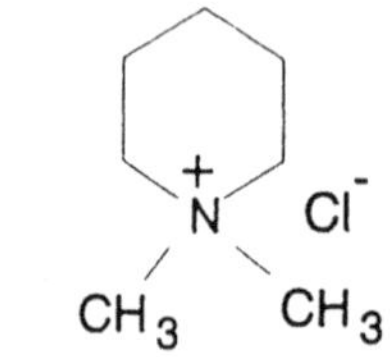

mepiquat chloride

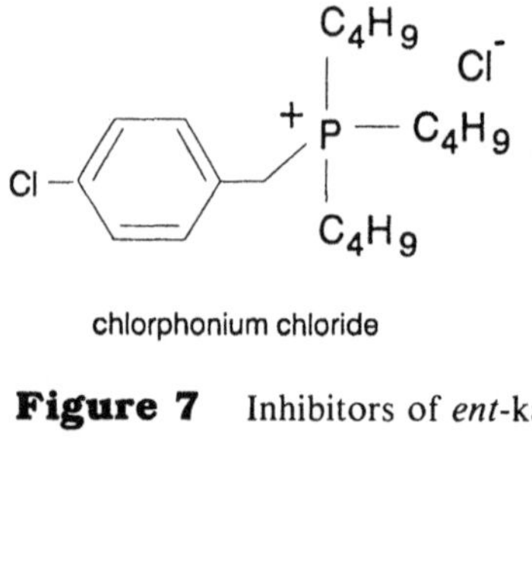

chlorphonium chloride

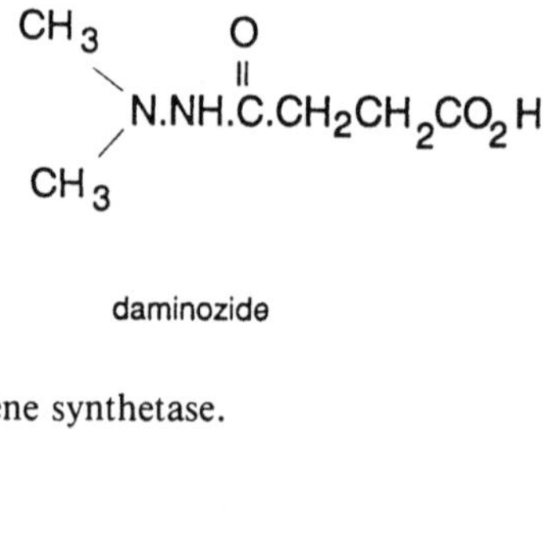

daminozide

Figure 7 Inhibitors of *ent*-kaurene synthetase.

The use of CCC as an antilodging agent extends to a range of temperate cereals, but especially the wheat crops of Northern Europe, where the application of large amounts of nitrogen, coupled with high rainfall, can cause severe lodging. In addition to strengthening the stems and reducing height, CCC applications were shown to enhance yield, thus ensuring its use in commercial practice. Whilst it is generally recognized that CCC is less effective on barley than on wheat, certain formulations have been marketed which are claimed to have a more consistent effect on barley [80].

Although CCC has been tested on a wide range of other species, including field, fruit, vegetable, flower, and ornamental crops and grasses, very few uses have been found for the compound on a commercial scale. It is only in the production of certain ornamental plants, where it is applied to reduce internode length and hasten flowering that CCC has found a major use, apart from that on cereals.

Mepiquat chloride is also used widely with cereals, either alone or in conjunction with ethephon (see above). However, the major use of this retardant is to limit undesired vegetative growth in cotton [81]. A reduction in height and breadth is coupled with slower closing of the canopy in treated plants allowing more light to reach the lower parts and better air circulation in the stand. Bolls on treated plants often mature earlier without a reduction in lint quality.

Daminozide (N-dimethylaminosuccinamic acid) is a growth retardant active on a wide range of plant species [82]. The compound is used to control growth in a range of ornamentals, but by far the most important application is in the control of vegetative growth, enhancement of fruit bud initiation, prevention of preharvest fruit drop, and promotion of firmness and quality in top fruit [30]. Other crops which have shown benefit from the use of daminozide include cherries, grapes, peaches, nectarines, plums, cantaloupes, tomatoes, peas, and brussels sprouts. In recent years its use has been questioned due to findings in the United States, which were difficult to substantiate, that daminozide is carcinogenic.

A new and extremely potent class of GA-biosynthesis inhibitors was launched in 1970 with the description of ancymidol by Tschabold et al. [83]. This compound prevents gibberellin biosynthesis by inhibiting the oxidation of *ent*-kaurene to *ent*-kaurenoic acid by *ent*-kaurene oxidase [84]. Many inhibitors of *ent*-kaurene oxidase are now known and they include pyrimidines, 1,2,4-triazoles, imidazoles, pyridines and a norbornanodiazetine. Examples of each type are given in Figure 8. They all possess an N-containing ring in which an N atom is sp^2 hydridized [85]. As with the related sterol biosynthesis inhibitors (SBIs) [86], a lone electron pair on the N atom is thought to coordinate with the heme Fe on the cytochrome P-450-dependent oxygenases that catalyze these reactions.

Some retardants have weak fungicidal activity and vice versa. In the case of the triazoles, which are usually produced as racemic mixtures, retardant and fungicide activity can sometimes be separated. Thus paclobutrazol consists of a mixture of the 2S,3S and 2R,3R enantiomers, the former being the more effective growth retardant and the latter possessing fungicidal activity [87]. The 2S,3S compound was found to be about 30 times more effective than its enantiomer as an inhibitor of *ent*-kaurene oxidation in a cell-free *C. maxima* endosperm preparation [88].

Several cytochrome P-450 inhibitors such as tetcyclacis and paclobutrazol inhibit the growth of plant cells in culture [89,90]. In these cases, since cell division could be restored with plant sterols, phytosterol biosynthesis was implicated as the site of action. The target enzyme in the phytosterol bio-

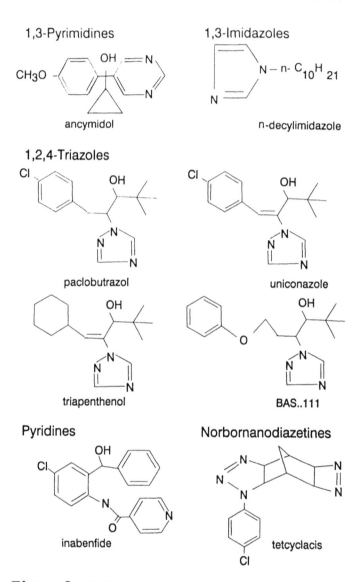

1,3-Pyrimidines

ancymidol

1,3-Imidazoles

n-decylimidazole

1,2,4-Triazoles

paclobutrazol

uniconazole

triapenthenol

BAS..111

Pyridines

inabenfide

Norbornanodiazetines

tetcyclacis

Figure 8   Inhibitors of *ent*-kaurene oxidase.

synthetic pathway is obtusifoliol 14α-demethylase, which was shown to be more strongly inhibited by triazole fungicides than by related growth retardants [91]. Thus, inhibition of sterol biosynthesis may be a cause of growth retardation by some triazoles, particularly the fungicides. However, GA biosynthesis is generally much more sensitive to growth retardants than is sterol biosynthesis. On the other hand, certain triazole herbicides, which cause considerable growth retardation, are potent inhibitors of obtusifoliol 14α-demethylation [92].

The triazoles and related compounds are known to inhibit other cytochrome P-450 mediated reactions, including the oxidation of ABA to phaseic acid [93]. Elevated ABA levels in treated plants could explain the reduced water loss sometimes observed in such plants [94]. Some triazoles inhibit herbicide metabolism and act as synergists [95].

The extreme potency and persistency of *ent*-kaurene oxidase inhibitors has limited the use of some compounds for certain applications where carry-over to subsequently sown crops has been shown to have detrimental effects on yield [96]. It is in tree shaping of top fruit and in the control of growth of other woody species that the triazoles, and particularly paclobutrazol, have found their major use. Paclobutrazol, which is marketed as "Clipper" for horticultural applications, is used in the United States extensively to stop the growth of branches on roadside trees, which would impede overhead power lines if left untreated. The savings in pruning costs amount to several million dollars per year.

The lack of systemicity of paclobutrazol, in common with other triazoles, does not seriously limit its use, but has led to the introduction of novel application methods, including trunk injections, to ensure that the compound is carried in the xylem stream to growing points on the trees. This has been found to be more reliable than applying the compound around the base of trees, where dose rate cannot be determined accurately due to effects of different soil types. An added benefit of the former type of application is that it avoids contact with nontarget plants and reduces environmental residues.

Paclobutrazol and some of the other triazoles, including uniconazole, triapenthanol, BAS 110..W, and BAS 111..W have found application in different small grains and in rice. Some of the other compounds, which lack the persistency of paclobutrazol, are potentially useful with oilseed rape, which has become a major break crop in Europe over the last decade. Stem and branch length and canopy structure can be manipulated very effectively with the triazoles, reducing lodging and fungal attack, and increasing light penetration into the canopy [97]. Significant yield increases have been obtained and there is scope for further improvement if the problem of seed shedding at the time of harvest can be overcome.

Triazoles and the norbornanodiazetines (e.g., tetcyclacis) have been tested for their effect on stimulating cold resistance of cereals. They have been shown to increase the overwintering capacity of a number of fall-sown oat varieties, resulting in yield increases [98]. The exact mechanism for the effect has not been established.

The final stages of gibberellin biosynthesis are catalyzed by a series of soluble, 2-oxoglutarate-dependent dioxygenases [99]. Recently, a new class

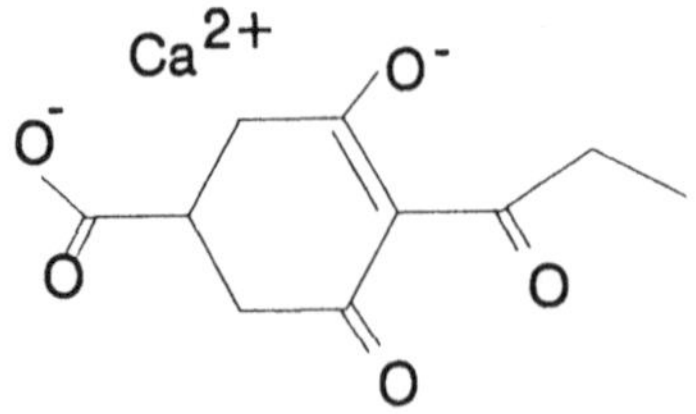

Figure 9   Prohexadione calcium, an inhibitor of soluble gibberellin hydroxylases.

of retardants, derivatives of 4-acyl-3,5-dioxocyclohexanecarboxylic acids, have been shown to inhibit these enzymes, particularly the 3β- and 2β-hydroxylases [100,101]. In some cases, at least, the inhibitors are competitive with respect to 2-oxoglutaric acid [101]. A typical member of this group, prohexadione calcium, produced by Kumiai Chemical Industry, Co., inhibits primarily the 3β-hydroxylation of GA_{20} to GA_1 (see Figure 9) [100]. The acylcyclohexanedione inhibitors are less persistent than the triazoles, are systemic and effective with a broad range of species. They are likely to become important growth retardants in the future.

INHIBITORS OF AUXIN TRANSPORT

The polar, basipetal transport of IAA from its site of synthesis at the shoot apex occurs from cell to cell and is generally accepted to be mediated by specific protein carriers on the plasma membrane. Influx carriers, primarily sited at the apical region of the cell and efflux carriers, at the basal region have been proposed [102]. Many compounds that inhibit polar auxin transport are known and the structures of several are given in Figure 10. The earliest to be discovered was 2,3,5-triiodobenzoic acid (TIBA). Derivatives of fluorene-9-carboxylic acid, such as 2-chloro-9-hydroxy-9H-fluorene-9-carboxylic acid (chlorflurecol-methyl), known as morphactins [103], and the phytotropins [104], typified by N-1-naphthylphthalamic acid (NPA), bind to the efflux carrier and so inhibit the transport of IAA across the plasma membrane, although in a noncompetitive manner. Thus, they bind to the carrier on a different site from IAA, perhaps on a separate subunit [102]. There is weak polar transport of TIBA, but not of NPA [105] and they appear to have different binding sites [106].

Many of the physiological effects of TIBA, the morphactins and phytotropins can be readily explained in terms of reduced IAA transport. Thus the stimulation of branching in fruit trees, an early use of TIBA, indicates

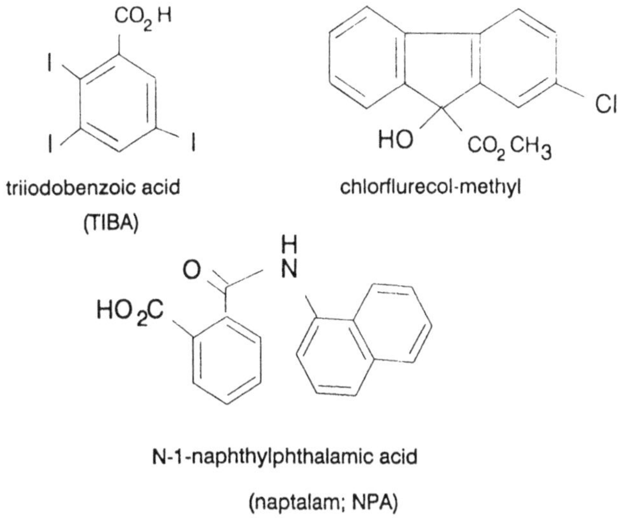

Figure 10 Inhibitors of polar auxin transport.

loss of apical dominance. Growth inhibition would occur when IAA does not reach its subapical target cells and fruit abscission would result from reduced auxin movement down the pedicel. The auxin transport inhibitors have proved to be very valuable experimental tools, but, despite a great deal of early interest, they have not been used extensively as growth regulators.

NONHORMONAL GROWTH INHIBITORS

Maleic hydrazide (Figure 11) is the best-known example from a group of compounds that inhibit cell division in the apical meristem. The mode of action of maleic hydrazide, whose uses include inhibition of axillary bud outgrowth in tobacco and of sprouting in potatoes and onions, and as a growth retardant in ornamental horticulture, is not well understood. A major effect appears to be inhibition of DNA synthesis and, perhaps indirectly, RNA synthesis, although the molecular mechanism for this inhibition is unclear [107]. It has been suggested that maleic hydrazide might induce lesions in heterochromatin [107], or that it may act as a pyrimidine antagonist. It was found to inhibit the uptake of [^{14}C]uracil into sieve elements of willow [108], although such an effect has not been confirmed in many cases.

Dikegulac sodium (sodium 2,3:4,6-bis-O-(1-methylethylidene)-α-L-xylo-2-hexulofuranosonate) (Figure 11), an inhibitor of apical growth used in

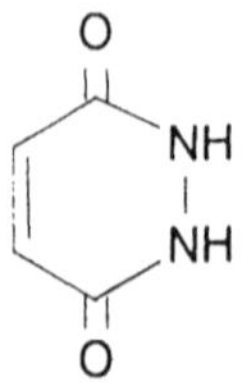

maleic hydrazide

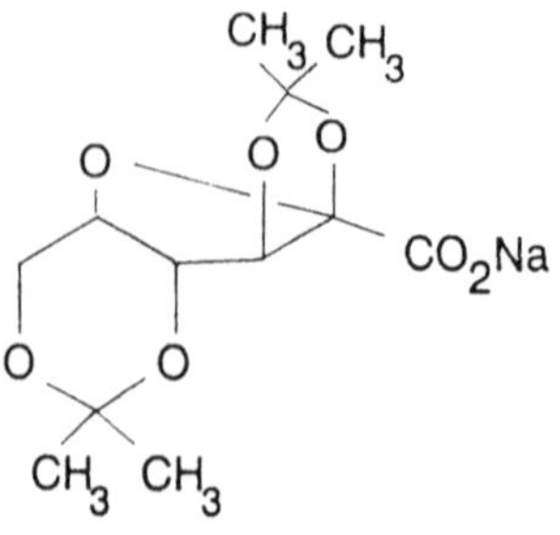

dikegulac sodium

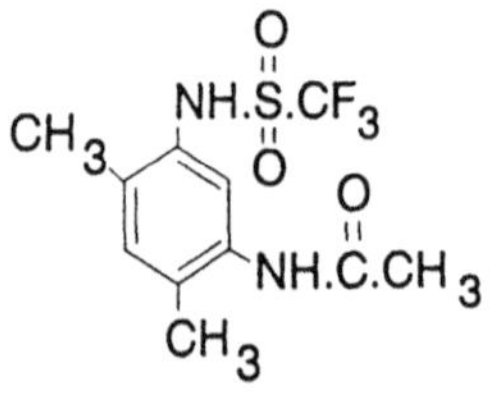

mefluidide

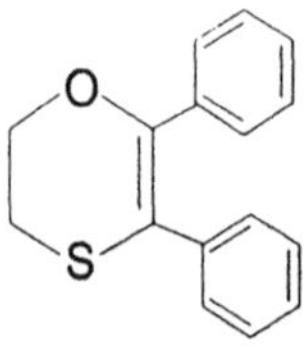

UBI-P293

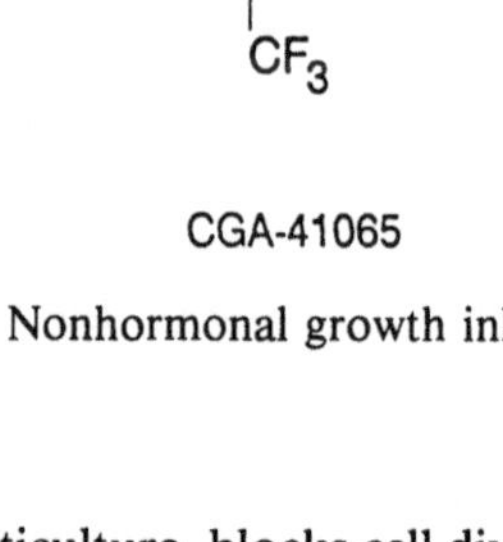

CGA-41065

Figure 11   Nonhormonal growth inhibitors.

ornamental horticulture, blocks cell division by an apparently different mechanism. It inhibited uptake of amino acids into *Solanum nigrum* cells grown
in culture and eventually caused general leakiness [109]. Thus it acts at the
plasmamembrane, as might be anticipated from its detergent-like chemical
nature.

A number of other growth regulators are known to affect meristematic
growth, although the biochemical mechanisms are not fully understood.

Mefluidide (N-[2,4-dimethyl-5-[(trifluoromethyl)sulfonyl]phenyl]acetamide) (Figure 11), which is used to regulate the growth of grass and broad-leaf vegetation, acts synergistically with several growth regulators and herbicides and may alter cell membrane function [110]. Destruction of cell membranes is probably the major effect of the hydrocarbon contact pinching agents. The most effective are the C_8 and C_{10} fatty alcohols and C_{10} fatty acid methyl ester, for which the major use is to control suckers in tobacco [111]. The cuticles of young buds are penetrated preferentially by these compounds so that the buds are more sensitive than more mature tissues. Other contact or localized growth inhibitors, such as UBI-P293 (2,3-dihydro-5,6-diphenyl-1,4-oxathiin) and CGA-41065 (N-ethyl-N-(2-chloro-6-fluorobenzyl)-2',6'-dinitro-4'-trifluoromethyl aniline) (Figure 11), restrict bud growth although the bud is not killed [111].

HERBICIDES AS GROWTH REGULATORS

There is quite extensive use of herbicides to induce abscission, desiccation, or ripening, all of which are senescence processes. Citrus fruit abscission agents such as 5-chloro-3-methyl-4-nitro-1-pyrazole and ethanedial dioxime (Figure 12), are thought to cause superficial damage to the fruit, which, in response, produce wound ethylene [112]. Cotton defoliants, such as organo-

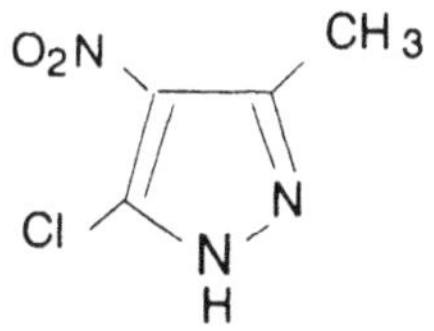

5-chloro-3-methyl-4-nitro-
1-pyrazole

HON=CH.CH=NOH

ethanedial dioxime
(glyoxime)

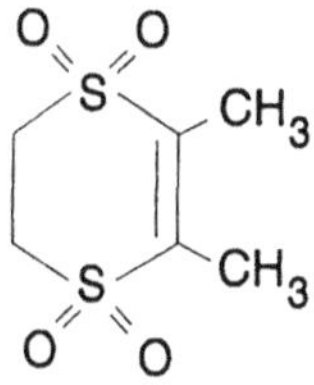

dimethipin

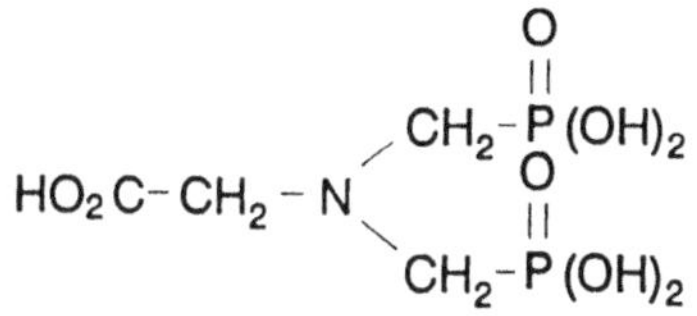

glyphosine

Figure 12   Herbicides used as abscission, desiccation, or ripening agents.

phosphate compounds and dimethipin (2,3-dihydro-5,6-dimethyl-1,4-dithiin-1,1,4,4-teraoxide) (Figure 12), presumably have a similar mechanism. They are used prior to mechanical harvesting, as is the desiccant arsenic acid; in this case desiccation occurs so rapidly that there is no leaf abscission. Glyphosine and the more potent herbicide glyphosate are used to ripen sugarcane and produce significant increases in yield. The most clearly defined biochemical effect of glyphosate is the inhibition of 5-enolpyruvylshikimic acid-3-phosphate synthase, an enzyme involved in the biosynthesis of aromatic amino acids and related metabolites [113]. Glyphosine is a less effective inhibitor of this enzyme.

FUTURE PROSPECTS FOR PLANT GROWTH REGULATORS

Currently, the compounds that cause defoliation and desiccation occupy 40% of the total growth regulator world market of over $1 billion, followed by ethephon (25%), chlormequat chloride (12%), and mepiquat chloride (8%) (W. Rademacher, BSAF AG, personal communication). Although mature products dominate the market and face little competition, there is some scope for expansion through widening the range of crop species to which PGRs are applied. Developing countries also represent areas of potential development, as does the increasing use of PGRs in amenity horticulture. It is, however, becoming increasingly more difficult to find new markets and to extend PGR usage in existing ones. Environmental considerations are of ever greater concern, such that registration of new products can be prohibitively expensive and investment in new PGRs is often given low priority. For this reason several major agrochemical companies have discontinued the production of new compounds for the PGR market, although the biological activity of compounds from other programs may still be assessed in screens. It is unlikely, therefore, that new compounds will be developed for anything other than major world crops.

REFERENCES

1. H. B. Tukey, *Plant Regulators in Agriculture*, John Wiley, New York (1954).
2. S. H. Wittwer, *Outlook Agric.*, *6*: 205 (1971).
3. R. F. Weaver, *Plant Growth Substances in Agriculture*, W. H. Freeman, San Francisco (1972).
4. A. J. Haagen-Smit, W. B. Dandliker, S. H. Wittwer, and A. E. Murneek, *Amer. J. Bot.*, *33*: 118 (1946).
5. F. Kögl and A. J. Haagen-Smit, *Proceedings Kon. Ned. Akad. Wet.*, *34*: 1411 (1931).
6. F. E. Gardner, P. C. Marth, and L. P. Batjer, *Science, 90*: 208 (1939).

7. G. Schneider and J. V. Enzie, *Proceedings American Society of Horticultural Science, 42*: 167 (1943).
8. B. O. Phinney, *The Biochemistry and Physiology of the Gibberellins. Vol. 1* (A. Crozier, ed.), Praeger Press, New York, p. 19 (1983).
9. C. O. Miller, F. Skoog, M. H. von Saltza, and F. M. Skoog, *J. Amer. Chem. Soc., 77*: 1392 (1955).
10. F. B. Abeles, *Ethylene in Plant Biology*, Academic Press, New York (1973).
11. R. C. deWilde, *Hort. Sci., 6*: 364 (1971).
12. J. W. Cornforth, B. V. Milborrow, G. Ryback, and P. F. Wareing, *Nature, 205*: 1269 (1965).
13. J. V. Jacobsen and P. M. Chandler, *Plant Hormones and Their Role in Plant Growth and Development* (P. J. Davies, ed.), Martinus Nijhoff Publishers, Dordrecht, The Netherlands, p. 164 (1987).
14. J. V. Jacobsen and L. R. Beach, *Nature, 316*: 275 (1985).
15. J. A. Zwar and R. Hooley, *Plant Physiol., 80*: 459 (1986).
16. T. M. Ou-Lee, R. Turgeon, and R. Wu, *Proceedings of the National Academy of Sciences (U.S.), 85*: 6366 (1988).
17. K. Skiver, F. L. Olsen, J. C. Rogers, and J. Mundy, *Proceedings of the National Academy of Sciences (U.S.), 88*: 7266 (1991).
18. R. Hooley, M. H. Beale, and S. J. Smith, *Planta, 183*: 274 (1990).
19. E. G. Jefferys, *Adv. Appl. Biol., 13*: 283 (1970).
20. B. O. Phinney, *The Biosynthesis and Metabolism of Plant Hormones, Society for Experimental Biology, Seminar Series 23* (A. Crozier and J. R. Hillman, eds.), Cambridge University Press, London, England, p. 17 (1984).
21. J. R. Bearder, *The Biochemistry and Physiology of Gibberellins, Vol. 1* (A. Crozier, ed.), Praeger, New York, p. 251 (1982).
22. W. Rademacher and J. E. Graebe, *Biochem. Biophys. Res. Commun., 91*: 35 (1979).
23. G. M. Palmer, *J. Inst. Brewing, 80*: 13 (1974).
24. A. J. Christadoulou, R. J. Weaver, and R. M. Pool, *Proceedings of the American Society of Horticultural Science, 92*: 301 (1968).
25. P. W. Goodwin, *Phytohormones and Related Compounds: A Comprehensive Treatise. Vol. 2* (D. S. Lethem, P. B. Goodwin, and T. J. V. Higgins, eds.), Elsevier/North-Holland Biomedical Press, Amsterdam, The Netherlands, p. 175 (1978).
26. M. V. Bradley and J. E. Crane, *Bot. Gaz., 123*: 154 (1962).
27. Y. Vercher, A. Molowny, C. López, J. L. García-Martínez, and J. Carbonell, *Plant Sci. Letters, 36*: 87 (1984).
28. M. A. Katob and W. W. Schwabe, *J. Hort. Sci., 46*: 89 (1971).
29. G. K. Goldwin, *Growth Regulators in Horticulture* (R. Menhennet and M. B. Jackson, eds.), British Plant Growth Regulator Group Monograph *13*, p. 71 (1985).
30. G. V. Hoad, *Plant Growth Regulator Potential and Practice* (T. H. Thomas, ed.), British Crop Protection Council, Croydon, England, p. 123 (1982).
31. S. P. Monselise, *Scientia Hort., 11*: 151 (1979).
32. M. W. Williams and E. A. Stahly, *J. Amer. Soc. Hort. Sci., 94*: 17 (1969).

33. C. W. Coggins Jr., H. Z. Hield, and S. B. Boswell, *Proceedings of the American Society of Horticultural Science, 76*: 199 (1960).
34. C. W. Coggins Jr. and H. Z. Hield, *Proceedings of the American Society of Horticultural Science, 81*: 227 (1965).
35. S. J. Wertheim, *J. Hort. Sci., 57*: 283 (1982).
36. P. H. Moore, R. V. Osgood, J. B. Carr, and H. S. Ginoza, *J. Plant Growth Reg., 1*: 205 (1982).
37. R. P. Pharis and C. G. Kuo, *Can. J. For. Res., 7*: 299 (1977).
38. R. P. Pharis and R. W. King, *Ann. Rev. Plant Physiol., 36*: 517 (1985).
39. G. V. Hoad, *The Biochemistry and Physiology of Gibberellins. Vol. 2* (A. Crozier, ed.), Praeger, New York, p. 57 (1983).
40. M. Los, C. A. Kust, G. Lamb, and R. E. Diehl, *HortSci., 15*, 22 (1980).
41. M. K. Upadhyaya, A. I. Hsiao, and M. E. Bonsor, *Plant Physiol., 80 (Suppl.)*: 145 (1986).
42. M. Lal and V. S. Jaiswal, *Plant Growth Regul., 7*: 29 (1988).
43. H. Yoshii and H. Imaseki, *Plant Cell Physiol., 22*: 369 (1981).
44. J. A. Roberts and G. A. Tucker, *Ethylene and Plant Development*, Butterworths, London, England (1985).
45. J. A. Maynard and J. M. Swan, *Austral. J. Chem., 16*: 596 (1963).
46. S. F. Yang, *Plant Physiol., 44*: 1203 (1969).
47. H. T. Hartmann, W. Reed, and K. Opitz, *J. Amer. Soc. Hort. Sci., 101*: 278 (1976).
48. D. J. Osborne and J. A. Sargent, *Ann. Appl. Biol., 78*: 83 (1974).
49. J. Jung and W. Rademacher, *Plant Growth Regulating Chemicals, Vol. 1* (L. G. Nickell, ed.), CRC Press, Boca Raton, FL, p. 253 (1983).
50. G. W. Bates and M. H. M. Goldsmith, *Planta, 159*: 231 (1983).
51. R. E. Cleland, *Plant Hormones and Their Role in Plant Growth and Development* (P. J. Davies, ed.), Martinus Nijhoff Publishers, Dordrecht, The Netherlands, p. 132 (1987).
52. J. L. Key, *Bioessays, 11*: 52 (1989).
53. R. M. Napier and M. A. Venis, *TIBS, 16*: 72 (1991).
54. M. A. Venis, E. W. Thomas, H. Barbier-Brygoo, G. Ephritikhine, and J. Guern, *Planta, 182*: 232 (1990).
55. S. S. Miller, *Hort. Rev., 10*: 309 (1988).
56. F. G. Gustafson, *Amer. J. Bot., 24*: 102 (1937).
57. L. C. Luckwill, *J. Hort. Sci., 28*: 14 (1953).
58. F. Bangerth, *Physiol. Plant., 76*: 608 (1989).
59. R. Aloni, *Ann. Rev. Plant Physiol., 38*: 179 (1987).
60. M. W. Williams, *Hort. Rev., 1*: 270 (1979).
61. C. S. Walsh, H. J. Swartz, and L. J. Edgerton, *HortSci., 14*: 704 (1979).
62. A. E. Hitchcock and P. W. Zimmerman, *Contrib. Boyce Thomp. Inst., 12*: 497 (1942).
63. J. L. Key, *Ann. Rev. Plant Physiol., 20*: 449 (1969).
64. T. J. Guilfoyle, C. Y. Lin, Y. M. Chen, R. T. Nagao, and J. L. Key, *Proceedings of the National Academy of Sciences (U.S.), 72*: 69 (1975).
65. L. G. Chen, C. M. Switzer, and R. A. Fletcher, *Weed Sci., 20*: 53 (1972).

66. S. K. Nandi, D. S. Letham, L. M. S. Palni, O. C. Wong, and R. E. Summons, *Plant Sci., 61*: 189 (1989).
67. K. Koshimizu and H. Iwamura, *Chemistry of Plant Hormones* (N. Takahahi, ed.), CRC Press, Boca Raton, Florida, p. 153 (1986).
68. L. R. Burch and R. Horgan, *Phytochem., 28*: 1313 (1989).
69. J. C. Thomas and F. R. Katterman, *Plant Physiol., 81*: 681 (1986).
70. J. M. Chatfield and D. J. Armstrong, *Plant Physiol., 80*: 493 (1986).
71. K. Skriver and J. Mundy, *Plant Cell, 2*: 503 (1990).
72. W. Rademacher, R. Maisch, J. Leissegang, and J. Jung, *Plant Growth Regulators for Agricultural and Amenity Use. BCPC Monograph No. 36* (A. F. Hawkins, A. D. Stead, and N. J. Pinfield, eds.), BCPC Publications, Thornton Heath, England, p. 53 (1987).
73. G. Asante, L. Merlini, and G. Nasini, *Experientia, 33*: 1556 (1977).
74. J. W. Dicks, *Recent Developments in the Use of Plant Growth Retardants.* Monograph 4 (D. R. Clifford and J. R. Lenton, eds.), British Plant Growth Regulator Group, Wantage, England, p. 1 (1980).
75. D. T. Dennis, C. D. Upper, and C. A. West, *Plant Physiol., 40*: 948 (1965).
76. T. J. Douglas and L. G. Paleg, *Plant Physiol., 54*: 238 (1974).
77. A. Duriatti, B. N. Pierrette, P. Benveniste, F. Schuber, L. Delprino, G. Balliano, and L. Cattel, *Clinic. Pharm., 34*: 2765 (1985).
78. P. Hedden, *Plant Growth Substances 1988* (R. P. Pharis and S. B. Rood, eds.), Springer Verlag, Berlin, p. 322 (1990).
79. J. E. Graebe, G. Böse, E. Grosselindemann, P. Hedden, H. Aach, A. Schweimer, S. Sydow, and T. Lange, *Progress in Plant Growth Regulation* (C. M. Karssen, L. C. Van Loon, and D. Vreugdenhil, eds.), Kluwer Academic Publishers, Dordrecht, The Netherlands, p. 545 (1992).
80. M. J. Samson, *News Bulletin, British Plant Growth Regulator Group, 3*: 10 (1979).
81. W. T. Thomson, *Agricultural Chemicals. Book III. Miscellaneous Chemicals*, Thomson Publications, Fresno, CA (1979).
82. H. M. Cathey, *Ann. Rev. Plant Physiol., 15*: 271 (1964).
83. E. E. Tschabold, H. M. Taylor, J. D. Davenport, R. E. Hackler, E. V. Krumkalns, and W. C. Meredith, *Plant Physiol., 46* (suppl.): 19 (1970).
84. R. C. Coolbaugh, S. S. Hirano, and C. A. West, *Plant Physiol., 62*: 571 (1978).
85. W. Rademacher, H. Fritsch, J. E. Graebe, H. Sauter, and J. Jung, *Pestic. Sci., 21*: 241 (1987).
86. T. Kato, *Chemistry of Plant Protection 1* (G. Hung and H. Hoffman, eds.), Springer, Berlin, p. 1 (1986).
87. B. Sugavanam, *Pestic. Sci., 15*: 296 (1984).
88. P. Hedden and J. E. Graebe, *J. Plant Growth Regul., 4*: 111 (1985).
89. K. Grossmann, E. W. Weiler, and J. Jung, *Planta, 164*: 370 (1985).
90. P. A. Haughan, J. R. Lenton, and L. J. Goad, *Phytochem., 27*: 2491 (1988).
91. M. Taton, P. Ullmann, P. Benveniste, and A. Rahier, *Pestic. Biochem. Physiol., 30*: 178 (1988).
92. R. S. Burden, C. S. James, D. T. Cooke, and N. H. Anderson, 1987 British Crop Protection Conference—Weeds, p. 171 (1987).

93. J. A. D. Zeevaart, *Plant Growth Substances 1988* (R. P. Pharis and S. B. Rood, eds.), Springer, Berlin, Germany, p. 233 (1990).
94. N. K. Asare-Boamah, G. Hofsta, R. A. Fletcher, and E. B. Dumbroff, *Plant Cell Physiol., 27*: 383 (1986).
95. M. S. Kemp, L. V. Newton, and J. C. Caseley, "Factors Affecting Herbicidal Activity and Selectivity." Proceedings of the European Weed Research Society Symposium, Wageningen, The Netherlands, p. 121 (1988).
96. C. A. Stutte, "Bioregulators: Chemistry and Uses." A. C. S. Symposium Series No. 257 (R. L. Org and F. R. Rittig, eds.), p. 23 (1984).
97. R. D. Child, D. R. Butler, and D. E. Evans, *Proceedings Plant Growth Regulator Society of America, 1989*, Plant Growth Regulator Society of America, Ithaca, New York, p. 173 (1989).
98. H. M. Anderson, *Crop Research, 29*: 29 (1989).
99. J. E. Graebe, *Ann. Rev. Plant Physiol., 38*: 419 (1987).
100. I. Nakayama, Y. Kamiya, M. Kobayashi, H. Abe, and A. Sakurai, *Plant Cell Physiol., 31*: 1183 (1990).
101. D. L. Griggs, P. Hedden, K. E. Temple-Smith, and W. Rademacher, *Phytochem., 30*: 2513 (1991).
102. P. H. Rubery, *Plant Hormones and their Role in Plant Growth and Development* (P. J. Davies, ed.), Martinus Nijhoff, Dordrecht, The Netherlands, p. 341 (1987).
103. G. Schneider, *Naturwissenschaften, 51*: 416 (1964).
104. G. F. Kateckar and A. E. Geissler, *Plant Physiol., 66*: 1190 (1980).
105. K. S. Thomson, R. Hertel, S. Muller, and J. E. Tavares, *Planta, 109*: 337 (1973).
106. M. R. Sussman and M. H. M. Goldsmith, *Planta, 151*: 15 (1981).
107. L. D. Noodén, *Plant Cell Physiol., 13*: 609 (1972).
108. D. Coupland and A. J. Peel, *Planta, 103*: 249 (1972).
109. S. Zilkah and J. Gressel, *Planta, 145*: 273 (1979).
110. K. J. Tautvydas and T. G. Hargroder, *Proceedings Plant Growth Regulator Society of America*, Plant Growth Regulator Society of America, Lake Alfred, Florida, p. 13 (1985).
111. G. L. Steffens, *Plant Growth Substances 1979* (F. Skoog, ed.), Springer, Berlin, Germany, p. 397 (1980).
112. R. E. Holm and W. C. Wilson, *J. Amer. Soc. Hort. Sci., 102*: 576 (1977).
113. N. Amrhein, B. Deus, P. Gehrke, and H. C. Steinrücken, *Plant Physiol., 66*: 830 (1980).

7

Impact of the Greenhouse Effect on Plant Growth and Crop Productivity

Sylvan H. Wittwer

Michigan State University
East Lansing, Michigan

INTRODUCTION

Climate and the weather can be proclaimed as the most determinant factors both for plant growth and for crop productivity. The world's present agricultural production system is based primarily on the needs of annual crops. There are only a few exceptions—bananas, coconuts, other tree fruits, grapes, nuts, and a few vegetables, such as asparagus, rhubarb, and others under special cultural conditions.

Agricultural productivity varies dramatically from year to year. Extremes in weather, rather than averages, affect agriculture. Crops and livestock are sensitive to weather over relatively short periods of time, and annual averages do not convey short-term differences. Interannual variabilities have

---

The theme for this paper was set in 1980 with the publication of the paper "Carbon Dioxide and Climate Change: An Agricultural Perspective" in the *Journal of Soil and Water Conservation* *35* :116–120 [1]. It was further amplified by an international conference sponsored in 1982 with a focus on research imperatives related to the direct (biological) effects of rising levels of CO_2 in plants [2]; in 1990 with the report, "Implications of the Greenhouse Effect on Crop Productivity" in *HortScience* *25*(12): 1560–1567 [3]; a report entitled "Climate Change and Agriculture," prepared for a symposium to be sponsored by the University of Missouri; and finally a paper entitled "In Praise of Carbon Dioxide" in *Policy Review* Number 62, Fall 1992: 4–9 [4].

a greater impact on agricultural productivity than does any projected climate change. There is no evidence that with projected climate changes there will be increases in interannual variabilities.

Farmers have always coped with climate change. The magnitudes of interannual variabilities have already exceeded those of projected climate change. Year-to-year fluctuations in climate are several orders of magnitude greater than the slow secular change of climate in the imaginary scenarios being proposed. This has been dramatically illustrated for the summer months in the U.S. corn belt between the years 1988 (dry and hot) and 1992 (wet and cold). Temperatures differed in the 2 years by as much as 8°F during July and August in some states. The most critical unknown is the pattern of drought frequency. Drought occurs someplace on the earth every year. It also occurs someplace in the U.S. corn belt almost every year. Annual droughts are also commonplace in the major agricultural areas of India, China, and Russia and in many African nations.

The big environmental news story for the year 1988 was the so-called warming of the globe or projected climate change, known as the "greenhouse effect." The protagonist was carbon dioxide. The message was of doom and gloom, global warming was portrayed as having catastrophic consequences for the biosphere. True, an enormous global and geophysical experiment is underway. We are all participants in the experiment and will be exposed to the results of it. There are no exceptions. It is projected that the consequences or results will impact the environment in which we all live, the climate, our water resources, the total biological productivity of the earth, and crop production. The variable by which humanity is currently conducting this global experiment on itself, is the rising level of atmospheric carbon dioxide and other so-called greenhouse gases, which include methane, dinitrogen oxides, and chlorofluorocarbons. The result is the greenhouse effect and a projected global warming [5–11].

It is now generally agreed that the atmospheric CO_2 concentrations have increased since the early part of the 19th century by nearly 30%. During this span, fossil fuels have made a net contribution of 130 to 180 teragrams, while deforestation and expanding agriculture another 100–200 teragrams. Human activities have also led to release of other radiatively active gases listed above. Meanwhile, the atmosphere has gained only 60 teragrams of the above, the remainder having gone somewhere. The ocean is a big unknown sink, as is also the biosphere. Estimates of the ocean's sink capacity leave a large quantity of released carbon dioxide unaccounted for.

The combined and cumulative projected effects of greenhouse gases other than CO_2 on the earth's climate are equivalent to an increase in atmospheric CO_2 of about 55 ppm above the now observed concentration of 360 ppm, which is increasing at the rate of 2 ppm/year. It is projected that the combined effects of increasing concentrations of CO_2 and other trace gases might lead to a equivalent of doubling at the current level of 355 ppm of CO_2 as early as the 2040s [12].

Extending the projections further as derived from very uncertain climate models, the global mean equilibrium surface temperature will increase by 1.5 to 4.5°C, if the atmospheric CO_2 concentration doubles. (Values outside this range, especially on the low end, should not be excluded.) With a projected global warming of 1.5 to 4.5°C, the sea level might rise 20 to 140 cm from thermal expansion of water. Major ice sheets, however, would not be expected to melt within the next century. It is further projected that doubling of atmospheric concentrations of CO_2 and other greenhouse gases could profoundly affect global ecosystems, agricultural and forest productivity, water resources, and sea ice. Further, it is now believed by some that during the first half of the next century, a rise in global mean temperature could occur, which is greater than in our history [6,7,8,12–15].

Interest was heightened in a greenhouse effect by the drought and heat wave across the continental United States in May and June of 1988. By July, there were cover stories in news weeklies. It was the lead topic for broadcast news programs, and magazine write-ups appearing on what were presumed to be correlations between the heat wave and drought with a greenhouse effect. All this was coupled with the testimony of an overzealous scientist before the U.S. Congress that he was 99% sure the greenhouse effect was already here [16].

There was little if any scientific content in most of the stories which were visually portrayed — damaged crops, dried up rivers, sweltering cities, and burning forests. All of this rippled off into a plethora of local, national, and international conferences, symposia, and colloquia.

Meanwhile, it has been authoritatively established that no greenhouse effect was involved in the U.S. drought and heat wave of 1988. It was the result of an out of position jet stream, which directed storms into Canada rather than across the mid-United States [17,18]. Nevertheless, there now exists the assertion by a very large segment of both the scientific and political communities that the greenhouse effect is more than a phenomenon to be described but a world reality, and that global warming has already set in [19].

This paper will address the implications of the greenhouse effect, or more specifically, possible global climate changes induced by rising atmospheric levels of CO_2 and other greenhouse gases on crop productivity. Also considered will be the direct or biological effects of rising atmospheric levels of CO_2 and possible climate–biological interactions on plant growth and crop productivity.

In the biological sense, CO_2 should not be considered an air pollutant, whereas, much less is known of the effects on biological productivity of methane, dinitrogen oxides, and the chlorofluorocarbons. In 1988, there was no "hotter" topic than the greenhouse effect, but in 1989 and again in 1990, the topic cooled a bit, though it was kept very much alive by press reports, environmental activists, and some climatologists. The central and eastern United States were contrastingly cooler and wetter than in 1988. During June, July, and August of 1992, temperatures in the central U.S. corn belt were up to 5°C cooler than in 1988.

CLIMATIC EFFECTS

The most overriding variable projected to affect crop productivity arising from increasing levels of atmospheric carbon dioxide and other gases, is a warmer earth. Some warming appears inevitable, but how much is the big question. There is a high degree of extremism associated with this issue [20–23]. Based on real world observations thus far, however, it may be considerably less than the very crude computer models indicate and almost everyone also religiously appears to believe [24]. More recently (1990s), the dissenters have become more numerous and vocal, with little support, however, from the media and press [25]. The projections are based upon very crude predictive climate models. According to the majority that interpret them, there will be a global mean equilibrium surface temperature increase of 1.5 to 4.5°C if the atmospheric CO_2 doubles from a not too well-defined level. Globally, the climate averages could warm by 3 ± 1.5°C by the end of the next century. The change would be greatest at the poles and least at the equator [6, 15–17,26–32].

Second, projected increases in temperature would predictively, but presumably, induce widespread changes in precipitation patterns. For many food and fiber crops and other plant species, a change in water availability would have a greater impact on both the magnitude and stability of crop production than temperature changes. Regional scale climatic changes have been modeled with even less confidence than global climate (temperature) changes [33,34]. This is particularly true of any future precipitation patterns [23]. The modeling science is yet far too weak to support major policy decisions either as to global or regional warming or changes in precipitation patterns as now being and having been proposed by many climate change advocates and environmental activists [6,7,8,15,31].

Projected changes, if they occur in precipitation and temperatures under high atmospheric CO_2 levels, will favor irrigated crop production. Regionally, major increases in irrigated acreage will occur for the United States in the nothern plains and the Mississippi delta. Similar expansions could be expected for the northern and central provinces of China, many of the states of India, and most all southern republics of the former U.S.S.R.

In the United States, the corn and winter wheat belts will shift northward and grain sorghum and millet will take over the current wheat and corn producing areas.

A third global-warming-induced variable for crop productivity, and it could be the most important for temperate zone agriculture, would be both an increase in the length and intensity of the growing season [1,35], or growing degree units [36]. Seasonal and interannual variabilities in rainfall and snow cover, length of the growing season, and thermal variability in growing degree days (heat sums), are the climatic concern of farmers. Stability or dependability of production is as important as the magnitude of production itself.

The prospects of climate change from increasing atmospheric levels of CO_2 should not unduly frighten agriculturists and farmers. Seasonal and interannual variabilities in climate have always made agriculture uncertain and a gamble. The purchase of a farm, acquisition of machinery, including irrigation equipment, choice of a fertilizer formula, crops to be grown, seed varieties, and development of markets, all depend upon next year's weather resembling that of previous years [35,37].

Fortunately, the past century provides evidence that farmers in the world's agriculture and their research establishments can cope with, and even improve during climate change. Over the past 100 years, for example, the high plains in the United States became the wheat belt during the moist period, then a dust bowl during a dry period. Agriculture, through migration and technology, was able to adapt. The course of yields for the major food crops (rice, wheat, maize, potatoes) in the United States, Western Europe, India, China, Japan, Egypt, and the United Kingdom during the past century is a continuous upward trend, although there are significant yearly fluctuations, some of which were induced by war. During the last century, one sees only upward trends attributable to new technology applied during both warm and cold, and wet and dry periods. Interannual variations in climate for a particular location may equal or even exceed those projected for long-term trends associated with a presumed greenhouse effect.

The impacts of new technology will continue. Apprehension about a more prolonged climate trend in the future and a greater change, can be balanced by the expectation that research, perhaps spurred by the opportunity of more CO_2 for photosynthesis and a reduced water requirement, can continue to increase crop yields. For the past 70 years, the variation in weather, as reflected by the wheat fields of the United States, Russia, China, India, Argentina, Canada and Australia, has allowed nations with good weather to supply grains to nations with bad weather. Worldwide droughts have been nonexistent in modern times.

While there is some assurance that United States and world agriculture can cope with the moderate climatic changes projected, such a change will add one more problem that agriculture must handle during the coming decades [1]. Other problems facing agriculture include shortages of water, water quality, arable land, and fossil energy, and sustainability of current production practices. Some agricultural practices will continue to threaten the environment and must deal directly with food safety and human health concerns. At the same time, there is the need to maintain dependable production at high levels [38]. A projected climate change makes comprehensive agricultural research even more critical if we are to deal effectively with that change.

A specific research initiative related to possible climate change, would be to mount immediately a major effort to alleviate environmental stress on crop production through genetic improvement and genetic engineering, chemical treatments, and management practices. This would be beneficial both as a buffer against short-term effects of climate variability, including droughts, which have always existed; and as a means of combating long-term climatic change, and to capitalize on existing genetic resources for enhancing production.

An extensive review of the CO_2-climate issue by a joint committee of the World Meteorological Organization and International Council of Scientific Union (ICSU) reached a most perceptive conclusion in 1983, that still holds [12,32]. It was stated that by simple comparison of the overall temperature and CO_2 increase for the last 100 years, one must conclude that the climate (temperature) sensitivity to CO_2 is at the lower limit of mathematical climate computer model prediction. In other words, global temperatures in the past 100 years have risen less than half as much (0.3–0.7°C) as the same models which now predict a warming of 1.5–4.5°C by the middle of the 21st century. This is based on the more than 25% rise in atmospheric CO_2 which has already occurred, along with major increases in other greenhouse gases. Furthermore, these simultaneous increases in greenhouse gases, other than CO_2, should now be the equivalent, as far as any warming is concerned, to the halfway point of doubling of CO_2, but this is not at all substantiated by model projections [20,39].

The global warming predicted by climate models has not come to pass. More warming has occurred in the southern hemisphere than in the north, which is 90% covered by water, even though water is supposed to take longer to warm than land. The models also claim that the far north latitudes should warm the most. Yet Alaska had record cold in the winter of 1988. This country, presumably, has the best geographical distribution of temperature and precipitation records of anywhere in the world, and from the most extensively and strategically located weather stations on earth. Yet for the continental United States the record shows no climate change during the past 100 years. While the United States cannot be extrapolated as the entire earth, it produces 10% of the world food grains and 25% of the feed grains and beef. It is a major producer of agricultural products.

All this strongly suggests a possible 1.5°C, or lower, increase in global mean temperature with a doubling of the current atmospheric CO_2 level and other gases, not the 4.5°C or as much as 8°F as is predicted by some over-zealous environmental activists. These observations confirm what has happened thus far in real world climate change statistics [21-23,39]. Also, it has been concluded that a considerable fraction of the observed temperature variations during the past 100 to 150 years is related to causes other than CO_2 [40,41]. The fact that model projected temperature changes from a doubling of atmospheric CO_2 have an estimated uncertainty factor of 2, lends further doubt as to the validity of climate models for an accurate assessment of what the future may hold. These facts and conclusions are vitally important considerations for nations when considering current and future policies and actions to be taken as to energy resources and conservation, biological productivity, crop production, water resources, human health and welfare, in general.

Thus, a different view of global warming, with considerable scientific credibility is now emerging [21,22,42-44], supporting those clearly articulated much earlier [34,39]. It has now been concluded by some scientists, formerly strong advocates of a global warming threat [44], that climate models are crude simplifications of the myriad complex physical processes taking place in the atmosphere and oceans. They cannot prove that emissions of greenhouse gases will significantly alter the earth's climate. The causes of global warming may be even less certain than the trend itself.

Moreover, as Hare and associates [45] have emphasized, there are as yet no grounds for viewing a CO_2 warming, if it occurs, as an unmitigated disaster. We can speculate, as many have done, as to gains as well as losses, or winners and losers. Plants, in particular, may profit from the fertilizer and physical effects of CO_2 enrichment [46]. There may even be a net gain. It has been particularly unfortunate that on the issue of the greenhouse effect and global warming, many scientists and others have held to only one side of the issue and quoted only that which seems to support their argument [5-8,15,16,24]. This may be laudable for lawyers, policymakers, and environmental advocates but not for scientists [46].

DIRECT BIOLOGICAL EFFECTS

Greenhouse-Grown Plants

The beginnings of CO_2 enrichment for improvement of plants grown in greenhouses took place over 100 years ago [47]. As early as 1888, the benefits of

CO_2 fertilization were recognized and reported for practical greenhouse culture in Germany, a few years later in England, and about 80 years ago in the United States. Favorable results were first reported for food crops, especially lettuce, tomatoes, and cucumbers, and then for flowers and ornamentals.

Many reviews, conferences, assessments, and compilations of literature have since been assembled, primarily of crops grown in controlled or partly controlled environments with the atmospheres variously enriched with CO_2. Wittwer and Robb [48] made a very thorough assessment of past work and added data of their own, primarily for greenhouse grown tomatoes and lettuce. They related the results to commercial practices of growing crops to maturity under enriched CO_2 atmospheric levels. The results were substantial increases in yields of marketable products and an improvement in quality. Meanwhile, Strain and Cure [49] assembled extensive bibliographies of literature concerned with the direct effects of atmospheric CO_2 enrichment on plants and ecosystems. Kimball and associates followed in 1983, 1985, and 1986 [50–52] with an assemblage of 770 observations on agricultural yield enhancement of crops grown in greenhouses and subjected to elevated levels of atmospheric CO_2. The overall yield increase was estimated at 32%. The U.S. Department of Energy, with other concerned federal and state agencies, including the U.S. Department of Agriculture, has sponsored extensive research programs and published numerous research reports on the direct effects of subambient and superambient carbon dioxide levels on plant growth, photosynthesis, respiration, and water relations of a wide variety of commercially important crops in the United States [4,26, 53–63].

The primary purpose of a 1982 international conference [2] was to identify researchable issues relating to, first order, or direct biological effects of rising atmospheric carbon dioxide on plant productivity. An inseparable linkage of the biological effects of photosynthetic efficiency [64–66], water use efficiency [63], and biological nitrogen fixation [67] with the climatic resources of sunlight, temperature, and moisture, was recognized [63,68,69].

The subject focus of this paper is, in part, a follow-up of that conference. More thorough documentation of the direct effects of CO_2 on plant productivity appeared in 1985–1987 with a series of reports by the U.S. Department of Energy [53,55,57,61] and my reviews, [47] in 1985, and in 1992 [4]. These were accompanied by two volumes assembled and edited by Enoch and Kimball in 1968 on Carbon Dioxide Enrichment of Greenhouse Crops covering the status and CO_2 sources [70] and physiology, yield, and economics [71].

There is and has been for 35 years, a very active working group on CO_2 nutrition as a unit of the Commission on Protected Cultivation of the International Society for Horticultural Science. This working group sponsors CO_2 symposia on the direct or positive effects of elevated levels of atmospheric CO_2 for increasing the productivity and quality of greenhouse grown crops. They also publish an annual newsletter with an update on latest developments with respect to CO_2 fertilization for tomatoes, cucumbers, peppers, lettuce, potted plants, and flowers. One of their reports [72] is on summer CO_2 enrichment for greenhouse tomatoes resulting in 12–13% increase in yields when levels of atmospheric CO_2 are maintained at 335 ppm or above.

The two most significant direct effects of elevated levels of atmospheric carbon dioxide for enhancement of crop yields, are on photosynthesis and water use efficiency.

Photosynthetic Efficiency

There is little doubt that current atmospheric levels of CO_2 are suboptimal for photosynthesis when other factors affecting plant growth (light, water, temperature, nutrients) are optimal [70,73]. Net photosynthesis is the sum of gross photosynthesis minus photorespiration, and the latter may be of such magnitude that up to 50% of the CO_2 newly fixed into carbohydrate may be oxidized back to CO_2. With the currently rising levels of atmospheric CO_2 and those yet projected, photorespiration rates may be expected to decrease [66]. Increases in biomass are generally observed with atmospheric CO_2 enrichment. This may not always come from an increase in net photosynthesis. Higher than normal levels of atmospheric CO_2 induce greater water use efficiencies with C_4 plants, such as corn. This increase in water use efficiency has been largely responsible for the resultant stimulations in plant growth [59,63,69,74–76].

The most verifiable direct effect of high levels of CO_2 on plants is an increase in leaf and canopy photosynthetic rates. The increase in photosynthesis with increasing atmospheric CO_2 will continue up to about 1000 ppm [47,48,50,52,68,77,78]. The most obvious result is that plants grow faster and get bigger. There are differences among species. C_3 species have a qualitatively greater photosynthetic response to elevated CO_2 levels than the C_4 species [53–55,60,61,68,74,75,78]. There are usually increases in leaf areas, weight per unit area, leaf thickness, stem height, branching and seed and fruit number and weight. Organ size may increase along with root:top ratios. The C:N ratio increases. Most important of all, yields of the marketable product will most likely increase. Aside from overall increases in growth and earlier maturity, particularly noteworthy have been the direct effects of elevated levels of atmospheric CO_2 on enhancement of tuber growth in potatoes [79], root growth in sweet potatoes [80], biological nitrogen fixation in soybeans [67,81–83], and on root:top ratios [84]. Generally, the harvest index,

meaning those parts of the plant which command some economic importance, is increased. With a doubling of the current ambient CO_2 concentration (360 pm) worldwide, agricultural yields may average an increase of 32%
[50,51]. It is likely that some of the substantial increases in global crop productivity witnessed during the past 100 years, may be attributable to the rising level of atmospheric CO_2. A tentative increase of 5–10% in productivity
may already be ascribed to this variable. Important progress has been made
during the past 10 years on the direct effect of CO_2 enrichment on crops.
The enhanced growth and yield from more CO_2 for some plants is now widely
recognized. Numerous tests have been conducted on a few major crop plants
(rice, corn, wheat, soybeans, cotton, potatoes, tomatoes, sweet potatoes,
and forest tree species), usually under a controlled or partly controlled environment. There is, however, still a wide gap in knowledge as to the responses in open fields of the major food crops, as well as for native species
and ecosystems to CO_2 enrichment [85].

Water-Use Efficiency

What agriculture needs more than anything else is water. This resource,
including both water quantity and quality, is becoming the most limiting
of all natural resources, exceeding that of both land and energy. Water for
agriculture is becoming increasingly critical for all five of the most popu-
lated nations on earth—China, India, the former U.S.S.R., the United States,
and Indonesia. It is by far the most important in the Middle East, Egypt,
the Sudan, and the countries of northern and Sub-Saharan Africa. One of
the most critical unknowns in crop productivity is the pattern of drought
frequency [35]. Drought is what farmers fear above all. This is true in the
U.S. corn and wheat belts, North China, Ukraine, the Middle East, Sub-
Saharan Africa, and the semiarid tropics of India, Pakistan, Chile, Australia,
and Brazil.

Water is now and will increasingly become a critical and limiting resource for global agricultural productivity. Of the fresh water resources con-
sumed annually in the United States, 80–85% go to crop irrigation. One-
third of the world's food supply is now grown on 18% of the crop land that
is irrigated.

An important aspect of increasing CO_2 concentrations on plants is that
the leaf stomata tend to partly close. This increases the resistance to trans-
pirational water loss, with decreases in leaf transpiration rates, and an increase in water use efficiency.

Water stress is the single most limiting factor for crop productivity.
Evidence now accumulated shows that improved water use efficiency at elevated levels of atmospheric CO_2 associated with the greenhouse effect, is an
important finding for both agriculture and ecology. The implications are

many. They include protection or greater resilience of crops from drought and other water-related stresses, and a decrease in the quantity of water required for crops to mature and produce a harvestable product. There is the potential for greatly improved biological productivity. A modified water use requirement could reduce water requirements for irrigation. An indirect effect would be an extension of the distribution and boundaries into semi-arid and desert areas for specific crops now constrained by water supplies. As a global antitransparent in biological productivity, elevated levels of atmospheric CO_2 could reduce overall evaporative water loss and increase water availability for use in agriculture and industry [86,87].

That high CO_2 levels in the atmosphere alleviate water stress in plants has been confirmed [59,61,69,73,88–93]. Yields of water-stressed wheat at high CO_2 levels were as large or larger than those from well-watered wheat at normal CO_2 [94–96]. Elevated levels of CO_2 will increase water-use efficiency in many plant species. The most striking changes in water use efficiency and reduction in water use requirements up to 1983 were summarized by Pearcy and Bjorkman [65]. Elevated atmospheric CO_2 affected stomatal closure and a reduction of transpiration rates. These effects progressed as levels of atmospheric CO_2 rose. As stomatal conductance decreased in response to increased CO_2, transpiration rates decreased proportionally, providing other variables were constant. The ratio of CO_2 taken up in photosynthesis to the water lost in transpiration can be termed "photosynthetic water use efficiency." Water use efficiency also may be expressed in terms of the amount of biomass gained for the amount of water lost in a given period of time [71].

The direct effect of the atmospheric CO_2 concentrations on photosynthesis of C_4 plants is one of increasing photosynthetic water use efficiency. Stomatal conductance declines with increasing CO_2 concentration in both C_3 and C_4 plants, but the C_3 plants would likely benefit more from an increase in atmospheric CO_2. Rogers et al. [75] demonstrated that with soybeans, high atmospheric CO_2 not only promoted greater growth but prevented the onset of severe water stress under conditions of low water availability.

Plants of a C_3 photosynthetic pathway, may benefit in dry matter production from high CO_2 in three ways. First, there is an enhancement of leaf expansion; second, an increase in the photosynthetic rate per unit leaf area, and finally, there is an increase in water use efficiency.

There are still many unknowns concerning the consumptive use of water in crop production as it may be affected by rising levels of atmospheric CO_2 and possible accompanying climate changes. Speculations as to landscape-scale consequences are premature in the absence of actual field data for unconfined plants continually exposed to elevated levels of CO_2. As suggested by Kimball [50] and Acock et al. [53], plants are probably going to be larger and have a greater leaf area in the future high CO_2 world. This will tend to increase transpiration. They will also, to compensate, likely have a larger, more vigorous root system to extract more water from the soil [84]. A CO_2 induced decrease in transpiration will make more thermal energy available for soil evaporation. Consequently, the amount of consumptive water use (and requiring irrigation) that will be reduced by twice the current level of atmospheric CO_2 very likely will be less than the potential 33% reduction in leaf transpiration. Reductions up to 10% might reasonably be expected. This could still be very significant in world agriculture and for plant growth. Plants, whose relative stomatal closure in high CO_2 is more than their leaf area increase, are more likely to have a reduction in water use. If warmer temperatures occur, however, as is projected by some, along with the rise in global CO_2 concentration, we may also see a partially compensating rise in transpiration [97].

The above observations strongly suggest that a substantial research effort should be directed toward the direct effects on plants of a rising level of atmospheric CO_2. The rise in CO_2 is real, the effects on photosynthesis, water use efficiency, plant growth, and crop yields are real. We should determine the short- and long-term effects of a doubled atmospheric CO_2, and the beneficial effects on water supply and increased water-use efficiency of plants [73,98].

Irrigated Crop Production

Projected changes in precipitation and temperatures under the greenhouse effect and global warming will likely favor irrigated crop production. Regionally, in the United States, major increases in irrigated acreage will occur in the northern plains and Mississippi Delta. This will also be true for China, India, and other nations, for the more humid areas that are not now irrigated but devoted to the production of staple grain crops and seed legumes. This trend, however, is already occurring in the United States, China, and elsewhere. Corn and wheat belts will shift to areas that are now somewhat cooler and wetter [35]. The production of grain sorghum and millet will take over what are now predominately some of the drier wheat and corn producing areas. For the United States, this will be the more westerly parts of the grain (corn and wheat) belt. It is to be anticipated and hoped that for the future, models from atmospheric science will be coupled with those now being developed by the plant scientists and agricultural economists [46]. By such means, greater sensitivity of plants and crops to climate change can then be explored. The results will depend both upon the severity of climate change and the compensating direct effects of CO_2 on crop yields. Simulations heretofore suggest that irrigated acreages will expand and regional patterns of crop production will shift.

PLANT GROWTH AND CROP PRODUCTIVITY-ADAPTABILITY TO THE PRESENT CLIMATIC RESOURCES OF THE EARTH

Most agriculturally important crops are adaptable to some climate change (Table 1). There is no place on earth too hot and humid to grow rice, cassava, sweet potatoes, bananas, or plantains. They are all major staple food crops and rank as number one in many agriculturally developing countries. A global warming would extend the currently set climatic boundaries both north and south. Maize is the third major food crop, and second only to rice and wheat, and number one in the United States and for many agriculturally developing nations. It is grown in more diverse areas of the earth than any other crop. This includes the lowland humid and high elevation tropics and throughout the temperate zones. Commercial production has moved 800 km further north in the United States during the past 50 years, and continues as a major southern state crop. Soybeans and field beans, two of the world's leading legumes, can be grown successfully from the equator to beyond lat. 45° N, or 40° S. Winter wheat, which is generally much more productive than spring wheat, has moved 360 km further north in Russia and China. The U.S. winter wheat zone could also be moved 360 km farther northward by using a new level of winter hardiness, now genetically available [1,35]. The projected global warming, if it occurs, especially during the winter months and high latitudes, could greatly speed up the process, or may not make necessary the introduction of new genetic material.

There are some places now too hot and humid where wheat cannot be grown but genetic material is now on hand, at the International Wheat and Maize Development Center in Mexico, and in China and elsewhere, to make that soon possible. It is projected that a doubling of CO_2 and projected climate change would increase production of the major food and fiber crops in North America [99-102], and likely holds for northern Europe, Russia, and the northern provinces and autonomous regions of China and several of the states of India, as well as the most southern countries of South America. Other crops, of more horticultural interest, such as potatoes, sweet corn, green beans, tomatoes, celery, cabbage, onions, head lettuce, broccoli, and strawberries are currently grown or could be at some season of the year in every state in the United States. Both the most southern and northern boundaries for successful cotton, corn, rice, peanut, sugar beets, sugarcane, and watermelons are being extended by the use of plastic film covers. This is occurring most successfully and extensively in Japan, Korea, and China. In any event, if there is to be a global warming within the magnitudes projected, this might be preferable to another ice age, also predicted about a decade

Table 1 Current Geographical Distributions of Food Crops and Possible Changes with Projected Greenhouse Effects

Food crops	Current geographical distribution	Possible changes with global warming
Rice	Lat. 40° S to lat. 45° N. and throughout the tropics	Both southern and northern boundaries extended
Wheat	Lat. 45° S to lat. 60° N but not in the lowland humid tropics	Both southern and northern boundaries extended for winter wheat, cannot be grown in the humid tropics
Maize	Lat. 42° S to lat. 60° N and throughout the tropics	Boundaries both north and south extended
Grain/sorghum and pearl millet	Lat. 40° S to lat. 45° N and throughout tropics	Production extended into areas that are now the drier wheat and corn fields
Barley, oats, and rye	In temperate zones from lat. 50° S to 60° N	Production extended both north and south with reductions in warm temperate zones
Soybeans and field beans	Lat. 40° S to 45° N and through tropics and semi-tropics	Production extended both north and south
Chickpeas, cowpeas, and pigeon peas	Lat. 30° S to lat. 35° N and semitropics	Production extended both north and south
Potatoes	Lat. 48° S to 60° N and irrigated semitropics	Production extended both north and south and in semitropics
Sugarcane	Lat. 30° N and lat. 30° S	Boundaries may be extended both to north and south
Sugar beets	Lat. 35° to 63° N 30° to 45° S also in India and China in some of the same locations as sugarcane	Boundaries extended north and south
Sweet potatoes	Lat. 35° S to 40° N and throughout humid tropics, none are too hot	Boundaries may extend north and south
Cassava, yams	Lat. 22° S to 25° N and humid tropics, none are too hot	Boundaries slightly extended north and south
Coconuts, plantains, and bananas	Lat. 20° S to lat. 25° N hot humid tropics	Boundaries slight extended north and south
Deciduous fruits, Grapes and small fruits	Temperate zones High elevations in tropics	Boundaries slightly extended north and south
Citrus, other tropical fruits	Tropics and semitropics	Boundaries slightly extended north and south

Table 1 (Continued)

Food crops	Current geographical distribution	Possible changes with global warming
Tender vegetables, tomatoes, peppers, eggplant, melons, cucumbers, squash, snap beans	Frost free periods 60–120 days	Boundaries may be slightly north and south extended
Hardy vegetables, cabbage, kale, carrots, cauliflower, parsnips, beets, asparagus, and onions	Temperate zones to lat. 50° S to 65° N	Production extended to more areas north and south

ago, especially with most of the world's food production now concentrated between lat. 30° and 50° N and S [24].

The future adaptability of agricultural crop production can also be indexed by already observed rapid rates of change. Hybrid maize production in Iowa increased from 5 to 95% of the total acreage between 1935 and 1940. The acreage of high-yielding wheat varieties in India went from nothing to 82% of the total in the decade between 1967 and 1977. Over 80% of the cultivated land devoted to rice in the Philippines is now planted to high-yielding varieties. Rice production in Indonesia doubled in a 5-year span from 1980 to 1985. Within the decade of the 1960s, hybrid corn became a major crop in northern Europe, and during the 1970s soybeans became a major crop in Brazil. Sunflowers in the Red River Valley of the northern United States, oil palm in Malaysia, and Canola (rape seed) oil production, are now exponentially expanding in Canada. Sunflowers, potatoes, sweet corn, onions, carrots, cabbage, and many other vegetable crops can be grown from Texas to Minnesota, even Alaska, and in practically every agriculturally developing country, either in the northern or southern hemispheres during some season of the year (Table 1). Successful asparagus production with high yields and superior quality is now operational from the tropics and semitropics in Taiwan and China to the countries of northern and western Europe.

One of the most important, perhaps the most important biosphere of biological productivity, that will be affected by both the direct biological and climatic effects of greenhouse warming and closely allied to agriculture is our forests. Little is known concerning either the direct biological (photosynthetic, water-use efficiency) or climatic effects. The long life span of trees, whether for food or forestry, makes their response to both climate and the direct effects of CO_2 very different from annual agricultural crops which constitute our basic food supply.

It is claimed by some that forest trees or forests are more climatically sensitive than agricultural crops because trees planted now will mature during a period of anticipated climate change [46]. This assertion is open to question [98]. The most weather sensitive portion of a tree's life is the first year, the establishmentarian phase.

Trees may live for hundreds of years. This suggests that they can and do withstand great climatic interannual fluctuations. The forester also has the option of harvesting a tree whenever it seems appropriate. Important trees, such as aspen, red maple, Douglas fir, and Ponderosa pine, over the latitudinal range are found from Canada to Mexico. With the model-projected changes in climate, however, forest succession would be modified in many parts of the world, probably more in the most northern forests of America, Europe and Asia, rather than in the midlatitudes and the tropics. The rate of succession, however, would be very uncertain and the time cycle probably centuries, not decades. Planted or managed forests would speed up the transition [98,103].

Making the Minnesota climate that of Texas, which would be an extreme rendition of even the most exaggerated predictive climate scenario, but already artistically portrayed by environmental activists and a responsive press, would not eliminate many crops. A warming, as projected, might encourage successful commercial production of mangos, papayas, lichee, passion fruit, bananas, and pineapple in what are now our most southern states. Tropical and subtropical fruits could become more important, both imported and those grown domestically.

Although neither the severity nor reality is known, a warming trend of some magnitude could occur coincident with increased variability of precipitation. These changes could exceed past interannual variations, especially in temperate zones. The unknowns also outweigh the knowns. CO_2 models, however, have not been validated by observational data. There is no evidence that when and if the climate warms, the variability of climate will increase [98].

Nevertheless, a wise option would be a prudent course of preparing for the worst. This would mean an accelerated effort in energy conservation, designing energy sources other than fossil fuels, reforestation, tree plantings, and the promulgation of conservation tillage. Such actions would be desirable resource conservation efforts and should be vigorously pursued, independent of any greenhouse warming.

REGIONAL PROJECTIONS: CLIMATIC PATTERNS AND CROP RESPONSES

There is no controversy, however, over the fact that atmospheric carbon dioxide has increased by more than 25% since 1850 because of fossil fuel combustion and changes in land use (mostly deforestation). Levels of other so-called trace greenhouse gases, such as methane, dinitrogen oxides, and chlorofluorocarbons have also increased by even larger percentages. It is claimed by some that the combined effects of trace greenhouse gases will equal that of CO_2. There is, however, great controversy as to whether a global warming's first signal has already been detected [13,16,30]. There is a similar controversy, already referred to, as to the magnitude of such a global warming, if it occurs, during the next century or thereafter. Estimates and projections range from a negative change to 1.5° or 6°C or even higher [6, 10,15,16,24,30,39,40,42,44,46,102,103]. Forecasts of regional climatic change with distribution of variables, such as cloud covers, soil moisture, or precipitation patterns have even greater uncertainties [44,98]. It is within such a backdrop of uncertainties that we review some of the reports that relate to implications of the greenhouse effect on crop productivity. Some of the possible impacts of climate change on agriculture are summarized in two recent volumes initiated by the International Institute for Applied Systems Analysis. The first is entitled, "Assessments in Cool Temperate and Cold Regions" [104]; the second "Assessments in Semi-arid Regions" [105]. Waggoner has summarized some of the potential effects on the United States, and the impacts on western and other water reserves of a climate change [106]. These were a follow-up of earlier papers by the author [1,48,107] and succeeded by an additional review [47] and other projections [35,108–110]. Meanwhile, comprehensive reports have appeared on CO_2 climate change in U.S. agriculture [111–113].

These volumes, along with others, emphasize that there will be both negative and positive effects from a CO_2-induced climate change, and that the direct effects of more CO_2 in the atmosphere are, for most part, beneficial. More CO_2 in the air will cause leaves, as a result of the narrowing of stomata, to assimilate more carbon and lose less water. But if there is less rainfall in the American grain belt, as some very uncertain models suggest, the net or integration of both climatic and biological factors on yields are not known [97]. For overall effects on plants of the global changes in atmosphere CO_2 and in climate foreseen for the year 2000 and beyond, some will be positive and some will be negative. The important message, however, is that the effects, both the direct, climatic, and those pertaining to water resources will be manageable [106,114]. Actions by farmers and by scientists can be done or taken bit by bit and they extend present policies rather than require changes in the way we live [114].

While regional-scale climate changes have not been modeled with any degree of confidence, there are many press reports and those circulated in slick magazines, newspaper headlines, and Sunday supplements with speculative projections as to shifts and changes in crop productivity during the next 50 to 100 years. Irrespective of such popularized reports, we can suggest the following: Cereal crop production in Europe will not be affected as significantly as elsewhere. Cold marginal regions, such as Canada, Alaska, Iceland, Scandinavia, New Zealand, Argentina, Russia, and north China should benefit considerably from higher crop yields associated with higher temperatures and longer growing seasons [115]. The direct effects of CO_2 enrichment should increase yields and tolerance to drought stress, although the effects may vary considerably according to plant species. A figure of 15 to 32% for yield increases has been suggested [51], but such estimates are still very speculative and hypothetical.

It is reasonable that a warmer climate will increase the length of the growing season, the growing degree units, and frost free periods in both the northern and southern hemispheres. This could be significant for most horticultural crops whether they be flowers, fruits, vegetables, or ornamentals. The most pronounced effects would be near the poles, and specifically the North Pole. The risk of freezing temperatures in Texas, Florida, and California and other western and southern states in the United States and in parts of other countries in both northern and southern hemispheres should be reduced under new climatic regimes that might be induced by higher levels of atmospheric CO_2. The production of winter wheat should be extended further north and south in the two hemispheres. If there is to be a warmer and drier climate in the U.S. grain belt, sorghum and millet could become more important crops in the west and south. It has been suggested that the U.S. corn belt might shift 175 km northeasterly for each 1°C rise in temperature [116], and wheat production would shift eastward [35].

There are many more speculative projections as to the greenhouse effects on crop productivity. For the United States, farmers in Minnesota might see yields of corn and soybeans doubled, while in other more southern areas of the corn belt, production may not be possible. Michigan, New York, and Pennsylvania, and southern Ontario could become even more important than they now are in fruit and vegetable production with more serious consideration of irrigation. Crop production in the southern states could shift to more citrus. Yields of major crops in the great plains could drop if irrigation water becomes limiting. Similar projections within nations could be made for China, India, Russia, northern Europe, Spain, Brazil, Argentina, Chile, South America, Australia, and New Zealand. A northern migration of agriculture in the northern hemisphere or a southern migration in the southern hemisphere would increase the use of irrigation and fertilizers on sandy soils, which may create or worsen ground water problems. Higher crop yields would require greater amounts of fertilizer and water. This was demonstrated over 25 years ago for vegetables grown in greenhouses and exposed to elevated atmospheric levels of CO_2 [48]. There may be significant effects on U.S. agricultural trade, especially with Russia and China. Global warming would help Russia boost wheat production by perhaps as much as 50% if a climate much like that of southern Canada were to occur. Crop production in China should be greatly enhanced, providing additional water was available, with the northern migration of soybeans, winter wheat, rice, corn, and cotton. Finally, assuming the climate models have a semblance of accuracy, there should be little change in the tropics. Thus the agricultural crop productivity impacts on most developing countries in Central and South America, Africa, India, Southeast Asia, Indonesia, and the islands of the Pacific should be minimal. Some regions and crops are climatically more vulnerable than others. North America is strategically critical to the stability of world food supplies. However, resources for crop production are usually most critical in agriculturally developing countries [117].

Projected CO_2 induced climate changes have caused much speculation, and some research has been initiated on crop–pest relationships and changing strategies relating to crop protection. Weeds, insects, nematodes, and diseases inflict substantial losses on crops grown for food and in forests and on range lands. A possible climate change, accompanying elevated atmospheric carbon dioxide levels, will impact plant–crop–pest relationships for agricultural and other ecosystems as well as have direct biological inputs [118–121]. Plants will accumulate more carbohydrates and grow faster with elevated levels of carbon dioxide. Nutritional levels of carbohydrates and nitrogen will change, along with feeding habits of insects. This has been demonstrated by a few isolated studies. A major factor in global warming for temperate zone agriculture, could be greater survival, through overwintering and persistence of plant diseases and insects.

Pest control is or should be entering a new era with integrated management. Pests will change with the weather, and new actors will also enter [37, 110,120]. Of special interest and concern will be crop–weed interactions.

Of all crop pests, weeds are the most damaging. The current increases in atmospheric CO_2 concentrations and projected warming will affect the growth and productivity of crops and associated weeds. Weeds compete directly with crops for water, sunlight, space, essential nutrients, and atmospheric CO_2. Estimated losses combined with the costs of weed control in the United States alone, annually, exceed $20 billion. Of the 20 most important food crops, 16 have a C_3 photosynthetic pathway. C_3 plants include rice, wheat, barley, oats, rye, soybeans, field beans, mung beans, cowpeas, chickpeas, pigeon peas, potatoes, sweet potatoes, cassava-yams, sugar beets, bananas-plantains, and coconuts. Most all fruits and vegetables and all forest tree species are C_3 plants; the exceptions are corn, sorghum, millet, sugar-cane, and some tropical grasses, which have C_4 pathways. On the other hand, of the world's 18 most noxious weeds, 14 (primarily the grasses) have the C_4 photosynthetic pathway. Conversely, for the few C_4 plants that are im-portant food crops (corn, sorghum, millet, sugarcane), many of their major weeds are C_3 plants. In fact 19 of the 38 major weeds of corn in the U.S. are C_3 plants.

   C_3 and C_4 plants respond differently to elevated levels of atmospheric CO_2, the greater response being with the C_3. It could be good fortune, with some major exceptions, that rising levels of atmospheric CO_2 will generally favor food crop production over that of weed growth [37]. This could be of special benefit for horticultural crops and forest tree species, almost all of which are C_3 plants. For the moment, we do not know the effects of a pro-jected warming and a higher CO_2 on the productivity of C_3. C_4, and Crassu-lacean Acid Metabolism (CAM) plants (most commonly represented by the pineapple) under real world conditions of crops and weeds, limited water, adverse temperatures, air pollution, restricted sunlight or limited soil nu-trients [35,47,51,118,119].

CONCLUSIONS

We know for certain that the carbon dioxide levels are rising globally. The increase has been an overall near 30% with the advent of industrialization. In the last 75 years, there has been an approximate 23% increase, going from about 290 up to the current level of 360 ppm. The rate of increase is now 2 ppm/year. We are also reasonably sure that one of the prime causes of the rise is the release of CO_2 from the combustion of fossil fuels. The pro-gressively greater destruction of tropical and other forests are also a con-tributing factor as is the cultivation of land in crop production, resulting in the irreversible oxidation of soil organic matter.

   Currently, there is a widely prevailing perception that the threat from a greenhouse warming of the globe is growing progressively greater and it will become more costly and inconvenient to stabilize the future climate re-sources of the earth [6,7,8,15,31,122]. Counter to this concept is that most CO_2 emission reduction measures would save money, protect the environ-ment, conserve the natural resources of land, water, and energy, and im-prove the quality of life [1,46,50,53,55,60,70,71,99,114,123].

Optimal Levels of Atmospheric CO_2 and Other Greenhouse Gases

At this moment no one can say, or really knows, the earth's optimal CO_2 and other greenhouse gas levels for the most favorable total or accumulative effects on the environment, on crop productivity, on natural resources, on total biological productivity, or on human health or society. I have personally witnessed a rise of atmospheric CO_2 from about 290 to near 360 ppm during my lifetime. Within this three-quarters of a century, there has been no verifiable climate change, either for hot or for cold; for dry or wet, nor any catastrophic climatic shifts with major impacts on crop productivity. The summer drought and heat wave of 1988 in the United States was not caused by an induced greenhouse effect but a shift in the jet stream. By contrast, the year 1992 for the corn belt of the United States will be labeled as one of the coolest and wettest, when total agricultural production reached an all-time record. Meanwhile, it is difficult to accept the suppositions made repeatedly in the U.S. press, in scientific and quasi-scientific reports, in testimonies before congressional committees, and conclusions and recommendations emanating from national and international conferences and symposia, that global temperatures may go up by 4.5°C (8°F) in an additional 75 years or less, with major crop production dislocations. However, innovative, but perhaps well-meaning scientists have learned that frightening the public gets results.

We have been on a treadmill of ecological gloom ever since Earth Day, over 20 years ago. We have been hit with one doomsday prediction after another: It was said that Lake Erie was "dead" or dying. They said: DDT was killing all ocean life. The population "bomb" was set to explode and worldwide famine was just around the corner. Now it's the greenhouse effect with resultant rising sea levels, sunken cities, and destruction of the ozone layer [6,7,8,15,31].

If there are to be climatic changes as a result of the greenhouse effect, there will be impacts on crop productivity. But here, there will be both winners as well as losers. Higher CO_2 levels, such as a doubling accompanied by warmer weather would also have beneficial effects. There should also be milder winters, longer growing seasons, and more growing degree units. Above all, there would be increased efficiency of photosynthesis and greater water use efficiency in plants and food crops. The additive effects could well increase yields of crops by over 10% and could extend the boundaries of crop production, especially the major food and horticultural crops, now limited by insufficient moisture and cold temperatures.

Some warming appears inevitable, although it could, based on real world data, be much less than the still crude computer models suggest and a handful of overzealous prominent scientists and environmentalist promulgated in the hothouse atmosphere of a 1988 summer heat wave, which quickly flowered luxuriantly into an accepted fact [124].

Climate Modeling

There is still much debate about the credibility and accuracy of the projections for global warming. Examples of critical looks being taken and their scientific underpinnings, have recently emerged. One such report, "Scientific Perspectives on the Greenhouse Problem," from the George C. Marshall Institute, Washington, DC [41], shows that temperature and solar activity have followed a parallel course. Solar variability may be a significant part of the explanation for the post-1880 0.5°C rise in temperature, and the greenhouse effect projected for the 21st century will also be relatively small, perhaps 1°C rather than 4.5°C or higher. Any greenhouse warming of such a magnitude could be partially offset by the natural cooling expected in the 21st century, and widely projected only a decade ago, and by increased cloud cover. Such a balance of man-made and natural forces of climate change would not be catastrophic, being only a degree or less of the present temperature.

Climate modeling is intellectually stimulating but very flawed. This is not unknown to those working on models. People, however, dealing with advisory and policy issues and the media will often take the worst scenarios and magnify them in Sunday supplements, news headlines, and the contents of slick magazines. Stories become much more interesting if they portray that the mid-Western United States will turn into a desert, that palm trees will be growing in Minnesota, and that coastal cities will sink with a rise in sea levels, than if they speculate about the effects of a degree increase or decrease in temperature. One important fact remains, the CO_2 models for greenhouse warming have not been validated against observational data in the real world. Until now there seems to have been very little interest in doing so. In fact, Idso [39] has presented evidence that climate models may have overestimated global surface temperature increases from greenhouse warming by a full order of magnitude.

There are, however, some important facts to remember. The effects of climate change on food and agriculture are the only readily identifiable global impacts of significant magnitude on future living standards. Virtually all agriculture everywhere is outdoors, although there are some horticultural exceptions. Agriculture depends on sun and rain or irrigation. It is sensitive to temperature. It is subject to both the beneficial and harmful activities of insects, diseases, microorganisms and weeds, all of which in turn are affected by weather and climate [109,110].

Irrespective of agriculture's substantial dependence on weather and climate for the future, it is not yet possible, even with today's crops, technologies, and distribution of agriculturally diverse activities over the earth, to assess the aggregate impact of projected climate change, nor even to be sure of the arithmetic sign, whether it be plus or minus. There is a prevailing presumption among some that any climate change, independent of what the change is, has a disadvantageous expectation. There is also the rightful presumption that the direct effect of rising levels of atmospheric CO_2 could have a globally positive affect on both photosynthetic and water use efficiency and thus total crop and biological productivity. Results of research assembled over the past decade now provide good information on the CO_2 direct (biological) effects, and there is a rich agrometerological literature available for people who want to construct better climate models.

Agricultural scientists have been noticeably silent concerning the "greenhouse effect." We ought to join the chorus for increased research on sources of energy alternative to fossil fuels, encourage tree plantings, reforestation, and conservation tillage and above all, support basic and applied research on greate resilience of crops to climatic and other environmental hazards and competing biological systems [108,125], including integrated pest management. A tremendous opportunity exists for all the basic food crops, horticultural crops, and forest tree species. Increased support of research in these areas was among the top priorities recommended in 1977 in the World Food and Nutrition Study [125] and at the International Conference on Crop Productivity Research Imperatives at Boyne Highlands, Michigan in 1975 [126] and again in 1985 [127]. The recommendations prepared by these studies remain, as yet, to be implemented. Support is also needed for research on the direct effects of rising levels of atmospheric CO_2 and other greenhouse gases on crop productivity which has received minimal effort, compared to that going into climate research and computer modeling. Climate research, among agricultural scientists should not be neglected.

Actions to Be Taken

We ought to proceed with the following actions immediately, independent of scare tactics of a greenhouse warming.

First, I refer to energy conservation and the development of energy resources, independent or alternative to the use of fossil fuels. We need to reinstate the energy conservation tactics of a decade ago—more efficient gasoline engines and improved mileage for automobiles, tractors, trucks, and other farm machinery, the development of engines with greater fuel economy; a renewed research program for improved utilization of solar, wind, hydropower, and atomic energy; biological nitrogen fixation as an alternative to chemical fixation (which now requires large infusions of fossil energy); and finally, integrated pest management to reduce the use of chemical pesticides. Such measures would not only greatly temper the generation of atmospheric CO_2 but would reduce crop production costs, result in environment improvements, preserve nonrenewable resources for future generations, and initiate more sustainable alternative agricultural production systems.

Secondly, there should be a massive global program for reforestation and tree planting and a ban on further destruction of tropical forests in all agriculturally developing countries. It is estimated that tropical rain forests are disappearing at the rate of 8 million hectares per year [15]. The exploitation of high-value species, such as mahogany, is proceeding at an accelerated rate in Central America. Reforestation would also tend to stabilize the levels of atmospheric CO_2, and add to, rather than diminish one of the great natural resources of the earth, now vital for firewood, timber, logs, buildings, furniture, soil stabilization, control of soil erosion and sedimentation, and as a food source (agroforestry). Again, reforestation with high-value species should proceed as a world-wide effort, independent of the frightening shadow of a global greenhouse effect. Currently, there is a striking positive correlation between the rate of deforestation and national foreign debts in most all agriculturally developing countries.

Third, the presumed CO_2 induced climate change could also be partially averted or at least delayed, by soil conservation, conservation tillage, and other tillage practices, which would reduce the oxidation of organic matter, practically eliminate soil erosion, and make cropping and plant harvest an annual event in the tropics. I refer to a combination of conservation tillage and alley cropping.

Finally, I refer to the urgency for research support for stabilizing crop production through greater resistance to biological and environmental stresses, such as pests, weather aberrations, short-term droughts, temperature extremes, aluminum toxicity, and related nutrient deficiencies of acid soils and those of high salinity. Climatic stresses frequent almost all crops in all countries [125]. Making or developing plants (crops) more resistant to environmental stresses would have worldwide application and interest to meet the exigencies of the ever-present interannual climate variations. Such an initiative would be of benefit in crop production independent of a projected greenhouse warming. It would buffer short-term effects of variability and combat long-term climate change. It would be a step toward the catch phrase "LISA," or low-input sustainable agriculture on a global scale.

In conclusion, the overall positive benefits and negative outputs—climatic, biological, social, economic, and political—of the rising level of atmospheric CO_2 and other gases, still need resolution. So far as the overall agricultural outlook and that of total biological productivity is concerned, a CO_2-induced climate change will not be catastrophic [1,2,4,5,11,24,35,37, 44-47,49,50,57,90,98,100,111,112,114,124].

REFERENCES

1. S. H. Wittwer, *J. Soil and Water Conservation, 35*: 116–120 (1980).
2. E. R. Lemon, *CO₂ and Plants*, Westview Press, Boulder, CO (1983).
3. S. H. Wittwer, *HortSci., 25*: 1560–1567 (1990).
4. S. H. Wittwer, *Policy Rev., 62*: 4–9 (1992).
5. W. T. Brooks, *Forbes*: 144: 96–102 (1989).
6. C. Flavin, *Slowing Global Warming: A Worldwide Strategy*, Worldwatch Paper 91, Washington, DC (1989).
7. F. Lyman, *The Greenhouse Trap*, Beacon Press, Boston, Massachusetts (1990).
8. B. McKribben, *The End of Nature*, Random House, New York (1989).
9. R. Revelle, *Scient. Amer., 247*: 35–43 (1982).
10. S. H. Schneider, *Global Warming: Are We Entering the Greenhouse Century?* Sierra Club Books, San Francisco, California (1989).
11. S. H. Wittwer, *The Greenhouse Effect*, Carolina Biol. Supply Co., Burlington, North Carolina (1988).
12. UNEP, *An Assessment of the Role of Carbon Dioxide and Other Greenhouse Gases in Climate Variations and Associated Impacts*, United Nations Environmental Program/World Meteorological Organization/International Council of Scientific Unions. The Villach Conference. WMO, Geneva (1985).
13. M. C. MacCracken and F. M. Luther (eds.), *Detecting the Climatic Effects of Increasing Carbon Dioxide*, Carbon Dioxide Res. Div., U.S. Dept. Energy, DOE/ER-0235, Washington, DC (1985).
14. M. C. MacCracken and F. M. Luther (eds.), *The Potential Climatic Effects of Increasing Carbon Dioxide*, Carbon Dioxide Res. Div., U.S. Dept. Energy, DOE/ER-0237, Washington, DC (1985).
15. World Resources, *Climate Change: A Global Concern*, World Resources Institute, Washington, DC, pp. 11–31 (1990).
16. R. A. Kerr, *Science, 244*: 1041–1043 (1989).
17. S. H. Schneider, *Climate Change, 13*: 113–115 (1989).
18. K. E. Trenberth, G. W. Branstator, and P. A. Arkin, *Science, 242*: 1640–1645 (1988).
19. P. J. Crutzen and M. O. Andreae, *Science, 250*: 1669–1675 (1990).
20. R. C. Balling Jr., *The Heated Debate*, Pacific Institute for Public Policy. San Francisco, California (1992).
21. H. W. Ellsaesser, *Atmosfera, 3*: 3–29 (1990).
22. R. S. Lindzen, *Bul. Amer. Meteor. Soc., 71*: 288–299 (1990).
23. N. D. Strommen, *National Geographic Research and Exploration*, 8: 10–21. Washington, DC (1992).
24. V. Smil, *Curr. Hist., 88*: 9–12, 46–48 (1989).
25. *Global Climate Change: A New Vision for the 1990's.* Laboratory of Climatology, Arizona State University, Tempe, Arizona (1990).

26. W. C. Clark (ed.), *Carbon Dioxide Review*, Oxford University Press, New York (1982).
27. J. E. Hansen, D. Johnson, A. Lacis, S. Lebedeff, P. Lee, D. Rind, and G. Russell, *Science, 213*: 957–966 (1981).
28. S. Manabe and R. Stouffer, *Nature, 282*: 491–493 (1979).
29. S. Manabe and R. T. Wetherald, *J. Atmos. Sci., 37*: 99–118 (1980).
30. S. Manabe, R. T. Wetherald, and R. J. Stouffer, *Climate Change, 3*: 347–386 (1981).
31. M. Shepard, *EPRI J., 11*: 5–15 (1986).
32. WMO, *Report of the WMO (CAS) Meeting of the Experts on the CO_2 Concentrations from Pre-Industrial Times to I.G.Y.*, WCP 53, World Meteorological Organization, Geneva, Switzerland (1983).
33. S. H. Schneider, *Climate Change, 13*: 113–115 (1989).
34. N. E. Landsberg, *The Resourceful Earth*, Blackwell, New York, pp. 272–315 (1984).
35. W. L. Decker, V. Jones, and R. Achutuni, *Characterization of Information Requirements for Studies of CO_2 Effects: Water Resources, Agriculture, Fisheries, Forests and Human Health* (M. R. White, ed.), National Technical Information Service, U.S. Dept. Commerce, Springfield, Virginia, pp. 69–93 (1985).
36. C. Rosenzweig, *Climate Change, 7*: 367–389 (1985).
37. P. E. Waggoner, *Changing Climate*, National Academy Press, Washington, DC, pp. 383–418 (1983).
38. Council for Agricultural Science and Technology, *Long-Term Viability of U.S. Agriculture*, CAST Report No. 114, Ames, Iowa (1988).
39. S. B. Idso, *Carbon Dioxide and Global Change: Earth in Transition*, IBR Press, Tempe, Arizona (1989).
40. D. A. Lashof, *Climate Change, 14*: 213–242 (1989).
41. F. Seitz, R. Jastrow, and W. A. Niernberg, *Scientific Perspectives on the Greenhouse Problem*, George C. Marshall Institute, Washington, DC (1989).
42. A. B. Abelson, *Science, 243*: 461 (1989).
43. R. W. Spencer and J. R. Christy, *Science, 247*: 1558 (1990).
44. P. D. Jones and T. M. L. Wigley, *Scient. Amer., 263*: 84–91 (1990).
45. F. K. Hare, *Climate Impact Assessment*, SCOPE 27 (R. W. Kates, J. H. Ausubel, and M. Berberian, eds.), John Wiley, New York, pp. 37–68 (1985).
46. V. W. Ruttan (ed.), *Resource and Environmental Constraints on Sustainable Growth in Agricultural Production: Report on a Dialogue.* Staff paper Dept. Agriculture and Applied Economy, University of Minnesota, St. Paul, pp. 90–33 (1990).
47. S. H. Wittwer, *Crit. Rev. Plant Sci., 2*: 171–198 (1985).
48. S. H. Wittwer and W. Robb, *Econ. Bot., 18*: 34–56 (1964).
49. B. R. Strain and J. D. Cure, *Direct Effects of Atmospheric CO_2 Enrichment on Plants and Ecosystems: A Bibliography with Abstracts*, National Technical Information Service, U.S. Dept. Commerce, Springfield, Virginia (1986).
50. B. A. Kimball, (*WCL Report 14*). U.S. Water Conservation Laboratory, Phoenix, AZ (1985).

51. B. A. Kimball, *Direct Effects of Increasing Carbon Dioxide on Vegetation* (B. A. Strain and J. D. Cure, eds.), National Technical Information Service, U.S. Dept. Commerce, Springfield, Virginia, pp. 185–204 (1985).
52. B. A. Kimball, *Physiology, Yield and Economics*, Vol. II. CRC Press, Boca Raton, Florida, 105–115 (1968).
53. B. Acock and L. H. Allen, Jr., *Direct Effects of Increasing Carbon Dioxide on Vegetation* (B. R. Strain and J. D. Cure, eds.), National Technical Information Service, U.S. Dept. Commerce, Springfield, Virginia, pp. 53–97 (1985).
54. L. H. Allen, *Modification of the Aerial Environment of Crops* (B. J. Barfield and J. F. Gerber, eds.), American Society Agricultural Engineering, St. Joseph, MI, pp. 500–519 (1989).
55. J. H. Allen Jr., K. J. Boote, J. W. Jones, P. H. Jones, R. R. Valle, B. Acock, H. H. Rogers, and R. C. Dahlman, *Response of Vegetation to Carbon Dioxide*, U.S. Dept. Agriculture and U.S. Dept. Energy, Washington, DC (1987).
56. L. H. Allen Jr., *J. Environ. Qual., 19*: 15–34 (1990).
57. J. M. Callaway and J. W. Currie, *Characterization of Information Requirements for Studies of CO_2 Effects*: *Water Resources, Agriculture, Fisheries, Forests and Human Health* (M. R. White, ed.), National Technical Information Service, U.S. Dept. Commerce, Springfield, Virginia, pp. 23–67 (1985).
58. R. W. Carlson and F. A. Bazzaz, *Environmental and Climatic Impact of Coal Utilization* (J. Singh and A. Deepak, eds.), Academic, New York, pp. 609–623 (1980).
59. R. W. Carlson and F. A. Bazzar, *Oceologia, 59*: 50–54 (1982).
60. J. T. Baker, L. H. Allen, Jr., K. T. Boote, P. Jones, and J. W. Jones, *Crop Science, 29*: 98–105 (1989).
61. J. D. Cure, *Direct Effects of Increasing Carbon Dioxide on Vegetation* (B. R. Strain and J. D. Cure, eds.), National Technical Information Service, U.S. Dept. Commerce, Springfield, Virginia, pp. 99–116 (1985).
62. B. A. Kimball and S. T. Mitchell, *J. Amer. Soc. Hort. Sci., 104*: 515–520 (1979).
63. B. R. Strain and F. A. Bazzaz, *CO_2 and Plants* (E. R. Lemon, ed.), Westview, Boulder, Colorado, pp. 177–222 (1983).
64. W. J. S. Downton, O. Bjorkman, and C. Pike, *Carbon Dioxide and Climate*: *Australian Research* (G. I. Pearman, ed.), Australian Academy of Science, Canberra, Australia (1981).
65. R. W. Pearcy and O. Bjorkman, *CO_2 and Plants* (E. R. Lemon, ed.), Westview, Boulder, Colorado, pp. 65–105 (1983).
66. N. E. Tolbert and I. Zelitch, *CO_2 and Plants* (E. R. Lemon, ed.), Westview, Boulder, Colorado, pp. 21–64 (1983).
67. M. R. Lamborg, R. W. F. Hardy, and E. A. Paul, *CO_2 and Plants* (E. R. Lemon, ed.), Westview, Boulder, Colorado, pp. 131–176 (1983).
68. D. N. Baker and H. Z. Enoch, *CO_2 and Plants* (E. R. Lemon, ed.), Westview, Boulder, Colorado, pp. 107–130 (1983).
69. M. R. White, *Characterization of Information Requirements for Studies of CO_2 Effects*: *Water Resources, Agriculture, Fisheries, Forests and Human Health*, U.S. Dept. Energy, DOE/ER-0236, Washington, DC (1985).

70. H. Z. Enoch and B. A. Kimball (eds.), *Carbon Dioxide Enrichment of Greenhouse Crops. Vol. 1. Status and CO₂ Sources*. CRC Press, Boca Raton, Florida (1986).

71. H. Z. Enoch and B. A. Kimball (eds.), *Carbon Dioxide Enrichment of Greenhouse Crops. Vol. 11. Physiology, Yield, and Economics*. CRC Press, Boca Raton, Florida (1986).

72. I. M. Mortensen (chairman), International Society Horticultural Science, Agricultural University of Norway, *Aas, CO₂ N/Letter*, Norway, Sweden (1986).

73. C. B. Osmond, O. Bjorkman, and D. J. Anderson, *Econological Studies, 36*, Springer Verlag, New York, pp. 419–425 (1980).

74. H. H. Rogers, G. E. Bingham, J. E. Cure, W. W. Heck, A. S. Heagle, D. W. Israel, J. M. Smith, K. A. Surano, and J. F. Thomas, *Report No. 001*, U.S. Dept. Energy and Dept. Agriculture, Washington, DC (1980).

75. H. H. Rogers, N. Sionit, J. D. Cure, J. M. Smith, and G. E. Bingham, *Plant Physiol., 74*: 233–238 (1984).

76. H. H. Rogers, J. F. Thomas, and G. E. Bingham, *Science, 220*: 428–429 (1983).

77. P. R. Hicklenton, *CO₂ Enrichment in the Greenhouse*, Timber Press, Portland, Oregon, Growers Handbook Series, Vol. 2, 58 pp (1988).

78. S. H. Wittwer, *Carbon Dioxide Enrichment of Greenhouse Crops. Vol. I. Status and CO₂ Sources*. CRC Press, Boca Raton, Florida (1986).

79. R. N. Arteca, B. W. Poovaiah, and O. E. Smith, *Science, 205*: 1279–1280 (1979).

80. N. C. Bhattacharya, P. K. Biswas, S. Bhattacharya, N. Sionit, and B. R. Strain, *Crop Sci., 25*: 975–981 (1985).

81. G. A. Finn and W. A. Brun, *Plant Physiol., 69*: 327–331 (1982).

82. R. S. F. Hardy and U. D. Havelka, Contributions to the scientific literature, Section IV, *Biology*, E.I. Dupont de Nemours, Wilmington, Delaware (1975).

83. D. A. Phillips, K. D. Newell, S. A. Hassall, and C. E. Filling, *Amer. J. Bot., 63*: 356–362 (1976).

84. H. H. Rogers, C. M. Peterson, J. N. McCrismmon, and J. D. Cure, *Plant & Cell Environ., 15*: 749–752 (1992).

85. B. R. Strain, *New Biolog., 4*: 87–89 (1992).

86. CDIC Communications, *Carbon Dioxide Information Center*, Oak Ridge National Laboratory, Oak Ridge, Tennessee (1985).

87. W. E. Riebsame, *Climate Change, 13*: 69–97 (1988).

88. U. N. Chauduri, R. B. Burnett, M. B. Kirkham, and E. T. Kanemasu, *Agr. For. Met., 37*: 109–122 (1986).

89. P. Jones, J. W. Jones, and L. H. Allen, Jr., *Trans. ASAE, 28*: 2021–2028 (1985).

90. B. A. Kimball, R. R. Mauney, J. W. Radin, F. S. Nakayama, S. B. Idso, D. L. Hendrix, D. H. Akey, S. G. Allen, M. G. Anderson, and W. Hartung, *Response of Vegetation to Carbon Dioxide*, Ser. 039. U.S. Dept. of Energy and the USDA U.S. Water Conservation Laboratory and Western Region Cotton Res. Laboratory, Phoenix, Arizona (1986).

91. A. M. Salomon and D. C. West, *Characterization of Information Requirements for Studies of CO₂ Effects: Water Resources, Agriculture, Fisheries, Forests and Human Health* (M. R. White, ed.), National Technical Information Service, U.S. Dept. Commerce, Springfield, Virginia, pp. 145–169 (1985).

92. N. Sionit, H. Hellmers, and B. R. Strain, *Crop. Sci., 20*: 687–690 (1980).
93. N. Sionit and D. T. Patterson, *Crop Sci., 25*: 533–537 (1985).
94. R. M. Gifford, *Austral. J. Plant Physiol., 6*: 367–378 (1979).
95. R. M. Gifford, *Search, 10*: 316–318 (1979).
96. U. D. Havelka, V. A. Wittenbach, and M. G. Boyle, *Crop Sci., 24*: 146–150 (1984).
97. P. Martin, N. J. Rosenberg, and M. S. McKenney, *Climate Change, 14*: 117–151 (1979).
98. W. F. Reifsnyder, *Agricultural and Forestry Meteorology, 47*:349–371 (1989).
99. C. Rosenzweig, *Climate Change, 7*: 367–389 (1985).
100. C. Rosenzweig, *Env. Prot. Agency J., 15*: 9–10 (1990).
101. B. Smit, L. Ludlow, and M. Brklacich, *J. Env. Quality, 17*: 519–527 (1988).
102. D. S. Wilks, *Climate Change, 13*: 19–42 (1988).
103. C. E. Cooper, *Carbon Dioxide Review* (W. C. Clark, ed.), Oxford University Press, New York, pp. 299–333 (1982).
104. M. L. Parry, T. R. Carter, and N. T. Konijn (eds.), *The Impact of Climatic Variations on Agriculture, vol. 1: Assessments in Cool Temperate and Cold Regions*, Kluwer Academic, Boston (1988).
105. M. L. Parry, T. R. Carter, and N. T. Konijn (eds.), *The Impact of Climatic Variations on Agriculture, Vol. 2: Assessments in Semi-arid Regions*, Kluwer Academic, Boston, Massachusetts (1988).
106. P. E. Waggoner (Ed.), *Climate and Water*, John Wiley, New York (1990).
107. S. H. Wittwer, *Carbon Dioxide Review* (W. D. Clark, ed.), Oxford University Press, New York, pp. 320–324 (1982).
108. National Academy of Sciences, A Report of the Committee on Climate and Weather Fluctuations and Agricultural Production, *Climate and Food*, Board on Agriculture and Renewable Resources, Commission on Natural Resources Council, Washington, DC (1977).
109. National Academy of Sciences, *Carbon Dioxide and Climate: A Scientific Assessment*, National Resources Council, Washington, DC (1979).
110. National Academy of Sciences, *Policy Implications of Greenhouse Warming*, National Academy Press, Washington, DC (1992).
111. R. M. Adams, C. Rosenzweig, R. M. Post, J. T. Ritchie, B. A. Carl, J. P. Glyer, R. B. Curry, J. M. Jones, K. J. Boote, and E. H. Allen, Jr., *Nature, 245*: 219–224 (1990).
112. L. H. Allen, Jr., R. M. Peart, J. W. Jones, R. B. Curry, and L. K. Boote, *Preparing for Climate Change* (J. C. Topping, ed.), The Climate Institute, Washington, DC, p. 186–191 (1989).
113. Council for Agricultural and Science Technology, *Preparing U.S. Agriculture for Global Climate Change*, Ames, Iowa (1992).
114. P. H. Abelson, *Science, 247*: 1529 (1990).
115. B. Smit, M. Brklacich, R. B. Stewart, R. McBride, M. Brown, and D. Bond, *Climate Change, 14*: 153–174 (1989).
116. J. E. Newman, *Biometeorology, 7*: 129–142 (1980).
117. P. A. Oram, *Climate Change, 7*: 129–152 (1985).
118. D. T. Patterson and E. P. Flint, *Weed Sci., 28*: 71–75 (1980).

119. D. T. Patterson and E. P. Flint, *Weed Sci., 30*: 389–394 (1982).
120. F. A. Bazzaz and E. D. Fajer, *Scient. Amer., 266*: 68–74 (1992).
121. H. W. Polley, H. B. Johnson, B. D. Marino, and H. S. Mayeux, *Nature, 361*: 61–64 (1993).
122. J. B. Smith and D. A. Tirpak (ed.), *The Potential Effects of Global Climate Change on the United States*. U.S. Environmental Protection Agency, Washington, DC (1988).
123. *Rocky Mountain Institute, 6*: 1–3 Old Snowmass, Colorado (1990).
124. J. R. Lang, *Barrons, 60*: 6–7,20,22 (1989).
125. National Academy of Sciences, *World Food and Nutrition Study, The Potential Contributions of Research*. A report of the steering committee NRC study on world food and nutrition of the Commission on International Relations, National Resources Council, Washington, DC (1977).
126. A. W. A. Brown, T. C. Byerly, M. Gibbs, and A. San Pietro (eds.), Crop Productivity-Research Imperatives, An International Conference. Michigan Agricultural Experimentation Station, E. Lansing, MI, and Charles F. Kettering Foundation, Yellow Springs, Ohio (1975).
127. M. Gibbs and C. Carlson (eds.), *Crop Productivity-Research Imperatives Revisited*, An international conference. Michigan Agricultural Experimentation Station, E. Lansing, Michigan, and Charles F. Kettering Foundation, Yellow Springs, Ohio (1985).

8

Cell and Tissue Culture for Plant Improvement

D. S. Brar and Gurdev S. Khush

International Rice Research Institute
Manila, Philippines

INTRODUCTION

Aseptic culture of cells, tissues, and organs in defined nutrient medium has been an important tool in basic and applied research aimed at plant improvement. The principles of tissue culture were implicated as early as 1838–1839 in cell theory advanced by Schleiden and Schwann but these were not clearly formulated until German plant physiologist Haberlandt, the pioneer of plant tissue culture, suggested in his address to the German Academy in 1902 that it should be possible to cultivate artificial embryos from vegetative cells. Since that day, the concept of totipotency (the potential of a cell to develop into a whole plant) has been demonstrated in a large number of plant species. It is now possible to culture in vitro large populations of totipotent cells and to regenerate plants from cultured somatic cells, pollen, and protoplasts.

The ability to culture plant cells under defined conditions primarily resulted from the experiments of White [1,2], Gautheret [3], and Nobecourt [4]. These findings set the stage for strong interest in tissue culture research and established procedures for in vitro propagation of germplasm and eradication of viruses through meristem/shoot tip culture [5], auxin-cytokinin manipulation to regulate organogenesis [6], somatic embryogenesis [7], large-scale culture of cells [8], large-scale enzymatic isolation of protoplasts [9], anther culture for production of haploids [10,11], regeneration of plants from single cells [12], and the uptake of DNA by plant cells [13]. The most commonly used media for culture of plant cells are Murashige and Skoog's medium MS [14] and B5 [15] with modifications in phytohormone levels.

During the last two decades, procedures have been refined for the culture of somatic cells, pollen, and protoplasts and for regenerating plants from such cultured cells. The techniques of meristem culture and in vitro propagation are commercially used to produce virus-free stocks of potato, sweet potato, and cassava and to achieve large-scale multiplication of orchids, and several other ornamental and horticultural species. Embryo rescue has successfully been used to overcome incompatibility and to produce numerous interspecific and intergeneric hybrids in cereals, legumes, forages, and horticultural species. Haploids have been produced in more than 200 plant species using anther culture. Several elite breeding lines and improved varieties have been developed through anther culture in wheat, rice, tobacco, and other crops. Cell cultures have been subjected to various toxic conditions such as pathotoxins, salts, and amino acid analogues. Several useful variants possessing resistance to diseases, increased tolerance for salt, and improved nutritional quality have been obtained. A number of somaclonal variants (without any exposure to toxic conditions during in vitro culture) have been isolated; notable examples include resistance to eyespot and Fiji virus disease in sugarcane, resistance to early and late blight in potato, and resistance to *Helminthosporium* in maize. Somatic embryogenesis is being explored for developing artificial seed technology for large-scale, true-to-type multiplication of hybrid varieties and other high-value crops. Although somatic embryogenesis has been achieved in more than 150 plant species, production of artificial seeds for commercial planting is still in its infancy. Cryopreservation of cultured cells, meristem, shoot tips, and plant organs is becoming increasingly important for long-term storage of germplasm.

Major advances have already been made in the establishment of embryogenic cell suspensions and in protoplast culture and plant regeneration of recalcitrant species such as cereals (rice, wheat, maize) and some tree species. Protoplast fusion has been used to obtain novel somatic hybrid plants among several sexually incompatible species and to produce cybrids difficult to obtain through conventional methods. Refined tissue culture procedures have made it possible to introduce foreign DNA and cloned genes into cultured cells, protoplasts, and plant organs from diverse biological systems and to regenerate transgenic plants in more than 45 plant species with new genetic properties [16,17,18]. Transgenic plants carrying agronomically useful genes for herbicide resistance, insect resistance, virus protection, fungal resistance, and male sterility have become available. Advances in molecular biology have further widened the scope of tissue culture in plant improvement.

Table 1   Some Applications of Cell and Tissue Culture Techniques in Plant Improvement

Technique	Application	Future research priority(ies)
In vitro propagation	Rapid and large-scale true-to-type multiplication of elite germplasm of ornamentals, horticultural, tree species, and other rare genotypes	Efficient propagation techniques applicable to wide range of genotypes particularly for horticultural and woody tree species such as coconut, oil palm, rubber tree, teak tree, pines, eucalyptus, female papaya, etc. Automated propagation system to reduce cost
Meristem/shoot tip culture	Large-scale production of virus-indexed (free) foundation stocks Clonal propagation of germplasm In vitro conservation and international exchange of germplasm	Efficient plant regeneration from smaller meristems to eliminate viruses Rapid indexing techniques for detection of viruses
Embryo culture	Overcoming incompatibility barriers and production of interspecific and intergeneric hybrids Overcoming seed dormancy and self-sterility of seeds	Optimization of conditions for culturing hydrid embryos aborting at very early stages of development
In vitro pollination	Production of hybrids which are otherwise difficult to produce through embryo rescue	Optimization of conditions for effective in vitro fertilization in distant crosses
Anther and microspore culture	Reducing breeding cycle of cultivars through production of homozygous lines (doubled haploid) in the immediate generation Fixation of certain gene combinations from heterozygous source materials Genetic transformation using microspores Useful gametoclonal variants	High anther culturability and plant regeneration applicable to a wide range of elite genotypes
Somaclonal variation	Isolation of useful variants in well-adapted, high-yielding genotypes lacking in a few desirable traits	High frequency of plant regeneration from long term callus cultures

Table 1 (Continued)

Technique	Application	Future research priority(ies)
Somaclonal variation (continued)		Extensive laboratory and field testing to isolate stable somaclonal variants of practical importance
In vitro selection at cellular level	Induction and selection of useful mutants at cellular level for disease resistance, tolerance for abiotic stresses, and improved nutritional quality	Refined procedures for high frequency of plant regeneration from cell cultures exposed to various biotic and abiotic stresses
		Optimization of selective conditions capable of discriminating epigenetic variants from desired mutants
Somatic cell hybridization	Production of somatic hybrids between otherwise sexually incompatible species	Efficient plant regeneration from protoplasts of elite genotypes
	Transfer of CMS in elite lines and production of cybrids and organelle recombinants	Efficient systems for selection of hybrid, and cybrid cells and plants
		Development of procedures for induced recombination and chromosome exchanges between genomes of parental species
		Optimization of donor-recipient method to transfer CMS into well adapted cultivars
		Production and utilization of novel somatic hybrids in vegetatively propagated crops
Protoplast and cell culture mediated DNA transformation	Introduction of foreign genes from diverse biological systems and production of transgenic plants with new genetic properties	Efficient protoplast-mediated transformation systems
Cryopreservation	Long-term preservation of useful germplasm (cell lines, meristems, plant organs) and strengthening of germplasm banks	Overcoming loss in viability and regeneration ability of cryopreserved cells, meristems, and other plant organs

Table 1 (Continued)

Technique	Application	Future research priority(ies)
Somatic embryo-genesis	True-to-type mass multiplica-tion of heterotic F_1 hybrids and other elite lines through artificial seeds Source of embryogenic proto-plast culture Genetic transformation	Optimization of conditions to obtain mature somatic em-broids with high conversion rate (embryo to plant) under field conditions

A number of cell and tissue culture techniques are available and these offer enormous potential in plant improvement (Table 1). Some of these techniques and their application in plant improvement are discussed.

IN VITRO PROPAGATION

The findings that organogenesis is regulated by hormones [6] and that plant cells are totipotent [12] laid the foundation for clonal propagation of plants through tissue culture techniques. The discovery regarding the role of cyto-kinins in shoot morphogenesis was an important step in clonal propagation [6]. The technique consists of rapid and large-scale propagation of plants from cultured organs and cells. Plants may be propagated through (1) mul-tiplication of existing (axillary) shoot meristems, and (2) formation of ad-ventitious buds either directly from organ explants or indirectly from callus or suspension cultures. The explants from desired plants are grown on a nu-trient medium enriched with cytokinins to get en masse production through organogenesis or somatic embryogenesis. The plant hormones stimulate the development of multiple axillary buds. The newly formed buds are re-cultured to produce more buds. Eventually, buds are allowed to grow into shoots which are then rooted in auxin-containing medium to produce plants. The production of artificial seeds through somatic embryogenesis is also be-ing explored as another approach for mass propagation of useful germplasm (see later section).

Murashige [19] described in vitro propagation as a three-stage procedure which normally requires change of culture medium or growth conditions between stages. These stages involve (1) establishment of tissue in vitro; (2) production of multiple shoots; and (3) root formation and conditioning of propagules prior to transfer to the greenhouse. In several cases, media and cultural conditions are not altered between stage 1 and stage 2.

The first successful application of tissue culture to propagation was the in vitro multiplication of orchids by Morel in the early 1960s. The findings of Morel and Martin [5] demonstrated that dahlia plants obtained from cultured shoot meristems were virus-free. This led to the widespread culture of orchid meristem for propagation. Now, several orchid species, strawberry, asparagus, Iris, carnation, and other ornamental plant species are regularly multiplied through this technique. More than 22 genera of orchid can now be propagated by tissue culture. In vitro orchid propagation is used routinely in the trade. Millions of strawberry plants can likewise be produced from a single mother plant in 1 year. Several industries in the United States, France, West Germany, Republic of China, Belgium, and other countries multiply ornamental plants for commercial use by tissue culture techniques. Worldwide, there are probably more than 300 laboratories involved in the multiplication of ornamental plants [20]. Many of the newer commercial laboratories have production capabilities of up to 200,000 in vitro plantlets per week [21]. Several private companies, such as the Indo-American Hybrid Seeds in Bangalore, India, have developed facilities for in vitro propagation and exported millions of plants to Europe and released several hundred thousands of bananas and cardamoms to farmers in India. The National Chemical Laboratory in Poona, India, is attempting to multiply germplasm from elite mature teak trees through tissue culture. In teak, the worth of a tree is judged after 10–20 years, and clonal propagation from such "elite" trees would be highly desirable.

In spite of the success in commercial propagation of ornamental and horticultural species, progress in the multiplication of woody plants has rather been low. In a few cases, mature woody plants have been multiplied in vitro; examples include the multiplication of *Eucalyptus, Tectona grandis* [22], *Seqioua sempervirens, Populus tremuloides, Pinus pinaster, P. radiata,* and *P. caribaea* [23,24]. Some "scion" cultivars of pome and stone fruits can also be micropropagated [25]. In vitro culture of nodal segments and development of axillary buds on medium supplemented with 1–4 mg/liter benzylaminopurine (BAP) has been successfully used in micropropagation of coffee. Rooting of buds can be induced on BAP-free medium. Using this method, more than 100,000 clonal plants have been multiplied for distribution to Latin American countries [26]. The limited success in propagation of woody plants has been achieved mainly by the use of appropriate explant and pretreatment. Propagation via adventitious bud formation and somatic embryogenesis in woody species could be of tremendous value in mass propagation.

The bottlenecks of in vitro propagation include (1) low frequency of regenerated plants (in many cases, propagation is sporadic and thus impractical for mass propagation); (2) plant regeneration being limited to only specific genotypes thus making several elite lines and useful materials difficult to multiply; (3) frequent variation in tissue culture-propagated plants (propagated germplasm is hence not true to type); and (4) its high cost.

Several developing countries have recently refined their facilities (e.g., automatic shade system, mist chambers, and movable benches for hardening of tissue culture plants), for more efficient in vitro propagation. The automated plant tissue culture system designed for mass propagation of germplasm can work with both organogenic and embryogenic cultures [27]. It separates propagules by size and dispenses masses of propagules into culture vessels for growth and development. In 8 hours, the bioprocessor can produce sufficient material for 100,000 propagules. A microprocessor-controlled transplanting machine automates another labor-intensive work that of removing plantlets from culture vessels and placing them in nursery trays. The machine transfers about 8000 plantlets per hour. Further modifications in automated systems can enhance the efficiency of in vitro propagation of germplasm.

Future perspectives of clonal propagation include large-scale multiplication of elite and mutant strains of various crop plants, particularly horticultural and tree species in which the natural rate of multiplication is low, are highly heterozygous, have longer generation time, and are difficult to improve by conventional breeding procedures. Efficient techniques are required for mass propagation of coconut, oil palm, date palm, pineapple, rubber tree, teak tree, and for increasing apple root stock, female lines of papaya, male sterile asparagus, and others. Commercial propagation of elite genotypes of fruit crops, woody trees, and forest species through tissue culture could lead to a breakthrough in the productivity of these crops. There is an urgent need to reduce propagation cost and to develop efficient automated methods for mass propagation of useful germplasm.

MERISTEM/SHOOT TIP CULTURE

The term *meristem culture* denotes in vitro culture of the apical dome along with a portion of the subjacent tissue containing one to several young leaf primordia. The meristem, measuring about 0.1 mm in diameter and 0.25 to 0.30 mm in length is located at the extreme tip of a shoot. These "meristemtips," when cultured in vitro on a suitable nutrient medium under optimal conditions, grow, and differentiate into whole plants. In some species, each meristem in a culture grows into one plantlet. In others, it results in multiple plantlets, thereby providing an efficient means of mass propagation. Plants regenerated by meristem culture result in the recovery of identical progeny because the constituent cells of the shoot apical meristem are less differentiated than the other cells.

As early as 1949, Limmaset and Cornuet [28] demonstrated that the virus was unevenly distributed in systemically infected plants and that the virus titer decreased as the vegetation point is reached. The pioneering work of Morel and Martin [5] demonstrated regeneration of healthy plants through meristem culture of virus-infected dahlias. Since then, meristem culture has been widely used in clonal propagation and in the production of virus-free plants in a large number of plant species [29]. The technique consists of dissecting meristems (0.2–0.5 mm) and growing them in test tubes on nutrient medium. Meristem/tip culture-derived plants are allowed to develop and grown in insect-proof screenhouse when ready. At this stage, plants are indexed for the presence of viruses. Various techniques such as grafting, sap transmission, serology, electron microscopy, and enzyme-linked immunosorbent assay (ELISA) are used to detect viruses. Meristem culture provides foundation stock of virus-free plants. However, these plants are not immune to reinfection by the same or other causative agents. Plants found free of viruses serve as a source of virus-free propagation materials. Thus, the nuclear stock of healthy mother plants should be fully protected against reinfection. Only propagules directly propagated from mother stocks should be distributed for commercial field planting.

Meristem culture has great potential in (1) the production of virus-free germplasm; (2) mass propagation of desirable genotypes; (3) facilitating international exchange of germplasm; and (4) cryopreservation of germplasm of recalcitrant crop species. It has been used to eliminate several viruses such as mosaic virus (cassava, raspberry, sugarcane, cauliflower, garlic); cucumber mosaic virus (banana); crinkle, vein banding latent A and C viruses, mottle (strawberry); internal cork, feathery mottle (sweet potato); virus X, Y, S, A, M, leaf roll, and spindle tuber virus (potato). The technique is now commercially used to produce virus-free plants in potato, sweet potato, dahlia, strawberry, carnation, chrysanthemum, orchids, etc. More than 10% of the potato area in China is planted to virus-free potatoes produced through meristem culture [30]. Virus-free plants of gladiolus, orchids, and carnation are now commercially produced in China. International centers like CIAT in Colombia and IITA in Nigeria routinely use meristem culture to produce and distribute virus-free germplasm of cassava, potato, and sweet potato to various national programs [31,32]. Several countries have well-defined programs to propagate and distribute virus-free plants for commercial purposes. Indexed plants can be shipped in vitro, overcoming the most stringent plant quarantine barriers.

The causes of meristem devoid of viruses are not well understood. Earlier hypotheses that (1) higher level of hormones in meristems inactivated the virus or (2) virus movement from cell to cell did not keep pace with the faster rates of cell division in meristems, do not hold true. For example,

Walkey [33] found that viruses such as tobacco mosaic virus (TMV), cowpea mosaic virus (CMV), and potato virus × (PVX) are present in high concentrations in the apical meristem cells. Virus-free plants have been regenerated from meristem culture, however, with the viruses presumably lost during in vitro culture due to a combination of unknown effects.

The efficiency of meristem-tip culture to produce virus-free germplasm is enhanced when combined with thermotherapy, which involves keeping the infected plants at 37°C for several weeks to inactivate the virus. Later, meristem-tip explants are cultured to regenerate virus-free plants. Kartha and Gamborg [34] regenerated 100% mosaic virus-free plants from cassava meristems isolated from infected cuttings grown at 35°C for 30 days. After heat treatment, large meristem tips (0.8 mm) could be cultured. Roca [31] used a combination of thermotherapy and meristem culture in eliminating the frog skin disease of cassava.

Chemical suppression of plant viruses has also been reported. For example, ribavirin reduces the concentration of CMV and alfalfa mosaic viruses in cultured tissues. In combination with thermotherapy and meristem culture, compounds with antiviral properties could be used to eliminate viruses which are difficult to eradicate through the usual meristem culture technique.

The success of meristem culture is judged by the percentage of plantlets regenerated and the absence of viruses. It varies from species to species and with the kind of viruses and size of tips cultured. Some viruses are more easily eliminated than others. Potato viruses are eliminated in the following order of increasing difficulty: potato leaf roll virus V (PLRV), potato virus A (PVA), potato virus Y (PVY), potato virus X (PVX), and spindle tuber viroid [35]. In general, the larger the meristems, the higher the regeneration, but the number of virus-free plants is inversely proportional to the size of the meristems cultured. The characteristics of primordia on the tips are better criteria of successful meristem culture than is length or volume of the tips. For example, most of the potato viruses can be eliminated by culturing shoot tips with one or two primordia.

The main bottlenecks of meristem culture are (1) lower frequency of plant regeneration from small-sized meristem; (2) high proportion of regenerated plants having virus infection; (3) lack of rapid virus-indexing techniques; and (4) limited facilities for multiplication of germplasm from meristem-derived plants under insect-proof conditions. Culturing of meristems of smaller size (0.1–0.2 mm) leads to reduced plant regeneration; but where size of the meristem is larger (more than 0.4 mm), virus elimination becomes a problem. Several vegetatively propaged crops such as cassava, potato, sweet potato, sugarcane, and others are infected with a range of viruses that cause huge crop losses. Refined procedures are needed to culture meristems of smaller sizes in such crops.

As a means to produce virus-indexed plants, the technique is simple and promising. The efficiency of meristem culture could be further improved if combined with heat treatment and also if the meristem culture facility is accompanied with extensive vegetative propagation of the material under protected conditions, availability of indicator plants, and efficient virus detection techniques such as nucleic acid spot hybridization, latex agglutination, and ELISA.

EMBRYO, OVARY, AND OVULE CULTURE

Embryo Culture

To widen the gene pool of crop plants, breeders search for variability from diverse sources. The wild species of crop plants are an important reservoir of useful genes for resistance to diseases, insect pests, tolerance for abiotic stresses, male sterility, and quality traits. However, several crossability barriers hinder the transfer of useful genes from wild species into crop plants [36,37].

Embryo culture is an important technique to overcome incompatibility in such interspecific crosses. It is also useful technique to accelerate seed germination and overcome seed dormancy and self-sterility of seeds and to produce haploids through chromosome elimination from distant crosses such as *Hordeum vulgare* × *H. bulbosum.* Abortion of embryos at different stages of development is a characteristic feature of wide crosses. Incompatibility between the embryo and endosperm is the major problem resulting in the abortion of hybrid embryos. The technique involves excision of hybrid embryos and culturing of such embryos in a defined nutrient medium to enable them to grow into seedlings. Laibach [38] has demonstrated that embryos of nonviable seeds of *Linum perenne* × *L. austriacum* could be cultured in a nutrient medium and reared to maturity. The technique has since then been successfully used for rescuing abortive embryos and for producing interspecific hybrids in a large number of plant species that include cereals, legumes, vegetables, and horticultural and tree species [39]. Several intergeneric hybrids have been produced through embryo rescue in cereals — *Triticum* × *Hordeum, Triticum* × *Elymus, Triticum* × *Secale, Hordeum* × *Secale*, and others. *Triticale*, a hybrid of *Triticum* and *Secale* species, is a classical example of a man-made cereal produced through embryo rescue technique.

Several international centers routinely use this technique to produce interspecific hybrids. For example, CIMMYT in Mexico has an extensive program on wide crosses (maize × *Tripsacum*, maize × sorghum, wheat × barley, wheat × *Elymus*, and wheat × several allied species of *Triticum*). Similarly, ICRISAT in India is using this technique to incorporate genes for disease resistance from wild species to cultivated peanut varieties. At IRRI, we have produced a number of interspecific hybrids in rice through embryo rescue [40]. Following embryo rescue, genes for resistance to brown plant-hopper, blast, and bacterial blight have been transferred across the cross-ability barrier from wild *Oryza* species into elite breeding lines of rice [41,42].

Embryo rescue has also been used extensively to produce haploid seed-lings from the crosses of *H. vulgare* × *H. bulbosum*. Chromosome elim-ination in such a cross results in the production of haploids [43]. The tech-nique involves making a cross between *H. vulgare* and *H. bulbosum*. The pollinated spikes are sprayed with 75 ppm GA_3 for 3 consecutive days. After 12 days of pollination, embryos are dissected and cultured on the nutrient medium in the dark at 22°C. One or 2 weeks later, embryos are transferred to a lighted cabinet (12 h) at 22°C. Up to 50–60% of the cultured embryos give rise to haploid plants. The preferential loss of *bulbosum* chromosomes has been demonstrated cytologically and by the use of gene markers. The bul-bosum method has been used to produce haploids in barley and eight varieties have so far been released for commercial cultivation (Table 2).

In Vivo/Vitro Embryo Culture

In vivo/vitro embryo culture is a modification of the embryo rescue tech-nique and is used when embryo abortion starts at the very young stages of development. In this technique, the endosperm of the female parent serves as a nurse tissue for the hybrid embryo. Normal endosperm from the female parent is so placed on the surface of the nutrient medium that it is not in direct contact with the hybrid embryos. Sometimes, the endosperm extract is added to the nutrient medium. Immature endosperm is known to pro-vide critical nutrients to the developing embryo and helps induce germina-tion of the immature embryo. Kruse [56] used this technique to produce *Hordeum* × *Triticum* hybrids. The success rate in *Hordeum* × *Secale* crosses increased by 30–40% through nurse tissue technique as compared with 1% achieved through the traditional embryo rescue technique.

A slightly modified technique was used by Williams and De Lautour [57]. It involves the insertion of a hybrid embryo into a healthy endosperm dissected from a normally developing ovule and their transfer to the nutrient medium. The normal embryo is pressed out of the sac of the endosperm and the hybrid embryo is inserted into the exit hole of this endosperm. These implanted embryos are cultured on nutrient medium to produce hybrid seed-lings. This technique has been used to obtain interspecific hybrids in *Tri-folium, Lotus,* and *Ornithopus* [57].

Regenerating Plants from Callus of Hybrid Embryos

In most distant crosses, hybrid embryos are manipulated to produce F_1 plants directly. In some crosses, however, the hybrid embryos fail to differentiate

Table 2 Some Examples of Improved Varieties Developed Through Anther Culture and Bulbosum (Chromosome Elimination) Methods of Haploid Breeding

Technique	Crop	Variety	Reference
Bulbosum method	Barley	Mingo	[44]
		Rodeo,	[45]
		Craig, Gwylan,	[46]
		Etienne, Winthrop, TBR	
		579-5, TBC 555-1	
Anther culture	Wheat	Florin	[47]
		Huapei-1, Lunghua-1	[48]
		Jingdan 2288	[49]
		Zing Hua 1, Zing Hua 3,	[30]
		Zing Hua 5	
	Rice	Huayu-1, Huayu-2, Tanfong-1,	[50]
		Xin Xiu-1	[48,49]
		Hua Han Zhao	[51]
		Zhe Keng 66	[52]
		Zhong Hua 8, Zhong Hua 9	[53]
		Zhong Hua 10, Zhong Hua 11	[30]
		Hwacheongbyeo	[54]
		Hwajinbyeo, Suweon-384	[55, H.P. Moon
		(Hwaseonchalbyeo)	(pers. comm.)]
		Hwaseongbyeo, Hwaryeongbyeo,	
		Joryeongbyeo	
	Tobacco	Tanyu-1, Tanyu-2	[50]
		Tanyu-3	
		HaiHua-19, HaiHua-29	[49]
		HaiHua-30, HaiHua-31	

into plants. To overcome this barrier, the undifferentiated embryos are induced to proliferate as callus on a culture medium and hybrid plants are regenerated from the callus. Thomas and Pratt [58] employed this technique to increase the chances of recovering F_1 plants from the cross *Lycopersicon esculentum* × *L. peruvianum*. In the cross *Nicotiana suaveolans* × *N. tabacum*, viable seeds are produced, but the F_1 seedlings die soon after emergence at the first- or second-leaf stage. Lloyd [59] cultured the young cotyledons on a callus-inducing medium and from the callus regenerated several shoots which were grown to maturity. Normal hybrid plants from the partially necrotic F_1 seedlings of *Trifolium repens* × *T. uniflorum* have been obtained via the callus phase [60]. Poysa [61] could not get hybrid plants from the cross of *Lycopersicon esculentum* × *L. peruvianum* following embryo rescue, ovule culture, and through the use of immunosuppressants and hormonal treatments. The barrier was overcome through embryo–callus culture technique.

The embryo culture technique has great potential to overcome hybrid incompatibility and produce hybrids among widely divergent species. Further efforts are required to culture young embryos which abort at the earlier stages of development. Use of GA$_3$ and immunosuppressants like L-lysine, E-amino caproic acid, and recognition pollen could increase the efficiency of embryo culture.

In certain cases, it is difficult to obtain seed germination in vivo. In such situations, embryo culture has been used successfully to overcome the problem of poor seed germination in *Colocasia esculenta, Musa balbisiana*, and *Pinus amandii* × *P. koraiensis*. Certain early-maturing varieties in *Prunus* do not produce viable seed. Before fruit ripening, the embryo and endosperm tissue stop developing. Embryo culture could overcome this problem. Skene and Barlass [62] showed that *Persea americana* (avocado) embryos excised from seeds that abscise before maturity can be grown in a nutrient medium supplemented with cytokinin to yield multiple shoots suitable for growing onto seedling root stocks. There are many species that exhibit seed dormancy which is often localized in the seed coat and or in the endosperm. Embryo culture has been used to overcome this problem in Brussels sprouts, *Rosa* species, apple, oil palm, and Iris [39].

Ovary Culture

In many wide crosses, embryo abortion starts at the very early stages of development. These small embryos are difficult to excise and culture. Moreover, the nutritional requirements of younger embryos are more exacting than those of old embryos. To overcome these problems ovaries or ovules are cultured. The technique consists of excising ovaries 2–15 days after pollination (depending on the cross combination) and culturing them on nutrient medium. The calyx, corolla, and stamen are removed. Before culturing, the tip of the distal part of the pedicel is cut off and the ovary is implanted with the cut end inserted into the medium. The nutrient medium is generally simple, consisting mostly of inorganic salts. However, auxins, cytokinins, yeast extract, and casein hydrolysate can support continued growth of excised ovaries in culture. After the embryos become visible, they are excised aseptically and cultured on nutrient medium to obtain hybrid seedlings following all the steps of embryo rescue previously described.

Following ovary culture, interspecific and intergeneric hybrids have been produced in *Brassica campestris* × *B. oleracea* [63], *B. nigra* × *B. oleracea* [64], *Diplotaxis erucoides* × *B. napus* [65], and *Moricandia arvensis* × *B. oleracea* [66]. Inomata [63] excised ovaries 4 days after pollination of *Brassica campestris* with *B. oleracea* pollen and cultured them on white's medium containing mineral salts, vitamins, and casein hydrolysate. After 36 days of culturing, the embryos from these ovaries were excised and cultured on the same medium under similar conditions. The plantlets arising from the cultured embryos were transferred to the soil. Similarly, Takahata [66] produced hybrids between *Moricandia arvensis* and *Brassica oleracea* through ovary culture. The ovaries were excised 3-8 days after pollination and cultured on MS medium free of hormones. The embryos from these ovaries were excised after 3-4 weeks of ovary culture and transferred to the same medium. The embryos failed to develop into plantlets. Shoots were therefore induced from hypocotyl segments. The plants were regenerated from such shoots.

Hybrids have also been produced between wheat × maize through ovary (spikelet) culture from 2-4 day old pollinated spikelets. The maize chromosomes get eliminated after hybridization, however, giving rise to haploid wheat plants [67-69].

Ovule Culture

Like ovary culture, ovule culture is sometimes helpful in overcoming the barriers which affect the growth of the zygote during earlier stages of development. In this procedure, ovaries are harvested 1-12 days after pollination (depending on species), surface-sterilized, and cut open with a sterilized scalpel. The fertilized ovules are scooped out and placed as evenly as possible on the nutrient medium. The growth of the ovule is monitored by changes in size and developmental stage of the enclosed embryo and endosperm. The nutrient medium generally consists of inorganic salts, auxins, and cytokinins, and, in other cases, complex growth substances which promote development of embryo such as coconut water, casein hydrolysate, and yeast extract. With ovule culture, the hybrid embryos can be cultured at earlier stages than excised embryos.

Interspecific hybrids have been obtained through ovule culture in species crosses of *Abelmoschus, Lolium* × *Festuca, Nicotiana, Impatiens, Gossypium,* and *Brassica* [70]. Cotton is the classical exmaple of a crop where ovule culture has been used to produce a series of interspecific hybrids. Stewart and Hsu [71] cultured ovules 2-4 days after anthesis. After 8-9 weeks of culture, all ungerminated ovules were dissected and the embryos were transferred to the same medium used for germinating seedlings. This technique, commonly referred to as in ovulo-embryo culture, overcomes the incompatibility between the A genome diploid cottons and the AD genome tetraploid cottons. Hybrid plants have also been produced through ovule culture from crosses of *Gossypium hirsutum* × *G. australe, G. barbadense* × *G. australe,* and *G. arboreum* × *G. australe.* Ovule culture has also been used in obtaining hybrids between cultivated and wild species of *Nicotiana, Glycine,* and *Lens.* Reed and Collins [72] used ovule culture to produce hybrids of cultivated tobacco *Nicotiana tabacum* with three wild species *N. nesophila,*

N. stocktonii, and *N. repanda*. Arisumi [73] found that in producing inter-specific hybrids in *Impatiens*, embryo rescue was successful only when well-developed embryos were cultured, 1–2 weeks beyond the late heart stage. In contrast, rescue by ovule culture was possible with much younger, globular-stage embryos.

The reasons for the recovery of hybrids in some wide crosses by ovary or ovule culture rather than by embryo culture are not well understood. The embryos developing within the ovule may have more favorable chemical and physical environment for growth and development than embryos cultured outside the ovule. It is also possible that tissues of the ovary and the ovule serve various functions such as giving nutrition and protection of the en-closed embryos.

IN VITRO FERTILIZATION

Any manipulation of excised material and paternal tissue to obtain pollen tube penetration to the embryo sac to accomplish fertilization is referred to as in vitro fertilization. It is an important technique for overcoming the bar-riers in stigma and style that inhibit pollen tube growth and embryo abor-tion in the early stages of development. In vitro fertilization followed by culturing of fertilized ovules to maturity is a promising approach and may be a viable alternative to parasexual or somatic cell hybridization. Stewart [70] and Tilton and Russell [74] have reviewed the usefulness of in vitro fer-tilization technique for overcoming incompatibility barrier to produce wide hybrids.

Various methods are used for in vitro fertilization. The whole gynoecia are excised and placed on a medium for 24–48 hours followed by dusting of pollen on the stigma. Kanta et al. [75] were the first to attempt in vitro pollination technique in *Papaver somniferum*. Since then, a number of re-ports on this technique have been published [70]. However, the technique has been used to a very limited extent in wide crosses. Zenkteler [76] reported the production of intergeneric hybrid plants through in vitro pollination of *Melandrium album* × *Silene schafta* and *M. album* × *Viscaria vulgaris*.

Dhaliwal and King [77] attempted hybridization between maize × maize and maize × teosinte (*Zea mexicana*). The styles of maize were removed and the exposed nucellus of caryopsis on cultured sections of scape were dusted with pollen of *Z. mexicana*. Seed set was about 5%. No seed devel-oped after application of sorghum pollen to maize styles. Application of sorghum pollen directly to maize ovules resulted in a rapid and massive growth of nucellar tissue without any indication of embryo and endosperm develop-ment. Slusarkiewicz-Jarsina and Zenkteler [78] obtained hybrid plants from in vitro pollination of ovules of *Nicotiana tabacum* with *N. knightiana* pollen. Embryos produced callus, which later differentiated into shoots and formed plants. Refaat et al. [79] produced hybrid plants through in vitro pollination of *Gossypium hirsutum* with *G. barbadense* pollen.

It is now possible to isolate and fuse sperm and egg cells in vitro [80]. In vitro fertilization provides another opportunity to produce genetically engineered plants by injecting exogenous DNA into nuclei of gametes and zygotes as has been done with protoplasts.

ANTHER AND MICROSPORE CULTURE

Anther Culture

A number of techniques to produce haploids in crop plants are available. Of these techniques, anther culture is one of the most efficient. Since the first successful production of haploids from anther culture of *Datura innoxia* [10,11], haploids have been produced through anther culture in more than 200 plant species [81]. Several reviews have been published on the production and utilization of anther culture-derived haploids [81–83]. Numerous studies have reported about the factors affecting plant regeneration from anther culture of crop plants. Various media, pretreatments, stage of the pollen, genotype, and growth condition of donor plants affect culturing ability of anthers [82,83]. Haploids have been produced from anther culture of various crops such as cereals, legumes, and forages. It has also been possible to produce haploids from woody plants such as the rubber tree, apple, litchi, and poplar [48].

Besides anther culture, the chromosome elimination technique involving *Hordeum vulgare* × *H. bulbosum* was also proven to be an efficient technique for production of haploids in barley [43,84]. The technique, called bulbosum method, leads to the production of haploids through selective elimination of bulbosum chromosomes from the cross of *H. vulgare* × *H. bulbosum*.

The haploids produced through anther culture and chromosome elimination methods offer promise in various crop improvement programs. The following are some of the applications:

Production of Homozygous Lines and Development of New Varieties/Elite Breeding Lines

The technique consists of culturing anthers on semisolid and liquid medium. The cultured anthers produce callus or, in some cases, form embroids. These calli or embroids are regenerated into haploid plants. The chromosome number of regenerated haploid plants is doubled through colchicine application to produce dihaploid (DH) lines. Spontaneous chromosome doubling is also quite common in haploids. The seeds of DH lines are multiplied and evaluated for their agronomic performance. The technique is important in developing true breeding lines in the immediate generation from any segregating population, thereby shortening the breeding cycle of new varieties. In conventional breeding, after a cross is made between desired parents, the segregating populations from F_2 to F_7 are grown to ultimately develop true breeding homozygous lines. On the other hand, anther culture-derived lines can be multiplied and evaluated in the immediate generation.

Anther culture has successfully been used to develop improved varieties of wheat, rice, and tobacco (Table 2). In rice alone, more than 100 varieties and breeding lines have been developed in China through anther culture [48,82,83,85]. Notable examples include Huayu-1, Huayu-2, Tanfong 1, and Xin Xiu in rice; Hupei-1, Lunghua-1, and Jingdan 2288 in wheat; and Tanyu-1, Tanyu-2, and Tanyu-3 in tobacco. Nakamura et al. [86] developed promising lines of tobacco through anther culture. These lines are more resistant to bacterial wilt and have a milder smoking quality in comparison with other varieties. Similarly, four other varieties (Haihua-19, Haihua-29, Haihua-30, and Haihua-31) have been developed through anther culture breeding in tobacco [49]. In rice, a gene (Pi-zt) for blast resistance from Toride No. 2 was introduced into a local variety in 2 years through anther culture breeding, resulting in the development of Zhong Hua 8 and Zhong Hua 9 [53]. Huayu-1 and Huayu-2 took only 5 years from initial crossing to being released as new varieties [84]. Huayu 1 is high-yielding, resistant to bacterial blight, and has wide adaptability. Similarly, Xin Xiu has been grown in 10,000 ha in eastern China. Lee et al. [54] developed japonica rice variety Hwacheongbyeo through anther culture of F_1 of a cross between Suweon 298 and Milyang 64. This variety is resistant to brown planthopper, blast, and bacterial leaf blight and is recommended for cultivation in the southwestern plain and southwestern coastal area of Korea. Moon et al. [55] used anther culture and developed a variety, Hwajinbyeo, from the cross Milyang 64 × Iri 53. This variety is resistant to stripe virus and bacterial leaf blight. In wheat, anther culture-derived variety Florin was released in France [47]. The superior wheat varieties Jing Hua-1, Jing Hua-3, and Jing Hua-5 developed in China occupy more than 650,000 ha [30]. Several institutes in Canada and Germany have used anther culture and developed improved strains of *Brassica*, rye, and potato [87,88]. Through anther culture breeding, improved varieties can be developed up to 4 years/crop seasons earlier than conventional breeding methods.

Enhanced Selection Efficiency

The DH lines offer a unique opportunity to improve selection efficiency for various traits since there is no dominance variance in the population [89]. For traits such as yield which have low heritability and where selection efficiency in early generations (F_2–F_4) is low with conventional breeding procedures, doubled haploids can give more additive genetic variation expression through elimination of dominance genetic variation. Anther culture-derived plants are genetically identical from one generation to the next, and their selection efficiency is thus likely to be higher when dominance variation is significant. Breeding at the DH level in tetraploid species such as potato is much more efficient.

Recessive mutations can also be detected in the same generation. The haploid method is based on gamete instead of sporophyte selection and the probability of obtaining desired genotype in haploid ($1/2^n$) is much higher than that in diploids ($1/4^n$) (n = number of genes controlling a particular character). Haploids are useful in inducing mutations where genes are in the hemizygous condition.

Dihaploids provide useful mapping population for analysis of restriction fragment length polymorphism (RFLP). Seeds of such populations can be distributed to workers in other laboratories, and populations can be grown repeatedly in many different environments, greatly facilitating the addition of both DNA markers and genetic loci controlling traits of agronomic importance. In rice, mapping population is being developed from the DH lines produced from (IR64 × Azucena), an indica/japonica cross (S. McCouch, IRRI, personal communication).

Haploids can be applied advantageously in the recurrent selection procedures to breed for disease and pest resistance governed by polygenes. Wenzel et al. [90] have outlined the procedures for the use of haploids in combination breeding as an example for the transfer of barley yellow mosaic virus resistance from the Japanese spring barley (YM1) cultivar to the German winter barley variety (IGRI).

Gametoclonal Variation

Repeatedly subcultured callus lines derived from anther culture could also give rise to a variation referred to as gametoclonal variation. Such variation has been reported in several species and can be used in crop improvement. A bell pepper variant was identified from anther culture-regenerated plants. From that, a variety (Bell Sweet) with only a few seeds has been developed — an average of 9 seeds per fruit compared with 330 seeds of control variety Yolo Wonder [91]. Recently, a new rice variety "Dama" in Hungary has been released through gametoclonal variation [92]. A large number of DH somaclones of haploid origin were produced from three rice varieties. The DH somaclones were evaluated for different agronomic traits including increased resistance to *Piricularia*. Among various selections, one promising somaclone was selected and released as a variety in 1992.

The main bottlenecks observed in anther culture include low frequency of regenerated haploid plants, and quite often, production of haploids from only a limited number of genotypes. In rice, for example, japonica rices are known to be more amenable to anther culture than indica rices. In a global context, however, indica rices are far more important than japonicas. Another bottleneck is the high frequency of albino plants. In vitro androgenesis is strongly genotype-dependent. It has now become possible to map genes governing anther culturability. Cowen et al. [93] found in maize that high anther culture response (as measured by formation of embryolike structure) is conditioned by two major recessive genes which are epistatic and two minor genes. RFLP analysis showed the two major loci to be located on chromosomes 3 and 9. Such studies widen the scope of introducing high anther culture ability into elite breeding lines of crop plants and thus increase the efficiency of anther culture breeding. Future research needs to be focused on developing anther culture techniques that are not genotype-specific and that can give rise to a large number of pollen-derived plants covering a wide spectrum of gametic combinations.

The bulbosum method (*Hordeum vulgare* × *H. bulbosum*) of producing haploids is used in several countries. In this method, any variety can be used for haploid breeding. To illustrate, Ciba-Geigy in Canada developed the first doubled-haploid cultivar Mingo [44] and since then, seven more varieties (Rodeo, Craig, Gwylan, Etienne, Winthrop, TBR579-5, TBC555-1) have been released using the same method [46].

Microspore Culture

Microspores provide another opportunity to produce genetically engineered plants through direct DNA transfer methods and through induced mutations. The low frequency of plant regeneration from isolated microspore culture is one of the major bottlenecks in using this technique. Microspore in culture can undergo repeated divisions to form calli or embroid-like structures. Microspore embryogenesis has been observed in anther/microscope cultures in more than 100 plant species [94]. As an example, one of the most efficient systems for microspore embryogenesis and plant regeneration is observed in *B. napus* [95]. The technique consists of macerating the anthers using a blender or grinder. The macerate is filtered through $40 \times 100 \ \mu m$ nylon mesh and filtrate centrifuged at $100 \times g$ for 5 min. The microspore pellet is resuspended gently in B5 medium [15] and centrifuged 3 more times. The microspores are cultured in filter-sterilized medium. Embryos at the cotyledonary stage begin to appear 14 days after microspore isolation. Plantlets ready for transfer to soil become available approximately 8 weeks after microspore isolation. In *B. napus*, more than 80% of the microspore-derived plants are haploids. Chromosome doubling can be achieved by immersing roots in 0.2% colchine for 2–4 hours.

Plants have been regenerated from microspores of barley, maize, wheat, and rice. Datta et al. [96] produced more than 200 green plants through microspore embryogenesis of indica and japonica rices. Cold pretreatment and modified MS medium [14] containing reduced nitrogen and ficoll seem to favor somatic embryogenesis and green plant regeneration in rice. To get a high frequency of embryos from microspore culture of *Brassica*, it is important that donor plants be grown at low temperatures (10°C day/5°C night, 16 h light) and buds are cultured near first pollen mitosis in B5 medium [15] with $CaCl_2$ increased to 750 mg/liter and supplemented with 13% sucrose [95].

Though microspores are important materials both for mutagenesis and genetic transformation, these have been used on a limited scale because of poor plant regeneration. Some success has been achieved in *Brassica* to produce herbicide-resistant plants through microspore mutagenesis [97]. Recently, Fennell and Hauptmann [98] used electroporation and polyethylene glycol (PEG) treatment for delivery of DNA into maize microspores and obtained a high level of chloramphenicol acetyltransferase (CAT) activity.

SOMACLONAL VARIATION

Somaclonal variation refers to the variation arising through tissue culture in regenerated plants and their progenies. Larkin and Scowcroft [99] reviewed the occurrence of somaclonal variation in various plant species and detected it in tobacco, potato, tomato, wheat, rice, *brassica*, sugarcane, oats, and others. Similarly, somaclonal variation occurred for a series of agronomic traits such as disease resistance, plant height, tiller number, and maturity and for various biochemical traits. Thus, somaclonal variation appears to be a rich and novel source of genetic variability for various traits.

Application of somaclonal variation in crop improvement has been discussed in several publications [100,101]. The technique consists of growing callus or cell suspension cultures for several cycles and regenerating plants from such long-term cultures. Screening for somaclonal variants (as opposed to selection) involves making observations or measurements on a large population of cell lines or regenerated plants until individuals with a new phenotype are detected. Examining large populations of regenerated plants is the most common approach to screening or identifying somaclonal variants.

Some useful somaclonal variants have been isolated, notably resistance to Fiji virus, eye spot, and downy mildew diseases in sugarcane [102], resistance to blight in potato [103], resistance to *Helminthosporium* in maize [104], and resistance to *Fusarium oxysporum* in celery [105]. Shepard et al.

[103] analyzed 10,000 somaclonal variants arising from mesophyll proto-plasts of potato variety Russet Burbank. Variation in tuber shape, yield, maturity, and resistance to *Alternaria solani* and *Phytophthora infestans* was recorded in such somaclones. Heath-Pagliuso et al. [105] regenerated plants from cell suspension and developed the UC-T3 somaclone of celery which is resistant to *Fusarium oxysporum*. This resistance was inherited by the progeny. The findings support the observations that the tissue culture cycle itself can generate genetic variation and that the frequency of soma-clonal variants is more or less similar to those where in vitro selection and mutagenesis of cultured cells were used. It is thus possible to develop disease- and insect-resistant germplasm even if suitable selective agents are lacking.

A few cultivars have been released through somaclonal variation. Sugar-cane cultivar Ono which is resistant to Fiji disease was developed from sus-ceptible cultivar Pindar [106]. A cultivar of sweet potato Scarlet having yield and disease resistance characteristics similar to those of the parent cultivar but with darker and more stable skin color was produced [107]. In tomato, two promising varieties have been developed from cell culture – DNAP-9 having high solid and DNAP-17 with resistance to *Fusarium* race 2 [91]. Some somaclonal variants of anther culture origin have been released as new cultivars in sweet pepper and rice [91,92]. In general, most of the soma-clonal variants have not advanced beyond the laboratory or greenhouse phase, possibly because the selected material has no practical value. This may be explained by unwanted genetic variation or it maybe that the trait obtained was not a novel one.

Several factors such as genetic background, explant source, medium composition, and age of culture affect somaclonal variation. The exact mech-anism of somaclonal variation is not understood. The probable causes in-clude changes in chromosome number and cryptic changes associated with chromosome rearrangement, transposable elements, somatic gene arrange-ments, DNA amplification, and deletion, etc. It is possible that different processes cause such variation in different species or one or several factors operate simultaneously during in vitro culture, resulting in somaclonal var-iation. Phillips et al. [108] hypothesized that mutational events are directly or indirectly related to the modification of DNA, specifically DNA hypo/ hypermethylation. Hirochika [109] reported activation of transposable ele-ments in tissue culture of tobacco. The results on activation of Tnt2 sug-gest that transposable elements are responsible for at least part of the soma-clonal variation.

The main advantage of somaclonal variation is that it involves limited tissue culture facilities and it can be carried out in any tissue culture labora-tory in many developing countries. It could be a good supplement to con-ventional program which aim to overcome specific defects in otherwise well-adapted and high-yielding genotypes. Some bottlenecks encountered are (1) poor plant regeneration from long-term cultures of various cell lines; (2) regeneration being limited to specific genotypes which may not be of much interest to breeders; (3) several somaclonal variants being unstable and not inherited by the progenies; and (4) some somaclonal variants being associated with other undesirable features of aneuploidy, sterility, and others. Breeders will continue to look for useful somaclonal variants in well-adapted cultivars. More research is needed to understand the mechanism of somaclonal variation and to isolate stable somaclonal variants of practical importance.

Somaclonal Variation to Enhance Alien Gene Introgression

The technique appears to be particularly important in enhancing variation in interspecific crosses particularly where the parental genomes of the two species show little or no homology. Under such situations, chromosome breakage and reunion could result in new combinations and in the transfer of alien chromosome segments into cultivated species. A tissue culture cycle of the hybrid material (F_1, monosomic alien chromosome addition or substitution lines, somatic hybrids) could enhance the frequency of genetic exchanges. In hybrids of *Hordeum vulgare* × *H. jubatum*, enhanced variation in isozyme pattern and chromosome pairing was observed in contrast to the original hybrid which was asynaptic [110,111]. Larkin et al. [112] reviewed the usefulness of cell culture to enhance alien introgression in wide crosses.

Tissue culture of interspecific hybrids of barley resulted in regeneration of 1315 plants. Karyotype analysis of 82 plants showed these changes in 46 plants—aneuploidy, deletions, extra C-bands, and extra euchromatic bands [113]. Following the cell culture of wheat × rye hybrids, chromosomal exchanges have been observed between IR and 4D and 3R and 2B chromosomes [114]. Larkin et al. [115] obtained chromosomal exchanges through tissue culture of wheat-rye monosomic addition lines. The derived lines showed introgression of cereal cyst nematode resistance from rye to wheat. Similarly, tissue culture of monosomic addition lines of wheat-*Thinopyrum intermedium* showed introgression of barley yellow dwarf virus resistance into wheat. These examples demonstrate that cell culture can induce chromosomal exchanges between distantly related genomes. The technique appears equally promising to obtain chromosomal exchanges and derive useful progenies from asymmetric somatic hybrids produced through protoplast fusion among widely divergent species.

IN-VITRO SELECTION AT THE CELLULAR LEVEL

The ability to manipulate large populations of homogeneous cells provides the opportunity for in vitro selection of useful mutants. It is much easier to screen 10^6–10^7 cells in small petri dishes than to screen a similar number of whole plants in the field. The latter requires more space and time. In vitro selection at the cellular level involves growing of millions of cells (callus, cell suspension, protoplasts) in tissue culture, followed by exposure of such cultured cells to toxic concentrations of various selective agents such as culture filtrate, pathotoxins, salts, amino acid analogues, etc. Cell selection involves imposing a selection pressure on population of cells so that only rare individuals with a specific phenotype can survive or grow. This is followed by selection of tolerant cells from the exposed cultures in successive subcultures, regeneration of plants from the surviving cells, and eventual testing of the regenerated plants and their progenies for the desired trait. Collin and Dix [116] have reviewed various cell selection strategies used to isolate variant cell lines and plants. Some useful mutants have been identified through in vitro selection (Table 3) and are discussed below.

Disease Resistance

In vitro selection for disease resistance was first demonstrated in 1973 when tobacco plants resistant to wildfire disease (*Pseudomonas tabaci*) were obtained by selecting cell lines tolerant of methionine sulfoximine — an analogue of the wildfire toxin [117]. Since then, several useful mutants for resistance to diseases have been selected using toxin and or culture filtrate as selective agents (Table 3). Behnke [124] used culture filtrate of *Phytophthora infestans* (late blight pathogen) for selection in potato callus cultures grown under toxic conditions. From 41,040 original calli, 153 were found to be growing on toxic medium. Of these, only 36 calli survived. The selected calli showed resistance to the four pathotypes and this resistance was maintained in the regenerated plants. In vitro selection for resistance to *Helminthosporium* T toxin in maize has been well-demonstrated [119]. Larkin and Scowcroft [120] exposed callus cultures of sugarcane cultivar Q101 to *Helminthosporium sacchari* toxin. A total of 480 culture-derived plants (somaclones) were characterized. Several clones tolerant of the toxin could be identified with or without selection pressure. Hartman et al. [125] selected resistance to Fusarium wilt of alfalfa using a culture filtrate of *Fusarium oxysporum*. Resistance was expressed in cultures on inhibitory medium and also in the regenerated plants. Pathotoxins have been used to screen rice for resistance to brown spot (*Helminthosporium oryzae*) disease [122]. Chawla and Wenzel [123] used 3000 calli of barley and 2000 calli of wheat for selection against *Helminthosporium sativum*. The selection resulted in 6–17% surviving calli. Some of the regenerants were found to be resistant to the pathogen.

Protein Quality

Improvement in protein quality requires increases in the concentration of limiting amino acids such as lysine, threonine, and tryptophan. In vitro

Table 3 Some Examples of In Vitro Selection for Isolation of Useful Variants

Selection agent	Trait	Crop	Cell culture	Regenerated plants	Inheritance in progeny	Reference
					Trait expression[a]	
Methionine sulfoximine (*Pseudomonas syringe*)	Pathotoxin resistance	Tobacco	+	+	+	[117]
Partially purified tabatoxin (*Pseudomonas syringae* pv *tabaci*)		Tobacco	+	+	+	[118]
Toxin fraction from culture filtrate (*Alternaria alternata*)		Tobacco	+	+	+	[118]
HmT-toxin (*Helminthosporium maydis* race T)		Maize	+	+	+ (Maternally inherited)	[119]
Partially purified Hs-toxin (*Helminthosporium sacchari*)		Sugarcane	+	+	+ (Vegetative propagation)	[120]
Victoria toxin (*Helminthosporium victoriae*)		Oats	+	+	+	[121]
Culture filtrate (*Helminthosporium oryzae*)		Rice	+	+	+	[122]
Culture filtrate (*Helminthosporium sativum*)		Wheat, barley	+	+	−	[123]
Culture filtrate (*Phytophthora infestans*)		Potato	+	+	−	[124]
Culture filtrate (*Fusarium oxysporum* f. sp. *medicaginis*)		Alfalfa	+	+	+	[125]
Culture filtrate (*Xanthomonas campestris* pv *pruni*)		Peach	+	+	−	[126]

Amino acid analogs	Improved protein quality				
5-methyltryptophan	Rice	+	+	+	[127]
S-(2-amino ethyl)-L-cysteine (AEC)	Rice	+	+	+	[128, 129]
AEC	Pearl millet	+	+	−	[130]
Feedback inhibitors					
Lysine + threonine	Maize	+	+	+	[131, 132]
Threonine	Barley	+	+	−	[133]
Lysine	Tobacco	+	+	+	[134]
	Salt tolerance				
NaCl	Tobacco	+	+	+	[135, 136]
Seawater	Rice	+	+	−	[137]
NaCl	Rice	+	+	−	[138]
	Rice	+	+	+	[139]
	Oats	+	+	+	[139]
	Poncirus trifoliata	+	+	−	[140]
Aluminum	Metal tolerance				
	N. plumbaginifolia	+	+	+	[141]
	Rice	+	+	+	[142]
	Herbicide resistance				
Picloram	Tobacco	+	+	+	[143, 144]
Chlorsulfuron	Tobacco	+	+	+	[145]
Terbutryn	*N. plumbaginifolia*	+	+	+	[146]

[a] + = positive expression. − = not known.

selection has been attempted involving the use of amino acids or their analogs to select cell lines and plants which produce increased concentrations of these amino acids. The biosynthesis of such amino acids is controlled by feedback regulation. Three selection agents have been used to identify mutants overproducing aspartate-derived amino acids; the lysine analog AEC, a combination of lysine plus threonine (LT), and the tryptophan analog 5-methyltryptophan (5-MT). AEC-resistant cell lines have been isolated in rice [127–129] and pearl millet [130]. The regenerated plants contained higher levels of lysine than controls. Similarly, increase in lysine and threonine was obtained in maize [131,132].

Salt Tolerance

Limited progress has been made in enhancing salt tolerance level through conventional breeding. Several attempts have been made to improve salt tolerance through tissue culture. Salt-tolerant cell lines have been isolated in several species such as tobacco, *Capsicum*, and alfalfa. But only limited progress has been achieved in isolating plants with increased level of salt tolerance in tobacco and rice (Table 3). The regenerated plants, particularly the progenies, do not exhibit a high degree of salt tolerance. One of many reasons cited is that the polygenic nature of salt tolerance and the physiological adaptation of cultured cells limit the usefulness of this technique to isolate useful salt-tolerant variants. In vitro selection has also been used to obtain plants with increased herbicide resistance [143–146].

The findings on in vitro selection at the cellular level show that it is possible to isolate cell lines tolerant of selective agents and to obtain several kinds of useful somaclonal variants. However, none of these variants have been used in practical plant improvement.

Certain problems need to be addressed before in vitro selection can be routinely used for developing useful germplasm. Some of these include: (1) plant regeneration from cultured cells is usually low in many crop plants and moreover restricted to only a few genotypes, thus limiting its application; (2) cell lines selected under toxic conditions generally show poor plant regeneration; (3) in vitro selection is under epigenetic effect in many cases; (4) correlation between in vitro selection at the cellular level and resistance to pathogen and other biotic and abiotic stresses at the plant level may not be very strong; and (5) appropriate selective agents are lacking for many fungal, viral, and bacterial diseases and insect pests.

SOMATIC CELL HYBRIDIZATION

Somatic cell hybridization involves isolation, culture, and fusion of protoplasts from different species and regeneration of somatic hybrid plants. It is an alternative to sex "parasexual hybridization." The technology provides enormous potential to: (1) produce somatic hybrid "novel plants" between otherwise sexually incompatible species; (2) produce cybrids and exploit the cytoplasmic variability (mitochondrial recombination); (3) transfer cytoplasmic male sterility (CMS) into adapted cultivars through fusion of X-irradiated protoplasts of the CMS parent; (4) study compatibility between nuclei and cytoplasms of different species; and (5) in the uptake of chloroplasts and isolated chromosomes into protoplasts. The protoplasts or "naked" cells can be isolated enzymatically from any plant organ or cultured cell. Upon culturing in nutrient medium, these protoplasts regenerate cell walls and subsequent divisions result in callus formation and plant regeneration. In certain species, plant regeneration from protoplasts can be achieved through somatic embryogenesis. Since the first demonstration of plant regeneration from mesophyll protoplasts of tobacco [147], protoplasts have been successfully cultured and regenerated into plants from more than 300 species [S. Roest, personal communication, 148]. Vasil and Vasil [149] reviewed the successful culture and regeneration of plants from protoplasts of recalcitrant cereals such as wheat, maize, rice, pearl millet, sugarcane, sorghum, and barley. The development of embryogenic cell suspensions has been the key factor for success in plant regeneration from protoplasts of cereals and many other species such as sandalwood, white spruce, etc. [150]. Plant regeneration has also been obtained from protoplasts of woody plants such as citrus, coffee, *Eucalyptus* spp., white spruce, *Pinus taeda, Populus, Prunus,* and *Pyrus* species [148,151].

Production of Somatic Hybrids

Several methods are available to induce fusion between protoplasts of different species and to select hybrid cells and regenerate somatic hybrid plants from such cells [152]. Polyethylene glycol (PEG) is the most commonly used chemical to induce fusion among protoplasts [153], however, electrofusion is also employed to induce protoplast fusion [154]. The hybrid cells can be selected mechanically or on the basis of biochemical markers. These hybrid cells are cultured into suitable nutrient media to regenerate somatic hybrid plants. Precision for sorting fusion products (hybrid cells) have increased using flow cytometry [155].

Carlson et al. [156] were the first to produce somatic hybrid between *Nicotiana glauca* and *N. langsdorffii.* Since then somatic hybrids have been produced through protoplast fusion in numerous plant species [157]. Most somatic hybrids produced among divergent species are asymmetric. Numerous somatic hybrids have been produced between species of the same genera. Some examples of intergeneric somatic hybrid plants produced through protoplast fusion are given in Table 4. Notable examples on intergeneric somatic hybrid plants include Arabidobrassica (*Arabidopsis thaliana* +

Table 4   Some Examples of Intergeneric Somatic Hybrid Plants Produced Through Protoplast Fusion

Hybrid combination	Reference
Solanum tuberosum + *Lycopersicon esculentum*	[158, 159]
Arabidopsis thaliana + *Brassica campestris* (*Arabido brassica*)	[160]
Datura innoxia + *Atropa belladonna*	[161]
Aegopodium podagraria + *Daucus carota*	[162]
Nicotiana tabacum + *Hordeum vulgare*	[163]
N. plumbaginifolia + *Atropa belladonna*	[164]
N. tabacum + *Daucus carota*	[165]
N. tabacum + *Hyoscyamus muticus*	[166]
N. tabacum + *Petunia hybrida*	[167]
Physalis minima + *Datura innoxia*	[168]
Moricandia arvensis + *Brassica oleracea* (*Brassicomoricandia*)	[169]
Sinapsis turgida + *B. oleracea*	[169]
Sinapsis turgida + *B. nigra*	[169]
Brassica juncea + *Diplotaxis muralis*	[170]
B. napus + *Eruca sativa* (*Erucobrassica*)	[171]
B. juncea + *Eruca sativa* (*Erussica*)	[172]
B. napus + *Raphanus sativus*	[173]
Citrus sinensis + *Poncirus trifoliata*	[174]
C. sinensis + *Severinia disticha*	[175]
C. sinensis + *Murraya paniculata*	[176]
C. aurantifolia + *Feroniella lucida*	[177]
C. aurantifolia + *Swinglea glutinosa*	[177]

Brassica campestris), pomatoes and topatoes (potato + tomato), *daucus carota* + *Aegopodium podagraria, Brassicomoricandia* (*Brassica* + *Moricandia*), and *Eruco brassica* (*Brassica napus* + *Eruca sativa*). Recently, somatic hybrids have also been produced in sexually incompatible woody genera such as citrus [174–177] and *Prunus* [178]. The somatic hybrid cells and regenerated plants have been characterized on the basis of chromosome behavior, isozyme pattern, plant morphology, and RFLP analysis [168,173, 178–180].

Although a number of somatic hybrids between distantly related species have been produced during the last 20 years, the practical value of these hybrids has been extremely limited. The greatest value of somatic hybrids appears to be in combining cytoplasm of one species with the nucleus of the other to obtain cytoplasmic male sterile combinations. Somatic hybridization can also be employed for intervarietal transfer of cytoplasmic male sterility. Somatic hybrids of vegetatively propagated crops may be useful if found superior to either parental species or adapted to certain ecological niches. Recently, tetraploid potato hybrids produced through protoplast fusion of diploid (2n = 2x = 24) *Solanum tuberosum* and *S. phureja* have been found to be promising under field conditions [180]. One hybrid clone (35-4) had higher yield than the commercial variety "Bintje." It is interesting to note that this technique is included in commercial breeding programs in the Netherlands. A large number of somatic hybrids need to be examined to find those of value in agriculture. Lelivelt and Krens [173] produced beet cyst nematode (BCN)-resistant somatic hybrids through protoplast fusion between *Brassica napus* and *Raphanus sativus*. Although the hybrid had a high level of resistance, it showed reduced fertility and could not be backcrossed with *B. napus*. More efforts are needed to transfer the BCN resistance of *R. sativus* to *B. napus*.

Protoplast fusion offers promise for crop improvement. No new varieties, however, have been produced using this fusion. The main bottlenecks include the following: (1) somatic hybrids are amphiploid and thus need to be backcrossed to the cultivated crop to transfer a few useful genes and develop new varieties. Since somatic hybrids are usually sterile, it may be necessary to use back fusion or embryo culture to isolate stable combinations; (2) lack of chromosome pairing and recombination between genomes of parental species hinders the transfer of useful genes into cultivated varieties. Alien chromosome substitution is likely to affect yield adversely, and it may be necessary to induce mitotic recombination prior to plant regeneration to ensure intergenomic gene transfer. The progenies derived from asymmetric hybrids with chromosomal exchanges between divergent genomes hold promise in crop improvement.

Partial genome transfer is considered an alternative approach to gene transfer through protoplast fusion. Ramulu et al. [181] have reviewed the procedures for isolation and fusion of microprotoplasts. Microcell fusion is a routine technique for partial genome (one or a few chromosomes) transfer in human and other mammalian cell systems. Spangenberg et al. [182] described details of various procedures for protoplast–cytoplast microfusion. It is now possible to transfer a single chloroplast and establish a new plastid population. The functional DNA can be introduced into the cytoplasm and expressed in the organelle compartment.

Production of Cybrids and Organelle Recombinants

The production of cytoplasmic hybrids (cybrids) following protoplast fusion is one of the most exciting applications of protoplast fusion. In cybridization, the nuclear genome of one parent is combined with the organelles of a second parent. In effect, organelles are transferred from one parent to the other in a single step. Cybrids have been produced in several species such as *Nicotiana*, rice, and citrus through protoplast fusion. The cybridity is confirmed on the basis of mt DNA restriction patterns, morphological traits, and isozyme and cytological tests.

In several cases, the mtDNA of the somatic hybrids differ from both parents. Nonparental mtDNA are interpreted to be the result of genetic recombination of the parental mitochondrial genomes [183,184]. Vardi et al. [185] produced novel alloplasmic (cybrid) plants between *Citrus aurantium* and *Microcitrus* and found complete sorting out of chloroplasts. Intergeneric fusion resulted in mitochondria with novel DNA, however, indicating recombination between the organelle genomes of the two species. Intergenomic recombination between mitochondrial genomes of *Brassica* has also been reported in the regenerated somatic hybrid plants [186].

During the process of cell division, the organelle genomes begin to sort out to yield cell lines with one or the other parental organelle type. Sorting out of chloroplasts and mitochondria occur independently or may cosegregate when selective pressure is exerted on early fusion products. For chloroplasts, only one parental type is usually found in somatic hybrid plants, although a mixture of chloroplasts may persist in the progeny of fusion products. Recombination of chloroplast DNA rarely occurs [187]. It is possible to recover plants from protoplast fusion that contain chloroplasts from one parent and mitochondria from the other parent. Protoplasts provide unique opportunity to produce cybrids and recombine cytoplasmically inherited traits [183,187,188].

Transfer of Cytoplasmic Male Sterility

Transfer of CMS to elite breeding lines requires 5–7 repeated backcrosses. The success in protoplast fusion has made it possible to transfer CMS within several months. Some notable examples of the successful transfer of CMS by protoplast fusion in plants are given in Table 5. Donor-recipient protoplant fusion is an efficient procedure for the selective formation of cybrids and for the transfer of CMS. Protoplasts of the donor CMS line are exposed to high doses of irradiation and fused with the iodoacetamide-treated protoplasts of the recipient line. Irradition inactivates the nucleus while chemical treatment with iodoacetamide prevents cell division. As a result, metabolically complementary cells are capable of developing into plantlets after fusion treatment. This method does not require use of selectable cytoplasmic marker of the donor. The donor-recipient method has been used in *Nicotiana, Petunia, Brassica,* and rice to transfer CMS into elite lines (Table 5).

Aviv and Gallun [193] restored fertility in a CMS line of *N. sylvestris* following protoplast fusion of X-irradiated fertile *N. sylvestris*. Pelletier et al. [188] produced *Brassica-Raphanus* cybrids, having nucleus of *Brassica*

Table 5   Some Examples Showing Transfer of Cytoplasmic Male Sterility (CMS) and Other Traits Through Protoplast Fusion

Recipient parent	Donor parent	Trait transferred	Reference
Nicotiana sylvestris	*N. tabacum*	CMS	[189]
Petunia hybrida	*P. axillaris*	CMS	[190]
Nicotiana plumbaginifolia	*N. tabacum*	CMS	[191]
Brassica napus	*Raphanus sativus*	Triazine resistance	[188]
N. tabacum	*N. plumbaginifolia*	Triazine resistance	[192]
N. sylvestris	*N. plumbaginifolia*	Tentoxin resistance	[193]
Brassica napus	*B. napus*	CMS	[194]
O. sativa	*O. sativa*	CMS	[195, 196]
Daucus carota	*D. carota*	CMS	[197]

napus, chloroplasts of atrazine-resistant *B. campestris*, and mitochondria that confer male sterility from *Raphanus sativus*. Yang et al. [195] produced cybrid plants in rice by electrofusing gamma-irradiated protoplasts of A-58 CMS and iodoacetamide-treated protoplasts of the fertile cultivar Fujiminori. Cybrids had peroxidase isozyme of the fertile parent but had four plasmid-like DNAs (B1, B2, B3, and B4) from the sterile A-58 CMS parent in their mitochondrial genomes. Protoplast fusion has been successfully used to transfer CMS from indica rice Chinsurah Boro into 35 japonica rice varieties ([196], T. Fujimura, Mitsui Toatsu Chemical Inc., Japan 1992, personal communication).

PROTOPLAST- AND TISSUE CULTURE-MEDIATED GENETIC TRANSFORMATION AND PRODUCTION OF TRANSGENIC PLANTS

Genetic transformation refers to the transfer of foreign genes isolated from plants, bacteria, or animals into a new genetic background. It requires production of normal fertile plants which express the newly introduced gene. The process involves several distinct stages, namely insertion, integration, expression, and inheritance of the new DNA. Foreign genes could be transferred through vector-mediated (*Agrobacterium*) or direct DNA transfer methods using protoplast or other tissues. The genetic transformation techniques are essentially dependent on precise tissue culture methods. Some of the transformation techniques are discussed below.

Agrobacterium-Mediated DNA Transformation

Agrobacterium tumefaciens has provided an effective DNA delivery system for transformation of several dicot species. The development of simple cocultivation techniques such as the leaf disc transformation method [198] and use of selectable marker genes encoding for kanamycin resistance have allowed routine production of transgenic plants. In this method, DNA is transferred to plant cells using the natural gene transfer capacity of the soil bacterium *Agrobacterium*. Surface-sterilized leaf discs or other axenic explants (seedlings, hypocotyl, cotyledons, embryos, or shoot apex) are infected with *A. tumefaciens* and cocultured on regeneration media for 2 or 3 days. After transformation, the explants are transferred to the regeneration/selection medium. During the next 3 weeks, the transformed cells grow into callus or differentiate into shoots via organogenesis. Between 3 and 6 weeks, the shoots developed are grown for rooting for transfer to soil. The putative transformants are tested for integration of gene-using southern hybridization techniques. *Agrobacterium*-mediated DNA transformation has been used to produce a large number of transgenic dicot plants [16,17].

In certain species such as cereals, efficient DNA vectors for transformation are lacking; direct DNA transfer or protoplast-mediated transformation are commonly followed. Some reports indicate *Agrobacterium*-mediated transformation in cereals, but convincing evidences are still lacking. Potrykus [199] has made a critical assessment of *Agrobacterium*-mediated transformation in cereals.

Protoplast-PEG-Mediated Transformation

PEG has been an important fusion agent in somatic cell hybridization of various plant and animal species. It has been equally effective in delivering DNA into protoplasts and subsequently in obtaining transformed cells. Paszkowski et al. [200] transferred foreign genes in the form of plasmid DNA into plant protoplasts and obtained transformed tobacco plants. Since then, numerous reports on PEG-mediated protoplast transformation have appeared [201–203]. Wang et al. [202] obtained transgenic plants of tall fescue (*Festuca arundinacea*) by direct DNA transfer through PEG treatment of protoplasts. Datta et al. [203] produced herbicide-resistant IR72 through PEG-mediated transformation of rice protoplasts.

Protoplast-Electroporation-Mediated Transformation

Electroporation involves the application of high-voltage electrical pulses to solutions containing protoplasts and foreign DNA. The introduced DNA enters the cells through reversible pores created in the protoplast membrane by the action of the short electrical pulse treatments. Electroporation has been used to introduce foreign DNA into protoplasts and produce transgenic rice [204], maize [205], and other plants. Electroporation has also been used for delivery of DNA into maize microspores [98].

Transformation Through Microprojectile Bombardment of Cultured Cells and Plant Organs

Several plant species have been transformed by the use of high-velocity microprojectiles to deliver DNA directly into cultured cells and intact plant organs. The technique utilizes DNA-bearing particles that are accelerated to velocities that permit their penetration into cell walls and membranes. Genes can be introduced into a wide range of tissues including suspension cultures, callus cultures, and tissues isolated directly from plants or even tissues of whole seedlings, immature embryos, and others. Several plant species including tobacco, soybean, rice, and maize have been transformed by microprojectile bombardment. Vasil et al. [206] produced herbicide-resistant transgenic wheat through microprojectile bombardment of embryogenic callus. Transformed plants were resistant to *basta* and the *bar* gene segregated as a dominant Mendelian trait in R_1 and R_2 plants. Koziel et al. [207] produced transgenic maize through microprojectile bombardment of immature embryos. Plants expressing high levels of insecticidal protein derived from *Bacillus thuringiensis* were field-tested and they exhibited a high degree of resistance to European corn borer. Foreign DNA has also been introduced into chloroplasts of tobacco by microprojectiles [208].

Transformation via Microinjection of Cultured Cells

Microinjection refers to the delivery of foreign DNA into cells using a glass micropipette having an orifice of less than 1 μm. It is a useful technique for the precise delivery of DNA into target cells. In this method, location of DNA delivery can be controlled using a micromanipulator and the volume of injection can be controlled by the microinjection apparatus. Microinjection of DNA into individual cells using commercially available micromanipulators fixed to the movable stage of an inverted microscope has been used in transformation of tobacco protoplasts [209] and multicellular microspores of *Brassica* [210]. Schweiger et al. [211] have reviewed the potentials of microinjection in genetic manipulation of plants. Microinjection into vacuolated, nonembryogenic cells, and into protoplasts has not been very successful as much of the DNA is injected into vacuoles of these cells and is degraded. Microinjection of evacuolated protoplasts is being considered as an alternative approach to genetic transformation [212]. Microinjection requires technical skill, precision equipment, and low-density cell culture system with a high plant regeneration ability. The method has limited success as compared with the other transformation techniques.

Table 6 Recent Examples on Transgenic Plants Carrying Agronomically Useful Genes

Crop	Foreign gene	Gene transfer method	Useful trait(s)	Reference
		Herbicide resistance		
Rice	*bar*	Protoplast-PEG mediated	Tolerance for basta (phosphino-thricin)	[203]
Rice	*bar*	Microprojectile bombardment of suspension culture cells	Tolerance for basta (phosphino-thricin)	[213]
Wheat	*bar*	Microprojectile bombardment of embryogenic callus	Tolerance for basta	[206]
Tall fescue	*bar*	Protoplast-PEG mediated	Tolerance for phosphinothricin	[202]
		Insect resistance		
Tomato	*Bt*	*Agrobacterium*-mediated	Tolerance for lepidoptera insects	[214]
Cotton	*Bt*	*Agrobacterium*-mediated	Tolerance for cotton bollworm (*Heliothis zea*)	[215]
Populus	*Bt*	Microprojectile bombardment	Tolerance for lepidoptera pests	[216]
Maize	*Bt*	Microprojectile bombardment of immature embryos	Tolerance for European cornborer	[207]
Rice	*Bt*	Electroporation of protoplasts	Tolerance for striped stemborer and leaffolder	K. Shima-moto (pers. communication)
Tobacco	*p-lec* (pea lectin)	*Agrobacterium*-mediated	Tolerance for *Heliothis virescens*	[217]
Tobacco	*CpTi*	*Agrobacterium*-mediated	Tolerance for lepidoptera pests	[218]

Virus protection

Tomato	CP-TMV	*Agrobacterium*-mediated	Tolerance to tobacco mosaic virus	[219]
Rice	CP-stripe virus	Protoplast electroporation	Tolerance to stripe virus	[220]

Fungal resistance

Tobacco and rapeseed	Bean chitinase gene	*Agrobacterium*-mediated	Increased resistance to *Rhizoctonia solani*	[221]
Tobacco	Barley ribosome in-activating protein (RIP)	*Agrobacterium*-mediated	Increased protection against *R. solani*	[222]

Male sterility and fertility restoration

Tobacco, rapeseed	Ribonuclease (barnase) gene	*Agrobacterium*-mediated	Dominant genetic male sterility	[223]
Tobacco, rapeseed	Ribonuclease inhibitor (barstar) gene	*Agrobacterium*-mediated	Dominant male fertility restoration	[224]

Abiotic stress tolerance

Tobacco	Glycerol-3 phosphate acyl transferase	*Agrobacterium*-mediated	Decreased chilling sensitivity	[225]
Tobacco	*mt*ID	*Agrobacterium*-mediated	Tolerance for salinity	[226]

Transgenic Plants Carrying Agronomically Useful Genes

As previously discussed, a number of procedures have been used to insert foreign DNA and produce transgenic plants. Cocultivation of leaf disc and the Ti plasmid carrying foreign DNA (target gene) followed by selection of transformants on selective culture medium has been the major method of gene transfer in dicots. In monocots, protoplast-mediated transformation and, more recently, microprojectile bombardment of cultured cells and tissues have been commonly used to produce transgenic plants. Efficient tissue culture techniques are required to recover high frequency of transformants. Transgenic plants have been produced in more than 45 plant species such as wheat, rice, maize, oats, potato, cotton, tomato, *Brassica*, and soybean [16–18]. In majority of the cases, reporter genes of bacterial origin have been transferred. However, some genes of agronomic importance for insect resistance (*Bt, CpTi*), herbicide resistance (*aroA, bar, bxn*), virus protection (coat protein genes-TMV, CMV, PVX, AIMV), fungal resistance (bean chitinase), male sterility (ribonuclease-barnase), and tolerance to abiotic stresses (*mtld*) have also been transferred (Table 6).

Protoplast technology is an important prerequisite for direct DNA transfer to achieve genetic transformation in monocot species. In many species, however, plant regeneration from protoplast is poor and often limited to specific genotypes. There is thus a need to develop protoplast culture techniques which are applicable to a wide range of genotypes. Microprojectile bombardment of cultured cells, particularly plant organs such as immature embryos and meristems, hold great promise to enhance the efficiency of genetic transformation and should be applicable to a large number of genotypes and species recalcitrant to tissue culture.

SOMATIC EMBRYOGENESIS

Somatic embryogenesis is the process by which somatic cells develop through the stages of embryogeny to give whole plants without gametic fusion. Somatic embryogenesis has been reported in more than 150 plant species. Somatic embryos have been induced from a variety of plant tissues such as germinating seedlings, shoot meristems, young inflorescence, nucellus, leaf, anther, root, and others. The embryonic cells may develop directly from explant cells or indirectly with a number of unorganized, nonembryogenic mitotic cycles interposed between differentiated explant tissues and recognizable embryonic structures. The cells undergoing somatic embryogenesis could either be preembryogenic determined cells (PEDCs) or induced embryogenic determined cells (IEDCs). PEDCs are epigenetically embryonic at explanting (culturing time) whereas IEDCs are the product of an epigenetic switch to the embryonic state in culture. Once induced, IEDCs are functionally equivalent to PEDCs and both can be maintained and multiplied in the embryogenic state under appropriate culture conditions. Induction of the embryogenic state in differentiated explants often requires extensive proliferation through unorganized callus cycles. Major advances have been made to understand the role of auxins and cytokinins in the induction of embryogenesis from cultured cells, maturation of embryos, and germination of embryos into plantlets [227-229]. These advances have provided opportunity for producing artificial seeds.

Redenbaugh et al. [230] identified various stages in the production of somatic embryos and their utilization as artificial seeds: (1) selection of candidate crops based on both technological and commercial potential; (2) optimization of somatic embryogenesis system from cultured cells; (3) optimization of embryo maturation; (4) automation of embryo production; (5) production of mature synchronized embryos; (6) encapsulation of embryos with necessary adjuvants; (7) coating of encapsulated embryos; (8) optimization of greenhouse and field conditions for conversion of embryos to plants; and (9) delivery system for artificial seeds in terms of increased productivity. Considerable progress has been made to refine these stages. However, more research is needed to enable use of this technology in practical plant improvement.

The first application of artificial seed technology in breeding may be in alfalfa and celery where considerable progress has been made. Fujii et al. [229] could grow alfalfa plants from artificial seeds planted directly in the field. Various methods of protective covering of the embryos were used. Naked somatic embryos had an average of 13 and 9% conversion in the field under plastic and cloth covering, respectively, whereas the corresponding figures for encapsulated embryo were 5 and 14%. Directly planted embryos converted at 1% in the field. The successful conversion of coated and naked somatic embryos planted in the field indicates the possibility of using artificial seeds as supplement to natural seeds.

Synthetic seed technology via somatic embryogenesis would be very useful for multiplying true-to-type plants from various fruit and tree species, other recalcitrant seed-producing species, and for multiplying promising heterotic F_1 hybrids, male sterile lines, and others. The technique is still in its infancy. A major problem is the conversion of embryos (whole plant) planted in soil environment. The method may not be economical unless an efficient system is developed for direct field planting. Other problems need to be overcome before this technology is adopted: (1) lack of efficient procedures for induction of somatic embryogenesis from cultured cells of many genotypes; (2) poor synchronization of somatic embryo development; (3) incomplete somatic embryo development resulting in poor plant regeneration;

and (4) poor viability of somatic embryos following storage, handling, or transportation.

Improving somatic embryogenesis technology coupled with embryo desiccation and encapsulation may lead to the use of artificial seeds in the commercial production of high-value genotypes and as parental seed stocks of hybrid varieties.

CRYOPRESERVATION OF GERMPLASM

In vitro conservation of germplasm is appropriate where in situ conservation may prove inadequate. Such conservation is of particular value for vegetatively propagated plants, species with recalcitrant seeds, and other genetically engineered germplasm. The material could be cultured cells, meristems, somatic embroids, or plant organs. Cryopreservation is the storage of living biological material at ultralow temperature (normally at or near $-196°C$), the temperature of liquid nitrogen. Since all metabolic activities of plant cells are arrested at this temperature, the germplasm could be stored indefinitely in genetically unaltered conditions. This approach has opened up new possibilities for long-term storage of germplasm and for strengthening the gene banks. The technique consists of preserving the meristems, shoot tips, or cultured cells of the elite germplasm with the objective of maintaining viability for a longer period and without any genetic change. Various procedures for cryopreservation of plant cells have been described [231]. Meristems are more suitable as they grow out directly into plantlets. The method also facilitates the international exchange of germplasm. The first shoot tips to be successfully grown after being exposed to liquid nitrogen were those of carnation *Dianthus caryophyllus* [232]. Since then, meristems or shoot tips from a number of plant species such as pea, cassava, strawberry, potato, chickpea *Rubus*, and others have been cryopreserved with varying degrees of success (Table 7). Procedures for cryopreservation of alkaloid-producing cell cultures of periwinkle [244] and cell suspension of rice have been described [245]. The suspension culture of periwinkle established from liquid nitrogen-stored cells retained the capability of alkaloid synthesis and accumulation. There is a great need to preserve elite germplasm without any loss of viability or genetic change of the vegetatively propagated crops such as sugarcane, cassava, sweet potato, including other horticultural and tree species (cacao, coffee, oil palm, coconut, rubber tree) which are highly heterozygous, have long generation time, and are recalcitrant in producing good seed. The International Board for Plant Genetic Resources has also established a network to facilitate conservation and movement of vegetatively propagated germplasm.

Table 7   Some Examples of Cryopreservation of Plant Materials as a Means of Long-Term Storage of Germplasm

Species	Organ	Survival	Reference
Carnation	Shoot tips	15–30% callus, shoots	[232]
Apple	Buds	100% regrowth	[233]
Carrot	Somatic embryos and clonal plantlets	100% meristem growth	[234]
Cassava	—	Up to 86% (callus + leaves and plantlets)	[235]
Strawberry	Meristem	95% plant regeneration	[236]
Peanut	Meristem	23–30% regrowth	[237]
Pea	Shoot tip	73% plants	[238]
Potato	Shoot tip	60% survival	[239]
Oil palm	Adventitious somatic embroids	Normal plant development	[240]
Mulberry	Intact vegetative buds	79% bud + scale survival 26% shoot formation	[241]
Rubus	Meristem	40–63% regrowth	[242]
Brussels sprouts (*Brassica oleracea*)	Apices (1–2 mm)	85–93% survival	[243]

Some of the factors which need to be considered during cryopreservation include (1) the development stage and cellular activity of the starting material; (2) pretreatment to avoid injury due to low temperature or desiccation; (3) incubation in suitable cryoprotectants; (4) slow cooling rate and gradual reduction in temperature; and (5) identification of suitable environmental conditions and culture medium for normal recovery of cells and tissues. Further improvement are needed to overcome the problems of loss of cell viability, poor plant regeneration ability, and the occurrence of heritable changes during cryopreservation.

GENETICS OF TISSUE CULTURE RESPONSE

There exists voluminous literature on genotypic differences of tissue culture response as measured by callus formation, somatic embryogenesis, and plant regeneration ability in several plant species. Some information on the genetics of tissue culture response has recently become available. Kaleikau et al. [246,247], through aneuploid analysis, found that tissue culture response (callus formation and callus growth rate) in wheat was controlled by genes located on chromosome 2. Both major and minor genes affected tissue culture response. Crosses between monosomic series Wichita and a highly regenerable line ND7532 were analyzed. Segregation in the cross involving monosomic 2D showed a high frequency of regeneration (93.6%) and high callus growth rate (1.87 g/90 days), indicating that the genes for tissue culture response are located on chromosome 2D. Substitution of chromosomes from a low regenerable cultivar Vona showed that chromosome 2D possesses genetic factors which promote callus growth and regeneration.

Lazar et al. [248] cultured anthers and immature embryos from seven addition lines of rye in the wheat variety Chinese spring. The findings indicated that chromosome 4 contained a factor which promotes anther culture response and that rye chromosomes 6 and 7 possess genes which enhance immature embryo culture response. Nadolska-Orczysk and Malepszy [249] analyzed regeneration ability in three inbred lines of cucumber, their reciprocal hybrids, F_2 and BC_1 generations. The ability to regenerate plants from leaf explants of cucumber was found to be controlled by 2 pairs of dominant genes, characterized by complementary and probably additive interaction. There were no reciprocal differences in these crosses.

Through RFLP analysis, genes for anther culturability as measured by embryo-like structure have recently been located on chromosomes 3 and 9 of maize [93]. High anther culture response is conditioned by two recessive genes (which are epistatic), and two minor genes. In another study on maize, six chromosomal regions on chromosomes 1, 2, 3, 6, and 8 were found to be associated with the formation of embryo-like structure from microspores or the subsequent formation of regenerating callus from such embroids [250]. Genes for high culturability can be introduced into elite breeding lines and thus enhance the efficiency of tissue culture techniques in crop improvement.

FUTURE PROSPECTS

Considerable progress has been made to refine procedures for the successful culture and regeneration of plants from somatic cells, pollen, and protoplasts of a large number of plant species. It is now possible to manipulate through tissue culture some of the recalcitrant species such as cereals, legumes, and tree species. Techniques of in vitro propagation and meristem culture are commercially used in the production of virus-free germplasm and in the mass multiplication of several ornamental plants and other vegetatively propagated species. These techniques need to be extended to the multiplication of woody species like coconut, oil palm, rubber tree, teak tree, and other trees which are highly heterozygous, have long generation time, and are difficult to improve through conventional procedures, and whose worth is known only when they become old enough. Embryo rescue will continue to be an important technique to overcome incompatibility in wide crosses and to widen the gene pool of crop plants. However, emphasis is needed to develop procedures to culture young embryos which abort at the very early stages of development. The potential of anther culture has been demonstrated and a few improved varieties have also been developed; however, the technique has been used to a limited extent due to the low frequency of haploid plants and to regeneration being restricted to only a few genotypes. Emphasis should be given to develop protocols for regenerating haploid plants in a large number from a wide range of genotypes. Future research should focus on isolation of stable somaclonal variants of practical importance. Somatic hybrids from several species have already been produced but their use per se has been limited by high sterility and lack of chromosome pairing and recombination between parental genomes. Somatic hybridization offers promise to transfer CMS into elite breeding lines in the shortest time possible. Plant regeneration from protoplasts and selection of hybrid cells are major limitations which need to be overcome. Protoplast technology will continue to be an important means of direct gene transfer especially in cereals and other monocots until other alternative methods for transformation become available. Cryopreservation holds great promise for long-term storage of germplasm of recalcitrant species and other elite plant materials. The problems of loss in regenerability of cryopreserved meristems and cultured cells need to be solved.

Tissue culture techniques coupled with recent advances in molecular biology have opened new avenues for broadening the crop gene pool and for increasing the efficiency of various conventional plant improvement programs. The application of some of the tissue culture techniques in crop improvement has been quite limited. The main reasons seem to be genotype specificity of the tissue culture techniques and their expense. Quite often, cultivars or genotypes of breeding value are difficult to manipulate in tissue culture. Besides, many of the tissue culture techniques are unsuitable for manipulating the traits governed by polygens. Simple and reproducible tissue culture techniques that require low inputs and are applicable to a wide range of germplasm are needed for use in plant improvement.

ACKNOWLEDGEMENT

D.S.B. is supported by the Rockefeller Foundation Grant 92002 No. 9.

REFERENCES

1. P. R. White, *Plant Physiol., 9*: 585 (1934).
2. P. R. White, *Bul. Torrey Bot. Club., 66*: 507 (1939).
3. R. J. Gautheret, *C.R. Hebd. Seances Acad. Sc., 208*: 118 (1939).
4. P. Nobecourt, *C.R. Seances Soc. Biol. Ser. Fil., 140*: 953 (1939).

5. G. Morel and C. Martin, *C.R. Hebd. Seances Acad. Sc., 235*: 1324 (1952).
6. F. Skoog and C. O. Miller, *Symp. Soc. Exp. Biol., 11*: 118 (1957).
7. J. Reinert, *Ber. Dtsch. Bot. Ges., 71*: 15 (1958).
8. W. Tulecke and L. G. Nickell, *Science, 130*: 863 (1959).
9. E. C. Cocking, *Nature, 187*: 927 (1960).
10. H. Guha and S. C. Maheswari, *Nature, 204*: 497 (1964).
11. H. Guha and S. C. Maheswari, *Nature, 212*: 97 (1966).
12. V. Vasil and A. C. Hildebrandt, *Science, 150*: 889 (1965).
13. L. Ledoux, *Prog. Nucleic Acid Res. Mol. Biol., 4*: 231 (1965).
14. T. Murashige and F. Skoog, *Physiol. Plant., 15*: 473 (1962).
15. O. L. Gamborg, R. A. Miller, and K. Ojima, *Exp. Cell Res., 50*: 151 (1968).
16. H. Uchimiya, T. Handa, and D. S. Brar, *J. Biotechnol., 12*: 1 (1989).
17. D. S. Brar and H. Uchimiya, *Plant Tissue Culture: Applications and Limitations* (S. S. Bhojwani, ed.), Elsevier Science, Amsterdam, The Netherlands, p. 346 (1990).
18. A. Hiatt, *Transgenic Plants: Fundamentals and Applications*, Marcel Dekker, New York, p. 340 (1993).
19. T. Murashige, *Ann. Rev. Plant Physiol., 25*: 135 (1974).
20. T. Murashige, *Handbook of Plant Cell Culture*, Vol. 5 (P. V. Ammirato, D. A. Evans, W. R. Sharp, and Y. P. S. Bajaj, eds.), McGraw-Hill, New York, p. 3 (1990).
21. R. D. Hartman, *The Role of the Micropropagation Business in the Invention and Commercialization of Plants* (T. J. Mabry, ed.), Plant Biotechnology Institute, Austin, Texas, p. 193 (1988).
22. P. K. Gupta, A. L. Nadgir, A. F. Mascarenhas, and V. Jagannathan, *Plant Sci. Lett., 17*: 259 (1980).
23. D. I. Dunstan, *Can. J. Forest Res., 18*: 1497 (1988).
24. S. C. Halos and N. E. Go, *Plant Cell Tissue and Organ Culture, 32*: 47 (1993).
25. M. G. Mullins, *Plant Tissue and Cell Culture* (C. E. Green, D. A. Somers, W. P. Hackett, and D. D. Biesboer, eds.), Alan R. Liss, New York, p. 395 (1987).
26. V. M. Villalobos, *Biotechnology in Developing Countries* (A. Sasson and V. Costarini, eds.), Chayce Publishing Services, p. 247 (1989).
27. R. Levin, V. Gaba, B. Tal, S. Hirsch, D. Denola, I. K. Vasil, and B. Hall, *Biotechnol., 6*: 1035 (1988).
28. P. Limmaset and P. Cornuet, *Compt. Rend., 228*: 1971 (1949).
29. K. K. Kartha, *Plant Breed. Rev., 2*: 265 (1984).
30. L. Zhensheng, *Agricultural Biotechnology* (C. B. You and Z. L. Chen, eds.), China Science and Technology Press, Beijing, China, p. 1 (1992).
31. W. M. Roca, *Biotechnology in International Agricultural Research*, IRRI, Manila, Philippines, p. 3 (1985).
32. S. Y. Ng and S. K. Hahn, *Biotechnology in International Agricultural Research*, IRRI, Manila, Philippines, p. 29 (1985).
33. D. G. A. Walkey, *Frontiers of Plant Tissue Culture* (T. A. Thorpe, ed.), IPTAC, Calgary, Canada, p. 245 (1978).
34. K. K. Kartha and O. L. Gamborg, *Phytopathology, 65*: 826 (1975).
35. F. C. Mellor and R. Stace-Smith, *Phytopathology, 61*: 246 (1971).

36. D. S. Brar and G. S. Khush, *Handbook of Plant Cell Culture*, Vol. 4 (D. A. Evans, W. R. Sharp, and P. V. Ammirato, eds.), Macmillan, New York, p. 221 (1986).
37. G. S. Khush and D. S. Brar, *Distant Hybridization of Crop Plants* (G. Kalloo and J. B. Chowdhury, eds.), Springer Verlag, New York, p. 47 (1992).
38. F. Laibach, *Z. Botan., 17*: 417 (1925).
39. V. Raghavan, *Cell Culture and Somatic Cell Genetics of Plants*, Vol. 3 (I. K. Vasil, ed.), Academic Press, New York, p. 613 (1986).
40. D. S. Brar, R. Elloran, and G. S. Khush, *Rice Genet. Newsl., 8*: 91 (1991).
41. K. K. Jena and G. S. Khush, *Theor. Appl. Genet., 80*: 737 (1990).
42. A. Amante-Bordeos, L. A. Sitch, R. D. Dalmacio, N. P. Oliva, H. Aswidinnoor, and H. Leung, *Theor. Appl. Genet., 84*: 345 (1992).
43. K. J. Kasha and K. N. Kao, *Nature, 225*: 874 (1970).
44. K. M. Ho and G. E. Jones, *Can. J. Plant Sci., 60*: 279 (1980).
45. K. W. Campbell, R. I. Brawn, and K. M. Ho, *Can. J. Plant Sci., 64*: 203 (1984).
46. D. T. Tomes, *Progress in Plant Cellular and Molecular Biology* (H. J. J. Nijkamp, L. H. W. Vander Plas, and J. Van Aartrijk, eds.), Kluwer Acad. Pub., London, England, p. 23 (1990).
47. J. De Buyser, Y. Henry, P. Lonnet, R. Hertzog, and A. Hespel, *Plant Breed., 98*: 53 (1987).
48. Hu Han and Y. Hongyuan, *Haploids of Higher Plants In Vitro*. China Academic Publication, Beijing, Springer Verlag, Berlin, p. 211 (1986).
49. L. Chunling, *Agricultural Biotechnology* (C. B. You and Z. L. Chen, eds.), China Science and Technology Press, Beijing, China, p. 393 (1992).
50. H. Hu, T. Y. Hsi, C. C. Tseng, T. W. Ouyang, and C. K. Ching, *Frontiers of Plant Tissue Culture* (T. A. Thorpe, ed.), IAPTC, Calgary, Canada, p. 123 (1978).
51. Z. H. Zhang, *Review of Advances in Plant Biotechnology*, 1985–88 (A. Mujeeb-Kazi and L. A. Sitch, eds.), CIMMYT, Mexico and IRRI, Philippines, p. 31 (1989).
52. Z. Kang-Le, "Current Status and Future Prospects of Plant Biotechnology in Rice Improvement," Proceedings of the Rice Research Conference, Huangzhou, China (1989).
53. M. F. Li, P. C. Ni, Y. Q. Chen, and J. H. Shen, *Acta Agron. Sin., 9*: 173 (1983).
54. Y. Lee, M. S. Lim, H. S. Kim, H. T. Shin, C. O. Kim, S. H. Boe, and C. I. Cho, *Res. Rept. RDA (R), 31*: 27 (1989).
55. H. P. Moon, S. Y. Cho, R. K. Park, Y. H. Son, H. R. Kim, B. T. Jun, and K. H. Kang, *Res. Rept. RDA (R) 31*: 1 (1989).
56. A. Kruse, *Hereditas, 77*: 219 (1974).
57. E. G. Williams and G. De Lautour, *Bot. Gaz., 141*: 252 (1980).
58. B. R. Thomas and D. Pratt, *Theor. Appl. Genet., 59*: 215 (1981).
59. B. Lloyd, *Tobacco Sci., 19*: 4 (1975).
60. K. K. Pandey, J. E. Grant, and E. G. Williams, *Austral. J. Bot., 35*: 171 (1987).
61. V. Poysa, *Theor. Appl. Genet., 79*: 187 (1990).
62. K. J. M. Skene and M. Barlass, *Ann. Bot., 52*: 667 (1983).
63. Y. Inomata, *Jpn. J. Genet., 53*: 161 (1978).

64. N. Sarla and R. N. Raut, *Theor. Appl. Genet., 76*: 846 (1988).
65. R. Delourme, F. Eber, and A. M. Chevre, *Euphytica, 41*: 123 (1989).
66. Y. Takahata, *Euphytica, 46*: 259 (1990).
67. D. A. Laurie and M. D. Bennett, *Can. J. Genet. Cytol., 28*: 313 (1986).
68. D. A. Laurie and M. D. Bennett, *Theor. Appl. Genet., 76*: 393 (1988).
69. G. Chen, Z. C. Li, Q. C. Peng, Y. Z. Song, and Y. Z. Li, *Agricultural Biotechnology* (C. B. You and Z. L. Chen, eds.), China Science and Technology Press, Beijing, China, p. 557 (1992).
70. J. M. Stewart, *Environ. Exp. Bot., 21*: 301 (1981).
71. J. M. Stewart and C. L. Hsu, *J. Hered., 69*: 404 (1978).
72. S. M. Reed and G. B. Collins, *J. Hered., 69*: 311 (1978).
73. T. Arisumi, *J. Amer. Soc. Hort. Sci., 105*: 629 (1980).
74. V. R. Tilton and S. H. Russell, *Bioscience, 34*: 239 (1984).
75. K. Kanta, N. S. Rangaswamy, and P. Maheshwari, *Nature, 194*: 1214 (1962).
76. M. Zenkteler, *Internat. Rev. Cytol., 11B*: 137 (1980).
77. H. S. Dhaliwal and P. J. King, *Theor. Appl. Genet., 53*: 43 (1978).
78. A. Slusarkiewicz-Jarsina and M. Zenkteler, *Experientia, 39*: 1399 (1983).
79. M. Refaat, L. Rossignol, and Y. Demarly, *Z. Pflanzenzuchtg., 93*: 137 (1984).
80. R. Chasan, *Plant Cell, 4*: 369 (1992).
81. R. A. Morrison and D. A. Evans, *Biotechnol., 6*: 684 (1988).
82. S. K. Raina, *Adv. Agron., 42*: 339 (1989).
83. H. Hu and J. Zeng, *Handbook of Plant Cell Culture*, Vol. 3 (P. V. Ammirato, D. A. Evans, W. R. Sharp, and Y. Yamada, eds.), Macmillan, New York, p. 65 (1984).
84. K. J. Kasha, A. Ziauddin, and U. H. Cho, *Gene Manipulation in Plant Improvement* (J. P. Gustafson, ed.), Plenum Press, New York, p. 213 (1990).
85. Z. H. Zhang and Q. R. Chu, *J. Agric. Sci.* (China) *2* suppl.: 10 (1986).
86. A. Nakamura, T. Yamada, M. Kadotani, R. Itagaki, and M. Oka, *SABRAO J., 6*: 107 (1974).
87. W. A. Keller and K. C. Armstrong, *Z. Pflanzenzuchtg., 80*: 100 (1978).
88. G. Wenzel, B. Foroughi-Wehr, W. Friedt, and F. Kohler, *Biotechnology in International Agricultural Research*, IRRI, Manila, Philippines, p. 65 (1985).
89. J. W. Snape, *Review of Advances in Plant Biotechnology* 1985–88 (A. Mujeeb-Kazi and L. A. Sitch, eds.), CIMMYT, Mexico and IRRI, Philippines, p. 19 (1989).
90. G. Wenzel, F. Fadel, B. Foroughi-Wehr, U. Frei, A. Graner, C. Mollers, and T. Nguyen Quang, *Plant Tissue Culture and Gene Manipulation for Breeding and Formation of Phytochemicals* (K. Oono, T. Hirabayashi, S. Kikuchi, H. Handa, and K. Kajiwara, eds.), NIAR, Japan, p. 1 (1992).
91. D. A. Evans, *Trends Genet., 5*: 46 (1989).
92. L. E. Heszky and I. Simon-Kiss, *Hungarian Agricult. Res., 1*: 30 (1992).
93. N. M. Cowen, C. D. Johnson, K. Armstrong, M. Miller, A. Woosley, S. Pescitelli, M. Skokut, S. Belmar, and J. F. Petolino, *Theor. Appl. Genet., 84*: 720 (1992).
94. J. M. Dunwell, *Plant Tissue Culture and Its Agricultural Applications* (L. A. Withers and P. G. Alderson, eds.), Butterworth, London, England, p. 375 (1986).

95. B. Huang, *In Vitro Cell. Dev. Biol., 28*: 53 (1992).
96. S. K. Datta, K. Datta, and I. Potrykus, *Plant Sci., 67*: 83 (1990).
97. E. B. Swanson, M. J. Heergesell, M. Arnoldo, D. W. Sippell, and R. S. C. Wong, *Theor. Appl. Genet., 78*: 525 (1989).
98. A. Fennell and R. Hauptmann, *Plant Cell Rep., 11*: 567 (1992).
99. P. J. Larkin and W. R. Scowcroft, *Theor. Appl. Genet., 60*: 197 (1981).
100. D. A. Evans and W. R. Sharp, *Biotechnol., 4*: 528 (1986).
101. R. W. Van den Bulk, *Euphytica, 56*: 269 (1991).
102. D. J. Heinz, M. Krishnamurthi, L. G. Nickell, and A. Maretzki, *Applied and Fundamental Aspects of Plant Cell Tissue and Organ Culture* (J. Reinert and Y. P. S. Bajaj, eds.), Springer Verlag, Berlin, Germany, p. 3 (1977).
103. J. F. Shepard, D. Bidney, and E. Shahin, *Science, 208*: 17 (1980).
104. R. I. S. Brettell and D. S. Ingram, *Biol. Rev., 5*: 329 (1979).
105. S. Heath-Pagliuso, J. Pullman, and L. Rappaport, *HortSci., 24*: 711 (1989).
106. M. Krishnamurthi and J. Tlaskal, *Internat. Soc. Sugarcane Technol., 15*: 130 (1974).
107. J. W. Moyer and W. W. Collins, *HortSci., 18*: 111 (1983).
108. R. L. Phillips, S. M. Kaeppler, and V. M. Peschke, *Progress in Plant Cellular and Molecular Biology* (H. J. K. Nijkamp, L. H. W. Vander Plas, and J. Van Aartrijk, eds.), Kluwer, Dordrecht, The Netherlands, p. 131 (1990).
109. H. Hirochika, *Plant Tissue Culture and Gene Manipulation for Breeding and Formation of Phytochemicals* (K. Oono, T. Hirabayashi, S. Kikuchi, H. Handa, and K. Kajiwara, eds.), NIAR, Japan, p. 147 (1992).
110. T. J. Orton, *Theor. Appl. Genet., 56*: 101 (1980).
111. T. J. Orton, *J. Hered., 71*: 280 (1980).
112. P. J. Larkin, P. M. Banks, R. Bhati, R. I. S. Brettell, P. A. Davies, S. A. Ryan, W. R. Scowcroft, L. H. Spindler, and G. J. Tanner, *Genome, 31*: 705 (1989).
113. R. B. Jorgensen and B. Anderson, *Theor. Appl. Genet., 77*: 343 (1989).
114. N. Lapitan, R. G. Sears, and B. S. Gill, *Theor. Appl. Genet., 68*: 547 (1984).
115. P. J. Larkin, L. H. Spindler, and P. M. Banks, *Progress in Plant Cellular and Molecular Biology* (H. J. J. Nijkamp, L. H. W. Vander Plas, and J. Van Aaartrijk, eds.), Kluwer, London, England, p. 163 (1990).
116. H. A. Collin and P. J. Dix, *Plant Cell Line Selection: Procedures and Applications* (P. J. Dix, ed.), VCH Publishing, New York, p. 1 (1990).
117. P. S. Carlson, *Science, 180*: 1366 (1973).
118. P. Thanutong, I. Furusawa, and M. Yamamoto, *Theor. Appl. Genet., 66*: 209 (1983).
119. B. G. Gengenbach, C. E. Green, and C. M. Donovan, *Proc. Natl. Acad. Sci. USA, 74*: 5113 (1977).
120. P. J. Larkin and W. R. Scowcroft, *Plant Cell Tissue and Organ Culture, 2*: 111 (1983).
121. H. W. Rines and H. H. Luke, *Theor. Appl. Genet., 71*: 16 (1985).
122. D. H. Ling, P. Vidhyasekaran, E. S. Borromeo, and F. J. Zapata, *Theor. Appl. Genet., 71*: 133 (1985).
123. H. S. Chawla and G. Wenzel, *Theor. Appl. Genet., 74*: 841 (1987).
124. M. Behnke, *Theor. Appl. Genet., 55*: 69 (1979).

125. C. L. Hartman, T. J. McCoy, and T. R. Knous, *Plant Sci. Lett., 34*: 183 (1984).
126. F. A. Hammerschlag, *Theor. Appl. Genet., 76*: 865 (1988).
127. K. Wakasa and J. M. Widholm, *Theor. Appl. Genet., 74*: 49 (1987).
128. G. W. Schaeffer, *Environ. Exp. Bot., 21*: 333 (1981).
129. G. W. Schaeffer, F. T. Sharpe, and J. T. Dudley, *Theor. Appl. Genet., 77*: 176 (1989).
130. C. J. Boyes and I. K. Vasil, *Plant Sci., 50*: 195 (1987).
131. K. A. Hibberd and C. Green, *Proc. Natl. Acad. Sci. USA, 79*: 559 (1982).
132. T. J. Diedrick, D. A. Frisch, and B. G. Gengenbach, *Theor. Appl. Genet., 79*: 209 (1990).
133. B. J. Miflin, S. W. J. Bright, S. E. Rognes, and J. H. S. Kueh, *Genetic Engineering of Plants—An Agricultural Perspective* (T. Kosuge, A. Hollaender, and C. P. Meredith, eds.), Plenum Press, New York, p. 391 (1983).
134. I. Negrutiu, A. Reynearts, I. Verbruggen, and M. Jacobs, *Theor. Appl. Genet., 68*: 11 (1984).
135. M. W. Nabors, A. Daniels, L. Nadolny, and C. Brown, *Plant Sci. Lett., 4*: 155 (1975).
136. M. W. Nabors, S. E. Gibbs, C. S. Bernstein, and M. E. Meir, *Z. Pflanzenphysiol., 97*: 13 (1980).
137. Y. Yamada, M. Ogawa, and S. Yano, *Cell and Tissue Culture Techniques for Cereal Crop Improvement*, Science Press, Beijing, China, IRRI, Manila, Philippines, p. 229 (1983).
138. S. Yoshida and M. Ogawa, *Fertilizer Tech. Centre Technical Bulletin, 73*: 1 (1983).
139. M. W. Nabors and T. A. Dykes, *Biotechnology in International Agricultural Research*, IRRI, Manila, p. 121 (1985).
140. N. Beloualy and J. Bouharmont, *Theor. Appl. Genet., 83*: 509 (1992).
141. A. J. Conner and C. P. Meredith, *Theor. Appl. Genet., 71*: 159 (1985).
142. W. Abrigo, A. Novero, V. Coronel, G. Cabuslay, L. Blanco, F. Parao, and S. Yoshida, *Biotechnology in International Agricultural Research*, IRRI, Manila, Philippines, p. 149 (1985).
143. R. S. Chaleff, *Theor. Appl. Genet., 58*: 91 (1980).
144. R. S. Chaleff and M. S. Parsons, *Proc. Natl. Acad. Sci. USA, 75*: 5104 (1978).
145. R. S. Chaleff and T. B. Ray, *Science, 223*: 1148 (1984).
146. A. Cseplo, P. Medgyesy, E. Hideg, S. Demeter, L. Marton, and P. Maliga, *Mol. Gen. Genet., 200*: 508 (1985).
147. I. Takebe, G. Labib, and G. Melchers, *Naturewissenchaften, 58*: 318 (1971).
148. K. J. Puite, *Physiol. Plant., 83*: 403 (1992).
149. I. K. Vasil and V. Vasil, *Physiol. Plant., 85*: 279 (1992).
150. V. Vasil, *Biotechnol., 6*: 397 (1988).
151. S. M. Attree, D. I. Dunstan, and L. C. Fowke, *Biotechnol., 7*: 1060 (1989).
152. D. S. Brar, *Handbook of Plant Tissue and Cell Cultures* (A. R. Mehta and P. N. Bhatt, eds.), Academic Book Centre, Ahmedabad, India, p. 101 (1990).
153. K. N. Kao and M. R. Michayluk, *Planta, 126*: 105 (1975).
154. U. Zimmermann and P. Scheurich, *Planta, 151*: 26 (1981).

155. A. Lister, *Plant Cell Line Selection: Procedures and Applications* (P. J. Dix, ed.), VCH Publishing, New York, p. 39 (1990).

156. P. S. Carlson, H. H. Smith, and R. D. Dearing, *Proc. Natl. Acad. Sci. USA, 68*: 2292 (1972).

157. K. C. Sink, R. K. Jain, and J. B. Chowdhury, *Distant Hybridization of Crop Plants* (G. Kalloo and J. B. Chowdhury, eds.), Springer Verlag, New York, p. 168 (1992).

158. G. Melchers, M. D. Sacristan, and A. A. Holder, *Carlsberg Res. Commun., 43*: 203 (1978).

159. E. Jacobsen, P. Reinhout, J. E. M. Bergervoet, J. de Looff, P. E. Abidin, D. J. Huigen, and M. S. Ramanna, *Theor. Appl. Genet., 85*: 159 (1992).

160. Y. Y. Gleba and F. Hoffmann, *Naturewissenschaften, 66*: 547 (1979).

161. G. Krumbiegel and O. Schieder, *Planta, 145*: 371 (1979).

162. D. Dudits, GY. Hadlaczky, G. Y. Bajszar, C. S. Koncz, G. Lazar, and G. Horvath, *Plant Sci. Lett., 15*: 101 (1979).

163. D. A. Somers, K. R. Narayanan, A. Kleinhofs, S. Cooper-Blaud, and E. C. Cocking, *Mol. Gen. Genet., 204*: 296 (1986).

164. Y. Y. Gleba, S. Hinnisdaels, V. A. Sidorov, V. A. Kaleda, A. S. Parokonny, N. V. Boryshuk, N. N. Cherep, I. Negrutiu, and M. Jacobs, *Theor. Appl. Genet., 76*: 760 (1988).

165. D. Dudits, E. Maroy, T. Praznovszky, Z. Olah, J. Gyorgyey, and R. Cella, *Proc. Natl. Acad. Sci. USA, 84*: 8434 (1987).

166. I. Potrykus, J. Jia, G. B. Lazar, and M. Saul, *Plant Cell Reports, 3*: 68 (1984).

167. D. Pental, J. D. Hamill, A. Pirrie, and E. C. Cocking, *Mol. Gen. Genet., 202*: 342 (1986).

168. P. P. Gupta, O. Scheider, and M. Gupta, *Mol. Gen. Genet., 197*: 30 (1984).

169. K. Toriyama, T. Kameya, and K. Hinata, *Planta, 170*: 308 (1987).

170. G. Chatterjee, S. R. Sikdar, S. Das, and S. K. Sen, *Theor. Appl. Genet., 76*: 915 (1988).

171. J. Fahleson, L. Rahlen, and K. Glimelius, *Theor. Appl. Genet., 76*: 507 (1988).

172. S. R. Sikdar, G. Chatterjee, S. Das, and S. K. Sen, *Theor. Appl. Genet., 79*: 561 (1990).

173. C. L. C. Lelivelt and F. A. Krens, *Theor. Appl. Genet., 83*: 887 (1992).

174. T. Ohgawara, S. Kobayashi, E. Ohgawara, H. Uchimiya, and S. Ishii, *Theor. Appl. Genet., 71*: 1 (1985).

175. J. W. Grosser, F. G. Gmitter, and J. L. Chandler, *Theor. Appl. Genet., 75*: 397 (1988).

176. S. Shinozaki, K. Fujita, T. Hidaka, and M. Omura, *Jpn. J. Breed., 42*: 287 (1992).

177. R. Takayanagi, T. Hidaka, and M. Omura, *J. Jpn. Soc. Hort. Sci., 60*: 799 (1992).

178. S. J. Ochatt, E. M. Patat-Ochatt, E. L. Rech, M. R. Davey, and J. B. Power, *Theor. Appl. Genet., 78*: 35 (1989).

179. D. S. Brar, M. Ono, S. Kobayashi, H. Uchimiya, and H. Harada, *Protoplasma, 121*: 228 (1984).

180. W. M. Mattheij and K. J. Puite, *Theor. Appl. Genet., 83*: 807 (1992).

181. K. S. Ramulu, P. Dijkhuis, H. A. Verhoeven, I. Famelaer, and J. Blaas, *Physiol. Plant., 85*: 315 (1992).

182. G. Spangenberg, E. Freydl, M. Osusky, J. Nagel, and I. Potrykus, *Theor. Appl. Genet., 81*: 477 (1991).
183. G. Belliard, F. Vedel, and G. Pelletier, *Nature, 281*: 401 (1979).
184. E. Galun, P. Arzee-Genen, R. Fluhr, M. Edelman, and D. Aviv, *Mol. Gen. Genet., 186*: 50 (1982).
185. A. Vardi, P. Arzee-Gonen, A. Frydman-Shani, S. Bleichman, and E. Galun, *Theor. Appl. Genet., 78*: 741 (1989).
186. E. D. Earle, M. Temple, and T. W. Walters, *Physiol. Plant., 85*: 325 (1992).
187. P. Medgyesy, E. Fejes, and P. Maliga, *Proc. Natl. Acad. Sci. USA, 82*: 6960 (1985).
188. G. Pelletier, C. Primard, F. Vedel, P. Chetrit, R. Remy, P. Rouselle, and M. Renard, *Mol. Gen. Genet., 191*: 244 (1983).
189. A. Zelcer, D. Aviv, and E. Galun, *Z. Pflanzenphysiol., 90*: 397 (1978).
190. S. Izhar and J. B. Power, *Plant Sci. Lett., 14*: 49 (1979).
191. L. Menczel, F. Nagy, G. Lazar, and P. Maliga, *Mol. Gen. Genet., 189*: 365 (1983).
192. L. Menczel, L. S. Polsby, K. E. Steinback, and P. Maliga, *Mol. Gen. Genet., 205*: 201 (1986).
193. D. Aviv and E. Gallun, *Theor. Appl. Genet., 58*: 121 (1980).
194. L. Menczel, A. Morgan, S. Brown, and P. Maliga, *Plant Cell Rep., 6*: 98 (1987).
195. Z. Yang, T. Shikanai, K. Mori, and Y. Yamada, *Theor. Appl. Genet., 77*: 305 (1989).
196. J. Kyozuka, T. Kaneda, and K. Shimamoto, *Biotechnol., 7*: 1171 (1989).
197. L. Tanno-Suenaga, H. Ichikawa, and J. Imamura, *Theor. Appl. Genet., 76*: 855 (1988).
198. R. B. Horsch, J. E. Fry, N. L. Hoffman, D. Eichholtz, S. G. Rogers, and R. T. Fraley, *Science, 227*: 1229 (1985).
199. I. Potrykus, *Biotechnol., 8*: 535 (1990).
200. J. Paszkowski, R. D. Shillito, M. Saul, V. Mandak, T. Hohn, B. Hohn, and I. Potrykus, *EMBO J., 3*: 2717 (1984).
201. H. Uchimiya, H. Hirochika, H. Hashimoto, A. Hara, T. Masuda, T. Kasumimoto, H. Harada, J. E. Ikeda, and M. Yoshioka, *Mol. Gen. Genet., 205*: 1 (1986).
202. Z. Wang, T. Takamizo, V. A. Iglesias, M. Osusky, J. Nagel, I. Potrykus, and G. Spangenberg, *Biotechnol., 10*: 691 (1992).
203. S. K. Datta, K. Datta, N. Soltanifar, G. Donn, and I. Potrykus, *Plant Mol. Biol., 20*: 619 (1992).
204. K. Shimamoto, R. Terada, T. Izawa, and H. Fujimoto, *Nature, 338*: 274 (1989).
205. C. A. Rhodes, D. A. Pierce, I. J. Mettler, D. Mascarenhas, and J. J. Detmer, *Science, 240*: 204 (1988).
206. V. Vasil, A. M. Castillo, M. E. Fromm, and I. K. Vasil, *Biotechnol., 10*: 667 (1992).
207. M. G. Koziel, G. L. Beland, C. Bowman, N. Carozzi, R. Crenshaw, L. Crossland, J. Dawson, N. Desai, M. Hill, S. Kadwell, K. Launis, K. Lewis, D. Maddox, K. McPherson, M. Meghji, E. Merlin, R. Rhodes, G. Warren, M. Wright, and S. Evola, *Biotechnol., 11*: 194 (1993).

208. H. Daniell, J. Vivekananda, B. L. Nielsen, G. N. Ye, and K. K. Tewari, *Proc. Natl. Acad. Sci. USA, 87*: 92 (1990).
209. A. Crossway, J. V. Oakes, J. M. Irvine, B. Ward, V. C. Knauf, and C. K. Shewmaker, *Mol. Gen. Genet., 202*: 179 (1986).
210. G. Neuhaus, G. Spangenberg, O. M. Scheid, and H. G. Schweiger, *Theor. Appl. Genet., 75*: 30 (1987).
211. H. G. Schweiger, J. Dirk, H. U. Koop, E. Kranz, G. Neuhaus, G. Spangenberg, and D. Wolff, *Theor. Appl. Genet., 73*: 769 (1987).
212. R. J. Griesbach, *Plant Sci., 50*: 69 (1987).
213. J. Cao, X. Duan, D. McElroy, and R. Wu, *Plant Cell Rep., 11*: 586 (1991).
214. X. Delannay, B. Lavallee, R. Proksch, R. Fuchs, S. Sims, J. Greenplate, P. Marrone, R. Dodson, J. Augustine, J. Layton, and D. Fischhoff, *Biotechnol., 7*: 1265 (1989).
215. F. J. Perlak, R. W. Deaton, T. A. Armstrong, R. L. Fuchs, S. R. Sims, J. T. Greenplate, and D. A. Fischhoff, *Biotechnol., 8*: 939 (1990).
216. B. McCown, D. McCabe, D. Russell, D. Robinson, K. Barton, and K. Raffa, *Plant Cell Rep., 9*: 590 (1991).
217. D. Boulter, G. A. Edwards, A. M. R. Gatehouse, J. A. Gatehouse, and V. A. Hilder, *Crop Prot., 9*: 351 (1990).
218. V. A. Hilder, A. M. R. Gatehouse, S. Sheerman, R. Baker, and D. Boulter, *Nature, 330*: 160 (1987).
219. R. S. Nelson, S. M. McCormick, X. Delannay, P. Dube, J. Layton, E. Anderson, M. Kaniewska, R. Proksch, R. Horsch, S. Rogers, R. Fraley, and R. Beachy, *Biotechnol., 6*: 403 (1988).
220. T. Hayakawa, Y. Zhu, K. Itoh, Y. Kimura, T. Izawa, K. Shimamoto, and S. Toriyama, *Proc. Natl. Acad. Sci. USA, 89*: 9865 (1992).
221. K. Broglie, I. Chet, M. Holliday, R. Cressman, P. Biddle, S. Knowlon, C. Mauvals, and R. Broglie, *Science, 254*: 1194 (1991).
222. C. Logemann, G. Jach, H. Tommerup, J. Mundy, and J. Schell, *Biotechnol., 10*: 305 (1992).
223. C. Mariani, M. De Beuckeleer, J. Truettner, J. Leemans, and R. B. Goldberg, *Nature, 347*: 737 (1990).
224. C. Mariani, V. Gossele, M. De Beuckeleer, M. De Block, R. B. Goldberg, W. De Greef, and J. Leemans, *Nature, 357*: 384 (1992).
225. N. Murata, O. Nishizawa, S. Higashi, H. Hayashi, Y. Tasaka, and I. Nishida, *Nature, 356*: 710 (1992).
226. M. C. Tarczynski, R. G. Jensen, and H. J. Bohnert, *Science, 259*: 508 (1993).
227. S. A. Merkle, W. A. Parrott, and E. G. Williams, *Plant Tissue Culture: Applications and Limitations* (S. S. Bhojwani, ed.), Elsevier Science Publishing, Amsterdam, The Netherlands, p. 67 (1990).
228. K. Redenbaugh, *HortSci., 25*: 251 (1990).
229. J. Fujii, D. Slade, J. Aguirre-Rascon, and K. Redenbaugh, *In Vitro Cell. Dev. Biol., 28*: 73 (1992).
230. K. Redenbaugh, J. Fujii, and D. Slade, *Biotechnology in Agriculture* (A. Mizrahi, ed.), Alan R. Liss, New York, p. 225 (1988).

231. L. A. Withers, *Plant Cell and Tissue Culture* (J. W. Pollard, J. M. Walker, eds.), The Humana Press, Clifton, NJ, p. 39 (1990).

232. M. Seibert, *Science, 191*: 1178 (1976).

233. A. Sakai and Y. Nishiyama, *HortSci., 13*: 225 (1978).

234. L. A. Withers, *Plant Physiol., 63*: 460 (1979).

235. K. K. Kartha, N. L. Leung, and L. A. Mroginski, *Z. Pflanzenphysiol., 107*: 133 (1982).

236. K. K. Kartha, N. L. Leung, and K. Pahl, *J. Am. Soc. Hort. Sci., 105*: 481 (1980).

237. Y. P. S. Bajaj, *Indian J. Exp. Biol., 17*: 1405 (1979).

238. K. K. Kartha, N. L. Leung, and O. L. Gamborg, *Plant Sci. Lett., 15*: 7 (1979).

239. G. G. Henshaw, J. A. Stamp, and R. J. Westcott, *Plant Tissue Culture: Results and Perspectives* (F. Sala, B. Parisi, R. Cella, and O. Ciferri, eds.), Elsevier Science Publishing, Amsterdam, The Netherlands, p. 277 (1980).

240. F. Engelmann and Y. Duval, *Oleagineux, 41*: 169 (1986).

241. H. Yakuwa and S. Oka, *Ann. Bot., 62*: 79 (1988).

242. B. M. Reed and H. B. Lagerstedt, *HortSci., 22*: 302 (1987).

243. T. Harada, A. Inaba, T. Yakuwa, and T. Tamura, *HortSci., 20*: 678 (1985).

244. T. H. Chen, K. K. Kartha, N. L. Heung, G. W. Kurz, K. B. Chatson, and F. Constabel, *Plant Physiol., 75*: 726 (1984).

245. P. T. Lynch and E. E. Benson, *Rice Genetics II*, IRRI, Manila, Philippines, p. 321 (1991).

246. E. K. Kaleikau, R. G. Sears, and B. S. Gill, *Theor. Appl. Genet., 78*: 625 (1989).

247. E. K. Kaleikau, R. G. Sears, and B. S. Gill, *Theor. Appl. Genet., 78*: 783 (1989).

248. M. D. Lazar, T. H. Chen, G. J. Scoles, and K. K. Kartha, *Plant Sci., 51*: 77 (1987).

249. A. Nadolska-Orczyk and S. Malepszy, *Theor. Appl. Genet., 78*: 836 (1989).

250. Y. Wan, T. R. Rocheford, and J. M. Widholm, *Theor. Appl. Genet., 85*: 360 (1992).

9

Isolation and Characterization of Plant Genes

Dorothea Bartels

Max-Planck-Institut für Züchtungsforschung
Cologne, Germany

Gabriel Iturriaga

Instituto de Biotecnologia
Universidad Nacional Autonoma de Mexico
Cuernavaca, Mexico

INTRODUCTION

For more than a decade molecular biology has offered the tools to isolate plant genes and to analyze their structures. The first genes to be isolated and of which the primary structures were determined, were abundant genes encoding known gene products. Through rapid development of molecular techniques it is now possible to isolate genes existing as a single copy in the genome. Two general scientific motifs lead to the isolation of genes. Either some genetic, biochemical, physiological, or morphological information is available about a gene or its product, and on this basis it is desirable to isolate those particular genes. The other rationale is to describe the molecular basis of a process and isolate genes that are expressed in a particular situation or physiological stage. The strategy leads to the isolation of genes of which the function is unknown and needs to be revealed in due course.

In general, knowledge of the gene product is a prerequisite for gene isolation. Current molecular approaches are restricted by the information available about the gene product. Among the genes not easily accessible for isolation to date are mostly those that encode complex traits of tissues, organs, or whole plants. Important agronomic traits such as photosynthetic capacity, flowering requirements, disease resistance, stress tolerance, seed development, and germination ability are not characterized, because the

279

relevant gene products are unknown and also because these traits are most likely determined by more than one gene.

In the first part of this chapter the methodologies available for plant gene isolation are discussed and the examples described. In the second part the process made to isolate agronomically important genes is summarized for genes related to environmental stress, carbohydrate metabolism, storage proteins, and plant development as well as genes involved in transcriptional regulation.

MOLECULAR METHODS USED TO ISOLATE PLANT GENES

The strategy for gene isolation depends on the abundancy of the gene product as well as on the knowledge of the expression pattern or primary structure of the gene and encoded protein. In the first part of this section the molecular approaches will be introduced which allow the isolation of genes encoding a known gene product. The second part will deal with molecular techniques which can be applied to isolate genes of which the gene products are unknown.

Isolation of Plant Genes Encoding Known Products

DNA Probes

Specific eukaryotic genes or cDNAs have been isolated most frequently by screening genomic or cDNA libraries for sequences that cross-hybridize with specific nucleic acid probes, and could be either homologous or heterologous probes. Screening by nucleic acid hybridization is the most commonly used and reliable method of screening cDNA libraries, because a large number of clones can be examined, the cDNA clones need not to be full length, and it is not required that a biologically active protein is synthesized.

Homologous probes contain to a considerable extent the exact nucleic acid sequence of the gene of interest. They can be used to screen genomic or cDNA libraries under stringent hybridization conditions. Heterologous probes are only partially homologous to the gene which is to be isolated, and they will identify related but not identical clones. This procedure is applicable if the gene structure contains conserved domains; this is the case if the same gene has already been cloned from another species or alternatively if a related gene has been cloned from the same species. The degree of homology can never be predicted and often a series of trial experiments is necessary to determine whether there is sufficient nucleic acid sequence conservation. For instance, recently a gene from a green alga has been used to isolate a glucose transporter gene from *Arabidopsis* [1]. This gene has conserved regions among bacteria, mammals, fungi, and plants. Another example is the isolation of the heat shock protein 70 genes (HSP 70): the highly conserved nature of the HSP 70 genes has been utilized to isolate cross-hybridizing homologs from several plant species with the probe from *Drosophila* [2,3]. Since nowadays the synthesis of oligonucleotides is readily available, often synthetic probes are designed against conserved domains and used in screening procedures. This is preferred to using whole fragments in heterologous screens, because the hybridization is more specific. (The use of oligonucleotide probes is discussed p. 282).

Isolated Proteins Available

If the protein of interest can be purified to a sufficient amount, two main approaches exist to isolate the corresponding gene (Figure 1): First, antibodies raised against the protein are utilized in selecting the protein encoding transcripts. Alternatively, the amino acid sequence informations are determined and used to construct suitable nucleic acid probes for screening.

Expression Library Screening   Identification of cDNA clones from an expression library requires first to raise antibodies specifically directed against epitopes of the target protein. In a second step, expression libraries have to be constructed in either a plasmid or bacteriophage Lambda vector that carries a strong regulatable prokaryotic promoter. Because of the labor involved, phage expression vectors are preferred. Most cDNA expression libraries are constructed in lambda gt11 derivatives [4]. The lack of specificity of the antibody often leads to an experimental setback of this approach. Successful examples are the isolation of an anaerobically inducible gene encoding barley lactate dehydrogenase [5] and a gene from Zucchini (*Cucurbita*) coding for ACC (1-aminocyclopropane-1-carboxylate) synthase, the key enzyme for ethylene biosynthesis [6]. Both of these genes encode inducible enzymes. Purification of proteins is often difficult if the proteins are labile.

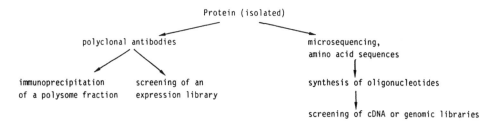

Figure 1   Overview of gene isolation strategies when the protein of interest has been isolated.

Therefore, taking advantage of the inducibility of the enzyme, antibodies were raised to only partially purified ACC synthase and were then purified by affinity chromatography with total proteins from uninduced tissue, a step which removes antibodies reacting with uninduced proteins. With the purified antiserum, a number of immunopositive clones were obtained.

Screening with Oligonucleotide Probes Derived from Amino Acid Sequences
Many cDNAs have been cloned by screening libraries with short synthetic single-stranded oligonucleotides corresponding to sequences of amino acids obtained from protein sequence determination [7]. Usually the number and length of oligonucleotides that are used in screening cDNA and genomic libraries are determined by the amino acid sequence information available of the proteins of interest. The sequence of the oligonucleotides is deduced from the amino acid sequence using the genetic code. Typically these probes are pools representing all possible codon choices for an amino acid sequence, as most amino acids are represented by more than one codon. Thus, a certain sequence of amino acids can be encoded by several different oligonucleotides. There is no definite way to design an oligonucleotide, but certain practical rules have been established in designing oligonucleotides [8–10]. This probe, together with optimized hybridization conditions, has led to the primary isolation of many genes that were previously inaccessible. Today, knowledge of the sequence of a few amino acids in a protein is sufficient for the isolation of cloned copies of the corresponding gene. For instance N-terminal amino acid sequences were the basis for the synthesis of S-allele-specific oligonucleotide probes which were used to identify S-associated proteins in *S. tuberosum*. The multiallelic S-locus controls the self-incompatibility, which prevents self-fertilization [11].

To obtain amino acid sequence of a protein, it is even not necessary to purify the protein to homogeneity. The protein microsequencing technique has been developed so far that it is sufficient for the synthesis of an oligonucleotide, if the protein can be resolved as a band or single spot in a protein gel [12]. A good demonstration for this technique is the isolation of a rice gene expressed in response to salt and drought stress [13].

If the protein has been purified, a decision must be made which way to proceed for gene isolation, via an expression library or via oligonucleotide probes. In the search for major auxin-binding proteins and the corresponding genes, both strategies were applied in parallel, which led to the independent isolation of identical cDNA clones from maize coleoptiles [14,15]. Both approaches yielded the molecular description of a gene for a putative auxin-binding protein. In general, nucleic acid probes are preferred because they detect all clones containing cDNA sequences, whereas antibodies react only with a subset of these clones in which the cDNA clone has been inserted in the correct reading frame and orientation in the vector so that the protein of interest will be expressed.

Immunoprecipitation of Polysomes   A different way to utilize antibodies in gene isolation is to immunoprecipitate polysomes as it was done to isolate a glutamine synthetase gene from *Phaseolus* [16]. There are not many successful examples for the application of this method.

Gene Isolation by Direct Genetic Selection

For many genes isolated by the methods described so far it is often difficult to prove that the clone contains the desired gene. In contrast to research with prokaryotic organisms a functional approach for gene isolation is rarely used in higher plants.

Genetic selection has been successfully exploited to isolate a number of prokaryotic cDNA clones. The isolation of a maize glutamine synthetase cDNA clone by direct genetic selection in *E. coli* demonstrates the feasibility of identifying eukaryotic cDNAs by directly screening cDNA libraries for functional rescue of known mutational defects in *E. coli* [17]. A prerequisite for this experimental approach is the availability of characterized null mutations in genes of interest and a selectable phenotype that is defined by the mutation. It should then be possible to identify and isolate specific cDNAs on the basis of their ability to rescue *E. coli* mutant cells when grown on a restrictive medium. Maize glutamine synthetase cDNA clones were isolated from a cDNA library by screening for functional rescue of an *E. coli* *gluA* mutant grown on agar medium lacking glutamine [17]. The great advantage of this method is that it allows to screen conveniently an enormous number of recombinant clones (10^8-10^9 cells can be screened on a single plate), and thus it is useful in isolating less abundant cDNAs.

Mutant Complementation   Similar to prokaryotes, in eukaryotes also a defect in a mutant can be overcome by expressing a complementary gene in the mutant phenotype. For *Petunia*, an experiment was described where in a mutant no pelargonidin derivatives were produced by the anthocyanin biosynthetic pathway, as a result all the flowers showed no pigmentation [18]. This effect is due to the substrate specificity of dihydroflavonol reductase in *Petunia*, that cannot reduce dihydrokaempherol which gets accumulated. This mutant genotype was transformed with the A1 gene of *Zea mays*, which encodes dihydroflavolol 4-reductase (DRF), the intermediate for pelargonidin biosynthesis. The expression of this gene in *Petunia* led to the appearance of brick-red flowers, as the maize DRF gene product could reduce dihydrokaempherol.

Polymerase Chain Reaction to Isolate cDNA Clones

Recently oligonucleotides either derived from conserved sequences or from amino acid sequences of proteins were used in combination with the polymerase chain reaction (PCR) to obtain cDNA clones [19-21]. cDNAs are generated by employing two oligonucleotide primers, one complementary to a 5′ end sequence on the ($-$) strand and the other to a downstream sequence on the ($+$) strand. In several cycles of denaturation, annealing, and extension, copies of the DNA lying between two primers are amplified (Figure 2). The necessary information required for this amplification is a short stretch of sequence within the mRNA to be cloned. From this sequence primers oriented in the 3' and 5' directions are synthesized, so as to provide specificity to the amplification step. Detailed protocols for this method are given in PCR protocols [22]. This is an efficient and powerful cDNA cloning technique particularly suitable for the isolation of rare transcripts, which was considered to be a difficult task. For the annealing reaction with the specific primer, hybridization conditions can easily be optimized in a series of parallel assays; something which is not really feasible if oligonucleotides are directly used to screen a library. The PCR technology has expanded very rapidly and will probably become the method of choice if the isolation of cDNA clones is based on oligonucleotides. It has been observed that the Taq-Polymerase makes errors in the synthesis of the DNA strand and therefore the clone isolated in a PCR reaction should be used to screen a library and select the corresponding cDNA clone.

Isolation of Plant Genes Encoding Unknown Gene Products

A major interest of plant biology is on development, cell differentiation, and cell response to environmental factors at the molecular level. Cell and tissue differentiation is the result of selective transcription of the genome, and therefore of differential expression of genes in space and time. Some of these genes are involved in determining plant productivity such as genes conferring resistance to stress, nutritional quality, fruit maturation, flowering time, senescence, and dwarfism among others. For these characters the phenotype is known, but rarely the gene products. Isolating these genes to study their regulation and their transfer to important crops is a major objective of plant biotechnology. Several approaches have been developed for selective isolation of genes expressed in a regulated fashion when their product is unknown.

Differential Hybridization

The first type of methods use probes derived from populations of mRNA molecules. The " $+/-$ " procedure is designed to detect cDNA clones derived from mRNAs which are present ($+$) in one cell type and absent ($-$) or expressed at significantly lower levels in another cell type. Poly (A)$^+$ RNAs extracted from two cell types are used as templates to generate radiolabeled first-strand cDNA probes that are hybridized separately to duplicate copies

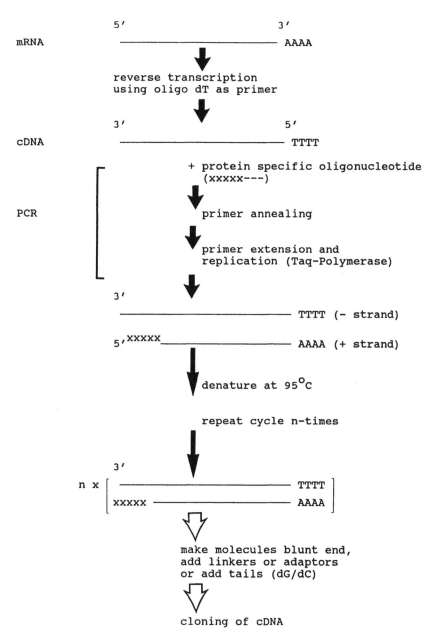

Figure 2   Experimental steps in cloning a transcript to which an oligonucleotide has been synthesized. The cDNA is amplified in a PCR polymerase chain reaction.

of the same cDNA library. Clones that hybridize to both probes correspond to genes that are expressed in both cell types, whereas clones that hybridize to only one probe correspond to mRNAs that are expressed in only one cell type. The limit of detection of this method accounts for mRNA representing approximately 0.1% of the total mRNA molecules (Figure 3). This strat-

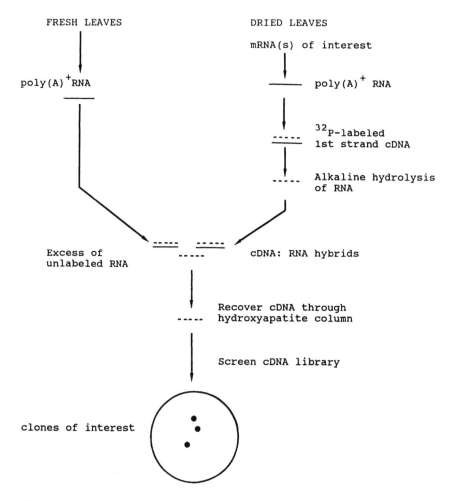

Figure 3 Screening by differential hybridization. Poly(A)$^+$ RNA from tissue under investigation is reverse transcribed into first strand cDNA and hybridized with an excess of poly(A) RNA from other tissue or with the same tissue treated under different physiological conditions. The nonhybridized cDNA is used to screen a cDNA library constructed from the tissue of interest. See text for more details.

egy was used to clone stigma-specific cDNAs encoding two self-incompatibil-ity S-locus alleles [24]. Several stress-induced genes have been also isolated by differential hybridization (see below), such as desiccation-induced genes [25,26] and cold-regulated genes [27].

A refined application of differential hybridization is subtractive hy-bridization. When a gene is differentially expressed but is of low abundance ($< 0.05\%$) the first strand cDNA used as a probe should be enriched by sub-stractive hybridization [8,23]. This procedure is based on the fact that a first strand cDNA population can be readily hybridized to complete kinetic ter-mination with excess of homologous mRNA. In other words, poly $(A)^+$ RNA extracted from one cell type is used as a template to make radio-labeled cDNA with high specific activity, which is then hybridized exhaustively to a tenfold excess of poly $(A)^+$ RNA extracted from a second cell type in which the gene of interest is not expressed. The cDNAs corresponding to mRNAs expressed in both cell types will form DNA:RNA hybrids that can be separated from single-stranded cDNA by chromatography on hydroxyapatite columns. This insoluble form of calcium phosphate binds more tightly double-stranded nu-cleic acids than single-stranded molecules. After two rounds of hybridization and chromatography, the remaining single-stranded cDNAs represent the enriched fraction of molecules coming from low abundant and differentially expressed mRNAs. The enriched cDNA can then be used to screen a cDNA library from the cell type expressing the gene(s) under investigation.

Alternatively, the enriched cDNA fraction can be used directly to gen-erate second-strand cDNA and then cloned into a plasmid vector. When only low amounts of radioactive first-stranded cDNA probe can be recov-ered after subtraction, the subtracted library has the advantage of directly amplifying this pool of molecules and therefore it constitutes per se the set of low abundant clones expressed in a cell-specific manner. These clones account for a tenfold reduction in the total type of different clones to be screened, they can be classified in groups by cross-hybridization, and checked for their expression by hybrid-release translation and Northern hybridization. In this way, genes expressed during flower induction by day-length in the mustard plant *Sinapsis alba*, could be isolated [28]. Similarly, a substractive library from the *Antirrhinum majus* flower mutant *Deficiens* allowed the isolation of the homeotic gene $defA^+$ which encodes a transcriptional ac-tivator [29].

Cloning of DNA-Binding Proteins

Gene isolation and sequencing is usually followed by the analysis of its pro-moter activity by fusion to reporter genes and expression in transgenic plants. This approach allows the dissection of the sequence upstream of the struc-tural gene in order to find out *cis*-acting elements by extensive 5' end deletions. In this way the minimal upstream sequence still able to confer tissue-specificity and promote transcription can be defined [30–32]. Nuclear proteins bind to the active sequences in the promoter region in order to turn on transcription; their presence is deduced by gel retardation and DNAse I footprinting assays [33,34]. Clearly, the next step is to clone the gene encoding the transcriptional activator protein. The strategy common nowadays is to construct an expression library with total cDNA from the active tissue. Fusion proteins will be expressed in the bacterial cells and blotted onto a nitrocellulose membrane. Positive clones are scored using a radio-labeled double-stranded oligonucleotide encoding the putative DNA-binding site. Detection is enhanced if concatemers of this sequence are used. This method was used to isolate transcriptional activators from tobacco which act upon the CaMV 35S promoter [35] a *trans*-acting factor from wheat embryos, acting on an ABA-inducible promoter [36] and transcriptional activators from tomato able to induce heat-shock genes [37].

Ultimately, the functionality of the protein must be demonstrated by in vitro transcription studies or in cotransfection experiments in which a plasmid carrying the full-length cloned cDNA under control of a constitutive promoter is introduced into plant cells together with a plasmid containing single or multiple copies of the DNA-binding sequence fused to a reporter gene. The ability of the protein encoded by the cDNA to activate transcription can be easily scored by the reporter activity. As examples, maize B regulatory protein *trans*-activated *Bronze 1* promoter [38]; *Fos* and *Jun* activated a wheat high molecular weight glutenin promoter [39], and maize *Opaque-2* protein switched on B32 promoter [40].

Map-Based Cloning of Genes by YAC Cloning and Genome Walking

Restriction fragment length polymorphisms (RFLPs) (for a review see [41]) reveal a source of genetic variation which has been proposed to be extremely useful in the isolation and description of genes whose products may not have been identified or which affect quantitative traits. If such a gene can be linked to a RFLP marker it is in principle amenable to cloning.

In combination with an RFLP map, YAC cloning (YAC = yeast artificial chromosome, see below) and genome walking provides a way to clone genes of which only the effect on a particular phenotype is known. The recently developed techniques involved in chromosome walking are currently being used to isolate genes defined by mutational analysis in several organisms, first of all in humans, nematodes, and *Drosophila*. This approach has allowed the isolation of a number of genes associated with the etiology of various human diseases [42,43].

For plants this technique is not yet advanced to a comparable extent because in general plants have a very large genome which complicates the application of cloning and genome walking. Among plants the small crucifer *Arabidopsis thaliana* is an exception because it has the least complex genome known among higher plants (estimated around 70,000 kb) and it has little repetitive DNA [44]. The second advantage is that a large number of well-characterized mutations affecting diverse and complex traits such as floral morphogenesis, hormone metabolism, plant height, or stress tolerance have been identified in *Arabidopsis* [45–47]. For these reasons, *Arabidopsis* has become a widely used experimental system for molecular and genetic studies [48], and it is at present the best model suited for gene isolation by chromosomal walking in plants.

At least four basic steps are involved in the map-based cloning of genes via chromosome walking. A prerequisite as a first step is the availability of a high-density genetic map saturated with molecular markers based on RFLPs. For *Arabidopsis*, two such linkage maps have been established [49,50]. Together these maps contain around 200 markers which means an average distance of 350 kb between two RFLP markers. The second step will be to transform the genetic map into a physical map applying Pulsed Field Gel Electrophoresis (PFGE) which permits the size fractionation of large DNA fragments (in the size range of 50–10,000 kb) and should allow to link a gene of interest to a particular DNA fragment. For long-range mapping and cloning of such DNA fragments yeast artificial chromosomes (YACs) were developed as cloning vectors (for a review see [51]). For *Arabidopsis*, several YAC libraries have been constructed [52]. The third step in a chromosome walking procedure is to identify YACs which contain the RFLP markers flanking the locus of interest. The next step will be to isolate end-clones of YAC inserts as hybridization probes and to use this probe to identify the clone closer to the desired locus. This walking step is repeated until a fragment containing the gene has been found. Fine genetic mapping is a possibility to determine whether the desired gene has been cloned. Finally, it is necessary to verify that the correct gene has been isolated: either in a plant transformation experiment where the desired phenotype is expressed or in a more elegant way by transformation of mutant plants and subsequent mutational complementation.

When this chapter was written no plant gene had been isolated by this approach, but it is expected that in the future several genes affecting interesting phenotypes will be identified. For *Arabidopsis* all the necessary tools are available. Many efforts are being made for agronomically important plants to use this technique for the isolation of genes of importance to plant breeding. For several crops progress has been made toward this objective: from major crop plants genetic RFLP maps are available (Table 1). For a number of cases these maps have been utilized to detect linkage between RFLP markers and disease resistant genes (Table 2).

Table 1 RFLP Maps Are Available from the Following Crop Plants

Crop plants	References	No.
Maize	Helentjaris (1987)	[165]
Lettuce	Landry et al. (1987)	[166]
Tomato	Zamir and Tanksley	[167]
Rice	McCouch et al. (1988)	[168]
Pepper	Tanksley et al. (1988)	[169]
Potato	Gebhardt et al. (1989)	[170]
Rape	Slocum et al. (1990)	[171]
Soybean	Tingey et al. (1990)	[172]
Wheat	Chao et al. (1989)	[173]
Barley	Gale et al. (1990)	[174]

Gene-Tagging by Transposable Elements and T-DNA

Another powerful technique to isolate genes encoding unknown products is gene-tagging. Transposable elements (TEs) are known to induce mutations which can be monitored by a change in the phenotype. Several TEs from maize have been well characterized; the *Ac/Ds* mobile elements and *En/Spm* being the most studied [53]. Transposable elements are characterized by causing genetic instability. They transpose in the genome during development; if they are inserted in a structural or regulatory gene, a mutated phenotype will be created. Some of these TEs are two-element systems of transposition. For instance, *Ds* (nonautonomous) requires the presence of *Ac* (autonomous) to be mobilized. In fact, *Ds* elements are naturally occurring *Ac* deletion mutants unable to encode a full-active transposase [53].

The particular unstable nature of TEs has been exploited to induce mutations which are then selected by segregation of the mutant phenotype after

Table 2 RFLP Markers Linked to Resistance Loci

Crop plant	Resistance mapped	References
Tomato	TMV-resistance	[175]
Tomato	Fusarium oxysporum	[176]
Tomato	Nematode resistance	[177]
Maize	Maize dwarf mosaic virus	[178]
Sugar beet	Nematode resistance	[179]
Potato	Potato cyst nematode	[180]
Barley	Mildew resistance	[181]

genetic crosses (Figure 4). If the TE is homologous to one or only a few sequences in the genome, then a correlation between the genomic location of the transposon in the DNA and the newly induced mutant phenotype may be fairly easily established using DNA hybridization techniques on segregating progenies. The DNA fragment which contains the mutated target gene must always cosegregate with the mutant phenotype.

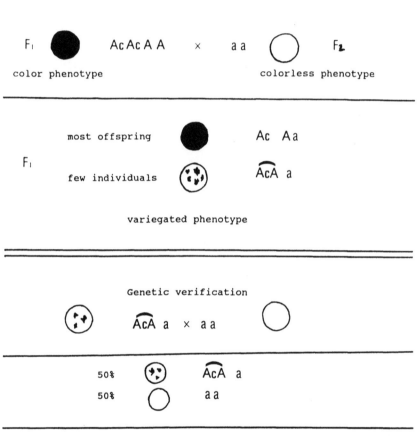

Figure 4 Transposon-tagging. In this example, a line of maize carrying Ac transposons and dominant homozygote for locus A (AcAcAA) is crossed with the recessive homozygote plant (aa). Normally, the F_1 offspring of this cross would result in all individuals having color phenotype, but due to Ac transposition (rare event) into the A locus (AcA) a few mutant plants can be recovered. A variegated phenotype results from Ac excision in some somatic cells. The mutant allele (AcA) should be inherited and segregate in half of the offspring population.

A transposon-induced mutant allele of a desired gene is physically isolated through the use of transposon-specific sequences as a hybridization probe in a genomic library from the mutant plant. The rationale is that flanking DNA sequence to the inserted transposon constitute the interrupted gene which can be cloned and sequenced. This DNA fragment is then used as a probe to isolate the intact gene from a wild type plant. This approach permitted the cloning of the *b2* [54], *al* [55], *c2* [56], *o2* [57], *kn-1* [58], *vp-1* [59], and *yl* [60] loci from maize. From *Antirrhinum majus* the *pallida* [61] and *floricaula* [62] loci were cloned.

For plants in which TEs have not been described, a transposon available from another plant can be utilized if a transformation and regeneration system has been established. For instance, the well-charactrized maize *Ac* transposable element (for a review see [54]), has been shown to transpose in *Agrobacterium*-transformed tobacco [63], *Arabidopsis*, carrot [64], potato [65], tomato [66], and *Petunia* [67]. The *En/Spm* system has been shown to transpose in transgenic tobacco [68] and potato [69]. The *Tam3* element has been shown to transpose in transformed tobacco [70]. As yet there are no reports of the isolation of genes using transposon-tagging in transgenic plants, but gene-tagging is regarded as a promixing method [71] for isolating key genes involved in directing development, conferring pathogen resistance, switching on photosynthesis or any other regulatory gene of low abundance encoding an unknown product.

Another way to induce mutations is the use of Ti-derived plasmids as a tool to select altered phenotypes. The T-DNA integration should correlate with morphological differences and this can be checked again by selfing tests. The analysis can be speeded up if the T-DNA contains a selectable marker. The physical isolation of the gene is basically the same as described above for transposon-tagging. For instance, the use of Ti plasmid-derived gene fusion vectors containing a promoterless reporter gene was used to select for expressed genes after the T-DNA was inserted next to active promoters [72,73]. Recently, a T-DNA mutagenesis strategy was successfully used to isolate dwarf [74], green-pale [75], and flower homeotic mutants [76] from *Arabidopsis*.

METHODS TO CONFIRM THE IDENTITY OF ISOLATED cDNA CLONES

A crucial step in gene isolation procedures is to prove that the desired gene has been cloned. Usually several pieces of evidence are obtained from the approaches listed below. It is, of course, essential that the experiments to prove the nature of a clone must not use the same test parameters that were employed to isolate the clone.

1. Determined DNA sequence of a clone corresponds to amino acid sequence derived from protein sequence.
2. DNA sequence homology to genes from other organisms or species.
3. Expression of the full-length cDNA clone in prokaryotic or eukaryotic cells and test for biological or enzymatic activity.
4. Hybrid-selected translation of the polypeptides synthesized from transcripts of the clone followed by immunoprecipitation with antibodies raised against the protein of interest.
5. Complementation of a mutant (see p. 283, Mutant Complementation)

The information which can be obtained from an isolated gene are summarized in a scheme (Fig. 5). This figure also demonstrates the interlink between the different experimental approaches.

Stress-Related Genes

The improvement of stress tolerance in plants is a difficult, but important challenge to biotechnology. The performance of plants can be affected by biotic and abiotic stresses (for a review see [77]). The phenomenon of stress tolerance is very complex and has not yet been fully understood [78]. The strategy adopted by molecular biology to reveal basic parameters of stress tolerance can be divided into four basic steps: (1) to discover differential gene action by comparing a stress and nonstress situation in suitable model systems; (2) to isolate genes correlated with stress or stress tolerance mainly done by differential hybridization; (3) to prove the direct role of the identified genes in the stress phenomenon by testing their effects in transgenic plants; (4) to use the genes to modify a crop or to assist in selection by monitoring stress levels. To date steps 1 and 2 have been applied to many stress situations. Consequently, a number of stress-induced genes have been isolated and characterized. These genes share the common feature of induction by some forms of physiological stress. Results from step 3 are just beginning to emerge.

Biotic Stress—Pathogenesis-Related Genes

Biotic stressors comprise pathogenic microorganisms and pests which adversely affect crop productivity. Gene activation is an important component of the resistance response of plants to pathogens. Many efforts are being made to isolate genes related to pathogen resistance. The resistance is based on multiple biochemical factors whereby transcriptional activation of specific defense genes is required. A number of genes involved in the defense response have been cloned and characterized (for a review see [79]; a list of cloned plant defense response genes is compiled in [80]).

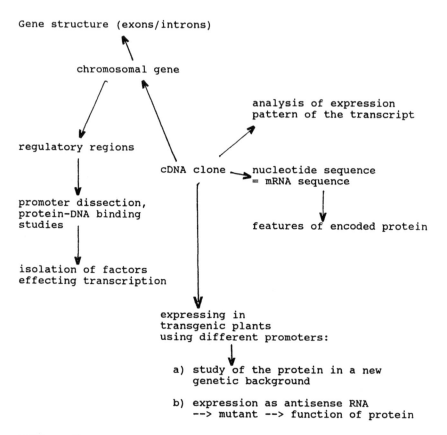

Figure 5   Information which can be obtained from an isolated gene.

The genes and their products can be grouped according to their functions or according to their structure as follows:

1. Products of general phenylpropanoid metabolism [81].
2. Phytoalexins [82].
3. Cell wall constituents [83,84].
4. Lytic enzymes [85].
5. Pathogenesis-related proteins [86,87].

The mode of isolation and characterization of the genes is given in the references.

Abiotic Stress-Induced Genes

The most important abiotic stress factors are temperature (heat, cold); light, water (drought, flooding); salt, heavy metals, and wounding. Most data originate from work on heat–shock genes which have recently been reviewed in excellent detail [77,78] and will not be treated here.

Anaerobic Stress   The plant anaerobic response is an adaptive mechanism that enables survival at low oxygen pressure for several days. Anaerobiosis mimics the stress situation encountered during flooding. In maize roots a set of about 20 anaerobic stress proteins have been identified [89,90]. The best studied gene system is the alcohol dehydrogenase (Adh) from maize. It was demonstrated that the anaerobic response of the Adh1 is most likely mediated by transacting proteins (transcription factors) which bind specifically to the anaerobic responsive element (ARE). These elements are required for anaerobic regulation. Promoter activity under anaerobic conditions is proportional to the number of complete ARE motifs in the Adh1 promoter as assayed in transient expression [91]. The fact that these elements are also present in other anaerobically inducible genes of maize (Adh2, sucrose synthase, aldolase) and other plants such as *Arabidopsis*, pea, and rice, suggests a general regulatory mechanism for these genes in anaerobic induction [92–97].

Genes Related to Water, Salt, and Cold Stress   The common effect of water, salt, and low temperature (cold) stress on a cell is that the water concentration is decreased and this leads to an increased osmotic pressure. Water stress is understood as water deficiency (dehydration or the extreme case desiccation). Although different sets of genes will be involved in these three stress reactions some of them are interlinked (salt or cold stress) and are part of a general reaction to osmotic stress (for a review see [98]). Some of the general features of these genes will be summarized here. The knowledge gathered about genes involved in osmotic stress is not as advanced as that for heat-shock or anaerobiosis. The progress which has been made in isolating genes relevant to water, cold, and salt stress is summarized in Tables 4, 5, and 6. Sources for gene isolation have been in vitro cell culture systems, agronomically important plants or model plants exhibiting extreme stress tolerance.

The most ubiquitous group of genes are those which code for the rab (rapid ABA induced) or dehydrin type of proteins [25,98]. These are characterized by a contiguous stretch of 8 or 9 serine residues and the conservation of a lysine-rich motif. The evolutionary conservation of these gene structures among diverse plant species must have been for functional constraints. These genes are expressed predominantly in mature embryos, osmotically stressed seedlings, and in dehydrated tissues of resurrection plants. Further biochemical and biophysical analysis must reveal the function of the rab proteins suspected of stabilizing cellular membranes during dehydration as

Table 3 Transcriptional Activator Genes

Genes or cDNA	Source	Target genes	Isolation method[a]	Protein domain[b]	References
C1	Maize	A1, Bz1, C2	TET	M	[151]
Lc	Maize	A1, Bz1, C2	HET	HLM	[152]
O2	Maize	22kD zeins B32, O2	TET	LZ	[154]
ASF-1	Tobacco	CaMV 35S	SWH	LZ	[36]
Hv1, Hv33	Barley	Unknown	HET	M	[153]
Zm1, Zm38, ZmC1	Maize	Unknown	HET	M	[153]
λ16-HBP-1	Wheat	Histone H3	SWH	LZ	[155]
OCSBF-1b	Maize	T-DNA ocs	SWH	LZ	[156]
Lp-HSF-24	Tomato	hs	SWH	LZ	[38]
λGC19	Wheat	Em	SWH	LZ	[37]
TFIID-1, TFIID-2	*Arabidopsis*	TATA box	HET	HLH	[214]
3AF1	Tobacco	rbcS-3A	SWH	ZF	[215]
Deficiens	*Antirrhinum*	Unknown	SUL	S	[31]
Agamous	*Arabidopsis*	Unknown	TDT	S	[76]

[a]TET = transposon-tagging; HET = heterologous DNA probe; SWH = South Western hybridization of expersion library; SUL = substractive cDNA library.
[b]M = myb-like; HLH = helix-loop-helix; LZ = leucine Zipper; ZF = zinc-finger; S = SRF-related.

suggested in models [99]. A general osmoprotective role is supported by the occurrence of these proteins in dry embryos and in resurrection plants, two plant systems displaying extreme desiccation tolerance.

A few cold and water stress-responsive genes are homologous to proteins with known functions (thiol and cysteine protease, alcohol dehydrogenase and aldehyde dehydrogenase, see Tables 4, 5), this may give a clue to which metabolic pathways are involved in the stress reactions; the listed examples here indicate pathways leading to the degradation of cell products. Most isolated genes are novel and their function is unknown. Until now, no pathways have been molecularly characterized that lead to the synthesis of compatible osmolytes. A beginning toward elucidating these pathways in plants is the cloning of the betaine aldehyde dehydrogenase, the last enzyme in the betaine biosynthetic pathway [100]. Betaine accumulates as a compatible osmolyte in some plants subjected to drought or salinity stress. Progress can be expected from using a genetic approach. Unfortunately, there are no mutants described for the structural genes listed in Tables 4–6.

A common feature of many stress-induced genes is that their transcripts are induced not only by stress but also by abscisic acid (ABA) [98]. This suggests a general role for ABA in the signal transduction pathway from sensing

Table 4 Isolation of Water Stress-Related Genes

Source of clone isolation	Clone	Comment	Inducing factors	References
Maize embryo	MA12	Identical	D, C, W, ABA	[182]
Maize dehydr. seedl.	M3	Identical	D, ABA	[183]
Barley dehydr. seedl.	B18	RAB group	D, ABA	[183]
Barley dehydr. seedl.	B17	RAB group	D, ABA	[183]
Barley dehydr. seedl.	B8	RAB group	D, ABA	[183]
Barley dehydr. seedl.	B9	RAB group	D, ABA	[183]
Rice embryos	RAB16A		D, ABA, S, C	[25]
Rice embryos	RAB16B		D, ABA	[184]
Rice embryos	RAB16C		D, ABA	[184]
Rice embryos	RAB16D		D, ABA	[184]
Cotton embryo	LeaD11		D, C, ABA	[185]
Crat. dried leaves	pcC6-19		D, S, ABA	[186]
Crat. dried leaves	pcC27-04		D, S, ABA	[186]
Wheat embryo	EM	EM group	D, ABA	[187]
Cotton embryo	LeaD11	EM group		[185]
Radish seed	p8B6	EM group	D	[188]
Rapeseed	pLea76		D, ABA	[190]
Cotton embryo	LeaD7			[185]
Cotton embryo	LeaD113		ABA	[185]
Cotton embryo	LeaD34		ABA	[185]
Maize embryo	pMAH9	homo to RNP	D, W, ABA	[189]
Barley seed	ASI		D, ABA	[191]
Pea wilted shoots	7a	p. homo to soybean nodulin	D, hs, ABA	[192]
Pea wilted shoots	15a	p. homo to cysteine protease	D	[192]
Pea wilted shoots	26g	Homo to aldehyde dehydro-genase	D	[192]
Arabidopsis dried pl.	pMAH9		D	[47]
Crat. dried leaves	pcC27-45		D, ABA, S	[186]
Crat. dried leaves	pcC13-62		D, ABA, S	[186]
Crat. dried leaves	pcC3-06	Homo	D, ABA, S	[186]
Cotton embryo	LeaD29	Homo	ABA	[185]

Abbreviations: ABA = abscisic acid, C = cold stress, Crat = Craterostigma plantagineum dried leaves, D = dehydration, homo = partial sequence homology, hs = heat shock, S = salt, W = wounding.

Table 5   Isolated Salt-Inducible Genes

Plant species	Clone name	Comments	Inducibility by other factors	References
Tomato	TAS 14	Similarity to Rab, dehydrin proteins	ABA	[193]
Tomato	TSW 12	Similarity to phospholipid carrier	Wounding	[194]
Tomato	pNP 24	Partial homology to thaumatin		[195]
Tobacco cells	Osmotin	Partial homology to thaumatin, maize amylase/ trypsin inhibitor, Pr protein of tobacco		[196]
Mesemb. cryst.	PEPCase			[197]
Rice	Salt		Desiccation ABA	[13]
Spinach		Coding for betainealdehyde dehydrogenase		[198]

osmotic stress to gene expression. To understand the complex network of gene activation is a major goal in this research area. For some species there are ABA mutants available which can be utilized. Pla et al. [101] investigated the differential regulation of the water stress related MA12 encoded protein and found that in the maize viviparous mutant the MA12 transcripts do not accumulate in water-stressed leaves whereas they are induced in the wild type. This approach together with promoter studies may shed light on the regulation of these genes. Promoter studies led to the identification of sequence elements involved in ABA activation of some of these genes [102, 103] and even a gene was isolated whose product recognizes an abscisic acid responsive element [36] (see also IIB4). Much progress in understanding the role of ABA in the stress response is expected from the combination of molecular and genetic analysis of the available ABA mutants in *Arabidopsis* [46].

Genes Involved in Carbohydrate Metabolism

Plant productivity is dependent on photochemical conversion and fixation of carbon dioxide into carbohydrates during photosynthesis, followed by a flow of energy from source to sink tissues. The photoassimilates are converted into sucrose which is then transported to the sink organs (tubers, roots or seeds) where it is transformed into the final reserve substances (for a review see [104]). Many steps in carbohydrate metabolism have been identified and analyzed at the level of enzyme activity. Several mutants affecting sugar concentrations or starch synthesis have been utilized to unravel the complex interaction of carbohydrate metabolites [105]. Many of these genes have been isolated from maize for the following reasons: First, around 70% of the dry weight of maize kernels is starch and is therefore an agriculturally important process of prime scientific interest. Second, endogenous transposable elements in combination with mutants made gene isolation possible without enzyme purification. It is not intended to give a general account of the molecular analysis of carbohydrate metabolism, but some features of recently isolated genes will be highlighted.

1. *Sucrose synthase* (EC2.U.1.13) is involved in starch metabolism of the developing endosperm, it catalyzes the reaction sucrose + UDP $\rightleftarrows$ UDP-glucose + fructose. In *Zea mays* it is encoded by the shrunken gene (Sh) on chromosome 9. The gene has been cloned and analyzed in detail, it comprises 5.4 kb and its structural gene is interrupted by 15 introns [93]. Recently, the transcriptional activity of the Sh promoter was studied in transient expression assays using maize protoplasts. It was shown that the activity of the Sh promoter was repressed by high levels of extracellular sucrose but not by glucose or fructose. It was, however, activated when the protoplasts commenced cell wall regeneration. By promoter deletion analysis a 26 bp region was identified to be sufficient for the expression behavior of the gene [106]. This discovery of a feedback control element in the promoter of the shrunken gene is a first step toward analyzing the transcriptional control of genes involved in carbohydrate metabolism.

2. D-glucose-1-phosphate adenyltransferase (Ec2.7.7.27) or ADP-glucose pyrophosphorylase catalyzes the reversible reaction: Glucose-1-phosphate + ATP $\rightleftarrows$ ADP-glucose + pyrophosphate, a regulatory step in the biosynthesis of starch in plants [107]. The enzymatic reaction is repressed by orthophosphate and activated by 3-phosphoglycerate. The enzyme has been purified from several sources like maize endosperm, rice endosperm, potato tuber, and *Arabidopsis* leaf. In all these cases the enzyme is supposed to be composed of two subunits encoded by two different genes [108]. Because of the importance of this enzyme several genes encoding ADP-glucose pyrophosphorylase from different species have been cloned either via isolated proteins, or via transposable elements in the Sh2 locus of maize [109–112]. A comparison of the deduced protein sequences revealed considerable sequence similarity on the amino acid level, including regions involved in substrate binding and activator binding [109]. A comparison between plant enzymes and the corresponding *E. coli* enzyme suggests a conservation of substrate binding sites, a feature that possibly can be expected for many genes of the carbohydrate metabolism.

3. Sucrose-6-phosphate synthase (EC2.4.1.14) catalyzes the synthesis of sucrose-6-phosphate out of fructose-6-phosphate and UDP-glucose, and it is therefore responsible for the biosynthesis of sucrose in plants. Molecular data about the corresponding gene are just beginning to emerge and cDNA clones have been isolated from wheat, *Arabidopsis*, and maize [108]. In maize the protein was first isolated and the information used to isolate the corresponding cDNA clone which encodes a polypeptide of 1068 amino acids. The polypeptide specified by the cDNA clone showed enzymatic activity and directed sucrose-6-phosphate synthesis in *E. coli*.

4. Starch branching enzyme (EC2.4.1.18) and UDP-glucose starch glycosyl transferase are two other enzymes in the starch metabolism of which the genes have been cloned. The UDP-glucose starch glycosyl transferase is encoded by the waxy locus in maize. The gene comprises a coding region of 3718 bp and is composed of 14 exons. The expression of the gene is tissue-specific and the protein is synthesized with an amyloplast-specific transit peptide [112]. Besides their metabolic importance, the starch branching enzymes (SBEI) are of particular interest, because in pea a transposable element like insertion in the gene leads to a wrinkled phenotype of the seeds [113]. The gene encoding an isoform of SBEI was isolated from an expression library using antiserum against SBEI. The isolated cDNA clone has significant homology to the glycogen branching enzyme of *E. coli*. When in the mutant phenotype the SBEI activity is absent, it affects starch, lipid, and protein metabolism of seeds [113].

Storage Protein Genes

Seed storage proteins act as reservoirs of reduced nitrogen to provide the required amino acids during germination and seedling growth. In accordance with their role, storage proteins accumulate at significant quantities and contain a high proportion of glutamine and asparagine [114]. In addition to carbohydrates, seed crops supply a large proportion of the protein required by man and animals. The nutritional problem with cereals and legumes is that they contain limited amounts of certain essential amino acids. In general, cereals are deficient in lysine and threonine, while legumes are deficient in cysteine and methionine [115]. Tuber storage proteins have a better amino acid balance but a much lower protein content [116,117]. It would be highly beneficial to genetically engineer major crops to produce proteins with a balanced essential amino acid content. This goal could be achieved by in vitro mutagenesis of existing storage protein genes to increase levels of essential amino acids or by construction of synthetic genes [118]. In addition, there are several naturally occurring genes which encode for methionine-rich proteins from maize, rice, Brazil nut, and sunflower which could also be used to improve grain quality [119].

The seed proteins have been widely studied and classified according to their solubility properties in four major categories [120]: albumins (soluble in water), globulins (soluble in salt solutions), prolamins (soluble in alcohol plus or minus disulphide reducing agents), and glutelins (soluble in dilute alkali). The main properties and nomenclature of the major storage proteins are summarized in Table 7.

The storage proteins are synthesized rapidly over a short period during seed development, their expression is both tissue- and time-specific, and it

able 6   Isolated Low-Temperature-Induced Genes

lant species source for lone iolation)	Clone name	Comments	Inducibility by other factors	References
omato	C14	Thiol protease protein degradation	High temperature	[199,200]
omato	C17			[199,200]
omato	C19			[199,200]
lfalfa	pSM784			[201]
lfalfa	pSM2201			[201]
lfalfa	pSM2358			[201]
arley	pT59			[27]
arley	pV60			[27]
arley	pAO29			[27]
arley	pAO86			[27]
arley	pAF93			[27]
rabidopsis	LTI		ABA	[47]
rabidopsis	kim1	Similarities to fish anti-freeze protein	ABA, water stress	[202]
rabidopsis	cor pHH7.2		ABA, water stress	[203]
rabidopsis	cor pHH28		ABA, water stress	[203]
rabidopsis	cor pHH29		ABA, water stress	[203]
rabidopsis	cor pHH67		ABA, water stress	[203]
ice embryo	Rab 21 (25)	Isolated as ABA inducible water stress-related clone	ABA, water stress	[204]
aize	Adh1 (91)	Cold inducibility shown in rice, Adh1 = key enzyme in ethanolic fermentation	Anaerobiosis	[205]

correlates with high levels of mRNA [115]. Therefore, the cDNA cloning of these genes was a relatively easy task in the beginning of the past decade, when recombinant DNA techniques began to be applied to plants. Biochemical data had shown that storage proteins are sequestered within protein bodies after being cotranslationally transported to the lumen of the endoplasmic reticulum [121].

Cereal Storage Proteins

Barley and Wheat Hordeins are endosperm proteins of the prolamine type of MW between 30 to 55 kD and comprise about 40% of the seed protein in barley. Two major groups, one sulphur-rich (B-hordeins) and one sulphur-poor (C-hordeins) account for most barley prolamines. A tandemly repeated octapeptide sequence rich in proline and glutamine was found in both B- and C-hordein deduced sequences, and it seems to promote a β-turn conformation [122]. A similar structure is a common feature of collagen, silk fibroin and keratin and it probably evolved from short primordial sequences by repeated duplications occurring as a result of errors in replication or repair, or through unequal crossing-over [122].

In wheat, prolamines have been classified in two groups according to their solubility in aqueous alcohols and molecular weight: Gliadins (α/β, γ and ω) of about 35 kD, and glutenins which have a disulfide-bonded subunit structure and MW between 30 to 70 kD. The latter are further subdivided into low molecular weight (LMW) and the high molecular weight (HMW) subunits [123]. Genes encoding members of these different protein types have been cloned and sequenced except for ω-gliadin. Like other cereals, each class of storage protein belongs to a multigene family, disposed in different chromosomes, with a characteristic absence of introns [124,125]. Comparison of the deduced amino acid sequence of wheat and barley prolamines reveal a common structural pattern of repetitive domains at the central part of the protein.

The bread-making quality of wheat is conferred by the viscoelastic nature of dough. Gliadins have been associated with viscosity and extensibility, and the glutenins with elasticity [126]. There is a direct correlation between allelic variation of HMW-glutenin genes and bread-making quality [123]. At least eight members of the HMW-glutenin gene family have been sequenced [126]. A model for the structure of the HMW subunit has been proposed based on the results of spectroscopic and hydrodynamic studies [127]. Accordingly, the β-turns present in the repetitive domain form a loose spiral, accounting for the rod-shaped structure of the whole protein. The nonrepetitive N- and C-terminal domains form α-helix structures with a possible globular conformation. The structure of HMW-glutenins is typical of elastomeric proteins, capable of undergoing large deformations under application of stress without braking. When the stress is removed the structure recovers. Long polymer chains are required to allow extensibility. The N- and C-terminal cysteins of HMW subunits comprise potential cross-linking sites to fulfill this function. The second feature of HMW subunit structure that could contribute to gluten functionality is the protein conformation, particularly that of the β-turn-rich repetitive domain. Recently, two particular HMW subunits, number 12 and 10, which were found previously to confer different degrees of elasticity, were shown to vary in their numbers of repeats in the central domain [128]. Subunit 10 has a higher proportion of hexamer and nonamer peptide units, and its presence in wheat grains correlates with good dough quality. A regular series of β-turns over a longer region could produce a stronger β-spiral with better elastic properties. If this accounts for different bread-making qualities, it can be predicted that modifications introduced by in vitro manipulation of the gene of subunit 10 could give rise to enhanced dough quality.

Maize   The synthesis of the prolamine fraction of maize, the zeins, is developmentally regulated and begins about 14 days after pollination [129]. Zeins are a family of proteins ranging from 10 to 28 kD of which the 22 kD and 20 kD represent 60–70% of the content of kernel protein bodies. There are approximately 100 copies of zein genes per haploid genome and several cDNA and genomic clones have been sequenced [130]. The deduced amino acid sequence of 22 kD zeins revealed a central portion which contains a 9–10-fold tandem repetition of a 20-amino acids block. There is no homology of the protein sequence with other storage proteins.

Rice   The major storage proteins present in rice seeds are the glutelins. These proteins, which may constitute about 80% of the total endosperm protein, accumulate during endosperm development at 4–6 days postanthesis. Rice glutelin is composed of subunit pairs, cross-linked by disulfide bridges, each of which comprise an acidic (28.5–30.8 kD) polypeptide and a basic (20.6–21.6 kD) polypeptide [131]. The two subunits are formed by posttranslational cleavage of the 57 kD polypeptide precursor probably during or after their accumulation in the protein bodies. Similar polypeptide structure and processing has been found in dicot globulins [132,133]. DNA sequencing of several glutelin genomic clones denoted the presence of three introns [134, 135], and it showed about 30% homology in the deduced protein sequence with 11 S globulins (soybean glycinin and pea legumin subunits).

Leguminous Storage Proteins
(Pea, Soybean, French Bean)

Generally, the storage proteins found in leguminous seeds are globulins and consist of two protein types with different physical and chemical properties: the glycosylated 7S (vicilin) and nonglycosylated 11S (legumin) types. Genes encoding legumin and vicilin have been sequenced and the coding sequences are interrupted by 3 introns [136–138]. Both proteins have an hexameric conformation, which aroses from posttranslational processing of their polypeptide precursors [133]. Legumin is important because it is the source of sulfur amino acids for animal consumption, in contrast to vicilin which is a sulfur-poor globulin. Legumin and vicilin display extensive primary and secondary structure homology, shared with phaseolin and conglycinin which has led to the view that 7S and 11S angiospermic storage proteins have evolved from a common ancestral gene [139].

Glycinins (leguminae type) are the predominant storage proteins in soybean seeds comprising up to 20% of the seed dry weight. They are produced in cotyledon cells, and sequestered in protein bodies. The mature protein is an oligomer of six similar subunits, each composed of two disulfide-linked polypeptides which result from posttranslational cleavage of a precursor during transport to the protein bodies, where the precursor is cleaved in its 40 kD acidic and 20 kD basic polypeptide chains [132]. The glycinin gene family has two types of submembers sharing 50% of homology [140], and each of them is composed of 4 exons containing sequences related to other legumes (see above). In the glycinin-coding region there are three variable regions which probably do not have a critical role in the assembly of subunits and therefore are regarded as potential targets to improve seed nutritional quality by protein engineering [140].

The other class of soybean storage proteins are conglycinins (vicilintype). They accumulate in storage vacuoles of cotyledons and embryos in trimer oligomers of various combinations formed by the two different subunits α/α' (72–76 kD) and β (52–54 kD) [141]. Conglycinins are glycosylated proteins which share homology with French beans phaseolin, pea vicilin, and convicilin [142]. The stretches of homology are spread in the amino acid sequence suggesting that these storage proteins did not evolve by splicing together exons.

In *Phaseolus vulgaris*, phaseolin (vicilin-type) comprises up to 50% of the protein in mature seeds, and it is represented by a group of three closely related glycoproteins (α, β, and γ) of 43 to 53 kD. The proteins are exclusively expressed in cotyledons [143].

Conglycinin and phaseolin are encoded by genes containing 6 exons. The homology between these two proteins is interrupted by a 174 amino acid sequence insertion in the first exon of α/α' conglycinin, which does not show transposonlike flanking sequences [142].

Tuber Storage Proteins

Tubers provide a major source of carbohydrates to human nutrition and although their relative content of protein is lower than in cereals, the nutritional value of their storage proteins is higher than that of cultivated gramineae. During potato tuber formation, several protein species accumulate in large amounts, the most abundant being patatin which comprises up to 40% of the total soluble protein [144]. Several patatin cDNA and genomic clones have been sequenced [145–147] and represent two types of submembers of the multigene family: highly expressed and tuber-specific genes (class I) and genes moderately expressed in tuber and roots (class II). In both types of genes the encoded product is a 43 kD glycoprotein which does not have homology to reported sequences of other storage proteins [146]. Remarkably, class I and class II patatins have esterase activity [147,148] of unknown function, this being the only case, so far reported, of a storage protein with enzymatic activity.

Regulatory and Developmental Genes

In contrast to animal development, plant organogenesis continues beyond the period of embryogeny, through the entire life span of the plant. Postembryogenic organogenesis occurs in the primary meristem, situated at the tip of each shoot and root. The plant body is formed through the activity of meristems that appear to be fixed or determined in specific developmental states. These states are stable but may be reprogrammed in response to signals received from the environment or from other parts of the plant [149, 150]. For instance, the regeneration of plants from single somatic cells demonstrates clearly that plant cells remain totipotent, that is, they contain all the genetic information required for the specification of the organism, but in the differentiated tissue they only express part of it. Therefore, it can be concluded that development and differentiation involve the selective control of gene expression. These phenomena are induced or regulated by plant hormones and environmental factors such as light quality, day length, temperature, gravity, etc. [149]. There is now clear evidence that responses to these signals also involve changes in gene expression (see below). Thus, it is imperative to understand the perception of these signals, the intermediate steps in the signal transduction mechanism, and the ultimate effects on gene expression, in order to understand plant development.

The ability of genes to respond to different stimuli lies mainly in their noncoding 5' end. Here, we will be dealing with the so-called master or regulatory genes involved in the selective activation or repression of groups of genes during development.

Several DNA-binding proteins have been characterized by their interaction to the 5' upstream sequence of well-known genes. It is thought that these proteins play a role in regulating gene transcription. In the past few years, an increasing number of DNA sequences encoding plant transcriptional activators has been reported. As summarized in Table 3, the methods

Table 7   Storage Protein Genes

Protein	Source	Protein class	Gene or cDNA isol. method[a]		Refs.
B-, C-, D-hordein	Barley	Prolamin	HST	HOM	[122,206]
α/β,-γ,-ω-gliadin	Wheat	Prolamin	HST	HET	[207]
LMW glutenins	Wheat	Prolamin	HST	HOM	[125,208]
HMW glutenins	Wheat	Prolamin	HST	HOM	[124,128]
20 kD-, 22 kD-zeins	Maize	Prolamin	HST	HOM	[130]
Glutelin	Rice	Glutelin	OLI	EXL	[134,135]
Legumin	Pea	Globulin	HST	HOM	[136,138]
Vicilin	Pea	Globulin	HST	HOM	[136,137]
Glycinin	Soybean	Globulin	OLI	HET	[140,209]
α/α'-, β-glycinin	Soybean	Globulin	HST	HOM	[210,211]
β-phaseolin	French bean	Globulin	HST	HOM	[212,213]
Patatin	Potato	Globulin	SUH	HOM	[145,146,147]

[a]HST = hybrid selected translation; HOM = homologous probe; HET = heterologous probe;
OLI = oligonucleotide probe; EXL = expression library; SUH = subtractive hybridization.

to isolate these genes include gene-tagging, South-Western, differential hybridization, or the use of heterologous probes. Remarkably, the encoded products, so far described for plant *trans*-acting factors, contain protein domains also found in yeast and animal cells, suggesting a conserved mechanism in eukaryotes for activating gene transcription.

The first plant transcriptional activator to be isolated was the *C1* locus involved in regulating the biosynthesis of anthocyanins in maize kernels. The DNA sequence of the *C1* gene encodes a protein homologous to *myb* oncoproteins in animals and a C-terminal acidic-domain probably involved in activation of transcription [151]. Another two genetic loci, *B* and *R*, are also involved in the regulation of anthocyanin pigmentation in maize. The deduced amino acid of gene *Lc* from the *R* gene family revealed a protein containing the characteristic helix-loop-helix domain of some animal transcriptional activators [152]. When a *B* allele was expressed after transformation in intact maize aleurones and embryogenic callus from a colorless mutant, it was able to transactivate the *bz1* gene which encodes an enzyme involved anthrocyanin biosynthesis [38]. A potential transcriptional regulator of *C1* is the *viviparous-1* (*vp-1*) gene [59]. Mutant *vp1* embryos show precocious germination due to reduced sensitivity to abscisic acid and have blocked anthocyanin synthesis. Using the DNA-binding domain coding sequence of *C1* gene as a probe several *myb*-like genes have been isolated from barley and maize [153]. The function of these genes still remains to be found.

The 22 kD zein storage protein genes are under control of the *Opaque-2* (*02*) locus which encodes a protein containing a domain similar to the leucine zipper motif identified in *fos* and *jun* animal proto-oncogenes and yeast GCN4 transcriptional activators [154]. Adjacent to this sequence, toward the amino terminus, a conserved cluster of basic amino acid residues is also present in *02* protein, which comprise the DNA binding domain. In transformed tobacco protoplasts, the *02* protein is able to *trans*-activate a maize endosperm-specific gene, otherwise inactive in these cells. In addition, this protein can autoregulate its own gene promoter, which is a mechanism also known to occur in *fos* [40]. Recently, several other transcriptional activators of the leucine zipper type have been isolated and shown to bind specifically to previously identified DNA motifs. For instance, tobacco TGA1a and TGA1b gene products bind to the CaMV 355 gene promoter [35]; wheat HBP-1 interacts with histone H3 promoter [155]; maize OCSBF-1 binds to T-DNA octopine synthase promoter; tomato MSF24 binds to the consensus upstream sequence of heat-shock genes [37]; and wheat EmBP-1 interacts with the DNA element known to confer abscisic acid-response to the Em gene during embryo development [36].

A good example of development is the pattern formation in flowers. Flowering occurs in response to a combination of endogenous and environmental triggers that change a self-perpetuating vegetative meristem into a terminal floral meristem [149]. Although a large collection of homeotic flower mutants has been described [157–159], for instance in *Arabidopsis thaliana* and *Antirrhinum majus*, no molecular information was available until recently. The *Deficiens (defA$^+$)* gene from *A. majus* causes homeotic transformation of petals into sepals and of stamens into carpels [29]. The DNA sequence of *defA$^+$* revealed a putative protein with homology to a conserved domain in SFR transcriptional activators which in mammals are known to active *c-fos* proto-oncogene. The SRF-like sequence in DEF A protein contains DNA binding and dimerization domains.

Mutations in the homeotic gene, *agamous* from *Arabidopsis*, result in the overall phenotype of a flower within a flower and the absence of stamens and carpels [52]. The deduced sequence from *agamous* shows homology to DEF A in the SRF domain but not elsewhere. The expression of *agamous* protein in transgenic mutant *Arabidopsis* restores the wildtype phenotype [76]. Another homeotic mutant, *floricaula (flo)* from *Antirrhinum*, seems to be blocked in an early acting gene which causes indeterminate influorescence meristems in place of floral meristems so that the inflorescence continues to proliferate without producing flowers [159]. The *flo* gene encodes a putative protein (FLO) containing a proline-rich N-terminus and a highly acidic central region but has no homology to other reported sequences [62]. Thus, although it is possible that FLO functions as a transcriptional activator, other roles cannot be excluded.

The genes responsible for other developmental mutants like *knotted* in maize which alters leaf morphology [58], *glabrous 1* in *Arabidopsis* which results in stem without trichomes [160,161], and a dwarf mutant of *Arabidopsis*, have been isolated but no DNA sequences are available yet. Again, it could turn out that the wild-type allele of these different mutants encode regulatory proteins.

As mentioned before, the helix-loop-helix, leucine zipper, and SRF transcriptional activators have a DNA binding and dimerization domain. If plant regulatory proteins as it has been shown in animals [162], can form dimers or heterodimers between different members of a *trans*-acting factor family, the diversity provided by all the possible combinations would offer an enormous scope for gene regulation. Thus, multiple oligomeric complexes could bind to the same DNA recognition sequence, and different combination of proteins that bind to a single site can lead to activation or repression of transcription.

In the future, it can be envisaged that whole sets of genes involved in biosynthetic pathways, organ development, nitrogen fixation, pathogen attack, or stress response, will be available for gene-transfer experiments. The correct and coordinate expression in transgenic plants of several structural genes will require concomitant expression of the appropriate regulatory genes.

FUTURE PERSPECTIVES

The number of plant genes that have been isolated and characterized is large and is increasing constantly. The molecular techniques have made it possible to isolate any structural genes of interest encoding a defined gene product. The more challenging task will be to isolate genes whose product is unknown. Most easily accessible will be genes of genetically well-characterized systems, first of all *Arabidopsis*. For *Arabidopsis* an enormous number of mutants [158] are available and with the progress of RFLP analysis, YAC cloning, etc., it should be possible to isolate any desirable gene from *Arabidopsis* in the near future. Hence, *Arabidopsis* will serve as a model system, and a second step will be to isolate the corresponding genes from agriculturally important plants. This may be achieved by using heterologous probes provided there is sufficient homology or the particular plant system needs to be investigated. All this work will lead to the analysis of the gene structure and description of expression patterns, but the major problem will be to understand the agriculturally desirable traits on the molecular level. To date very little information is available about enzymes and essential metabolites involved in desirable traits. To be able to manipulate important traits at the molecular level, it will be necessary to integrate the molecular biology into plant physiology. Only this approach will in the end lead to the successful application of genetic engineering in improving crop productivity.

As an example the authors in [163] discuss the manipulation of the glycine betaine pathway, implicated in drought and salt tolerance of some plants. They describe the complexity of steps if one wants to modify plant stress resistance by genetic engineering. For this chosen example they consider the different parameters which have to be taken into account like suitable choice of required target enzymes, correct compartmentation of enzymatic activity, appropriate regulation of the pathway and influence of the engineered pathway on general plant performance.

Being able to isolate the desired genes is the first step. Breeding crop plants by genetic engineering is based (1) on the isolation and manipulation of relevant genes (discussed in this chapter) and (2) on the availability of an appropriate plant transformation system. Considerable progress has been made in the development of gene transfer systems for higher plants [32,164]. The ability to introduce foreign genes into plants provides an unparalleled opportunity to modify and improve agriculturally important crops with relevant genes. The precise understanding on the molecular level of relevant agronomic traits is an essential step to a successful application of recombinant DNA technology to plant breeding. Traits controlled by one or a few Mendelian genes are the best candidates for molecular manipulations. Many important agronomical characters, for example, yield, are not easily related to defined genes. For these traits first a better understanding of their genetic control is essential. Despite these limitations genetically engineered crops are expected to be available for the market between 1993 and 2000.

ACKNOWLEDGMENTS

We gratefully acknowledge the assistance of Professor F. Salamini in compiling this review and M. Pasemann for typing the manuscript and for her patience.

REFERENCES

1. N. Sauer, F. Kriedländer, and U. Gräml-Wicke, *EMBO J., 9*: 30 (1990).
2. D. E. Rochester, J. A. Winter, and D. M. Shah, *EMBO J., 5*: 451 (1986).
3. J. K. Roberts and K. L. Key, *Plant Mol. Biol.* (in press).
4. R. A. Young and R. W. Davies, Proceedings National Academy of Sciences (U.S.), *80*: 1194 (1983).
5. D. Hondred and A. D. Hanson, Proceedings National Academy of Sciences (U.S.), *87*: 7300 (1990).
6. T. Sato and A. Theologis, Proceedings National Academy of Sciences (U.S.), *86*: 6621 (1989).
7. J. W. Szostak, J. J. Stiles, B. K. Tyoe, P. Chin, F. Sherman, and R. Wu, *Meth. Enzymol., 68*: 419 (1979).
8. J. Sambrook, E. F. Fritsch, and T. Maniatis, *Molecular Cloning: A Laboratory Manual*, Cold Spring Harbor Laboratory Press, New York (1989).

9. N. Touchot, P. Chardin, and A. Tavitian, Proceedings National Academy of Sciences (U.S.), *84*: 8210 (1987).
10. W. I. Wood, J. Gitschier, L. A. Lasky, and R. M. Lawn, Proceedings National Academy of Sciences (U.S.), *82*: 1585 (1985).
11. H. Kaufmann, F. Salamini, and R. D. Thompson, *Mol. Gen. Genet.* (in press).
12. Ch. Eckerskorn and F. Lottspeich, *Electrophoresis, 9*: 830 (1988).
13. B. Claes, R. Dekeyser, R. Villarroel, M. Van den Bulcks, G. Bauw, M. van Montagu, and A. Caplan, *Plant Cell, 2*: 19 (1990).
14. T. Hesse, J. Feldwisch, D. Balshüsemann, G. Bauw, M. Puype, J. Vandekerckhove, M. Löbler, D. Klämbt, J. Schell, and K. Palme, *EMBO J., 8*: 2453 (1989).
15. U. Tillmann, G. Viola, B. Kayser, G. Siemeister, T. Hesse, K. Palme, M. Löbler, and D. Klämbt, *EMBO J., 8*: 2463 (1989).
16. J. V. Cullimore and B. J. Miflin, *FEBS Lett., 158*: 107 (1983).
17. D. P. Snustad, J. P. Hunsperger, B. M. Chereskin, and J. Messing, *Genetics, 120*: 1111 (1988).
18. P. Meyer, I. Heidmann, G. Forkmann, and H. Saedler, *Nature, 330*: 677 (1987).
19. K. B. Mullis and F. A. Faloona, *Meth. Enzymol., 155*: 335 (1987).
20. S. J. Scharf, G. T. Horn, and H. A. Ehrlich, *Science, 233*: 1076 (1986).
21. M. A. Frohman, H. K. Dush, and G. R. Martin, Proceedings National Academy of Sciences (U.S.), *85*: 8998 (1988).
22. *PCR Protocols* (Innis, Gelford, Sninsky, White, eds.), Academic Press, New York (1990).
23. T. D. Sargent, *Meth. Enzymol., 152*: 423 (1987).
24. M. Trick and R. B. Flavell, *Mol. Gen. Genet., 218*: 112 (1989).
25. J. Mundy and N.-H. Chua, *EMBO J., 7*: 2279 (1988).
26. D. Bartels, K. Schneider, G. Terstappen, D. Piatkowski, and F. Salamini, *Planta, 181*: 27 (1990).
27. L. Cattivelli and D. Bartels, *Plant Physiol., 93*: 1504 (1990).
28. S. Melzer, D. M. Majewski, and K. Apel, *Plant Cell,* 953 (1990).
29. H. Sommer, J. P. Beltran, P. Huijser, H. Pape, W. E. Lönnig, H. Saedler, and Z. Schwarz-Sommer, *EMBO J., 9*: 605 (1990).
30. R. B. Goldberg, S. J. Barker, and L. Perez-Grau, *Cell, 56*: 149 (1989).
31. M. Thomas and R. B. Flavell, *Plant Cell, 2*: 1171 (1990).
32. N. Fenfey and N.-H. Chua, *Science, 250*: 959 (1990).
33. K. D. Jofuku, J. K. Okamuro, and R. B. Goldberg, *Nature, 328*: 734 (1987).
34. P. J. Green, M.-H. Young, M. Cuozzo, Y. Kano-Murakami, P. S. Silverstein, and N.-H. Chau, *EMBO J., 7*: 4035 (1988).
35. F. Katagiri, E. Lam, and N.-H. Chua, *Nature, 340*: 727 (1989).
36. M. J. Guiltinan, W. R. Marcotte, and R. S. Quatrano, *Science, 250*: 267 (1990).
37. K.-D. Scharf, S. Rose, W. Zott, F. Schöff, and L. Nover, *EMBO J., 9*: 4495 (1990).
38. S. A. Goff, T. M. Klein, B. A. Roth, M. E. Fromm, K. C. Cone, J. P. Radicella, and V. Chandler, *EMBO J., 9*: 2517 (1990).
39. P. Hilson, D. Froidmont, C. Lejour, S.-I. Hirai, J.-M. Jacquemin, and M. Yaniv, *Plant Cell, 2*: 651 (1990).
40. S. Lohmer, M. Maddaloni, M. Motto, N. Di Fonzo, H. Hartings, F. Salamini, and R. D. Thompson, *EMBO J.* (1991).

41. S. D. Tanksley, N. D. Young, A. H. Peterson, and M. W. Bonierbale, *Biotech.*, *7*: 257 (1989).
42. K. E. Davies and K. J. Robson, *BioEssays, 6*: 247 (1987).
43. J. M. Rommens, M. C. Januzzi, B.-S. Kerem, L. Drumm, G. Melmer, M. Dean, R. Rozmahel, J. L. Cole, D. Kennedy, N. Hidaka, M. Zsiga, M. Buchwald, J. R. Riordan, L. C. Tsui, and Collins, *Science, 245*: 1059 (1989).
44. R. E. Pruitt and E. M. Meyerowitz, *J. Mol. Biol., 187*: 169 (1986).
45. K. A. Feldman and M. D. Marks, *Mol. Gen. Genet., 208*: 1 (1987).
46. M. Koornneef, G. Reuling, and C. M. Karssen, *Physiol. Plant., 61*: 377 (1984).
47. The 5th International Conference on Arabidopsis, Vienna, Austria (1990).
48. E. M. Meyerowitz, *Cell, 56*: 263 (1989).
49. H. G. Nam, J. Girandot, B. den Boers, F. Moonan, W. Loos, B. Hange, and H. M. Goodman, *Plant Cell, 1*: 699 (1989).
50. C. Chang, J. L. Bowman, A. W. de John, E. S. Lander, and E. M. Meyerowitz, Proceedings National Academy of Sciences (U.S.), *85*: 6856 (1988).
51. S. Schlessinger, *Trends Genet., 6*: 231 (1990).
52. E. R. Ward and G. C. Jen, *Plant Mol. Biol., 14*: 561 (1990).
53. N. V. Fedoroff, *Cell, 56*: 181 (1989).
54. N. V. Fedoroff, D. B. Furtek, and O. E. Nelson, Proceedings National Academy of Sciences (U.S.), *81*: 3825 (1984).
55. C. O'Reilly, N. S. Shepherd, A. Pereira, Zs. Schwarz-Sommer, I. Bertram, D. S. Robertson, P. A. Peterson, and H. Haedler, *EMBO J., 4*: 877 (1985).
56. U. Wienand, U. Weydemann, U. Niesbach-Klösgen, P. A. Peterson, and H. Saedler, *Mol. Gen. Genet., 203*: 202 (1986).
57. M. Motto, M. Maddaloni, G. Pinziani, M. Brembilla, R. Marotto, N. Di Fonzo, C. Soave, R. Thompson, and F. Salamini, *Mol. Gen. Genet., 212*: 488 (1988).
58. S. Hake, E. Vollbrecht, and M. Freeling, *EMBO J., 8*: 15 (1989).
59. D. R. McCarty, C. B. Carson, P. S. Stinard, and D. S. Robertson, *Plant Cell, 1*: 523 (1989).
60. B. Buckner, T. L. Kelson, and D. S. Robertson, *Plant Cell, 2*: 867 (1990).
61. C. Martin, R. Carpenter, H. Sommer, H. Saedler, and E. S. Coen, *EMBO J., 4*: 1625 (1985).
62. E. S. Coen, J. M. Romero, S. Doyle, R. Elliott, G. Murphy, and R. Carpenter, *Cell, 63*: 1311 (1990).
63. B. Baker, J. Schell, H. Lörz, and N. Fedoroff, Proceedings National Academy of Sciences (U.S.), *83*: 4844 (1986).
64. M. A. van Sluys, J. Tempe, and N. Fedoroff, *EMBO J., 6*: 3881 (1987).
65. S. Knapp, G. Coupland, H. Uhrig, P. Starlinger, and F. Salamini, *Mol. Gen. Genet., 213*: 285 (1988).
66. J. I. Yoder, J. Palys, K. Alpert, and M. Lassner, *Mol. Gen. Genet., 213*: 291 (1988).
67. N. Houba-Herin, D. Becker, A. Post, Y. Larandelle, and P. Starlinger, *Mol. Gen. Genet., 224*: 17 (1990).
68. A. Pereira and H. Saelder, *EMBO J., 8*: 1315 (1989).
69. M. Frey, S. M. Tavantzis, and H. Saedler, *Mol. Gen. Genet., 217*: 1172 (1989).
70. C. Martin, A. Prescott, C. Listerff, and S. MacKay, *EMBO J., 8*: 1997 (1989).
71. L. Balcells, J. Swinburne, and G. Coupland, *TIBTECH, 9*: 31 (1991).

72. D. Andre, D. Colau, J. Schell, M. van Montagu, and J.-P. Hernalsteens, *Mol. Gen. Genet., 204*: 512 (1986).
73. T. H. Teeri, L. Herrera-Estrella, A. Depicker, M. van Montagu, and E. T. Palva, *EMBO J., 5*: 1755 (1986).
74. K. A. Feldmann, M. D. Marks, M. L. Christianson, and R. S. Quatrano, *Science, 243*: 1351 (1989).
75. C. Koncz, R. Mayerhofer, Z. Koncz-Kalman, C. Nawrath, B. Reiss, G. P. Redei, and J. Schell, *EMBO J., 9*: 1337 (1990).
76. M. F. Yanofsky, H. Ma, J. L. Bowman, G. N. Drews, K. A. Feldmann, and E. M. Meyerowitz, *Nature, 346*: 35 (1990).
77. D. Neumann, L. Nover, B. Parthier, R. Rieger, K.-D. Scharf, R. Wollgieh, and U. Zur Nieden, *Biol. Zentralblatt, 108*: 1 (1989).
78. J. Levitt, *Responses of Plants to Environmental Stresses*, Vol. 2, Academic Press, New York (1980).
79. D. Scheel, Proceedings of the 7th International Congress of Pesticide Chemistry (H. Frehse, ed.), VCH Publication, Weinheim, Germany (1990).
80. R. A. Dixon and M. J. Harrison, *Environmental Stress*, Vol. 28 (J. G. Scandalios, ed.), Academic Press, New York (1990).
81. K. Hahlbrock and D. Scheel, *Ann. Rev. Plant Physiol. Plant Mol. Biol., 40*: 347 (1989).
82. K. Hahlbrock, D. Scheel, In *Innovative Approaches to Plant Disease Control* (I. Chet, ed.), Wiley, New York, p. 229 (1987).
83. H. Kauss, T. Waldmann, W. Jeblick, G. Eulerf, R. Ranjewa, and A. Domard, *Signal Molecules in Plants and Plant Microbe Interactions* (B. J. J. Lugtenberg, ed.), Springer Verlag, Berlin, Germany, p. 107 (1989).
84. H. Buhlmann, S. Clausen, S. Behnke, H. Giese, C. Hiller, U. Reimann-Philipp, G. Schrader, V. Barkholt, and K. Apel, *EMBO J., 7*: 1559 (1988).
85. T. Boller, Oxford Surveys of Plant Mol. & Cell Biol., *5*: 145 (1988).
86. L. C. Van Loon, *Plant-Microbe Interactions. Molecular and Genetic Perspectives* (T. Kosuge, E. V. Nester, eds.), McGraw-Hill, New York, p. 198 (1989).
87. I. E. Somssich, J. Bollmann, K. Hahlbrock, E. Kombrink, and W. Schulz, *Plant Mol. Biol., 12*: 227 (1989).
88. R. T. Nagao, J. A. Kimpel, and J. L. Key, *Advances in Genetics (Genomic Responses to Environmental Stress)* (J. G. Scandalios, ed.), Vol. *28*: 235, Academic Press, New York (1990).
89. M. M. Sachs, M. Freeling, and R. Okimoto, *Cell, 20*: 761 (1980).
90. M. M. Sachs, T. H. D. Ho, *Ann. Rev. Plant Physiol., 37*: 363 (1986).
91. M. R. Olive, J. C. Walter, K. Singh, E. S. Dennis, and W. J. Peacock, *Plant Mol. Biol., 15*: 593 (1990).
92. E. S. Dennis, M. M. Sachs, W. L. Gerlach, E. J. Finnegan, and W. J. Peacock, *Nucl. Acids Res., 13*: 727 (1985).
93. W. Werr, W. B. Frommer, C. Maas, and P. Starlinger, *EMBO J., 4*: 1373 (1985).
94. E. S. Dennis, W. L. Gerlach, J. C. Walker, M. Lavin, and W. J. Peacock, *J. Mol. Biol., 202*: 759 (1988).
95. C. Chang and E. M. Meyerowitz, Proceedings National Academy of Sciences (U.S.), *83*: 1408 (1986).

96. W. J. Peacock, *J. Mol. Biol., 195*: 115 (1987).
97. Y. Xie and R. Wu, *Plant Mol. Biol., 13*: 537 (1989).
98. K. Skriver and J. Mundy, *The Plant Cell, 2*: 503 (1990).
99. L. Dure III, M. Crouch, I. Harada, T. H. D. Ho, J. Mundy, R. Quatrano, T. Thomas, and Z. R. Sung, *Plant Mol. Biol., 12*: 475 (1989).
100. K. F. McCue and A. D. Hanson, *Trends Biotechnol., 8*: 358 (1990).
101. M. Pla, A. Goday, J. Vilardell, J. Gomez, and P. Montserrat, *Plant Mol. Biol., 13*: 385 (1989).
102. W. R. Marcotte, C. C. Bayley, and R. S. Quatrano, *Nature, 335*: 454 (1988).
103. K. Yamaguchi-Shinozaki, M. Masanobu, J. Mundy, N.-H. Chua, *Plant Mol. Biol., 15*: 905 (1990).
104. M. Stitt and P. Quick, *Physiol. Plant., 77*: 633 (1989).
105. R. G. Creech, *Genetics, 52*: 1175 (1965).
106. C. Maas, S. Schaal, and W. Werr, *EMBO J., 9*: 3447 (1990).
107. J. Preiss, *Ann. Rev. Plant Physiol., 33*: 431 (1982).
108. *Journal of Cellular Biochemistry*, Suppl. 15A, Abstract 361 (1991).
109. M. R. Bhave, S. Lawrence, C. Barton, and L. C. Hannah, *Plant Cell, 2*: 581 (1990).
110. M. R. Olive, R. J. Ellis, and W. W. Schuch, *Plant Mol. Biol., 12*: 525 (1989).
111. M. K. Morrell, M. Bloom, V. Knowles, and J. Preiss, *Plant Physiol., 85*: 182 (1987).
112. R. B. Klösgen, A. Gierl, Z. Schwarz-Sommer, and H. Saedler, *Mol. Gen. Genet., 203*: 237 (1986).
113. M. K. Bhattacharyya, A. M. Smith, T. H. N. Ellis, C. Hedley, and C. Martin, *Cell, 60*: 115 (1990).
114. D. Spencer, Phil. Trans. R. Soc., London, *B304*: 275 (1984).
115. T. J. V. Higgins, *Ann. Rev. Plant Physiol., 35*: 191 (1984).
116. S. L. Desborough, *Potato Physiology* (P. H. Li, ed.), Academic Press, New York, p. 329 (1985).
117. M. Maeshima, T. Sasaki, and T. Asahi, *Phytochem., 24*: 1899 (1985).
118. J. M. Jaynes, M. S. Yang, N. Espinoza, and J. H. Dodds, *Trends Biotech., 4*: 314 (1986).
119. S. B. Altenbach and R. B. Simpson, *Trends Biotech., 8*: 156 (1990).
120. T. B. Osborne, *The Vegetable Proteins*, Longmans, Green, London, p. 1 (1924).
121. M. J. Chrispeels, Philosphy Transactions Royal Society, London, *B304*: 309 (1984).
122. B. G. Forde, M. Kreis, M. S. Williamson, R. P. Fry, J. Pywell, P. R. Shewry, N. Bunce, and B. J. Miflin, *EMBO J., 4*: 9 (1985).
123. P. I. Payne, *Ann. Rev. Plant Physiol., 38*: 141 (1987).
124. R. D. Thom,pson, D. Bartels, N. P. Harberd, and R. B. Flavell, *Theor. Appl. Genet., 67*: 87 (1983).
125. V. Colot, D. Bartels, R. Thompson, and R. Flavell, *Mol. Gen. Genet., 216*: 81 (1989).
126. P. R. Shewry, N. G. Halford, and A. S. Tatham, Oxford Surveys Plant Mol. Cell Biol., *6*, 163 (1989).

127. J. M. Field, A. S. Tatham, and P. R. Shewry, *Biochem. J., 247*: 215 (1987).
128. R. B. Flavell, A. P. Goldsbrough, L. S. Robert, D. Schnick, and R. D. Thompson, *Biotech., 7*: 1281 (1989).
129. M. Motto, N. Di Fonzo, H. Hartings, M. Maddaloni, F. Salamini, C. Soave, and R. D. Thompson, Oxford Surveys Plant *Mol. Cell Biol., 6*: 87 (1989).
130. G. Heidecker and J. Messing, *Ann. Rev. Plant Physiol., 37*: 439 (1986).
131. T. N. Wen and D. S. Luthe, *Plant Physiol., 78*: 172 (1985).
132. C. D. Dickinson, E. H. A. Hussein, and N. C. Nielsen, *Plant Cell, 1*: 459 (1989).
133. R. R. D. Croy, J. A. Gatehouse, I. A. Evans, and D. Boulter, *Planta, 148*: 49 (1980).
134. W. Higuchi and C. Fukazaawa, *Gene, 55*: 245 (1987).
135. T. W. Okita, Y. S. Hwang, J. Hnilo, W. T. Kim, A. P. Aryan, R. Laron, and H. B. Krishnan, *J. Biol. Chem., 264*: 12573 (1989).
136. R. R. D. Croy, G. W. Lycett, J. A. Gatehouse, J. N. Yarwood, and D. Boulter, *Nature, 295*: 76 (1982).
137. G. W. Lycett, A. J. Delauney, J. A. Gatehouse, J. Gilroy, R. R. D. Croy, and D. Boulter, *Nucl. Acids Res., 11*: 2367 (1983).
138. G. W. Lycett, R. R. D. Croy, A. H. Shorsat, and D. Boulter, *Nucl. Acids Res., 12*: 4493 (1984).
139. P. Argos, S. V. L. Narayana, and N. C. Nielsen, *EMBO J., 4*: 1111 (1985).
140. N. C. Nielsen, C. D. Dickinson, T.-J. Cho, V. H. Thanh, B. J. Scallon, R. L. Fischer, T. L. Sims, G. N. Drews, and R. B. Goldberg, *Plant Cell, 1*: 313 (1989).
141. V. H. Thanh and K. Shibasaki, *J. Agric. Food Chem., 26*: 692 (1978).
142. J. J. Doyle, M. A. Schuler, W. D. Godette, V. Zeuger, and R. N. Beachy, *J. Biol. Chem., 261*: 9228 (1986).
143. J. W. S. Brown, Y. Ma, F. A. Bliss, and T. C. Hall, *Theor. Appl. Genet., 59*: 83 (1981).
144. D. Racusen and M. Foote, *Food Biochem., 4*: 43 (1980).
145. G. A. Mignery, C. S. Pikaard, D. J. Hannapel, and W. D. Park, *Nucl. Acids Res., 12*: 7987 (1984).
146. M. Bevan, R. Barker, A. Goldsbrough, M. Jarvis, T. Kavanagh, and G. Iturriaga, *Nucl. Acids Res., 14*: 4625 (1986).
147. S. Rosahl, J. Schell, and L. Willmitzer, *EMBO J., 6*: 1155 (1987).
148. G. Iturriaga, *The Study of Protein Transport to the Secretory Pathway in Plants*. Ph.D. Thesis, Cambridge University, England, p. 156 (1989).
149. I. M. Sussex, *Cell, 56*: 225 (1989).
150. R. S. Poethig, *Science, 250*: 923 (1990).
151. J. Paz-Ares, D. Ghosal, U. Wienand, P. A. Peterson, and H. Saedler, *EMBO J., 6*: 3553 (1987).
152. S. R. Ludwig, L. F. Habera, S. L. Dellaporta, and S. R. Wessler, Proceedings National Academy of Sciences (U.S.), *86*: 7092 (1989).
153. A. Marocco, M. Wissenbach, D. Becker, J. Paz-Ares, H. Saedlerf, F. Salamini, and W. Rohde, *Mol. Gen. Genet., 216*: 183 (1989).
154. H. Hartings, M. Maddaloni, N. Lazzaroni, N. Di Fonzo, M. Motto, F. Salamini, and R. Thompson, *EMBO J., 8*: 2795 (1989).
155. T. Tabata, H. Takase, S. Takayama, K. Mikami, A. Nakatsuka, T. Kawata, T. Nakayama, and M. Iwabuchi, *Science, 245*: 965 (1989).

156. K. Singh, E. S. Dennis, J. G. Ellis, D. J. Llewellyn, J. G. Tokuhisa, J. A. Wahleithner, and W. J. Peacock, *Plant Cell, 2*: 891 (1990).
157. Zs. Schwarz-Sommer, P. Huijser, W. Nacken, H. Saedler, and H. Sommer, *Science, 250*: 931 (1990).
158. E. M. Meyerowitz, *Cell, 56*: 263 (1989).
159. R. Carpenter and E. S. Coen, *Genes Develop., 4*: 1483 (1990).
160. M. D. Marks and K. A. Feldmann, *Plant Cell, 1*: 1043 (1989).
161. P. L. Herman and M. D. Marks, *Plant Cell, 1*: 1051 (1989).
162. N. Jones, *Cell, 61*: 9 (1990).
163. K. F. McCue and A. D. Hanson, *Trends Biotechnol., 8*: 358 (1990).
164. J. Schell, *Science, 237*: 1176 (1987).
165. T. Helentjaris, *Trends Genet., 3*: 217 (1987).
166. B. S. Landry, R. V. Kesseli, B. Farrara, and R. W. Michelmore, *Genetics, 116*: 331 (1987).
167. D. Zamir and S. D. Tanksley, *Mol. Gen. Genet., 213*: 254 (1988).
168. S. R. McCouch, G. Kochert, Z. H. Yu, T. Y. Wong, G. S. Khush, W. R. Coffman, and S. D. Tanksley, *Theor. Appl. Genet., 76*: 815 (1988).
169. S. D. Tanksley, R. Bernatzky, N. L. Lapitan, and J. P. Prince, Proceedings National Academy of Sciences (U.S.), *85*: 6419 (1988).
170. C. Gebhardt, E. Ritter, T. Debener, U. Schachtschabel, B. Walkemeier, H. Uhrig, and F. Salamini, *Theor. Appl. Genet., 78*: 65 (1989).
171. M. K. Slocum, S. S. Figdore, W. C. Kennard, J. Y. Suzuki, and T. C. Osborn, *Theor. Appl. Genet., 80*: 57 (1990).
172. S. V. Tingey, J. A. Rafalski, and J. G. K. Williams, IV, Nato Advanced Study Institute, *Plant Mol. Biol.,* 185 (1990).
173. S. Chao, P. J. Sharp, A. J. Worland, E. J. Warham, R. M. D. Koebner, and M. D. Gale, *Theor. Appl. Genet., 78*: 495 (1989).
174. M. D. Gale, S. Chao, and P. J. Sharp, *Gene Manipulation and Plant Improvement*, Vol. 2 (J. P. Gustavson, ed.), Plenum Press, New York, p. 353 (1990).
175. N. D. Young, D. Zamir, M. W. Ganal, and S. D. Tanksley, *Genetics, 120*: 579 (1988).
176. M. Sarfatti, J. Katan, R. Fluhr, and D. Zamir, *Theor. Appl. Genet., 78*: 755 (1989).
177. S. Klein-Lankhorst, P. Rietvold, B. Machiels, R. Verkerk, C. Weide, C. Gebhardt, M. Koornneef, and P. Zabel, *Theor. Appl. Genet.* (in press).
178. M. D. McMullen and R. Louie, *Mol. Plant-Microbe Interact., 2*: 309 (1989).
179. C. Jung, M. Kleine, F. Fischer, and R. G. Herrmann, *Theor. Appl. Genet.* (in press).
180. A. Barone, E. Ritter, U. Schachtschabel, T. Debener, F. Salamini, and C. Gebhardt, *Mol. Gen. Genet., 224*: 177 (1990).
181. K. Hinze, R. D. Thompson, E. Ritter, F. Salamini, and P. Schulze-Lefert, Proceedings National Academy of Sciences (U.S.) (in press).
182. J. Vilardell, A. Goday, M. A. Freise, M. Torent, M. C. Martinez, J. M. Torne, and M. Pages, *Plant Mol. Biol., 14*: 423 (1990).
183. T. J. Close, A. A. Kortt, and P. M. Chandler, *Plant Mol. Biol., 13*: 95 (1989).
184. K. Yamaguchi-Shinozake, J. Mundy, and N. H. Chua, *Plant Mol. Biol., 14*: 29 (1989).

185. J. C. Baker, C. Steele, and L. Dure III, *Plant Mol. Biol., 11*: 277 (1988).
186. D. Piatkowski, K. Schneider, F. Salamini, and D. Bartels, *Plant Physiol., 94*: 1682 (1990).
187. W. R. Marcotte Jr, C. C. Bayley, and R. S. Quatrano, *Nature, 335*: 454 (1988).
188. M. Raynal, D. Depigny, R. Cooke, and M. Delseny, *Plant Physiol., 91*: 829 (1989).
189. J. Gomez, D. Sanchez-Martinez, V. Stiefel, J. Rigau, P. Puigdomenech, and M. Pages, *Nature, 334*: 262 (1988).
190. J. J. Harada, A. J. DeLisle, C. S. Baden, and M. L. Crouch, *Plant Mol. Biol., 12*: 395 (1989).
191. R. Leah and J. Mundy, *Plant Mol. Biol., 12*: 673 (1989).
192. F. D. Guerrero, J. T. Jones, and J. E. Mullet, *Plant Mol. Biol., 15*: 11 (1990).
193. J. A. Goday, J. M. Pardo, and J. A. Pintor-Toro, *Plant Mol. Biol., 15*: 695 (1990).
194. S. Torres-Schumann, J. A. Goday, and J. A. Pintor-Toro, *NATO-Abstract* (1990).
195. G. J. King, V. A. Turner, C. E. Hussey, E. S. Wurtele, and S. M. Lee, *Plant Mol. Biol., 10*: 401 (1988).
196. N. K. Singh, D. E. Nelson, D. Kuhn, P. M. Hasegawa, and R. A. Bressan, *Plant Physiol., 90*: 1096 (1989).
197. C. B. Michalowski, S. W. Olson, M. Piepenbrock, J. M. Schmitt, and H. J. Bohnert, *Plant Physiol., 89*: 811 (1989).
198. E. A. Weretilnyk and A. D. Hanson, Proceedings National Academy of Sciences (U.S.), *87*: 2745 (1990).
199. M. A. Schaffer and R. L. Fischer, *Plant Physiol., 87*: 431 (1988).
200. M. A. Schaffer and R. L. Fischer, *Plant Physiol., 93*: 1486 (1990).
201. S. S. Mohapatra, L. Wolfraim, R. J. Poole, R. S. Dhindsa, *Plant Physiol., 89*: 375 (1989).
202. S. Kurkela and M. Franck, *Plant Mol. Biol., 15*: 137 (1990).
203. R. K. Hajela, D. P. Horvath, S. J. Gilmour, and M. F. Tomashow, *Plant Physiol., 93*: 1246 (1990).
204. M. Hahn and V. Walbot, *Plant Physiol., 91*: 930 (1989).
205. P. J. Christie, M. Hahn, and V. Walbot, *Plant Physiol.* (in press).
206. B. G. Forde, M. Kreis, B. Bahramian, J. A. Matthews, B. J. Miflin, R. D. Thompson, D. Bartels, and R. B. Flavell, *Nucl. Acids Res., 9*: 6689 (1981).
207. C. C. Nimma, E. J.-L. Lew, M. D. Dietler, and F. C. Greene, Proceedings National Academy of Sciences (U.S.), *81*: 4712 (1984).
208. D. Bartels and R. D. Thompson, *Nucl. Acids Res., 11*: 2961 (1983).
209. M. A. Schuler, E. S. Schmitt, and R. N. Beachy, *Nucl. Acids Res., 10*: 8225 (1982).
210. T. Negoro, T. Momma, and C. Fukazawa, *Nucl. Acids Res., 13*: 6719 (1985).
211. J. J. Harada, S. J. Barker, and R. B. Goldberg, *Plant Cell, 1*: 415 (1989).
212. S. M. Sun, J. L. Slighton, and T. C. Hall, *Nature, 289*: 37 (1981).
213. J. L. Slighton, S. M. Sun, and T. C. Hall, Proceedings National Academy of Sciences (U.S.), *80*: 1897 (1983).
214. A. Gasch, A. Hoffman, M. Horikoshi, R. G. Roeder, and N.-H. Chua, *Nature, 390*: (1990).
215. E. Lam, Y.-Kano-Murakami, P. Gilmartin, B. Niner, N.-H. Chua, *Plant Cell, 2*: 857 (1990).

10

Regulation of Plant Gene Expression at the Posttranslational Level: Applications to Genetic Engineering

Allison R. Kermode

Simon Fraser University
Burnaby, British Columbia, Canada

INTRODUCTION

Of the various levels of gene regulation available to plants, e.g., transcriptional, posttranscriptional, translational, and posttranslational, the latter has only just recently become the focus of considerable study. Posttranslational processes include all events subsequent to the synthesis of proteins on cytosolic (or organellar) polyribosomes, to their final localization, and functional configuration within the cell. With the exception of a few proteins of chloroplasts and mitochondria, all proteins are synthesized on free or bound polyribosomes of the cytosol, and so must be transported to, or across, one or more membranes to reach their final subcellular location. Some very fundamental questions concerning these regulatory events are being addressed in plants. For example, what is the nature of the signal and sorting machinery which target proteins to different compartments of the cell? What are the mechanisms of signal "decoding" and of protein insertion into, or translocation across, the target membrane? What are the posttranslational modifications involved in protein maturity and the acquisition of a stable three-dimensional conformation? Also, what determines the half-life of a protein in a particular cell type? The maintenance of correct intracellular targeting of proteins is proposed to be particularly challenging in plants in which there is an additional organelle — the chloroplast — requiring extensive protein traffic from the cytosol and a high degree of cytoarchitecture to carry out its functions [1]. But the importance of the specificity of these processes to the cell cannot be underestimated, since it is the very basis of maintaining structural and functional integrity, enabling the various compartments of the cell to carry out their diverse metabolic roles.

The focus of this chapter will be on the various mechanisms involved in correct protein targeting in plant cells as well as some of the experimental approaches used to elucidate these events. A detailed examination of the components involved in the posttranslational modifications of proteins as well as the structure/architecture of the secretory pathway will not be emphasized. In general, detailed studies on protein transport and targeting in plant cells are at present in their infancy. Thus, where appropriate, reference is made to analogous mechanisms in other eukaryotic organisms, with the awareness that these "heterologous systems" may not fully reflect the processes in plants.

Plant genetic engineers have endeavored to manipulate metabolic pathways to improve plant productivity and to engineer seed crops with greater nutritional value or protein content. With respect to the latter, this has taken the form of attempts to enhance gene transcription and mRNA stability, or translatability. More recently a novel strategy has been toward enhancing protein stability and the "mis-" or "retargeting" of proteins into different subcellular compartments. Some of these studies will also be discussed.

PATHWAYS OF PROTEIN TRANSPORT IN PLANT CELLS

Major targets of protein transport in plant cells include the chloroplast, the mitochondrion, the nucleus and the peroxisome/glyoxysome. Proteins destined for these organelles are translated on "free" polyribosomes of the cytosol; concurrently, or shortly thereafter, they are imported into the appropriate organelle by direct recognition of specific targeting signals. In contrast, transport to the vacuole (or secretion from the cell) occurs along the endomembrane system (the ER, Golgi apparatus and transport vesicles); proteins destined for this "secretory" pathway are synthesized on polyribosomes associated with the ER and enter a common "gateway" by first translocating across the membrane of the endoplasmic reticulum (ER) into the lumen.

GENERAL PRINCIPLES OF PROTEIN TARGETING

Our understanding of the mechanisms of protein targeting is far from complete, particularly within plants, but present knowledge suggests that some general principles apply:

1. Targeting information resides in the protein itself. It may be in the form of a discrete signal (e.g., a specific amino acid sequence), but more commonly is encoded in a specific secondary or tertiary structure.

2. Targeting sequences are probably recognized and "decoded" by specific protein receptors; interactions with lipids of the membrane bilayer may also be important, perhaps playing a facilitory role in the initial binding process to receptors on the target membrane.

3. The types of signals ("topogenic" sequences) involved in protein targeting are limited and can be classified on the basis of the nature of the target membrane and, in some cases, on the mechanism of membrane translocation or integration (e.g., involvement of specific protein effectors). Examples are signal sequences, stop-transfer sequences, sorting sequences, and insertion sequences [2]. Not all membranes are translocation-competent and some translocate only specific proteins [3]. *Signal* sequences initiate translocation of proteins across specific cellular membranes; they interact with protein receptors/translocators, which effect unidirectional translocation. *Stop-transfer* sequences interrupt the translocation process that was previously initiated by a signal peptide and yield integration of proteins into translocation-competent membranes. *Sorting* sequences act as determinants for posttranslocational traffic of subpopulations of proteins, originating in translocation-competent donor membranes (and compartments) and going to translocation-incompetent receiver membranes (compartments). Finally, *insertion* sequences initiate unilateral integration of proteins into the lipid bilayer without the mediation of a distinct protein effector [2].

4. The mechanism of protein translocation across the target membrane may involve interactions of a variety of cytosolic and membrane components; it often requires an unfolded or "loosely folded" conformation of the protein undergoing translocation, the maintenance of which may involve assistance by another protein.

5. In some cases (but by no means all) the targeting signal is transient and cleaved following protein translocation; however, this step is generally not an obligatory component of the targeting/translocation process per se.

6. Proteins fold into their final three-dimensional structure (and in some cases, assemble into oligomers) shortly after translocation is completed; often these processes are assisted by a general group of proteins termed *molecular chaperones.*

7. While important exceptions exist, a striking feature of our understanding of the mechanisms of protein targeting is their universality among animals, plants, and eukaryotic microorganisms and even between prokaryotes and eukaryotes.

GENERAL METHODOLOGY TO STUDY PROTEIN TARGETING

Various methods have evolved over the years to study protein targeting in eukaryotic cells. Current approaches commonly involve the application of recombinant DNA techniques to engineer genes (e.g., to effect specific mutations or deletions in a given gene, or to construct chimeric genes). Transfer and expression of these modified genes in host cells or organisms allows the subsequent analysis of protein targeting in a heterologous environment (e.g., the subcellular localization of the genetically engineered proteins). In addition, where isolation and preservation of intact organelles is feasible (e.g., chloroplasts, mitochondria), protein targeting can be studied via in vitro systems. In animals and yeast, systems for reconstituted in vitro protein transport between two organelles (e.g., the ER and Golgi complex) have also been developed.

Because the targeting signal is part of the protein itself, engineering of the respective gene involves modifications to the coding region only (Figures 1 and 2). Two general approaches have commonly been undertaken:

1. *A "loss of function" approach* (Figure 1). This involves an analysis of the fate of mutants in which gene sequences (presumed to encode targeting signals of the corresponding protein) are deleted. The assumption here is that deletions in a targeting domain will cause the mutant protein to bypass the normal sorting reaction and hence, result in its mislocalization in

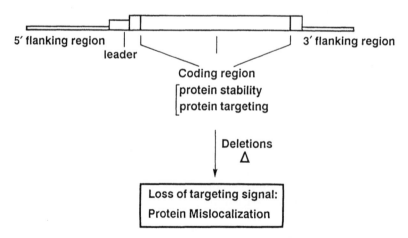

Figure 1   "Loss of function" approach to study protein targeting. Deletions in a targeting domain result in mislocalization of the mutant protein in the heterologous host cells.

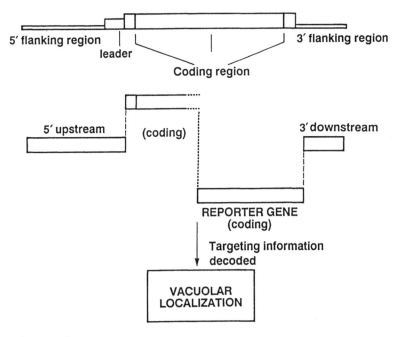

5′ flanking region

leader

Coding region

3′ flanking region

5′ upstream

(coding)

3′ downstream

REPORTER GENE
(coding)

Targeting information
decoded

VACUOLAR
LOCALIZATION

Figure 2 "Gain of function" approach to study protein targeting. The figure shows the approach applied to identify the amino acid sequences (or protein domain) sufficient to direct a reporter protein to the target organelle (e.g., the vacuole). Refer to text for further explanation.

the transgenic host cells (Figure 1). If a targeting domain is implicated by these studies, it may be further delimited or defined by mutational analysis of specific residues. This "loss of function" approach has certain limitations, since abolished targeting may be attributed to the deletion causing a non-specific protein conformational change (affecting transport), rather than to the removal of a sequence involved in targeting per se. Hence, this approach is often combined with another one.

2. *A "gain-of-function" approach* (Figure 2). Here, chimeric (fusion) genes are constructed, consisting of various lengths of coding sequence (from the gene encoding the targeted protein) linked to an intact reporter gene coding sequence. This "gain-of-function" approach allows identification of a protein targeting domain that is sufficient to direct the reporter protein into the organelle under study. For example, Figure 2 shows this approach applied to identifying the sorting signal of a vacuolar protein. One advantage of gene fusions is that the "reporter" or "passenger" protein may be easier to detect and localize in the heterologous host cells than the targeted

protein itself; ideally, the added sequences present in the chimeric protein should not affect (i.e., reduce) reporter protein detection (e.g., enzymatic activity). Cytosolic proteins are often preferred as reporters because they should not contain targeting signals; however, prudent choice of a reporter protein is necessary since they may be unable to cross certain membranes [4].

Specific methods have also been developed to identify the receptors involved in the recognition and decoding of specific sorting signals, as well as other components of the transport and targeting machinery; these are discussed in subsequent sections.

THE BIOSYNTHETIC TRANSPORT PATHWAY

As mentioned previously, some proteins synthesized in the cytosol (e.g., those subsequently routed to the vacuole or destined for secretion from the cell) are transported along the biosynthetic transport pathway [5,6] or "secretory" pathway (reviewed in [7]). The organelles of this pathway include the ER (rough and smooth), Golgi apparatus, *trans*-Golgi network (or partially coated reticulum), endosomes, a variety of transport vesicles, vacuoles, the tonoplast and the plasma membrane. As in other eukaryotic cells, the protein composition of these organelles in plant cells is highly dependent upon the tissue or developmental stage of the cell (i.e., strict temporal and spatial regulation of gene expression is operative); it may also be influenced by external environmental signals. For example, vacuoles are acidic compartments that perform a variety of functions dependent upon the physiological status of the plant cell. Within the reserve tissues of the seed, they perform a dual function; i.e., as temporary storage depots (during seed development, when large amounts of storage proteins and other specialized proteins are synthesized and accumulated), and as sites of macromolecular hydrolysis, during the postgerminative phase of the plant lifecycle, when seedling growth must be supported. It is during this latter phase that the vacuole accumulates several hydrolytic enzymes involved in reserve mobilization, and the plant cell vacuole is thought of as being equivalent to the animal lysosome. Many vegetative reserve tissues (e.g., tubers, roots, bark) accumulate storage proteins and lectins in their vacuoles (often in a seasonal-dependent manner) [8–12]; other vegetative tissues accumulate vacuolar storage proteins only when subjected to various stresses (e.g., nitrogen stress) [13]. Pathogen invasion induces a variety of enzymes (including β-glucanase and chitinase) which are subsequently directed to both vacuolar and extracellular locations [14,15].

Other major plant cell proteins transported along the "secretory" pathway are those destined for the plasma membrane and extracellular matrix (cell wall). Included in this latter category are the principal structural cell wall proteins, the hydroxyproline-rich glycoproteins (e.g., the extensins), as well as some arabinogalactan proteins (found in specialized mucilages and gums) (reviewed in [16,17]). The cell wall also appears to be a highly dynamic compartment; the abundance of several of its composite proteins is sensitive to external cues as well as being subject to the normal spatial/temporal controls [18]. Tissues in the cereal caryopsis (i.e., the aleurone layer and scutellum) have a glandular function (during postgerminative seedling growth) and secrete a large number of hydrolytic enzymes that play an important role in endosperm storage reserve mobilization.

Signal-Dependent and Signal-Independent Steps Along the Secretory Pathway

Evidence For a Default Pathway

Proteins destined for transport along the secretory pathway are synthesized on ribosomes associated with the ER. The first step of entry into the pathway is mediated by a N-terminal signal peptide on the nascent protein which directs a transmembrane translocation from the cytosol to the lumen of the ER [19], a step generally accompanied by signal peptide cleavage (see later discussion). The ER contains a mixture of proteins with multiple destinations. Some proteins will become permanent residents of the ER; others are exported to the Golgi apparatus for retention there, or for subsequent distribution to the cell surface or lysosome/vacuole. How this complex traffic pattern is organized, and which steps along the pathway occur by default (i.e., are signal-independent), or alternatively, require specific targeting signals, have been subjects for both speculation and debate (reviewed in [5, 20–26]). For example, there have been arguments for proteins requiring specific signals for export out of the ER; others have argued for nonselective export out of the ER, with positive signals being required for ER retention. The available evidence from eukaryotic systems indicates that each of the steps along the intracellular route leading to constitutive secretion from the cell (i.e., ER to Golgi complex, movement through the Golgi stacks, and movement from the Golgi complex to the cell surface) occurs by default: secretory and plasma membrane proteins are carried along by a nonselective "bulk-flow" process [5,7,23,25] (Figure 3). Proteins destined for targets other than the cell surface must contain additional positive topogenic information. Thus, more information is required for selective retention in the ER (or the Golgi complex) or for diversion from the bulk-flow pathway, by sorting along one of the branch pathways within the *trans*-Golgi network (e.g., to the animal lysosome or plant/yeast vacuole) (Figure 3).

There are several arguments in favor of a bulk-flow model in which (constitutive) secretion is the default fate for a protein containing no specific targeting information (in addition to the signal peptide). Although most

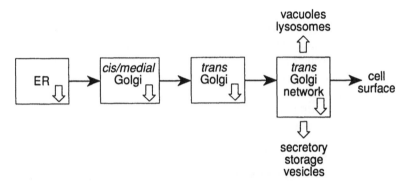

Figure 3 Current model of protein transport through the biosynthetic ("secretory") pathway. Solid arrows depict signal-independent transport and represent bulk-flow (the "default" pathway). Open arrows denote transport steps which are signal-mediated, including retention of resident proteins in the ER and Golgi complex, as well as transport to lysosomes/vacuoles, and sequestration of proteins into secretory storage vesicles. (Reproduced in modified from from Ref. 5. With permission.)

of the evidence comes from studies of animals and yeast, some recent studies in plants are also supportive of this view ([27–30], reviewed in [7,16]). Perhaps the most striking evidence for the existence of a default pathway in animal cells comes from the work of Wieland et al. [31] who devised a bulk phase marker of the secretory pathway. A synthetic tripeptide consisting of the glycosylation sequence Asn-Tyr-Thr, esterified to make it membrane permeable, can enter cells, reach the ER lumen where it becomes glycosylated (thus rendering it membrane-impermeant), and then be rapidly secreted, predominantly along the normal pathway (i.e., via the Golgi complex). It is assumed that the tripeptide contains no transport or retention signals and defines the bulk-flow rate of ER to cell surface movement and thereby the default pathway [24,25]. Moreover, the bulk flow rate defined by this marker is sufficiently rapid to account for the export of most secretory or membrane proteins, being close to the rate of transport of the most rapidly secreted proteins [31]. Although there are some valid criticisms regarding the appropriateness of the glycosylated tripeptide as a bulk phase marker (reviewed in [20,22,23]) there is currently no compelling reason to postulate the existence of a positive ER export signal for any protein [23].

The loss, or disruption of, a defined positive sorting signal on a protein (e.g., by mutation or deletion) generally leads to its transport along the default pathway (i.e., missorting and secretion). Similarly, when signal-dependent sorting to the animal lysosome or to specialized secretory vesicles (involved in regulated secretion) is saturated or incapacitated, the bulk-flow pathway to the cell surface seems to be the route taken by default. In I-cells that are unable to tag lysosomal hydrolases with the required sorting signals (i.e., mannose-6-phosphate residues) and in cells deficient in one of the mannose-6-phosphate receptors (see later discussion), a large fraction of lysosomal hydrolases are constitutively secreted [32,33].

Significant mistargeting and secretion also occurs in yeast cells when two proteins (normally targeted to the vacuole) are overexpressed, presumably because the capacity of vacuolar sorting machinery is surpassed [34, 35]. Removal of putative ER-retention signals (see later discussion) from ER-resident proteins causes them to be transported along the secretory pathway, even though they do not normally do so, and hence are not expected to have an export/transport signal [36–38]. Conversely, mutations in secretory or plasma membrane proteins which abolish transport and lead to their retention have subsequently been shown to affect transport out of the ER by indirect means (reviewed in [21,39,40]). Deletion or disruption of a specific export signal is not operative here; rather, these proteins are unable to undergo the normal folding/assembly processes required for subsequent transport out of the ER (see later discussion).

Disruptions of signal-dependent sorting in the *trans*-Golgi network by chemical treatments that affect the pH of the transport pathway (and presumably receptor-ligand interactions, e.g., chloroquine and monensin) also cause mistargeting and secretion [32,41–43]. Mistargeting and transport of proteins along the bulk-flow pathway occurs when animal cells are treated with the sodium ionophore monensin, which raises the pH of the *trans*-Golgi network [43]. A similar phenomenon occurs in plant cells; treatment of pea and jackbean cotyledons results in the transport of storage proteins and lectins (normally accumulated in the storage vacuole/protein body) to the plasma membrane and cell wall [44,45]. Not all storage proteins behave in this manner; pro-barley lectin and bean phytohemagglutinin are retained intracellularly following treatment with monensin [46,47]. The varying effects of this ionophore on the trafficking of secretory products, however, may reflect structural and functional cell type-specific differences in the organization of plant Golgi stacks [48].

Secretion is the outcome in plant cells when a protein contains no additional targeting information over that of the signal peptide [27–30]. In studies of chimeric proteins expressed in transgenic tobacco, a signal peptide (e.g., from the vacuolar proteins, phytohemagglutinin and patatin) is sufficient to direct a cytosolic "receptor" protein (e.g., pea albumin 2 and β-glucuronidase, or GUS) into the ER and leads to transport along the secretory pathway via a default/bulk-flow mechanism [27–29]. Likewise, in tobacco cells transformed by electroporation, three bacterial cytosolic enzymes, utilized as "reporters," enter the secretory pathway (when linked to the signal peptide of an extracellular pathogenesis-related protein, PR1) and are secreted [30]. Secretion in these transformed tobacco cells is relatively inefficient; however, the results are indicative that secretion can occur independently of active sorting, by non-specific migration through the secretory pathway.

Signal-Dependent Steps Along the Secretory Pathway

In contrast to the signal-independent export of proteins out of the ER (and intra-Golgi transport by bulk-flow), specific signals for retention have been described for some ER and Golgi proteins (Figure 3). Perhaps the best defined retention signal is the tetrapeptide His/Lys-Asp-Glu-Leu (HDEL or KDEL), which is present at the C-terminus of several soluble resident ER proteins, in animals, yeast and plants (see later discussion).

Sorting information is required to divert proteins from bulk-flow transport to the yeast cell vacuole (via sorting in the *trans*-Golgi network); recent evidence (reviewed in 7) suggests a similar requirement for transport of proteins to the plant cell vacuole. As mentioned earlier, a chimeric gene consisting of the signal sequence of the vacuolar seed protein, phytohemagglutinin (PHA) and the coding sequence of a cytosolic seed protein (pea albumin 2) has been expressed in seeds and suspension cultured cells of transgenic tobacco [7,27,28]. The signal peptide is necessary and sufficient to direct entry of the chimeric protein into the secretory pathway; however, it does not accumulate in the vacuole (e.g., in seeds [27]) and is efficiently secreted from suspension cultured cells [28].

In animal cells, proteins en route to the lysosome share a common pathway of intracellular transport with plasma membrane — and constitutively secreted — proteins which extends as far as the *trans*-Golgi network (reviewed in [49]). However, the fate of lysosomal proteins, which are all glycosylated in the ER, is determined earlier in the Golgi stack. Upon arrival in the *cis*-Golgi, a series of modifications by specific Golgi enzymes, ultimately result in a mannose-6-phosphate tag on lysosomal proteins which allows them to bind to mannose-6-phosphate receptors and hence be specifically sorted for delivery to the lysosome in the *trans*-Golgi network. Release of lysosomal proteins from their receptors occurs in an acidified (prelysosomal, endosomal) compartment; subsequently, the receptors recycle back to the Golgi (or to the cell surface, since these receptors are also involved in endocytosis), while the proteins are packaged into lysosomes.

In animal cells, there are two distinct pathways of secretion, i.e., constitutive and regulated [5,26,50,51]. In specialized cells involved in regulated secretion (which are also capable of secreting proteins constitutively), the regulated secretory products are concentrated and stored in dense vesicles (granules) in the cytosol. Fusion of these vesicles with the cell surface only occurs upon receipt of a specific extracellular signal (e.g., a hormone or neurotransmitter, depending on the cell type). This process, termed *exocytosis* [52], is thus regulated, and gives rise to the controlled secretion of the selected contents of the secretory storage vesicle (reviewed in [25,26]). By contrast, secretion via bulk-flow to the cell surface is constitutive and continuous. The signals which permit membership or entry of specific proteins into storage vesicles are not yet known; however, it is clear that their content is highly selected. A form of selective precipitation in the *trans*-Golgi network may be involved; "carrier-type" proteins have also been implicated in the mechanism [53].

In plants, the constitutive pathway of protein secretion is the principal pathway along which proteins are secreted to the cell exterior [7,16]; in general, there is a close temporal correlation between the synthesis of proteins and their secretion from the cell (e.g., cereal α-amylase) [54,55]. Regulated secretion occurs in the green alga, *Clamydomonas reinhardii*. Here the release of stored lysin (a cell wall degrading enzyme) is induced by gamete sexual signaling [56].

Signal-Dependent Translocation Across the ER Membrane   Translocation across the membrane of the ER represents the first signal-dependent step along the secretory pathway and is mediated by a N-terminal signal peptide on the nascent protein. Signal peptides generally function with a high degree of autonomy, being both necessary and sufficient to direct passenger proteins into the lumen of the ER, as shown by recent studies in transgenic tobacco (see earlier discussion). An examination of the derived amino acid sequences of plant secretory and vacuolar proteins shows that all of them have a N-terminal domain with the properties of other eukaryotic signal sequences [7,57,58]. Despite great sequence diversity between eukaryotic signal peptides, most have similar overall features i.e., a charged N-terminal region, a core of 10 to 15 hydrophobic amino acids and a more polar C-terminal sequence [57]. Translocation is mediated by receptorlike systems [19, 59,60] that involve proteins in the cytosol and on the target membrane; this complex cellular machinery is able to recognize signal peptides, despite their variation in amino acid sequence. Studies on the well-characterized mammalian ER have identified four distinct components of such a receptorlike system; these act sequentially and hence, greatly enhance the specificity of the system [reviewed in 3, 6, 61]. One component is the signal recognition particle (SRP), a ribonucleoprotein complex, consisting of a molecule of RNA and six polypeptide chains including a 54-kD protein with a GTP-binding domain. The 54-kD polypeptide subunit binds to the precursor's signal peptide sequence as it emerges from the ribosome, temporarily halting translation. Following movement of the ribosome to the cytoplasmic face of the ER, the ternary complex (consisting of the ribosome — nascent precursor — SRP) binds to a 72-kD integral ER protein (part of the SRP-receptor or docking protein). The ribosome binds to the ER membrane, the SRP is released, and the signal sequence interacts with a 35-kD integral ER glycoprotein, the signal sequence receptor (also equipped with a GTP-binding domain). These latter interactions allow translation to continue and somehow initiate translocation. At least two steps of the translocation event appear to be controlled by the hydrolysis of GTP, again contributing to the specificity of the system [61].

The ER-translocation process in plants is similar to that in animals and fungi. The existence of the SRP and SRP receptor in plants has now been documented. Moreover, efficient synthesis, translocation and processing of plant secretory proteins occurs in animal and fungal heterologous systems; thus, strong evidence exists for a common mechanism of protein translocation in eukaryotes [55,62,63].

Solutions to the enigma of how diverse signal peptide sequences/shapes can be recognized by, and accommodated within, the 54-kD polypeptide component of the SRP have been speculated upon (reviewed in [61]). Sequestration of signal peptide sequences may occur in a hydrophobic methionine-rich pocket formed by secondary structural domains (amphipathic α-helices) of the 54-kD SRP protein. More specifically, the unbranched methionine side chains lining this pocket provide projecting flexible "bristles" which recognize and accommodate only the common hydrophobic core of signal peptides, albeit diverse in primary sequence [61].

Protein synthesis (i.e., translation) and translocation of the protein generally occur simultaneously. Strong evidence argues against a *mechanistic* coupling between the two events; however, their temporal association may nonetheless play an important role, that of ensuring that nascent polypeptides maintain (or acquire) a translocation-competent conformation [3]. Binding of the SRP to the signal peptide and the arrest of translation until functional insertion into the ER occurs, would perform this important function [6]. A SRP-like component has not been detected in yeast [3]. Instead 70-kD heat-shock proteins (and probably additional proteins) are components of the ER-translocation machinery; in part, functioning to assist proteins to maintain a translocation-competent state [64–66].

Little is known about the mechanics of protein translocation; transport through a proteinaceous pore (rather than movement directly through the lipid bilayer) is suggested by studies in yeast [67].

In most cases, the signal peptide is removed (in a cotranslational manner) from the nascent polypeptide chain as it emerges into the ER lumen by a signal peptidase located on the inner face of the membrane. Cleavage generally occurs at the more polar C-terminal region of the signal sequence;

however, in cases where the signal peptide is located internally, or its hydrophobic core is integrated within a longer membrane anchor domain, cleavage does not occur [6]. In this latter case, the "stop-transfer" sequence halts translocation and remains embedded in the membrane functioning as the anchor for integral membrane proteins. Therefore, the presence of such a sequence determines whether a polypeptide is transported as a soluble or a membrane-bound protein [6]. There is not much information on the topogenic sequences of integral membrane proteins that are incorporated into the membranes (ER, Golgi, tonoplast, plasma membrane) of the secretory system of plant cells [7]. Analysis of the amino acid sequences of two secretory pathway proteins (a plasma membrane, proton-translocating ATPase, a vacuolar H^+-pyrophosphatase, and a seed-specific tonoplast protein) [68–71] reveals several membrane-spanning domains. The tonoplast intrinsic protein of bean seeds (α-TIP) is synthesized on the rough ER and its transport to the tonoplast is mediated by the secretory system; its C-terminal transmembrane domain seems to contain sufficient information for transport to the tonoplast [72] (see later discussion).

Interestingly, synthesis of TIP in seeds does not appear to be related (in a quantitative manner) to storage protein deposition and TIP is not found in leaves that accumulate vegetative storage protein. A role as a solute transporter in seeds has been suggested for this protein [69]; alternatively, its developmental regulation is indicative of a physiological function during late seed maturation, perhaps in the acquisition of desiccation-tolerance [73]. The vacuolar H^+-ATPases are thought to be an integral part of the organelles of the endomembrane system in plants (reviewed in 74–76); acidification of these compartments by the enzyme may be essential for protein sorting in plants, similar to its role in animal and yeast cells (see later discussion).

Signal-Dependent Retention of Proteins in the ER and Golgi Complex As mentioned previously, soluble resident proteins of the ER are distinguished by a carboxy-terminal tetrapeptide sequence. In animals, the preferred retention signal is KDEL; in yeast (e.g., *Saccharomyces cerevisiae*) it is HDEL [36,77]. Some of the more abundant soluble resident ER proteins of eukaryotes identified with this carboxy-terminal sequence include Binding protein (BiP), protein disulphide isomerase (PDI), glycosylation site binding protein (GSBP) and others [78–80]. In plants it has been identified at the carboxy-termini of several ER proteins (BiP, in tomato, maize and tobacco, an auxin-binding protein, in maize, and protein disulfide isomerase in alfalfa) [7,23, 81–87]. The carboxy-terminal tetrapeptide sequence is necessary for retention of proteins in the ER [36,88,89]. When the sequence is deleted from an ER-resident protein (e.g., BiP), secretion is the outcome [90]; conversely, addition of the tetrapeptide onto the carboxy-terminus of lysozyme (a protein that is normally secreted), or cathepsin D (a lysosomal enzyme) causes the proteins to accumulate in the ER [36,88]. Although some changes to individual amino acids can be tolerated, extending the sequence with random amino acids at the carboxy-terminus (or changing KDEL to KDAS), results in secretion [36,91]. Thus, retention is probably mediated by a specific receptor that recognizes (H/K)DEL only when it is present at the extreme carboxy-terminus of a protein. Interestingly, KDEL itself is not an efficient retention signal in yeast; conversely, HDEL functions poorly in animals [77]. Presumably both KDEL and HDEL are recognized in plant cells (tomato BiP has a carboxy-terminal HDEL; maize auxin-binding protein has KDEL).

Addition of KDEL or HDEL onto the carboxyterminus of some proteins retards transport, but is not sufficient for their absolute retention in the ER [92,93]. Retardation of export rather than total ER retention occurs with several mammalian secretory proteins that are modified at the exact carboxy-terminus to encode the sequence SEKDEL [92]. When the carboxy-terminal sequence of a plant vacuolar protein (phytohemagglutinin) is changed from LNQIL to LNKDEL and the derivative protein expressed in transgenic tobacco, a large proportion is localized in the ER and the nuclear envelope (which is continuous with the ER). However, a considerable amount of the PHA-KDEL is transported through the Golgi complex and reaches the protein storage vacuole/protein body [93]. Partial retention of these proteins in the ER may be due to a less than optimal display of the carboxy-terminal KDEL sequence, thus diminishing its recognition by the appropriate receptor [94] and allowing its escape from the normal retrieval mechanism (see subsequent discussion). Other structural features of the protein may be required for ER retention, perhaps to ensure optimal positioning of the tetrapeptide for receptor-mediated recognition [7,94]. Changing the carboxy-terminus of the vacuolar pea storage protein, vicilin to include KDEL (SEKDEL) results in a dramatic increase in the accumulation of vicilin in the leaves of transgenic tobacco and alfalfa [95,96]. The retention signal kept the protein from advancing to a compartment where it could be degraded (presumably the vacuole) and resulted in the formation of protein bodylike inclusions. These ER-derived, membrane-bound, electron dense structures (0.5–1.0 μm in diameter) contain vicilin and resemble the ER-derived protein bodies found in the endosperm cells of certain cereals such as maize and sorghum (see later discussion).

In addition to KDEL, the carboxy-terminal sequences HDEL and RDEL are sufficient for retention in the ER and nuclear envelope of the reporter enzyme phosphinothricin acetyl transferase (PAT) in stably and transiently transformed plant cells from a variety of species [97,98], indicating that the plant ER protein retention mechanism can recognize several sequences. However, changing certain sequences within KDEL (such as KEEL, SDEL, KDDL, KDEI, and KDEV), leads to partial or complete loss of ER retention.

Given that most proteins are transported by bulk-flow while certain proteins are selectively retained, it is pertinent to examine the mechanism of specific retention in the ER. The lumenal proteins are not associated with a membrane-bound receptor [23,90] nor is retention guaranteed by membrane anchoring; many membrane-bound viral proteins are still transported out of the ER [21]. It appears that for soluble proteins at least, ER retention mechanisms come into play at a later point in the secretory pathway; i.e., after the proteins have been transported out of the ER [23,88]. More specifically, luminal ER proteins that escape are continuously and specifically retrieved from a post-ER ("salvage") compartment (that is pre-Golgi or early Golgi) by a receptor; the receptor-ligand complexes are cycled back to the ER (by specialized vesicular transport; see subsequent discussion), where the KDEL-bearing proteins are released [23]. In yeast (*Saccharomyces cerevisiae*), the retrieval mechanism appears to be in the Golgi complex [77,99] and some very elegant studies in this organism [100–102] have identified a putative receptor involved in HDEL recognition. A genetic approach was utilized to identify the essential components of the "recycling" pathway; to this end, *Saccharomyces cerevisiae* mutants specifically defective in the HDEL-mediated ER retention system were isolated, allowing subsequent identification of two genes required in the process, i.e., *erd*1 and *erd*2 (for "ER retention defective"). Deletion of the *erd*1 gene causes a pleiotropic defect in part of the Golgi apparatus, which evidently results in inefficient retrieval from this organelle [101]. The *erd*2 gene encodes a 26-kD integral membrane protein that is required both for retention of ER proteins and perhaps indirectly, for normal traffic of proteins through the Golgi [100]. Strikingly, the abundance of this protein determines both the efficiency and capacity of the retention system: reduced expression leads to secretion of HDEL-tagged proteins; conversely, its overexpression improves their retention (both in wild-type cells and in other mutants). Moreover, *erd*2 determines the signal specificity of the retention system [102]; exchange of the *erd*2 gene from *Saccharomyces cerevisiae* (which only recognizes HDEL) for the corresponding gene from another yeast, *Kluyveromyces lactis* (which recognizes either HDEL or DDEL), allows equal recognition of DDEL and HDEL in *Saccharomyces cerevisiae*. These observations as well as the protein's subcellular localization (i.e., in a post-ER, Golgi-like compartment), make it an excellent candidate for the receptor that sorts lumenal ER proteins in yeast [100,102]. Recently, an antiidiotypic antibody approach has identified a 72-kD glycoprotein that is a good candidate for the KDEL receptor in animal cells [94]. An equivalent receptor in plants remains to be discovered.

Another pertinent question that has recently been addressed is how the ER membrane is maintained in the face of a heavy flow of vesicular traffic toward the Golgi stack [103,104]. The ER is a dynamic network of tubular membranes which extend along microtubules, continuously breaking, fusing, and rearranging [105]. Recent studies utilizing the antifungal antibiotic, brefeldin A suggest that there is an intermediate compartment to the ER transport pathway involving microtubules which potentially could restore membrane components and escaped proteins to the ER ([103]; reviewed in [104]). In the presence of the drug the Golgi complex breaks down and appears to merge with the ER; consequently, both components now interact with microtubules in the same way. Secretory and (H/K)DEL-tagged proteins may normally be sorted in the intermediate (salvage) compartment, such that the latter are returned via a retrograde pathway involving microtubules. Membrane vesicles destined to return to the ER may carry or acquire proteins which enable them to interact with microtubules; these proteins may normally be excluded from vesicles involved in nonselective (bulk-flow) transport during sorting in the salvage compartment [6,103,104]. Because the retrograde pathway into the ER depends upon microtubules, in contrast to the anterograde pathway out of the ER, this provides a potential means by which cells can regulate membrane transport along these two pathways [103]. The ERD2 protein in yeast not only functions as a receptor, but is also required for normal vesicular traffic through the Golgi complex; these two functions may be related [100]. For example, it may be required for the formation of the specialized vesicles involved in retrograde transport; whether it interacts directly with microtubules, remains to be determined. In mammalian cells, proteins localized in a pre-Golgi intermediate compartment may play a similar role (e.g., microtubule attachment proteins) [103, 104]. It is not clear whether any ER membrane proteins normally escape to the Golgi complex and require retrieval [100], but recycling of the SEC 12 protein (see later discussion), and perhaps other membrane proteins involved in vesicular traffic and membrane flow has been suggested [106]. *Erd2* and equivalent mammalian proteins might also be necessary to maintain the supply of such components in the compartments from which vesicles bud. A failure to retrieve these membrane proteins could be sufficient to reduce the flow of the secretory pathway to a trickle [100]. Whether a similar retrograde pathway and retrieval mechanism exist in plant cells remains to be determined.

It is noteworthy that other retention signals have been identified, primarily of viral proteins in infected animal cells. For the rotovirus VP7 protein (a 37-kD glycoprotein), the signal peptide, together with the first 60 amino acid residues of the mature protein forms the retention signal [107], although the mechanism of retention is not yet understood. The E19 protein of adenovirus is a membrane-anchored resident ER polypeptide which has a short tail protruding into the cytoplasm. A small stretch of amino acids on this exposed tail (i.e., the last six residues) is responsible for its retention in the ER [38]. Retention of the corona infectious bronchitis virus E1 glycoprotein in the Golgi stack is signal-dependent and an amino-terminal hydrophobic region has been implicated [108]. However, the E1 protein of another corona virus appears to require nearly all of the coding sequence for correct Golgi localization ([109]; reviewed in [110]).

Co- and Posttranslational Modifications Along the Secretory Pathway

Many proteins which are transported along the secretory pathway in plant cells undergo extensive modifications including glycosylation, folding, oligomer assembly, and proteolytic processing. Some of these processes have been studied in relation to their role in transport efficiency, protein stability, and protein targeting per se.

Glycosylation

Glycosylation is a modification which many vacuolar and extracellular proteins commonly undergo along the secretory pathway en route to their target oranelles. In the mature state, these vacuolar or extracellular glycoproteins often have both high mannose and complex asparagine (Asn)-linked glycans (reviewed in [7,16]; [111–113]). Glycans of the high mannose type (e.g., $Glc_3 Man_9 GlcNAc_2$) are assembled on dolichol lipid carriers in the ER lumen. These are transferred (in a cotranslational manner) to specific Asn residues (of the acceptor sequence Asn-X-Ser/Thr) on nascent polypeptides by oligosaccharyl transferase ([114]; reviewed in [115]). Two ER-resident enzymes (Glucosidase I and II) subsequently remove the terminal glucose residues from the high mannose glycan on the protein [116] (a notable exception is α-mannosidase of jackbean cotyledons which escapes this modification) [117]. The conversion of a high mannose glycan to a complex glycan takes place later in the secretory pathway (in the Golgi complex), and is the result of the sequential action of several glycosidases and glucosyl-transferases [111,112]. The complex glycans of plant glycoproteins differ substantially from those of mammalian glycoproteins, but may share some similar structural features with those present in insects and molluscs [118]. In plants, complex glycans are characterized by fucose and xylose sugar residues attached to the proximal GlcNAc and core mannose residues, respectively. This modification renders the glycoprotein resistant to the *Streptomyces* enzyme endoglycosidase H; hence this characteristic has been used as evidence for passage through the Golgi complex [7]. Apparently, glycans must be readily accessible to the appropriate enzymes in the Golgi apparatus; when inaccessible, they remain in a high mannose form as the protein moves through the Golgi complex [119].

The role of N-linked glycans in the transport and targeting of proteins to the plant cell vacuole has been examined (reviewed in [7]). An overwhelming amount of evidence indicates that glycans do not contain vacuolar targeting information. Not all vacuolar proteins are glycoproteins (e.g., 11S globulins, legumin, glycinin and others); moreover, some glycosylated precursors lose their glycans during posttranslational processing en route to the vacuole, e.g., concanavalin A, wheat germ agglutinin, and β-glucanase [120–122]. Tunicamycin inhibits N-linked glycosylation in the ER by interfering with the elaboration of glycans (GlcNAc-1-P) onto the dolichol lipid intermediate [123]. However, this antibiotic generally does not cause missorting of (normally glycosylated) vacuolar proteins, including pea vicilin and bean phytohemagglutinin [124,125]. Elimination of glycosylation sites by site-directed mutagenesis has been used as a more direct approach to study the role of glycosylation in protein transport. Elimination of the glycosylation sites of three plant vacuolar proteins by this means (e.g., phytohemagglutinin, patatin, and barley lectin) does not lead to mistargeting of the unglycosylated proteins in a heterologous host plant system [47,126,127]. Thus, in contrast to proteins of the animal lysosome, in which glycans play a pivitol role in their targeting (see earlier discussion), the glycans of plant vacuolar proteins do not contain targeting information; such is also the case for proteins of the yeast cell vacuole [128].

While N-linked glycans do not play a role in targeting per se, they may play a fundamental role in promoting correct protein folding and, as a consequence, enhance protein stability (e.g., via protection against proteolysis). Treatment of suspension-cultured plant cells with tunicamycin leads to a cessation of protein glycosylation and a decline in the accumulation of newly synthesized proteins in the culture medium and cell wall [129–131]; this is due primarily to an effect upon protein stability and not on protein synthesis [132]. Elimination of the glycosylation sites of phaseolin (a vacuolar storage protein in bean seeds) results in an increased susceptibility to proteolytic cleavage and a decreased stability of the protein in transgenic tobacco seed [133]. A general effect of glycosylation upon protein stability is indicated by similar studies of other seed vacuolar proteins (e.g., phytohemagglutinin and barley lectin) [47,127]. Transport efficiency may also be affected by glycosylation. In animal cells, transport of the vesicular stomatitis virus G protein to the cell surface is completely abolished when the protein's glycosylation sites are eliminated by site-directed mutagenesis. Conversely, the creation of additional glycosylation sites in the protein (at certain positions) promotes transport, as a consequence of having a positive effect upon protein folding [21].

Folding and Oligomer Assembly

Among the numerous posttranslational modifications that proteins of the secretory pathway undergo, one of the most important is the acquisition of a functional, three-dimensional structure and the assembly of polypeptide subunits into oligomers. With few exceptions, these processes take place in the lumen of the ER and in large part will determine a protein's stability and subsequent efficient transport (reviewed in [7,21,23,39]). Many plant vacuolar proteins are oligomers (commonly dimers, trimers, tetramers, or hexamers); characteristically, they undergo complete oligomerization in the ER [134, 135]. Transport of proteins out of the ER is subject to various rules of ER quality control; hence, this is often the rate-limiting step in transport along the secretory pathway [5]. Correct folding and assembly of proteins is necessary for their efficient exit from the ER and subsequent transport [23]. Thus, "sorting" of proteins at this step appears to be conformation-specific and failure to acquire the correct conformation can result in ER retention and/or degradation (see later discussion). Folding of proteins in vitro is often a slow and inefficient process; however, in vivo (i.e., within the lumen of the ER) most proteins achieve their final conformation within a very short time (i.e., within minutes). This efficiency in vivo is due, in part, to the presence of several resident proteins of the ER which play a role in assisting in both the folding and oligomer assembly of nascent proteins. In fact, it appears that a nascent polypeptide chain is accosted on the lumenal side of the ER membrane by a "welcoming" committee of enzymes and factors that facilitate and accelerate folding in several ways [21,23,39,64]. Perhaps the best characterized of these ER-"helper" proteins (termed *molecular chaperones*) is BiP (Binding protein, in mammalian cells also known as the glucose-regulated protein, GRP78) [80]. BiP is related to the cytosolic heat-shock proteins, but is a soluble protein which is targeted to the ER by a cleavable signal peptide; a carboxyterminal (H/K) DEL sequence mediates its retention [36,80,84–86]. Although the main function of BiP is probably to promote protein assembly, it may also serve to prevent export of misfolded proteins from the ER [23]. Increased synthesis of BiP is induced when abnormal proteins accumulate in the ER [136,137] and it preferentially associates with such proteins until they either fold correctly or are degraded. Thus by specifically retaining abnormally folded and incompletely assembled proteins, BiP contributes to the selectivity of transport [23]. Abnormal products retained in the ER by BiP include mutant and misfolded forms of proteins (e.g., influenza hemagglutinin) [138], chimeric proteins [139,140], and some unglycosylated forms of proteins (e.g., invertase) [141]. In an in vitro translation/translocation system, it binds to unoxidized (but not to mature, disulfide-bonded) prolactin [141]. A role in promoting protein folding per se is also suggested, perhaps by helping to prevent the aggregation of folding intermediates. More specifically, proteins such as BiP may act as reversible detergents, binding to hydrophobic surfaces, but intermittently utilizing ATP to change their conformation to a nonbinding state [142]. In this manner, unfolded proteins would be maintained in solution without sequestering them permanently in a nonfunctional complex; the proteins could thus avoid aggregation or precipitation but still be able to achieve their final tertiary and quaternary structures [23].

As intimated, secretory proteins may interact with BiP before they achieve their mature, correctly assembled, state [141]. In mammalian cells, BiP binds transiently in the ER to a portion of IgG heavy chains until assembly with the appropriate light chain occurs [143]. If no light chains are available, then the heavy chains remain associated with BiP until they are degraded. In a temperature-sensitive BiP mutant of yeast, BiP is essential for viability; import of proteins into the ER ceases within minutes of warming to the permissive temperature [144]. Thus, BiP interacts with a variety of nascent proteins and (in yeast cells, at least) may be required for some of them to complete their translocation into the ER [23], a process aided by additional proteins (e.g., SEC 61) which appear to be components of a delicate structure closely apposed to polypeptides that are moving through the membrane ([145,146]; reviewed in [147]).

The existence of BiP in plant cells has recently been demonstrated. It is present in tomato, tobacco, bean (*Phaseolus vulgaris*), and in the endosperm of maize and aleurone layer of barley [7,23,84–86,148,149]; a carboxy-terminal HDEL sequence typical of other soluble ER resident proteins is present. The ability of the tobacco BiP homologue to correct a temperature-sensitive BiP (*kar*2) defect in *Saccharomyces cerevisiae* [86] provides strong evidence that the functional role of plant BiP in the ER is similar to that of the yeast and mammalian BiPs. The level of tobacco BiP is elevated in tissues of germinating seedlings and in organs containing specialized secretory tissues, such as the anthers and stamens, in which the constituent cells likely have a large flux of protein through the secretory pathway [86]. ER isolated from aleurone layer cells of the barley seed contains a BiP cognate whose synthesis is up-regulated by GA_3 and down-regulated by ABA [149]. In maize endosperm, the abundance of BiP mRNA increases dramatically following tunicamycin treatment (as occurs in mammalian cells). Presumably, the antibiotic increases the amount of misfolded protein in the ER as a consequence of its inhibitory effect on protein glycosylation; thus, the requirement for BiP is increased. Maize BiP is also overproduced during endosperm development in plants carrying *fluory*-2, *defective endosperm-B30*, or *mucronate* mutations which are associated with a reduction in zein accumulation and altered protein body morphology [85,150].

Misfolded and misassembled proteins are produced continuously by the cell; the reactions involved in folding, covalent modification, and oligomeric assembly have a finite margin of error. Several studies (primarily in animals and yeast) suggest that the ER may be a major site of degradation of some of the proteins which fail to assume their final functional conformation (whether or not they are associated with BiP or a BiP-like protein) (reviewed in [39,151]). In animal cells, there are two major sites/pathways of degradation of misfolded/unassembled proteins, vis., the ER and the lysosome [152–154].

The restrictions in the transport of proteins out of the ER can be viewed as a late stage in a series of mechanisms utilized by the cell to ensure faithful expression of genetic information at the level of functional and correctly localized proteins [39]. The importance of oligomerization for protein transport and stability has recently been demonstrated for a plant storage protein; trimerization of phaseolin subunits is required for transport of phaseolin out of the ER in *Xenopus* oocytes [155]. This posttranslational control over gene expression is not exclusive to the ER; misfolded and unassembled proteins are specifically and rapidly degraded in the cytosol, mitochondria, chloroplasts, and the nucleus.

Many plant proteins transported along the secretory pathway (e.g., seed storage proteins) have disulphide bonds. The formation of these is not random, but is catalyzed rapidly in nascent chains by protein disulphide isomerase (PDI) (reviewed in [156]). This enzyme interacts with unfolded proteins in the ER and catalyzes thiol oxidation and disulphide exchange reactions [78]. In mammalian cells, PDI has several distinct roles. Besides existing as a free monomer, it is also an essential subunit of prolyl-4-hydroxylase (an enzyme that catalyzes the modification of prolyl residues); it also binds to the Asn-X-Ser/Thr acceptor sequence for N-linked glycosylation and is an important component of oligosaccharide transferase, essential for glycan transfer (i.e., as the glycosylation-site binding protein) [79]. Mammalian PDI can also be crosslinked to nascent immunoglobulin chains in vivo [157], which may be indicative of its role in correct protein folding, via direct interactions with newly synthesized secretory proteins. Whether the enzyme plays similar multiple roles in plant cells remains to be determined, although its importance for correct folding and disulphide bond formation has been established. All forms of PDI are soluble, or only loosely associated with the ER membrane, and are released by incubation at high pH. PDI-depleted microsomes can still import γ-gliadin (a nonglycosylated wheat storage protein), but the protein cannot achieve its correct disulphide-bonded state:

reconstitution of microsomes in the presence of purified PDI restores this function [158]. Alfalfa PDI has a C-terminal ER-retention signal, KDEL; synthesis of its corresponding mRNA is induced by tunicamycin, similar to BiP [87].

Plant secretory proteins (e.g., structural proteins of the cell wall such as extensins and other hydroxyproline-rich glycoproteins, or HRGPs) [17] may also undergo modification to specific amino acid residues en route to the cell surface. For example, the biosynthesis of these proteins involves the hydroxylation of prolyl residues, followed by the glycosylation of many of the hydroxyprolyl residues [159]. These reactions are carried out by 4-prolyl hydroxylase (an ER-associated enzyme) and a glycosyl transferase, associated with the Golgi complex [160,161]. It is not yet known whether the interesting protein homologies observed between 4-prolyl hydroxylase, PDI and glycosylation site binding protein of mammalian cells [79] are also present in plant proteins [7].

Proteolytic Processing

Seed storage proteins often undergo post-translational proteolytic processing; this may occur during their transit to the vacuole (i.e., along the secretory pathway), but more commonly occurs upon reaching their final destination (i.e., within the vacuole) [162]. In some cases, the polypeptide is simply cleaved into two or three smaller polypeptides, all of which remain together in the oligomer; in others, a domain is lost (e.g., from the amino-terminus, carboxy-terminus or middle of the polypeptide). The degree of processing is also variable; it may be complete or only partial, with some polypeptides remaining unprocessed [162]. In the vacuole/protein body of the developing soybean seed, glycinin (a legumin-type storage protein) occurs as a 12S hexamer made up of subunits that consist of an acidic and a basic polypeptide joined by a single disulphide bond [163]. Proglycinin subunits are assembled into 9S trimers in the ER and following their intracellular transport through the secretory system, arrive at the vacuole in that form. Within the protein body/vacuole, a specific cleavage occurs (to yield the acidic and basic polypeptides) and the trimers assemble into hexamers [135]. Assembly of monomers into trimers and then into hexamers can occur in vitro [134,135] and an in vitro system has been used to evaluate the effects of protein modification on oligomer formation [164]. Interestingly, the assembly of glycinin trimers into hexamers requires the proteolytic cleavage of proglycinin into its basic and acidic subunits.

Transport of Proteins Through the ER and Golgi Complex

All of the transport steps from the ER to the Golgi complex and transport through the Golgi itself are common to most proteins of the biosynthetic/

secretory pathway and occur via a signal-independent (bulk-flow) mechanism (see earlier discussion). Proteins destined for locations beyond the ER are transported in small membraneous vesicles which bud from the smooth regions of the ER membrane. These "transition" vesicles are directed to the Golgi apparatus (or stack) which is composed of flattened, membrane-enclosed saccules. In animal cells, the Golgi apparatus comprises three distinct compartments (each containing a characteristic array of protein-modifying enzymes), the *cis, medial,* and *trans* cisternae, plus the tubular *trans*-Golgi network [165,166]. Movement through the Golgi compartments is mediated by the budding and fusion of transport vesicles. Much of the present work in animals and yeast is focused upon the mechanisms of protein movement through the Golgi stacks, and how proteins are sorted according to destination, then packaged for delivery [6]. In both systems, signal-dependent sorting of proteins away from bulk-flow (i.e., to the lysosome/vacuole or to secretory vesicles) occurs in the *trans*-Golgi network, a complex tubular reticulum in which proteins are segregated into different transport vesicles to be dispatched to their final destinations (reviewed in [26,166]). Thus, it is in this compartment in which protein traffic signals are recognized by specific receptors that sort proteins according to their proper destination. For example, soluble lysosomal enzymes are diverted to lysosomes by a mechanism involving a protein sorting signal (a modification of a N-linked carbohydrate chain) and a membrane-bound receptor that recognizes the specific carbohydrate modification (reviewed in [49]). Likewise, within cells involved in regulated secretion, selected proteins undergo sorting and dense packing into special vesicles in the *trans*-Golgi network [167].

In mammalian cells, the vesicles that transport proteins from the *trans*-Golgi network to other compartments (via sorting-dependent mechanisms) appear to be covered with a protein coat consisting of clathrin [168] and a subset of characteristic polypeptides [169]. Small clathrin-coated vesicles function in both receptor-mediated endocytosis as well as in the receptor-mediated transport of proteins to the lysosome (reviewed in [5,170–172]). The endocytic and lysosomal transport pathways meet in the *trans*-Golgi network. The clathrin-coated vesicles facilitate the (mannose-6-phosphate)-receptor-mediated transport of lysosomal enzymes from the Golgi apparatus to a prelysosomal compartment. Here, they are sorted into lysosomes and the dissociated receptors are recycled back to the Golgi apparatus. An outstanding question is whether clathrin is involved in the export of constitutively secreted proteins and normal membrane proteins. One view is that clathrin contribution to intracellular traffic is quantitative rather than absolute; clathrin assembly concentrates and sorts receptors, facilitating their transport from one membrane to another [171]. It performs this function in both the import and export pathway, enhancing transport provided by bulk-flow in either direction. Localized control of clathrin assembly through diverse molecular signals from receptors and coated pit components allows its universal function to be applied to specific intracellular targeting [171]. A vesicle with only a partial clathrin coat is involved in the dense packing of proteins secreted via a regulated (and sorting-dependent) mechanism [173]. In the formation of hormone-containing secretory granules, clathrin-coated pits and vesicles remove membrane from the site of a condensing secretory bud that will eventually pinch off from the *trans*-Golgi network to form a secretory granule; the prophormone may be initially concentrated by a receptor-mediated process (reviewed in [171]). After granule acidification [26] and prohormone processing, the receptors may be removed by clathrin, resulting in granule condensation. This scheme again postulates that the role of clathrin is to concentrate and sort receptors [171]. A small non-clathrin-coated vesicle may be involved in bulk-flow transport [174,175]; in contrast to the clathrin coated vesicles, these carriers are found at all levels of the Golgi stack and contain protein at its prevailing (bulk) concentration in the parental Golgi cisternae [174].

In the cells of higher plants, the structural relations between the ER and Golgi apparatus (as well as intraorganellar transport pathways) are much less clear [7,16]. However, there is evidence that some features of Golgi structure and vesicular transport are similar to those of animals. For example, ultrastructural analysis of the individual Golgi stacks in suspension-cultured sycamore maple cells identified three morphologically distinct cisternae [176]. In plant cells, physical connections between the *trans*-Golgi and the partially coated reticulum (an organelle consisting of tubular membranes bearing clathrinlike coats over parts of their cytoplasmic surface) are indicative that the latter resembles the *trans*-Golgi network of animal cells [177]. The partially coated reticulum and clathrin-coated vesicles may participate in the sorting-dependent transport of proteins to the plant cell vacuole. For example, precursors of vacuolar seed storage proteins and lectins are found in clathrin-coated vesicles isolated from developing pea cotyledons [178–180]. Other findings are not supportive of a role for the partially coated reticulum in secretory protein transport [181] and further work is needed to clarify its role [7]. At this point, however, it is safe to presume that segregation of proteins en route to the plant cell vacuole (from those destined for constitutive secretion, via bulk-flow) occurs at a late Golgi step that relies on recognition of specific sorting signals by the appropriate receptors [182, 183].

Some insight into the mechanisms of vesicle budding, targeting and fusion have come from the use of in vitro transport systems in both animal and yeast cells. The reconstitution of vesicular transport between compartments has been achieved for transport between the cisternae of the Golgi stack [5]. In vitro transport of the vesicular stomatitis virus G protein from the *cis-* to the *medial*-compartments of the Golgi complex has been demonstrated; requirements for this transport include the presence of a crude cytosolic fraction, ATP, and proteins on the surface of the Golgi membranes [184–186]. Cytosolic factors (proteins) may be involved in collecting proteins for vesicular transport, as well as in mediating vesicle budding, transport, and fusion of the transport vesicle with the appropriate target organelle [5]. Remarkably, the cytosol from yeast [187] and plants [188] will substitute for animal cell cytosol in promoting transport in the Golgi stack and in forming coated vesicles. This interchangeability is indicative that the transport machinery is extremely similar, even in detail, in all eukaryotes [5]. More recently, techniques (previously utilized on rat liver) have been applied to isolate a population of vesicles derived from ER in a cell-free system from plants which likely represent an intermediate compartment in membrane transfer between the ER and Golgi complex [189].

The finding that proteins can transfer from one Golgi stack to another in a unidirectional fashion has led to the suggestion that vesicles may have a built-in targeting capacity; specifically, each type of vesicle has its own address marker which ensures that inappropriate transport does not occur. The target membrane must in turn have a complementary receptor that recognizes the address and allows docking of the vesicle [5,6]. Whether this type of specificity is involved in transport from the ER to the Golgi and in intra-Golgi transport remains to be demonstrated. Some proteins from mammalian intra-Golgi transport vesicles and from Golgi membranes which appear to fulfill these functions have been identified (reviewed in [6]). The working concept here is that during vesicle transit through the cytosol, the polypeptide configuration on the vesicle surface mediates vesicle docking and release of contents only when the correct membrane is encountered. Indeed, recent evidence is indicative that vesicle docking involves the formation of a complex that includes proteins from the cytoplasm as well as proteins permanently inserted in the vesicle and target membranes (reviewed in [190]). After reaching the *trans*-Golgi network, proteins are sorted according to their correct final destination, in this main distribution center of the cell.

Semiintact cells made permeable to macromolecules have been useful for studying ER to Golgi traffic in vitro [191,192]. In this system, there is efficient reconstitution of intracellular transport, provided ATP and a cytosolic fraction are added [191]. In vitro cell free systems have also been developed for yeast [193]. The secretory pathway in this organism has been defined genetically by a series of temperature-sensitive mutant strains blocked at various stages in protein transport [194–197]. At least 27 complementation groups (and thus gene products) are required for the transport of secretory proteins from their site of synthesis to the cell surface; about 12 are required for ER to Golgi transport alone. Additional gene products are necessary for localization of yeast hydrolases to the lysosomelike vacuole (see later discussion). In order to investigate the function of these gene products, researchers are cloning these genes and characterizing the proteins they encode (reviewed in [198]). Secretory pathway mutants can be complemented by unique expression plasmids that contain segments of wild-type yeast DNA. Open reading frames are identified (by DNA sequence analysis), gene fusions to β-galactosidase are constructed and gene product-specific antibodies are generated [5]. In this manner, some of the proteins required specifically for ER to Golgi transport have been identified and preliminarily characterized. For example, a 70-kD integral membrane glycoprotein (SEC 12) functions to promote protein transport to the Golgi complex; in the process of this transport, SEC 12 may itself migrate to the Golgi and function in subsequent transport events [106]. This type of analysis has also implicated GTP-binding proteins as regulatory elements in the secretory pathway. In very general terms, GTP binding and hydrolysis controls switching between two different protein conformations; the switch serves to propagate and amplify regulatory signals. A subset of these proteins (G proteins) may mediate the vectorial transport of individual vesicles, specifying the direction of vesicular transport between appropriate cellular compartments (reviewed in [198,199]). The yeast *ypt*1 gene encodes a raslike 23-kD protein that binds and hydrolyzes GTP; this protein likely resides in the Golgi complex and is involved in vesicular transport in that organelle [198,200]. Interestingly, there are mammalian homologues of the YPT 1 protein [201,202] and an antibody against the yeast protein detects a mammalian 23-kD protein localized in the Golgi complex [200]. The recent identification of two cDNA clones from maize encoding proteins highly related to *ypt* genes from yeast and mouse [203,204] suggests that the mechanisms of vesicular transport in higher eukaryotes are conserved. Using these cDNAs as probes, additional members of the *ypt* gene family have been isolated from maize and *Arabidopsis thaliana* [204]. The proteins predicted from the cDNAs show high similarity to other members of the ras family, particularly in the regions involved in GTP/GDP binding, GTPase activity and membrane binding. Their probable role in guiding vesicle transport in plants appears to be particularly important during pollen development, when these genes are expressed preferentially. Further experiments using reverse-genetic analysis with antisense constructs or dominant-interfering mutations, and cytological analyses using specific antibodies, may elucidate the precise roles of YPT proteins in vesicle trafficking and plant growth [204]. A defect in late secretory events occurs in the *sec* 4 yeast mutant. The *sec* 4 gene encodes a GTP-binding protein, most of which is tightly attached to the cytoplasmic faces of both post-Golgi secretory vesicles and the plasma membrane [205]. It may act primarily on the vesicle's surface to transduce an intracellular signal needed to regulate transport between the Golgi complex and the plasma membrane [205]. Another particle (SEC 8/SEC 15) is found in the cytosol and is peripherally associated with the plasma membrane but not with secretory vesicles. It may function as a downstream effector of the SEC 4 protein, serving to direct the fusion of vesicles with the plasma membrane [206]. It is not yet known whether the same GTP-binding proteins are involved in both constitutive and regulated secretion or whether particular ones are involved in the regulation of the targeting of secretory vesicles to the correct pathway. Their mode of action (e.g., in signal transduction or in other processes) in the secretory pathway is also unknown [198]. Secretory mutants in *Arabidopsis thaliana* are presently being sought [162]. Although specific defects in transport to the plant cell vacuole are the primary focus of these researchers, a more complete set of secretory mutants may help to identify specific proteins important in mediating the various transport steps of the biosynthetic/secretory pathway.

Targeting of Proteins to the Plant Cell Vacuole

Storage Protein Synthesis and Deposition in Seeds

Transport and targeting of proteins to the plant cell vacuole has been investigated most extensively in seeds, particularly in relation to the synthesis and deposition of storage proteins (reviewed in [7,48,207–210]). As mentioned previously, within the seed, vacuoles perform a dual function, i.e., as temporary storage depots (during seed development) and as sites of macromolecular hydrolysis (during postgerminative seedling growth). The synthesis and deposition of seed storage proteins is subject to strict temporal and spatial regulation; it only occurs in specific tissues/organs and only at specific times during seed development. Cell expansion and deposition of reserves (including storage proteins), follows the histodifferentiation stage of seed development, and occurs largely in the absence of further cell divisions (reviewed in [211]). It is during this time that the vacuoles of young storage parenchyma cells (particularly within the major reserve tissues of seeds) are transformed into protein storage vacuoles (protein bodies) which accumulate a variety of proteins (including storage proteins, acid hydrolases, plant defense proteins) and other reserve materials or metabolites (reviewed in [162]).

It is perhaps noteworthy to mention here that two distinct pathways of protein body formation are observed in seeds. One pathway (which is characteristic of most dicotyledonous seeds) involves the subdivision or fragmentation of a large central vacuole; the other occurs in some cereals and involves direct formation of protein bodies from the rough ER (reviewed in [212; 213–216] and references therein). In the former mechanism of protein body formation, storage proteins are packaged into protein bodies via transport through the Golgi complex (e.g., rice glutelins, wheat prolamines, legume globulins); with the latter mechanism, storage proteins are also synthesized on the rough ER, but are deposited directly within the lumen of this organelle where they remain and accumulate (e.g., prolamines of maize, rice, and sorghum). The subsequent discussion will focus primarily on the former pathway (i.e., vacuolar transport and targeting of storage proteins via the Golgi complex). Leaf and bark storage proteins accumulate in vacuoles which resemble the protein bodies of seeds [11,217].

Universality of Vacuolar Targeting Mechanisms

Because there is such a striking conservation of both general and specific features of the secretory pathway in eukaryotic cells, a pertinent question that has been addressed is whether plant vacuolar targeting signals are correctly processed in other eukaryotes. In mammalian cells, targeting of hydrolases to the lysosome (the vacuolar equivalent) is mediated by a mannose-6-phosphate sorting signal; this type of signal is not operative in plant vacuolar targeting (see earlier discussion). Plant vacuolar proteins expressed in animal cells (e.g., monkey COS cells, *Xenopus* oocytes) and insect cells, generally undergo normal co- and post-translational processing (e.g., signal peptide cleavage and glycosylation); however, they are not retained in any subcellular compartment, but rather are secreted by the bulk-flow pathway [218–221]. This further indicates that signals for targeting proteins beyond the Golgi apparatus in plant cells are different from those in animal cells. Not surprisingly, secretion is the outcome only for plant storage proteins which normally reach their target organelles via the Golgi complex; storage proteins exhibiting the alternative mechanism of transport (i.e., deposition into the ER, which subsequently becomes the protein body) are retained when expressed in animal cells. For example, injection of zein mRNA into *Xenopus* oocytes results in the synthesis and processing of zein proteins and their deposition into membrane-enclosed structures with the physical characteristics of protein bodies from maize [222,223]. Likewise, the wheat storage protein, α-gliadin is retained in *Xenopus* oocytes; in contrast γ-gliadin from wheat is secreted (to some extent) by these cells, which is indicative of the two different pathways transited by these related wheat storage proteins [218,224]. The mechanisms of retention of plant storage proteins like zein and α-gliadin in the foreign animal host cells (and in the natural plant host, for that matter) are as yet unknown; however, their failure to be secreted in animal cells is not due to improper recognition/processing of the signal peptide and inefficient sequestration into the membrane of the rough ER [218]. Thus, retention is likely a function of the mature protein. It is unlikely that a specific retention signal is recognized by the cell since these proteins have no C-terminal KDEL or HDEL. It may be that zeins are associated with proteins such as BiP that do have an ER retention signal [48,207]. Alternatively, the solubility characteristics of the protein may play a key role in retention, preventing subsequent transport, following translocation into the ER. Each member of the wheat gliadin family contains two separately folded regions that evolved independently as separate proteins and are thought to have later fused during the evolution of this family [225,226]. These are a N-terminal region composed of several tandem repeats rich in glutamine and proline and arranged in β-turn configurations and a C-terminal region arranged in α-helix and β-sheet configurations. Expression of wild-type γ-gliadin and two deletion mutant forms of the protein (missing either the N-terminal or C-terminal regions) in *Xenopus* oocytes has provided some information regarding the possible role of the different regions in protein sorting [224]. The N-terminal repetitive region of γ-gliadin led to retention and packaging of the protein into ER-derived protein bodies; conversely, the C-terminal region led to efficient secretion from *Xenopus* cells. Thus, sorting of wheat γ-gliadin is proposed to be determined by a balance between two opposing signals: (1) the effectiveness of the N-terminal region in ER retention and the initiation of ER-derived protein body formation, and (2) the counteracting effect of the C-terminal region to avert packaging within the ER and to promote the exit of the protein to the Golgi complex [224].

Some of the features of the plant vacuolar transport and targeting machinery are very similar to those of yeast. In those cases studied, when storage protein genes (e.g., phytohemagglutinin) are expressed in yeast cells under the control of the appropriate promoter, the majority of the protein is transported to the vacuole [227]. Thus, vacuolar targeting information is recognized by these two highly divergent species and yeast has been used as a model system to attempt to identify plant vacuolar targeting signals (see subsequent discussion).

Vacuolar Targeting in Yeast: Genetic and Biochemical Studies

Similar to the plant cell vacuole, the yeast vacuole is an acidic compartment which functions as a lytic site; it also serves as a storage depot for amino acids, phosphate, and inorganic ions. Numerous soluble hydrolytic enzymes reside in this organelle, including carboxypeptidase Y (CPY), proteinase A (PrA) and proteinase B (PrB). Its limiting membrane also contains a number of proteins and protein complexes, including α-mannosidase, alkaline phosphatase (AlP), dipeptidyl amino-peptidase B (DPAB), a proton-translocating ATPase and several permeases.

The soluble vacuolar hydrolases have been well characterized in terms of their biosynthesis. These proteins are synthesized as inactive precursors (preproteins) with N-terminal signal peptides that are proteolytically removed during translocation into the ER. Core glycosylation occurs in the ER; additional carbohydrate modifications take place during their transit through the Golgi complex. Sorting of vacuolar enzymes (from other proteins transiting the secretory pathway) occurs in a late Golgi compartment. Coincident with (or just prior to) their arrival in the vacuole, the modified precursors (e.g., proCPY, proPrA and proPrB) are activated by the proteolytic removal of a propeptide segment, a maturation process dependent upon the hydrolase PrA which is thought to activate itself autocatalytically [228]. Considerable progress has been made in characterizing the vacuolar sorting signals of these soluble hydrolases (reviewed in [229]). As mentioned earlier, in the absence of positive sorting signals for directed targeting away from bulk-flow, proteins traversing the yeast secretory pathway are delivered to the cell surface. This fact was exploited to design a screen for mutations in the CPY structural gene (pcr1) that result in secretion of CPY [230,231]. Another approach, utilized to determine elements sufficient for vacuolar targeting, was the construction of a nested set of carboxy-terminal deletions in CPY fused to invertase [232]. This latter approach demonstrated that the amino-terminal 50 amino acids of the CPY preprodomain (which includes a 20-residue signal peptide) are sufficient to sort invertase protein to the vacuole, while the first 30 amino acids direct high levels of secretion. Deletion analysis of the CPY gene (pcr1) in the propeptide-encoding region implicated a domain in the vicinity of amino acid 28 that is essential for efficient sorting of pro-CPY to the vacuole [231]. Deletions in this domain cause the mutant protein to bypass the Golgi sorting reaction and result in mislocalization of most of the proCPY to the cell surface. The analysis of point mutations in the pcr1 gene, encoding amino acid residues Lys 18 (in the signal sequence) to Leu 34, identified only four contiguous residues important for vacuolar sorting, $LQRP_{26}$ (Leu-Gln-Arg-Pro) [231,233]. Thus, this tetrapeptide sequence constitutes the core of an amino-terminal topogenic element, immediately following the signal peptide, that is necessary for the targeting of proCPY to the yeast vacuole [229]. Interestingly, the context in which the CPY tetrapeptide sequence is presented affects the efficiency of targeting, inferring the involvement of secondary structural elements in the sorting mechanism of this protein [233].

Subsequent analysis of the targeting signals of other vacuolar soluble hydrolases (e.g., PrA and PrB) revealed that the sorting mechanisms in yeast are much more complex than the initial simplistic model that emerged from the CPY studies. Here, the sorting domain is unlikely to be a short linear amino acid sequence. For PrA, it may be encoded by two signals, one in the mature protein and the other in the propeptide domain (e.g., between amino acids 61 and 76) [7,234]. There is no sequence similarity between the propeptides of CPY and PrA. Thus, currently no consensus sequence or common structural determinant has been demonstrated for the targeting mechanism of yeast vacuolar proteins, suggesting that a diverse array of factors may be operative in the sorting process. Although not clearly defined at present, the sorting domain of PrB likely resides in the mature portion of the protein, since the precursor undergoes an unusual early processing event in which the 280 amino acid propeptide is removed in the ER [235].

Although some commonalities exist, there may be independent mechanisms for the delivery of soluble and membrane proteins to the yeast cell vacuole. This appears to be the case for sorting of the vacuolar (type II) integral membrane protein, alkaline phosphatase [236]. Transport of this protein is less sensitive to some of the defects of the vacuolar sorting apparatus in mutants which exhibit dramatic missorting and secretion of the soluble hydrolases (e.g., CPY and PrA; see subsequent discussion). Dipeptidyl aminopeptidase B (DPAP B) also resides in the yeast vacuole as a type II integral membrane protein having short (amino-terminal) cytoplasmic and transmembrane domains (of 45 amino acids in total) and a large (800 residues) lumenal domain [237]. The cytoplasmic and transmembrane domains contain an element that is sufficient for vacuolar membrane targeting; however, a more detailed characterization of this topogenic signal must await mutational analysis [229].

It is likely that yeast vacuolar protein targeting signals are recognized by a sorting receptor(s), and the resultant receptor-ligand complexes are subsequently sorted into vesicles and transported to the vacuole by additional constituents of the vacuolar sorting machinery. Toward identifying the components of this sorting process, the application of several genetic selections has resulted in the isolation of a number of mutants that exhibit defects in vacuolar protein localization (primarily of the soluble hydrolases) and/or processing (reviewed in [229]). Instead of delivering vacuolar hydrolases to the vacuole, these *vacuolar protein sorting* (*vps*) defective mutants missort the enzyme precursors to the yeast cell surface [229,238–240]. Posttranslational glycosylation and secretion of proteins is normal in most mutants; thus, their defects are specific for targeting to the vacuole [239,240]. Genetic comparisons among the *vps* mutants demonstrate that they collectively define more than 47 unique complementation groups, indicating that delivery of proteins to the yeast cell vacuole represents a highly complex process requiring the coordinated participation of a relatively large number of gene products. Some of these components likely include sorting receptors, and proteins involved in the formation, interorganelle transport, targeting, and fusion of vesicles. Biochemical characterization of the *vps* mutants should not only identify cellular components directly involved in the specific segregation, packaging and delivery of vacuolar proteins, but also those gene products involved in the regulation, or control, of these processes. Progress toward this goal has recently been achieved with the demonstration that the *vps*15 gene encodes a novel protein kinase homologue that is essential for the delivery of soluble hydrolases to the vacuole [241]. The VPS15 protein is preferentially associated with the cytoplasmic face of a late Golgi or vesicle compartment and may regulate specific protein phosphorylation reactions required for efficient delivery and sorting of proteins to the yeast cell vacuole. For example, mutational alteration of the VPS15 protein kinase domain results in the biological inactivation of this protein and the secretion of multiple vacuolar hydrolases [241]. Protein phosphorylation has been implicated as a key regulator of protein sorting, specifically acting within pathways at branch positions, where proteins must choose between two or more different transport fates [241–243]. Thus, the proposed role of protein phosphorylation envisages its action as a molecular "switch" within intracellular transport pathways, such that proteins are actively diverted from a default pathway (i.e., secretion) to an alternative pathway (e.g., transport to the vacuole). This mechanism or level of control may be superimposed upon that carried out by GTP-binding proteins (vis., mediating the unidirectionality of secretory protein traffic, and perhaps designating a transport vesicle's final destination; see earlier discussion) [241]. In carrying out its role, the VPS15 protein appears to functionally interact with another protein (a phosphoprotein), that is encoded by the gene, *vps34* [241,244].

The generation and maintenance of a low lumenal pH with organelles of the secretory pathway is critical for vacuolar sorting (reviewed in [229]). Specifically, acidification is presumed to play a role in promoting the dissociation of vacuolar proteins from their sorting receptors, allowing the unbound receptors to return to the Golgi complex. Agents that abolish acidification of the vacuolar lumen (e.g., ammonium chloride, a classical lysosomotropic agent, or balifinomycin A1, a specific and potent inhibitor of vacuolar H^+-ATPases) also promote mislocalization of newly synthesized vacuolar proteins to the cell surface [245,246]. These agents presumably result in the recycling of bound receptors, causing a depletion of the pool of receptors available for sorting; secretion of newly synthesized vacuolar proteins by bulk-flow transport is the outcome. Two *vps* genes may encode components involved in the acidification process; *vps6* is a candidate for an H^+-ATPase subunit gene, while the *vps3* gene may encode a protein that participates in the assembly and localization of the vacuolar H^+-ATPase complex [229]. The availability of yeast cells lacking a functional vacuolar H^+-ATPase will allow a direct test of the role of acidification in the sorting of vacuolar hydrolases [229].

Future studies in yeast will include a detailed characterization of the *vps* genes and their encoded products. In conjunction with this, a recently developed in vitro reconstitution assay for vacuolar protein sorting [247] should aid in revealing the precise roles of these proteins in the sorting process (e.g., in receptor-mediated recognition/sorting, vesicularization, and vesicle targeting and fusion events). A clear understanding of these roles should in turn provide insights into how the various sorting components interact as well as the more general role of protein phosphorylation in intracellular protein sorting processes.

Identifying Plant Vacuolar Targeting Signals

Use of Transgenic Yeast as the Heterologous Host System   Because the plant vacuolar protein, phytohemagglutinin (PHA) is correctly processed and sorted to the yeast cell vacuole [227], transgenic yeast has been used as a model system to attempt to define the vacuolar targeting signal on this protein [248,249]. A PHA fusion protein (containing 43 amino-terminal residues of the mature protein, together with the signal peptide) is sufficient to redirect the secreted form of yeast invertase to the yeast vacuolar compartment [248]. Deletion analysis further localized the vacuolar sorting domain to an amino-terminal portion of mature PHA, between amino acids 14 and 23 [249]. Interestingly, this domain contains a yeast-like vacuolar targeting (tetrapeptide) sequence, $LQRD_{21}$, reminiscent of the one on CPY ($LQRP_{26}$), and conserved to some degree in other lectin proteins of legumes [7]. Site-directed mutagenesis to effect amino acid changes within this tetrapeptide sequence (in PHA-invertase fusion proteins) has demonstrated the importance of this short domain in the vacuolar sorting reaction in yeast. In particular, exchanging the aspartate at position 21, which introduces a glycan addition site into the sequence, results in considerable secretion of invertase activity (e.g., 64%). However, the effect of similar mutations within this short domain in full-length PHA, does not significantly affect targeting of the protein to the vacuole, indicating that there is additional vacuolar targeting information in PHA [7,249]. Thus, a second independent signal, present towards the middle of the polypeptide (i.e., carboxy-terminal to the first domain identified) may also be essential for correct sorting in yeast cells, notably similar to the requirements for PrA and PrB [7] (see earlier discussion). Perhaps more importantly, however, the same PHA-invertase fusion proteins that direct transport to the yeast cell vacuole are not successfully targeted to vacuoles in *Arabidopsis thaliana*. Therefore, the sorting determinant which contains sufficient information for vacuolar targeting in yeast, lacks the necessary information for efficient targeting in plants, suggesting that vacuolar sorting signals in these two organisms are dissimilar, perhaps in the extent to which other determinants in the mature protein are required for their receptor-mediated recognition. Thus, the approach which worked well in yeast does not work as well in plants and has not yet yielded results to define the vacuolar sorting signal of PHA in plants [7]. Further studies on various PHA-invertase chimeric constructs expressed transiently in plant protoplasts reveal that vacuolar sorting information is contained within an internal domain of PHA that is predicted to be exposed at the surface of the folded molecule [250]. The internal domain consists of 30 amino acids (amino acids 84 to 113 in mature PHA) and is capable of directing 50% of the reporter protein to the plant vacuole. It does not appear to share any sequence homology with the vacuolar targeting signals identified on other plant proteins (see subsequent discussion).

Fusion proteins of pea legumin and yeast invertase have yielded similar results; targeting information (for the yeast vacuole) appears to be contained in both the amino-terminal and carboxy-terminal portions of the legumin protein [251]. The vacuolar sorting signals of some yeast and plant proteins (e.g., legumin of field bean; [251]) are likely composed of regions on the surface of the protein which are commonly termed *signal patches*. Unlike signal peptides, signal patches will, in general, be formed from noncontinguous regions of the polypeptide chain that are brought together during protein folding; thus, they are conformation-dependent [5]. Although we presume it to be the case, we do not yet know with certainty whether the three-dimensional conformation of (intact or fusion) proteins is precisely the same in the heterologous yeast versus plant-host cells; there may be subtle differences, such that different protein determinants or features are more (or less) accessible for receptor-mediated recognition in these two systems. Alternatively (as intimated earlier) the sorting determinants per se which are critical for plant versus yeast vacuolar targeting may be different. A dependence on signal "patches" for correct sorting would obviously contribute to the difficulty of defining these domains experimentally, particularly via engineered (chimeric or deletion) proteins which contain (or lack) only contiguous amino acid sequences.

Use of Transgenic Plants as the Heterologous Host System   Several studies have demonstrated that plant vacuolar proteins are correctly targeted in other heterologous (plant) hosts (reviewed in [7]). In general, the vacuolar targeting signals on these proteins are recognized with a high degree of fidelity, regardless of the species, cell type, or organ, in which expression is directed, although transport efficiency may be variable and has not been systematically determined. Progress is being made toward identifying the molecular mechanisms regulating the vacuolar sorting of Gramineae lectins [182,252,253]. Barley lectin is synthesized as a preproprotein with a glycosylated carboxy-terminal propeptide that is removed, just prior to, or concomitant with, deposition of the protein in vacuoles [254]. The intact protein is correctly assembled, processed, and targeted to the vacuole in heterologous tobacco host cells [47]. However, expression of a mutant form of the lectin gene (in which the region encoding the short prodomain is deleted) results in missorting and secretion of the protein via the bulk-flow transport pathway [182]. The C-terminal propeptide on the lectin is also sufficient for vacuolar targeting; when the corresponding DNA sequence is fused to a gene encoding a secreted reporter protein (cucumber chitinase), the resultant fusion protein is redirected to the vacuole in transgenic tobacco with 70–75% efficiency [252,207]. Other Gramineae lectins, such as wheat germ agglutinin and rice lectin, contain similar proteolytically processed C-terminal domains (reviewed in [255]) and these probably serve the same function. Sporamin (a vacuolar storage protein in tuberous roots of sweet potato) undergoes a pattern of proteolytic processing which is similar to that of the cereal lectins. This protein is synthesized as a preproprotein with a short amino terminal propeptide domain which undergoes cleavage in a posttranslational manner, probably in the vacuole [256]. Studies of the fate of sporamin gene constructs in transgenic tobacco, show a dependence upon the amino-terminal prodomain for correct vacuolar targeting, similar to that demonstrated for the carboxy-terminal prodomain of barley lectin [257]. Thus, in these two vacuolar proteins (which have a similar pattern of precursor synthesis and posttranslational proteolytic processing) the prodomains appear to contain information essential and sufficient for vacuolar sorting. Subsequent studies analyzing the functional elements of the barley lectin C-terminal prodomain suggest that two independent short hydrophobic amino acid stretches are sufficient for correct vacuolar sorting. Neither charge nor glycosylation of the prodomain are necessary for targeting. The predicted secondary structure (i.e., an amphipathic α-helix) [258] is also not required for recognition by the vacuolar sorting apparatus; however, correct three-dimensional presentation of the domain appears to be essential [252, 253]. Comparison of the propeptide sequence of sporamin to other propeptide sequences (e.g., on barley aleurain, and potato cathepsin D inhibitor) reveals a common short region of hydrophobic amino acids with a hydrophilic residue, Arg, in the center (Asn-Pro-Ile-Arg-Leu-Pro); Asn and Ile are conserved, and Pro, Arg, and Leu can be substituted [209,259]. Progress is being made toward determining the critical residues in this region which are necessary and sufficient for vacuolar sorting [259].

Recent studies with some of the pathogenesis-related proteins of tobacco (e.g., chitinase, β-glucanase) has also led to the identification of the vacuolar sorting domains on these proteins. Homologous extracellular and vacuolar forms of these hydrolytic enzymes exist; acidic forms are generally extracellular, while their basic counterparts (likely encoded by different gene products) are vacuolar [14,260,261]. A basic chitinase of tobacco is synthesized as a higher molecular weight precursor, with three domains which are not present on the acidic form (e.g., an insertion in the middle of the polypeptide and short amino-terminal and carboxy-terminal domains) [262]. The carboxy-terminal tail of the basic tobacco chitinase appears to have significant vacuolar targeting information [263]; deletion of a carboxy-terminal heptapeptide from the basic enzyme leads to secretion of the mutant chitinase from transgenic tobacco cells. Conversely, a fusion protein (containing the carboxy-terminal domain of basic tobacco chitinase linked to an acidic [extracellular] cucumber chitinase) is retained to a substantial degree (i.e., 50%) in transgenic tobacco cells and is sufficient for vacuolar localization. A comparison of the C-terminal extensions of the lectins and vacuolar hydrolases (tobacco chitinase and β-glucanase) reveals no amino acid identities; however, a common feature is an abundance of hydrophobic amino acids, which may be important for recognition by the sorting machinery [207]. Aleurain, a vacuolar thiol protease in barley aleurone layer cells, is made as a 42-kD proenzyme (proaleurain). Similar to sporamin, aleurain is proteolytically processed into its mature form by the removal of a N-terminal propeptide [264] whose amino acid sequence shares identities with the sporamin pro-domain [48]. Two steps are required to form the mature 32-kD aleurain; the first yields a 33-kD intermediate; subsequent trimming results in the gradual loss of 1 kD. Two homologous thiol proteases, aleurain and EP-B (both having N-terminal propeptides of about 110 amino acids), are simultaneously expressed in aleurone layer cells but have different destinations (vacuolar vs. extracellular, respectively) [264,265]. Towards identifying the vacuolar sorting signal on aleurain, chimeric proteins (with switched prosequences from the two proteins) have been constructed and their fate (i.e., subcellular localization) examined in heterologous host plant cells [266,267]. Substitution of the propeptide of EP-B with the N-terminal propeptide of aleurain (containing the amino acids, SSSSFADSNPIR), results in redirection of about 50% of EP-B to the vacuole [266]. Shorter sequences within this domain (SSSSFADS and SNPIR) are also able to target pro-EP-B to the vacuole, albeit with lower efficiencies compared to the combined sequence. Efficient vacuolar sorting of aleurain may, therefore, be mediated by the combined action of small contiguous determinants [266,267].

Thus, significant insight into vacuolar sorting signals may be derived from comparisons of amino acid sequences of homologous extracellular and vacuolar proteins [7]. Subsequent studies utilizing this information, can then be geared to identifying those domains that are both necessary sufficient for correct targeting of the vacuolar form.

Some progress has been made toward identifying the targeting signals of tonoplast proteins in plants [72]. As mentioned earlier, in bean seeds the tonoplast intrinsic protein (α-TIP) is synthesized on the rough ER and its transport to the tonoplast is mediated by the secretory system. The C-terminal 48 amino acids of α-TIP, which include the sixth membrane spanning domain and the cytoplasmic tail, can redirect a reporter protein to the tono-plast in tobacco cells. A mutant form of α-TIP (in which the C-terminal cytoplasmic tail is deleted) is still targeted to the tonoplast; thus, the C-ter-minal transmembrane domain seems to contain sufficient information for transport to the tonoplast. Whether such transport occurs by bulk-flow or involves specific cellular machinery remains to be determined [72].

To fully understand plant vacuolar targeting mechanisms it will be nec-essary to undertake a detailed characterization of the sorting signals from various proteins with different functional and structural characteristics. For the above studies make it clear that the sorting signals may well vary widely from diffuse "signal patches," dependent upon correct protein con-formation, to more easily defined prodomains, which are removed when vacuolar transport is completed. Receptors mediating the recognition of these vacuolar targeting signals also await identification; experiments are underway to achieve this goal [183]. Another future goal is the isolation of plant mutants having specific defects in vacuolar protein targeting (similar to those in yeast); these will be valuable for identifying other components integral to the transport and sorting machinery [162]. As will be discussed in the next section, this basic research will be of value for more applied stud-ies geared toward the genetic engineering of plants with improved traits.

APPLICATIONS TO GENETIC ENGINEERING: MAXIMIZING LEVELS OF GENE EXPRESSION IN FOREIGN PLANT HOSTS

Subsequent to the advent of genetic engineering of plants, there has been an ongoing thrust towards refinement of the techniques for gene isolation and manipulation, plant host transformation, and efficient regeneration of whole plants from single transformed cells. Current attention is focused toward the direct application of this refined technology; in particular, its value as a tool for defining or elucidating regulatory mechanisms of plant gene expression, as well as its more practical potential—as a means of in-troducing agronomically useful traits or characters into crop and pasture plants. One major goal is to engineer seed crops for greater nutritional value, by introducing storage proteins modified to contain a more optimal balance of the essential amino acids (i.e., those amino acids which cannot be syn-thesized by animals and hence, must be supplied in their diet). Another re-lated goal (particularly important to the economy of Australia) is to obtain high levels of accumulation of certain sulfur-rich, rumen-resistant storage proteins within the leaves of pasture plants (e.g., lucerne and subclover), with the aim of increasing wool growth in sheep [95,96]. Other desirable traits which researchers are presently focused upon include the engineering of insect, disease, and herbicide resistance in crop plants, as well as the production of medicinally important substances in plants.

To be of agronomic value (particularly in relation to enhanced nutritive value), levels of the foreign protein must accumulate to significant levels in the host plant (e.g., to about 1–10% of the total cellular protein) [95,96]; it may also be desirable to target protein accumulation into a suitable subcellular compartment. The former objective is particularly challenging since levels of foreign proteins in transgenic plants (particularly within vegetative tissues) are generally low, although notable exceptions exist [268–271]. Thus, concerted efforts are being made to enhance gene transcription, and mRNA stability and translatability, to achieve high protein levels. While still a valuable pursuit, in general, such attempts have yielded variable success [95,96] and (so far) have led to insufficient increases in foreign protein accumulation to be of value. More recently, a novel strategy has been towards enhancing protein stability and the "mis-" or "retargeting" of proteins into different subcellular compartments.

Enhancing Protein Stability in Transgenic Host Plants

It has become evident that a lack of stability of the introduced protein in the foreign host environment is a major obstacle which must be overcome in order to obtain high levels of protein accumulation. Protein instability may be due to structural reasons (e.g., modifications to the coding region of a gene may yield an unstable protein, for various reasons); it may also arise as a consequence of the subcellular localization of the protein in the foreign host environment. Thus, success in achieving high protein levels in transgenic host plants will likely depend upon a close scrutiny of targeting signals, as well as the structural features of proteins which allow for their stable accumulation in cells. The requirement for a careful consideration of these factors is underscored by some recent studies attempting to introduce storage proteins which have been structurally modified to enhance their nutritional quality [272,273]. For example, a high-methionine β-phaseolin gene (modified to contain codons for 15 additional amino acids, including six methionine residues), yielded very low levels of expression in transgenic tobacco seeds, as a consequence of reduced protein stability and increased degradation. A great majority of the modified protein was degraded, either within the seed vacuole, or in transport vesicles en route to the vacuole [272]. Similarly, a gene with a modified CAT (chloramphenicol acetyl transferase) coding region (in which a 292-base pair coding sequence encoding 80% essential amino acids was inserted) gave very low levels of protein in transgenic potato tubers [273]. Protein instability may also occur as a result of deletions in the coding region of the vicilin gene, which yield highly reduced levels of protein (as compared to the intact vicilin gene) when expressed in the leaves and seeds of transgenic tobacco [274]. In the above cases, a decline in protein stability may occur as a consequence of the modification (i.e., insertion or deletion) having an effect upon protein folding; altered oligomeric assembly in the ER may also be operative.

Comprehending the rules that relate amino acid sequence to protein structure is not only fundamental to our understanding of such important biological processes as protein folding and the achievement of a functional three-dimensional structure, but will be essential to any attempts to modify the nutritional quality of proteins at the amino acid level. Tolerance to amino acid substitutions in relation to protein structure (and hence, stability) and function is just beginning to be understood (reviewed in [275–278]). Proteins appear to vary widely in their tolerance to amino acid changes. Zeins, the prolamin storage proteins of maize, are totally lacking in the essential amino acids lysine and tryptophan. In contrast to the outcome of modifying phaseolin, vicilin, and CAT genes (noted above), modifications to a zein gene (e.g., the introduction of lysine codons and lysine- and tryptophan-encoding nucleotides at several positions) does not affect the stability of the zein protein, when the in vitro transcribed mRNAs are expressed in *Xenopus* oocytes [279]. This result is promising in relation to the possibility of creating high-lysine corn by genetic engineering; stability of modified zein proteins in transgenic host plants now requires testing. Presumably, this protein can tolerate even severe modifications; a gross alteration (in which a 450 base-pair open reading frame from a simian virus 40-coat protein was inserted into the coding region) did not affect zein protein stability, although its ability to aggregate and form protein bodies was affected [279].

In general, seed storage proteins are significantly more stable in the seed of a foreign plant host than in the leaf, a concern for those attempting to achieve high levels of protein accumulation in vegetative organs [96,280, 281]. For example, despite similar levels of β-conglycinin mRNA in the leaves and seeds of transgenic petunia (utilizing a constitutive viral promoter to drive expression), the leaf cell environment yields a much lower level of protein accumulation [281]. A similar phenomenon occurs when pea vicilin is expressed in the leaves and seeds of transgenic tobacco; again reduced protein stability in the foreign leaf environment may be operative [96,282]. For example, much higher levels of vicilin are detected in younger tissues (leaves) than in older ones, which is indicative that the rate of breakdown exceeds the rate of synthesis in these older tissues [282]. The vacuolar targeting signals of seed storage proteins are recognized and processed correctly in the leaf cell environment, since correct vacuolar transport of the protein occurs with a high degree of fidelity, although transport efficiency in the leaf en-

vironment has not always been determined. As mentioned earlier, proteolytic processing of storage proteins commonly occurs upon their arrival in
the vacuole. Differential proteolytic processing of storage proteins (e.g.,
β-conglycinin, vicilin), yielding polypeptides of variable length, is evident
in these two tissues—i.e., the leaf and the seed [281,282]. Thus, the nature
of the endoproteolytic enzymes (and their substrate specificities) within the
leaf cell vacuole may differ from those inhabiting the seed storage vacuole;
the leaf cell vacuole (of some plant species) may also have a wider range of
proteases which are active and hence, "unfriendly" towards the (foreign)
seed storage proteins. Presumably, this latter factor would only be the case
in vegetative tissues which never accumulate vacuolar proteins (having a storage role) to any great extent. It is noteworthy that many of the proteases
that coinhabit the vacuole with storage proteins during seed development
are generally inactive at this stage, thus protecting storage proteins against
proteolytic attack. The relationship between the differential processing of
storage proteins and protein instability due to proteolytic breakdown may
need to be understood before the problem of low protein expression in the
leaf can be fully addressed. Reduced stability and differential processing
does not appear to be due to altered oligomeric assembly in the leaf; vicilin
assembly into trimers proceeds in a normal manner in this organ [282].

Since an instability of storage proteins in the leaf may, in part, be due
to their (correct) vacuolar localization, one novel approach to overcome
this problem is to "mistarget" proteins into subcellular compartments other
than the vacuole [95,96]. Removal of the signal peptide sequence of a protein should abolish its transport to the leaf cell vacuole; entry into the ER
lumen, and hence, the secretory pathway, is denied. A cytosolic localization
is predicted since this would occur by default. Vicilin accumulation in the
leaves of transgenic tobacco is increased by about five-fold when the DNA
sequence corresponding to the signal peptide sequence is removed [95]. This
modified vicilin gene results in a polypeptide of 50 kD; none of the normal
proteolytic processing products (characteristic of expression of an unmodified vicilin gene) are obtained, intimating that vacuolar transport is indeed
abolished. The fivefold increase in protein level is obtained from a level of
vicilin mRNA which is lower than that obtained with the unmodified gene.
Thus, the vicilin protein appears to be more stable as a consequence of its
new subcellular (cytosolic) localization; pulse-chase labeling and localization experiments are now required to confirm this [95].

As mentioned earlier, expression of a vicilin gene modified to encode
an ER-retention signal (SEKDEL) at the carboxy-terminus of the protein,
results in a dramatic (100-fold) increase in vicilin accumulation in the leaves
of transgenic tobacco and alfalfa, as compared to its unmodified counterpart [95,96]. This increase in protein is obtained without any change in vicilin

mRNA level.. Enhanced stability of the modified protein (which achieved a level of 2.5% of the total tobacco soluble leaf protein) may occur as a consequence of its new subcellular localization (i.e., predominantly in the ER), where presumably, it receives some protection from proteolysis [95]. Alternatively, there may be structural changes to vicilin as a consequence of its new carboxy-terminus which somehow render the protein more stable. Thus, targeting to a new subcellular locale may not be the sole factor involved in the enhanced protein stability. The stability of seed storage proteins in other subcellular compartments of the leaf (e.g., chloroplast, mitochondria, nucleus) will also be determined [283].

The targeting and stability of engineered zeins in transgenic yeast cells has been examined [284]. Unmodified zein is targeted to (and accumulates within) the ER lumen in transformed yeast cells, as predicted for this maize storage protein. However, a truncated zein protein (in which the signal peptide and the first 36 amino acids of the mature protein are deleted) is synthesized in the cytoplasm and massively accumulates in the mitochondria, nuclear encoded protein. Following synthesis in the cytosol and subsequent in maize endosperm cells. Thus, there appear to be distinct or separate domains of the zein polypeptide which, on the one hand, are responsible for membrane targeting; others may be responsible for the stability and aggregation of this storage protein [284].

Enhancing Posttranslational Processes: Folding and Oligomer Assembly

The design of polypeptide sequences with a functional and stable three-dimensional conformation is impeded by our limited knowledge of the rules that govern protein folding and oligomer assembly, and how these processes relate to a protein's ultimate stability in cells. Strategies must also be developed to predict and avoid degradation of recombinant proteins in heterologous host cells, particularly in cases where these proteins contain sequences which render them sensitive to proteases, specifically targeting them for degradation [285,286]. A novel way of overcoming some of the fundamental problems associated with protein design, may be the construction of proteins de novo with unnatural (but highly stable) chain architectures, by making use of the tools of synthetic chemistry (reviewed in [287]). It may also be possible to design modified storage proteins of desirable composition (e.g., rich in the sulfur-containing amino acid, cystine) which are stable and undergo correct folding and assembly in a manner reminiscent of natural (unmodified) seed storage proteins. Multiple cystine residues in a newly designed protein molecule may play an important role in stabilizing the protein; these have been shown to significantly stabilize the native structure of a protein [288] and may also render proteins resistant to proteolysis [289]. Transient associations of seed storage proteins with molecular chaperones resident in the ER, may also be a prerequisite for their correct folding and oligomeric assembly in foreign host cells (see earlier discussion). This may be achieved by ER-resident proteins which are endogenous in the foreign host cells; alternatively, it may require the introduction of genes encoding such "ER-helper" proteins from the donor genome (i.e., the plant from which the storage protein gene was derived). As mentioned earlier, glycosylation may play a general role in protein stability, primarily because of its promotive effect upon protein folding; increasing the number of glycosylation sites (and their strategic positioning within a designer protein, e.g., by site-directed mutagenesis) may increase both transport efficiency and stability.

One major system for selective protein degradation is the ubiquitin pathway in which proteins are committed to degradation by their ligation to ubiquitin (a highly conserved 76-amino acid polypeptide), as a multiubiquitin chain (reviewed in [290,291]). Although not yet tested in plant host systems, ubiquitin may be useful as a means of stabilizing proteins within heterologous hosts [291–293]. Interestingly, fusion of ubiquitin (by recombinant DNA techniques) to the amino-terminus of a protein can save it from being degraded in yeast [294], and results in increased expression, up to several hundredfold. The precise mechanism(s) responsible for such a dramatic increase in expression level is not clear. Ubiquitin may protect the amino-terminus of the fusion protein from proteolysis; alternatively it may somehow guide the protein into a new compartment where it receives greater protection from proteases. Enhanced translation of the fusion protein may also have contributed to the increased expression. It will be interesting to determine whether a similar fusion of the ubiquitin gene to the coding region of a seed storage protein yields a highly stable product in heterologous plant host cells.

PROTEIN TARGETING INTO AND WITHIN CHLOROPLASTS

The chloroplast is a complex organelle which carries out a wide range of metabolic processes; as well as housing the entire photosynthetic machinery, it is responsible for several vital biosyntheses (e.g., of fatty acids, phospholipids, and amino acids), the interconversion of carbohydrate intermediates, and the final steps in the assimilation of inorganic nitrogen and sulfate. It is a semiautonomous organelle, equipped with the machinery necessary for the transcription and translation of the limited number of proteins which are encoded by its own genome. Despite this synthetic capacity, the great majority of proteins required for its functioning are encoded by the nuclear genome; following their synthesis on free polyribosomes of the cytosol, they are imported into the chloroplast where further intraorganellar sorting may take place. The complexity of the organelle at the metabolic level is mirrored by a corresponding complexity at the structural level. It consists of three distinct membrane systems (the outer and inner envelope membranes and the thylakoid membrane) which enclose three distinct soluble subcompartments (e.g., the interenvelope membrane space, the stroma, and the thylakoid lumen). Thus, all six phases comprise targets for protein transport.

Protein transport into the stroma and thylakoids has been analyzed in some detail (reviewed in [295–298]) and is summarized in the next sections; reference is also made to Table 1. At present, little information is available on the mechanisms of protein targeting to the outer and inner envelope membranes or to the interenvelope membrane space.

Protein Targeting to the Stroma

Biogenesis and subsequent translocation of proteins across the chloroplast envelope membranes into the stromal compartment has been investigated in detail. The protein studied most extensively in this respect is the small subunit of ribulose bisphosphate carboxylase-oxygenase (SSU-Rubisco), a nuclear encoded protein. Following synthesis in the cytosol and subsequent translocation into the stroma, the SSU associates with its counterpart, the large subunit, LSU (which is synthesized inside the chloroplast) to form a large oligomeric complex comprised of eight subunits of each type [299]. The general principles derived from these studies on the SSU-Rubisco appear to be valid for the transport of similarly directed proteins, e.g., ferredoxin (Table 1) [295]. In large part, analysis has been by in vitro import systems [300] in which the transport of radiolabeled proteins (often constructed by recombinant genetic techniques and synthesized by in vitro transcription and translation) into purified intact chloroplasts is assessed. Verification of this import can involve treatment with external proteases; membranes protect imported proteins from degradation, while external proteins remain susceptible.

Proteins which must cross the outer and inner envelope membranes (i.e., stromal proteins and other proteins which undergo further intraorganellar targeting after reaching the stroma) are synthesized on cytosolic polyribosomes as precursors with a transient N-terminal sequence, the transit peptide. This transit peptide mediates both the initial binding of the precursor to the outer membrane of the chloroplast envelope as well as its posttranslational import. Coincident with translocation of the precursor, or shortly thereafter, a specific stromal peptidase effects transit peptide removal, yielding the mature protein. The proteolytic enzyme (180 kD), is chelator sensitive

Table 1   Targeting and Intraorganellar Transport of Some Nuclear-Encoded Chloroplast Proteins

Suborganellar target	Transported protein	Transport step(s)	Targeting signal(s)	Proteolytic processing
Stroma	SSU-Rubisco; ferredoxin	Translocation across outer and inner envelope membranes	N-terminal transit peptide; rich in hydroxylated amino acid residues serine and threonine, deficient in acidic residues; net positive charge	Stromal processing peptidase; cleavage of transit peptide to yield mature protein
Thylakoid membrane	LHCP	2 steps: Translocation across outer and inner envelope membranes; integration into thylakoid membrane	N-terminal transit peptide having common features similar to that of a stromal protein; thylakoid membrane integration signal in mature protein	Stromal processing peptidase; cleavage of transit peptide after membrane integration to yield mature protein
Thylakoid lumen	Plastocyanin; 33- and 23-kD proteins of photosynthetic O_2-evolving complex	2 steps: Translocation across outer and inner envelope membranes; translocation across thylakoid membrane	Composite transit peptide, first domain similar to the transit peptides of stromal proteins; second domain more hydrophobic; domains functionally independent	Stromal processing peptidase yields intermediate; thylakoidal processing peptidase yields mature protein

and is highly specific for imported precursors [301]. The precursor of SSU-Rubisco is processed to the mature size in two steps (involving the same enzyme, but different amino acid residues) and proceeds via an 18-kD intermediate [302,303].

The transit peptides of chloroplast proteins generally behave in an organelle-specific [304] and autonomous fashion; when attached to a heterologous (nonchloroplastic) passenger protein (via gene fusion experiments, followed by in vitro transcription/translation), they are capable of effecting unidirectional import of the passenger protein into the chloroplast. Successful import into the stroma was first demonstrated with the transit peptide of SSU-Rubisco linked to bacterial reporter proteins (chloramphenicol acetyl transferase, and neomycin phosphotransferase II) [305,306], and subsequently has been shown with a number of different passenger proteins [307–309], as well as with chimeric proteins having transit peptides from other chloroplast proteins [295,304,310–212]. Not all transit peptides may be equally efficient or competent in effecting translocation of a given passenger protein across the chloroplast membrane; likewise, different passenger proteins linked to the same transit peptide exhibit differences in the extent and efficiency of import [296]. Import of chimeric proteins can (in some cases) be increased when part of the mature chloroplast protein is added (in addition to the transit peptide), but this is not always a consistent result, and may, in part, be due to structural reasons (e.g., maintaining a particular secondary or tertiary configuration of the transit peptide which is conducive to import; see later discussion) [307,308,313,314]. More systematic and quantitative studies are needed to evaluate these results.

Attempts to define the functional domains of transit peptides (e.g., those sequences directly involved in receptor binding or translocation) have taken the form of two general approaches (reviewed in [296]). One approach has been the comparative analysis of the primary (and to a lesser extent, secondary) structures of transit peptides [296,315]. The other general approach has been to address structure-function relationships by generating and analyzing deletions (or specific changes to amino acid residues) in various regions of the transit peptide, e.g., those of ferredoxin and SSU-Rubisco [313,316–320]. With respect to the first approach, the amino acid sequences of the SSU-Rubisco transit peptides have now been determined from 48 genes representing 22 plant species [296]. From the statistical analyses of these and other transit peptides (e.g., those of other stromal and thylakoid membrane/lumen precursors), the general picture that emerges is that sequence similarities generally exist among transit peptides of the same precursor derived from different plant species. Few similarities are found among different precursors, even when the precursors are derived from the same plant species [296]. Despite a general lack of primary sequence similarities among transit peptides, some common features have emerged. They are rich in hydroxylated amino acids, serine, and threonine, and also have a number of small hydrophobic amino acids such as valine and alanine. A general deficiency in acidic amino acids is evident; overall they have a net positive charge [296,315].

Small deletions in the transit peptide of SSU-Rubisco generally do not abolish import into chloroplasts [316,317]. Mutagenesis experiments (to effect changes in specific amino acid residues in various regions of the transit peptide) indicate that N-terminal and C-terminal sequences may be essential for binding and/or uptake [295,317–319]. However, the general concept emerging is that the essential features of transit peptides are not found in their amino acid sequence, but rather in some higher-order structure. Thus, specific conserved secondary or tertiary structural features in the transit peptide may be more important in signal recognition/decoding by receptors or other components of the import/translocation machinery [295,296]. The presequences of several mitochondrial proteins have the capacity to form amphiphilic structures (see later discussion) and amphiphilicity may be one essential feature of their function [321,322]. It remains to be seen whether amphiphilic structures also play a role in chloroplast transit peptides (e.g., for initial membrane binding or receptor interaction). An alternative view is that chloroplast transit peptides are designed to be *devoid* of any regular secondary or tertiary structure [323]. If they are indeed "perfect random coils," a series of interactions with cytosolic and chloroplastic chaperones (see later discussion) may be all that is required for chloroplast targeting and import [323].

Protein Targeting Toward the Thylakoid Membrane System

The thylakoid membrane system encloses the thylakoid lumen and is itself completely surrounded by stroma. Thus proteins destined for either sub-compartment must undergo further intraorganellar targeting following their import into the stroma. Proteins of the thylakoid membrane or lumen utilize a targeting mechanism which consists of two independent steps; information is encoded for the sequential translocation across the outer and inner envelope membranes (info 1) and for the integration into (or translocation across) the thylakoid membrane (info 2) (Table 1) [312,324–326]. The most abundant thylakoid membrane protein imported from the cytosol is the light-harvesting chlorophyll binding protein (LHCP) of photosystem II. The transit peptide of this protein only directs import of precursor into the stroma; the signal for subsequent insertion into the thylakoid membrane is encoded in the mature protein [310,326,327]. Stable integration into the thylakoid membrane may involve complex interactions between a carboxy-terminus (putative) membrane-spanning domain and a number of other domains in the mature protein, which may have importance in protein refolding [326,328]; proteolytic cleavage to its mature size occurs only after membrane insertion is completed [329].

Nuclear-encoded thylakoid lumen proteins must transit all three chloro-plast membranes (e.g., the outer and inner membranes of the envelope and the thylakoid membrane) and hence have a uniquely complex import pathway (reviewed in [298]). Domain swapping and fusion experiments with such lumen proteins as plastocyanin (a small hydrophilic electron carrier) and the 33-kD and 23-kD proteins of the photosynthetic oxygen-evolving complex have provided strong evidence for two separate steps in the transport process [311,312,324,325]. Moreover, these two distinct targeting events (e.g., transport across the chloroplast envelope and subsequent translocation across the thylakoid membrane) are directed by a composite transit peptide, having two functionally independent domains. This bipartite structure is comprised of a N-terminal domain (which is structurally and functionally analogous to the transit peptides of stromal precursors) followed by a more hydrophobic domain (responsible for targeting across the thylakoid membrane) [295,296]. Interestingly, this latter domain strongly resembles the signal sequences of secretory proteins in bacterial and eukaryotic cells [330]. It is noteworthy that precursor maturation also occurs in two sequential steps, requiring different proteases (Table 1); one is located in the stroma and yields an intermediate protein form; the other is present as an integral membrane protein and generates the mature protein [324,325,331–334]. The thylakoid processing peptidase is capable of cleaving signal peptides of secretory proteins of both eukaryotic and bacterial origin ([295] and reference therein).

Cytochrome f is an example of a chloroplast-encoded protein which is targeted toward the thylakoid membrane system; its transport is presumed to be functionally equivalent to that of imported stromal intermediates en route to the thylakoids. Following its synthesis in the stroma on thylakoid-bound ribosomes, the precursor inserts partially into the thylakoid membrane by a C-terminal stop-transfer domain; a transient N-terminal sequence is then removed following membrane insertion [335].

Protein Targeting to the Envelope Membranes

Much less is known about the targeting of chloroplastic envelope proteins, in part because of their presence in low quantities compared to other chloroplastic proteins. Neither of the outer envelope membrane proteins studied so far (e.g., 6.7- and 14-kD proteins from spinach and pea, respectively) has a cleavable transit peptide [336,337]. Thus, proteins destined for the outer membrane of the chloroplastic envelope likely follow an import pathway distinct from that followed by proteins destined for other chloroplastic compartments and the necessary information must be located within the mature protein although its precise nature remains to be determined.

By contrast, import studies with different inner membrane proteins (e.g., a 37-kD protein and a phosphate translocator, both from spinach) indicate the presence of cleavable transit peptides [338,339]. Studies on the maize *bt-1*-encoded protein indicate that the transit peptides of inner membrane proteins function primarily as stromal targeting sequences; the specific information for subsequent targeting to the inner envelope is contained in the mature region of the protein [340].

Energy Requirements

Energy is required for both protein translocation across the chloroplast envelope and for efficient precursor binding to the outer envelope membrane [341,342]. Hydrolyzable ATP (inside the chloroplast) is the energy source utilized for these processes ([341,342]; reviewed in [296]); unlike protein transport within mitochondria, neither the electrical nor chemical components of a proton motive force (PMF) are involved in translocation across all three membranes in chloroplasts [341]. Confirmation of any ATP-dependent unfolding of precursors outside the chloroplast envelope [343] awaits further investigation [296].

Efficient integration into the thylakoid membrane requires ATP [326, 344,345]; energy is also likely required for translocation across the thylakoid membrane into the lumen [346].

Precursor Binding and the Identification of Receptors

The first step in the transport of precursor proteins into chloroplasts is a specific interaction between the precursor and outer envelope membrane; the components which mediate this binding, as well as subsequent translocation, likely involve both membrane lipids and intrinsic proteins [296]. There are several lines of indirect evidence for receptor-mediated binding. For example, studies showing diminished binding of precursors following protease (thermolysin) treatment (which specifically destroys outer membrane proteins), indicate that a protein component(s) of the chloroplast outer envelope is necessary for high affinity binding of precursors [347,348]. Moreover, precursor binding is ligand-specific; chloroplastic proteins lacking transit peptides do not bind to chloroplasts [296]. Likewise, binding is membrane-specific; chloroplastic precursors generally do not bind to nonchloroplast membranes (e.g., plasma membranes and erythrocyte membranes) ([296] and references therein). Several other characteristics of precursor binding are indicative of receptor involvement (e.g., binding sites are limtied and saturable; binding is rapid, specific and requires energy in the form of ATP) [342,348,349].

Two different proteins that may be components of the signal recognition and transport apparatus have been identified, utilizing different approaches. A heterobifunctional photoactivatable cross-linking reagent was used to implicate a 66-kD chloroplast surface protein [350]. Experiments utilizing anti-idiotypic antibodies (in this case, antibodies capable of recognizing other antibodies specifically directed against a synthetic analogue of the SSU-Rubisco transit peptide) have identified a putative (30-kD) import receptor, apparently located at contact sites between the outer and inner envelope membranes [351,352]. However, recent work suggests that this protein is identical to the phosphate-3-phosphoglycerate-phosphate translocator [353], suggesting a more indirect role in targeting.

ATP-dependent phosphorylation of a 51-kD envelope protein is correlated with precursor translocation into chloroplasts [354]. Whether it plays an integral role in the import process per se is unknown, however; at present, very little is known about the nature and composition of the translocation machinery.

Protein-lipid interactions may also be involved in the initial precursor binding to the chloroplast outer membrane; for example, they may mediate both the initial membrane insertion as well as facilitate transit peptide interaction with specific receptor proteins at the membrane surface [296].

Assembly of Oligomeric Protein Complexes: Role of Chaperonins

As mentioned earlier, the synthesis and assembly of Rubisco involves the interaction of two genetic systems viz., nuclear and chloroplastic. The large subunits of Rubisco are synthesized within the chloroplast; the nuclear-encoded small subunits are imported after synthesis in precursor form on cytosolic polyribosomes [355]. The assembly of the holoenzyme (which in its mature assembled form, consists of eight subunits of each type) occurs within the stromal compartment and appears to require the presence of another protein, the Rubisco subunit binding protein, more recently termed the *chloroplast chaperonin* (reviewed in [356–358]). This protein binds noncovalently to both newly synthesized large subunits and to imported small subunits. Convincing evidence for its role in assembly processes is available; for example, antiserum to the binding protein inhibits the transfer of newly synthesized Rubisco large subunits to the holoenzyme in stromal extracts ([356], and references therein). Interestingly, chloroplast chaperonin is related to the GroEL protein of *Escherichia coli*, a heat-shock protein that is essential for cell growth [259] and bacteriophage assembly [360]. In addition, GroEL is required for assembly of prokaryotic Rubisco synthesized in *E. coli* [361] and may also be involved in protein secretion [362,363]. Similar to the functions of the ER-resident molecular chaperones (discussed earlier), the chloroplast chaperonins are proposed to assist other polypeptides to maintain, or assume, a conformation required for their correct assembly (or localization), possibly via their ATPase activities [356,364]. This role is not limited to the assembly of Rubisco; rather, the chloroplast chaperonin appears to be involved in the assisted assembly and/or folding of a wide range of proteins in chloroplasts [365]. Similar chaperonin-type proteins (e.g., HSP 60) have been implicated in oligomeric protein assembly in mitochondria (see later discussion). Thus, assisted assembly of oligomeric protein structures is emerging as a general cellular phenomenon and is an important aspect of regulation of gene expression at the posttranslational level [356].

PROTEIN TARGETING INTO AND WITHIN MITOCHONDRIA

Mitochondria fulfill a variety of essential metabolic functions; in particular, being a major site for oxidative phosphorylation, they are often referred to as the "powerhouse" of the cell. The mitochondrial subcompartments each have a characteristic set of polypeptides and include the outer membrane, the intermembrane space, the inner membrane and the matrix. Like the chloroplast, the great majority of mitochondrial proteins (i.e., greater than 90%) are contributed by the nucleocytosolic system [366]. Precursor synthesis on free polyribosomes of the cytosol is followed by post-translational import into the organelle [367,368]. Much of our current understanding of the events required for organelle assembly and mechanisms of mitochondrial protein transport has derived from in vivo and in vitro studies in animals, yeast and *Neurospora* (reviewed in [369–372]). Little information is currently available on mitochondrial protein transport in plants; however, some of the general principles derived from studies of other eukaryotic systems may also prove to be applicable to plants.

Protein Targeting to the Matrix

Proteins which are destined for the mitochondrial matrix (as well as most of those which will ultimately reside in the inner membrane or intermembrane space) are synthesized as precursors containing targeting sequences (termed *presequences*) of 10 to 70 amino acids. In most (but not all) cases [373,374], these cleavable presequences are located at the N-terminus of the precursor. Much like the transit peptides of chloroplast precursors, sequence homology among mitochondrial presequences is lacking; common features include a high proportion of positively charged and hydroxylated residues and a general paucity of acidic amino acids [375,376]. Moreover, they also possess a high degree of autonomy, being both necessary and sufficient to direct nonmitochondrial passenger proteins into the mitochondrial matrix [377,378]. This has recently been shown for plant mitochondrial presequences in transgenic tobacco plants. The presequence of the mitochondrial β-sub-unit of ATP synthase (from tobacco) is capable of directing the bacterial protein, chloramphenicol acetyl transferase (CAT), into mitochondria, with high organellar specificity [379]. The nuclear-encoded precursor of mito-chondrial superoxide dismutase (SOD isozyme 3, a manganese-containing homotetrameric enzyme) [380] is translocated into isolated maize mitochondria [381]. Deletions in the transit peptide of maize SOD-3 (generated in vitro) show relative import efficiency to be highly correlated with deletion size [382]. A yeast mitochondrial presequence is capable of targeting a foreign protein into plant mitochondria in vivo [383]. The fusion protein consisted of the presequence of yeast mitochondrial tryptophanyl t-RNA-synthetase linked to bacterial β-glucuronidase (GUS); specific targeting and efficient import into the mitochondrial compartment of transgenic tobacco cells oc-curred, with no substantial misrouting. Proteolytic processing of the pre-cursor was equivalent in the two eukaryotic systems, with respect to both precision and fidelity; thus, the processing enzyme in plant mitochondria appears to recognize the same cleavage site within the presequence as the matrix protease from yeast [383]. It appears then, that the targeting of chim-eric proteins (expressed in heterologous plant hosts) is organelle-specific; dual targeting into mitochondria and chloroplasts is not a general phen-omenon. Exceptions do exist, however; for example, a yeast mitochondrial presequence (linked to a bacterial passenger protein) is recognized and can interact functionally with the protein translocation systems of both chloro-plasts and mitochondria in transgenic tobacco [384].

It is generally assumed that the essential feature of the presequence in the binding/translocation process is encrypted in a specific secondary or ter-tiary conformation (rather than in specific amino acid residues). Many tar-geting signals of mitochondrial precursors have the potential to form amphi-philic α-helices and β-sheets (i.e., secondary structural arrangements with the polar and nonpolar residues exposed to opposite faces) [321,322,385]. Such structures are said to have a hydrophobic moment [386], and react spontaneously with the surfaces of biological membranes [387]. This amphi-philicity is proposed to be important for the initial membrane insertion and/or the interaction with specific receptor proteins at the mitochondrial surface [370].

Most protein translocation into mitochondria occurs at sites where the outer and inner membranes are in close proximity [388]. It proceeds through a hydrophilic (proteinaceous) membrane environment; protein-lipid inter-actions may also occur [389]. The translocation complex seems to consist of two distinct translocation channels in the outer and inner membranes [390, 391]. Once translocated across the mitochondrial membranes, the N-terminal presequences of precursors are cleaved by a highly specific metal-dependent processing enzyme in the matrix [392]. This proteolysis is not coupled to membrane translocation, but is an essential process (e.g., in yeast; [393]), probably being required for proper assembly of imported proteins [370]. Proteolytic processing is a cooperative effort requiring two structurally related components which may recognize different structural elements of presequences and hence contribute to the high specificity of cleavage. These are the mitochondrial-processing peptidase (MPP) and the processing-enhancing protein (PEP) (identified in *Neurospora* and yeast) [393–396]. PEP may bind to presequences of incoming proteins (in a cotranslational manner) thus exposing the cleavage site towards MPP [370,394].

Protein Targeting to the Inner Membrane and Intermembrane Space

Proteins residing in the matrix reach their target compartment by translocation across the two membranes at contact sites; additional routing is required for correct localization of proteins of the inner membrane or intermembrane space. Soluble proteins of the intermembrane space (e.g., cytochrome b_2 of yeast mitochondria) as well as cytochrome c_1 of the bc_1 complex (which is largely exposed to the intermembrane space) are synthesized as cytosolic precursors with long complex presequences [397–399] having a bipartite structure. Their N-terminal parts exhibit the typical features of presequences of matrix-targeted proteins (and are functionally equivalent); the remaining C-terminal parts contain numerous hydrophobic residues preceded by one to four basic residues. This latter motif is thought to direct "export" of intermembrane proteins from the matrix back across the inner membrane [400,401]. However, another view is that it functions as a stop-transfer signal. According to the stop-transfer hypothesis, the N-terminal part of the presequence is imported into the matrix, but the intermembrane space targeting domain arrests further translocation through the inner membrane (reviewed in [372]). The composite presequences undergo cleavage in two steps executed by different processing peptidases [400,402–404]: the matrix MPP/PEP and an (as yet unidentified) membrane-associated peptidase at the outer surface of the inner membrane [400,403,404]. Cytochrome c_1 from potato also contains a transient presequence with a bipartite structure comparable to that described for the fungal proteins [405]; the targeting mechanism of the plant protein has not yet been determined.

The Rieske Fe-S protein, a peripheral component of the bc_1 complex at the outer surface of the mitochondrial inner membrane reaches its functional location on an import route via the matrix [370,406,407]. However, the presequence lacks a hydrophobic segment and contains information for targeting to the matrix only; information for its subsequent relocation to its final destination may be encoded in the mature protein [408,409]. Proteolytic processing is in the matrix only, and occurs via a two-step mechanism.

It is noteworthy that there are a few proteins of the mitochondria whose assembly and transport pathways are unique and do not conform to conventional intramitochondrial targeting routes (i.e., rerouting following transport into the matrix) [370]. The ADP- and ATP-carrier (AAC) of the inner membrane (and possibly some other structurally related proteins) are examples. The cytosolic precursor of AAC is made without a cleavable presequence [373]; targeting information resides in three internal segments [410,411]. Following its entry into the outer membrane (and subsequent transport into contact sites), there is lateral diffusion into the inner membrane. The signals that prevent translocation into the matrix and trigger integration into the inner membrane are unknown [370]. Cytochrome c also does not follow a reexport pathway; it reaches the intermembrane space by crossing the outer membrane only [412,413].

Other Components Involved in Protein Import and Assembly

Some of the components of the mitochondrial import machinery have been identified recently, mostly in *Neurospora crassa* and yeast (*Saccharomyces cerevisiae*). Precursor proteins must be maintained in a loosely folded conformation after synthesis, for their subsequent efficient translocation across mitochondrial membranes [414–416]. Constitutively expressed heat shock proteins of the HSP 70 family have been implicated in the stabilization of precursors in a translocation-competent state; by binding to precursors cotranslationally, they may help to keep targeting signals exposed [65,66,417, 418]. A N-ethylmaleimide (NEM)-sensitive factor associated with the cytosolic surface of mitochondria also appears to play a cooperative, facilitative role; along with ATP it is required for the release of proteins from HSP 70 [419,420].

The first specific step of the import pathway is the binding of precursors to the surface of the mitochondrial outer membrane [421,422]. In *Neurospora*, two outer membrane receptor/binding proteins (MOM 19 and MOM 72, 19 and 72 kD, respectively) have been identified [423]. Inhibition of precursor binding and import by antibodies recognizing the cytosolic domains of the proteins is indicative of their role as receptors participating in the specific recognition of precursors; also significant is their enrichment in the outer membrane at contact sites between inner and outer membranes, where most translocation takes place [423].

The subsequent entry of precursors into the translocation apparatus is thought to be facilitated by a common (integral) component in the outer membrane, the general insertion protein, GIP ([370], and references therein). Proteins of the matrix, inner membrane, and intermembrane space may be routed from GIP into contact sites for further translocation; precursors of outer membrane proteins (e.g., porin) may also insert into their target membrane directly via this protein [370]. Candidates for this (as yet unidentified) protein include an outer membrane protein from *Neurospora* (MOM 38) and a 45-kD protein from yeast [370,424,425].

HSP 60 (a large oligomeric protein complex located in the matrix) has been implicated as having an interal role in such posttranslocational events as protein folding and assembly in vivo [356,370,426–429]. This protein may have several functions. It may mediate the ATP-dependent folding of imported matrix proteins; folding may be prevented until the complete polypeptide is available [427]. A role in the maintenance of a translocation-competent state is also suggested, particularly for imported proteins which must undergo further intraorganellar targeting (e.g., insertion into, or translocation across, the inner membrane) [426]. A possible function in the translocation process per se has also been suggested; specifically it could interact with the extended N-terminus of a precursor reaching into the matrix, triggering complete protein unfolding and thus, effectively "pulling" it across the membranes [370]. Finally, HSP 60 of the mitochondrial matrix is highly homologous the GroEL protein of *E. coli* and to the Rubisco-binding protein of chloroplasts, and is also a member of the subgroup of "molecular chaperones" termed "chaperonins" (see earlier discussion). Thus, related to its role in mediating protein folding (e.g., recognizing structural motifs in unfolded or loosely folded polypeptide chains and repairing misfolded proteins), HSP 60 may also play an essential role in the assembly of large oligomeric proteins in the matrix [356,428].

Energy Requirements

Energy in the form of ATP is required in several steps along the mitochondrial import pathway including:

1. Release of precursors bound to HSP 70 or functionally related factors.
2. Transport of precursors from surface receptors to GIP.
3. Mitochondrial protein "export," e.g., release of imported proteins from HSP 60 (370 and references therein).

In contrast to chloroplast protein transport, some mitochondrial transport steps require a PMF in the form of an electrical potential ($\Delta\Psi$) (e.g., insertion of the targeting sequence into, or its translocation across, the inner membrane) [416,430]. The precise role of $\Delta\Psi$ is not understood, mainly because the mechanism of protein translocation and the components involved are still unknown [370].

PROSPECTS FOR ENGINEERING CHLOROPLAST AND MITOCHONDRIAL PROTEINS

A recent focus of genetic engineering in plants has been toward the ultimate development of systems for influencing the efficiency of plant growth and carbon fixation by photosynthesis, as well as the engineering of herbicide tolerance. This has involved the use of many novel approaches, including the manipulation of genes encoding important proteins (e.g., key enzymes of metabolic pathways) coupled with the exploitation of targeting mechanisms of proteins within both cells and organelles.

Herbicide Tolerance

Genetic engineering of herbicide tolerance into crop species is of significant interest to agricultural biotechnology. Particular interest in engineering tolerance to the herbicide glyphosate (N-[phosphono-methyl]glycine) stems from some of its desirable properties, e.g., its nontoxicity to animals and rapid degradation by soil microorganisms [431]. This herbicide functions by inhibiting the enzyme 5-enolpyruvylshikimate-3-phosphate (EPSP) synthase, and thus prevents plant growth by blocking the pathway of aromatic amino acid synthesis [432,433]. Some degree of tolerance to this herbicide has been achieved by introducing, into host plants, a bacterial EPSP synthase gene (which encodes a resistant enzyme). In order to direct correct chloroplastic localization of this mutant enzyme form, the DNA encoding the plant EPSP synthase transit peptide was linked to the bacterial enzyme coding region. The in vitro product of this chimeric gene is rapidly imported into chloroplasts, where it accumulates as a stable, glyphosate-resistant, enzyme [314]. Another strategy has been to introduce a plant EPSP synthase gene, under the control of the cauliflower mosaic virus (CaMV) 35S promoter; in this case, the inhibitory effect of glyphosate is counteracted by overproduction of the plant enzyme, targeted to the chloroplast by its own transit peptide [434].

A further approach to the engineering of herbicide tolerance involves a recent technology developed for the conversion of chloroplast (and mitochondrial) genes into nuclear genes and for the subsequent "retargeting" of the cytosolically synthesized protein into its respective "home" organelle (reviewed in [435]). This effective "relocation" of genes from organelle to nucleus is proposed to mimic the natural events occurring during evolution (i.e., the relocation of the majority of genes from an endosymbiotic organism to the host cell nucleus). "Reformatting" organellar genes for this "allotopic" expression (i.e., expression in a foreign or alien environment) involves several manipulations including:

1. Changing the open reading frame of the organellar gene (e.g., so that codon usage is compatible with the biases of the nucleocytosolic system).
2. Placing the open reading frame within a suitable nuclear transcription context (including appropriate promoter and terminator); a functional nuclear replicator sequence is provided to ensure the replication and maintenance of the restructured gene in the nucleus.
3. Addition of DNA encoding a N-terminal leader sequence to ensure correct targeting of the fusion protein into its "home" organelle, subsequent to its synthesis in the cytosol.

This strategy has been used in attempts to engineer resistance to the triazine herbicides (e.g., atrazine). This group of herbicides competes with thylakoid plastocyanin for binding with the quinone-binding protein (Q_B of photosystem II) and, thus, interrupts photosynthetic electron flow [436]. In situ, in thylakoid membranes, the Q_B protein of atrazine-sensitive plants, including the weed, *Amaranthus hybridus*, binds azidoatrazine; Q_B protein from an atrazine-resistant biotype does not bind the herbicide because of a single amino acid residue substitution [437]. The allotopic expression of the resistant form of Q_B protein from *Amaranthus* has been achieved in host tobacco cells [438]. The coding region of the donor (resistant) gene (*psb* A) was fused to transcriptional-regulation and transit-peptide-encoding sequences of a bonafide nuclear gene (i.e., those of the SSU-Rubisco gene). The transformed tobacco plants were tolerant to levels of atrazine that are toxic to nontransformed plants, although the atrazine-resistant phenotype was not sustained for long periods of plant growth [438]. Since expression of the endogenous ("selective") gene was not blocked in these plants, the variant Q_B protein was delivered to the organelles in which normal levels of the naturally encoded protein were produced. Thus, the transformants contained a mixed population of photosystem II complexes. Nonetheless, correct targeting of the variant Q_B form to the chloroplast could be demonstrated (by immunocytochemical means), and the recovery of atrazine-tolerant transgenic plants shows that the protein functions in photosynthesis [438].

Increasing Photosynthetic Productivity

Rubisco catalyzes the first step in the processes of both photosynthesis and photorespiration; since it is the balance between these two processes that ultimately controls plant productivity, this enzyme is a major target for genetic engineering (e.g., mutagenesis) with the aim of altering this balance for agricultural purposes [439]. Toward this goal, expression of Rubisco in a prokaryotic host has been undertaken; however, preliminary attempts have illustrated the importance of achieving the synthesis of proteins in the required active form in heterologous host cells [356]. So far little success has been achieved, in part due to a failure of synthesized Rubisco subunits to assemble properly in the bacterial host; hence an enzymatically active form is not produced [356,440]. Assembly may not occur because the chaperonins of these bacterial host cells (which would normally mediate protein assembly) are sufficiently different from those of plant cells, and hence are nonfunctional in Rubisco assembly. Thus a current aim is to express the cDNAs for the chloroplast chaperonin in the same *E. coli* cells that are expressing the genes for Rubisco from higher plants. In this way, Rubisco assembly should be rescued, permitting attempts to improve the properties of this agriculturally important enzyme [356].

Allotopic expression of the large subunit (LSU) of Rubisco has been achieved in *Oenothera hookeri* (evening primrose) ([435] and references therein). The transit peptide and transcriptional control sequences from the pea SSU-Rubisco gene were utilized for nucleocytosolic expression; the host was a plastome mutant containing a sigma mutation [441] in the Rubisco LSU (*rbc*L) gene of chloroplast DNA and, hence, expressing only a truncated version of the large subunit. Allotopic expression of full-length LSU was achieved, curing the sigma phenotype in transformants.

Since the technology of allotopic expression allows both controls over quantitative expression of the engineered gene and the subsequent correct delivery to the organelle of a protein of defined structure, it should have significant potential as a novel approach for basic investigation into the biogenesis, assembly, and function of a number of enzyme complexes in the chloroplast [435]. Its potential as a natural assay system for directly manipulating photosynthetic productivity (e.g., via changes to Rubisco subunits to diminish photorespiration and enhance CO_2 fixation) may also be exploited in the future.

Allotopic Expression of Mitochondrial Genes

This strategy may also prove to be useful for similar basic and applied studies of mitochondrial-encoded proteins. Successful allotopic expression of at least two mitochondrial genes has been achieved (so far in yeast) [435].For example, expression of a yeast gene, encoding subunit 8 of mitochondrial ATPase, was demonstrated in mutant host cells unable to synthesize the endogenous mitochondrial protein; transit peptide sequences (derived from the mitochondrial *Neurospora* subunit 9 gene) effected the required mitochondrial import [442]. In a similar manner, another mitochondrial gene (for the intron-encoded maturase of cytochrome b) was expressed allotopically in yeast cells [443]. Thus, there appears to be no intrinsic biological barrier to chloroplast or mitochondrial genes being encoded in the nucleus; the subsequent import of the synthesized protein into the target organelle is also possible, provided an appropriate presequence or transit peptide is utilized [435]. Other approaches to the manipulation and analysis of organellar genes could potentially make use of current techniques for direct delivery of organellar DNA into mitochondria and chloroplasts (e.g., via projectile techniques or possibly via DNA-protein conjugates) [444,445]. Although these approaches may have a more limited potential than allotopic expression strategies [435], future attempts to engineer organellar genes, and their encoded protein products, may exploit a combination of these avenues.

SUMMARY

It is apparent that posttranslational controls are an important aspect of the regulation of gene expression in all eukaryotes. In general, studies on the mechanisms of protein targeting in plant cells are in their infancy; much more information is presently available in other eukaryotic systems (e.g., animals, yeast, *Neurospora*). However, while important exceptions exist, a striking feature of the mechanisms and cellular machinery of protein targeting is their universality — among animals, plants, and eukaryotic microorganisms, and even between prokaryotes and eukaryotes. Moreover, this area will very shortly become a rapidly developing aspect of plant cell biology and studies are underway to characterize signals involved in the transport and targeting of proteins to all organelles of the plant cell (including the nucleus, peroxisomes and glyoxysomes) [446–454]. The specificity of protein targeting processes is the very basis of maintaining structural and functional integrity of the cell, enabling the various subcellular compartments to carry out their diverse metabolic roles.

Protein stability may also be influenced by subcellular localization; various factors appear to contribute to a protein's half-life in the cell. One of these is the ability of the protein to undergo normal folding and assembly processes and to assume a functional three-dimensional structure; failure to achieve this, results in the activation of the cell's disposal mechanisms to remove nonfunctional proteins or components. A consideration of these factors also has far-reaching implications for applied studies geared toward the genetic engineering of plants for agronomically useful traits or characteristics; in particular, where success will depend upon achieving high levels of accumulation of a foreign protein in a heterologous host. The ultimate challenge here is to design or manipulate polypeptide sequences (e.g., toward such desirable characteristics as enhanced nutritive value or more productive metabolic characteristics) but maintain a functional and stable three-dimensional conformation; this is impeded by our limited knowledge of the rules that govern protein folding and oligomer assembly and how these processes relate to a protein's ultimate stability in the cell. Thus, in addition to a characterization of protein targeting signals, more information is required about the structural features of proteins which allow for their stable accumulation in a particular subcellular compartment.

ACKNOWLEDGEMENTS

Dedicated to the memory of Kay V. Kermode who devoted her life to teaching and helping others.

REFERENCES

1. G. Della-Cioppa, G. M. Kishore, R. N. Beachy, and R. T. Fraley, *Plant Physiol.*, *84*: 965 (1987).
2. G. Blobel, *Proceedings of the National Academy of Sciences (USA)*, *77*: 1496 (1980).
3. K. Verner and G. Shatz, *Science*, *241*: 1307 (1988).
4. A. P. Pugsley, *Protein Targeting*, Academic Press, San Diego, California, p. 23 (1989).
5. S. R. Pfeffer and J. E. Rothman, *Ann. Rev. Biochem.*, *56*: 829 (1987).
6. G. Both, *Today's Life Science* (*Aug*): 12 (1990).
7. M. J. Chrispeels, *Ann. Rev. Plant Physiol. Plant Mol. Biol.*, *42*: 21 (1991).
8. D. R. Cyr and J. D. Bewley, *Planta*, *182*: 370 (1990).
9. M. J. Chrispeels and N. V. Raikhel, *Plant Cell, 3*: 1 (1991).
10. U. Sonnewald, D. Studer, M. Rocha-Sosa, and L. Willmitzer, *Planta, 178*: 176 (1989).
11. S. Wetzel, C. Demmers, and J. S. Greenwood, *Planta, 178*: 275 (1989).
12. K. Baba, M. Ogawa, A. Nagano, H. Kuroda, and K. Sumiya, *Planta, 183*: 462 (1991).
13. P. E. Staswick, *Plant Cell, 2*: 1 (1990).
14. M. Van den Bulcke, G. Bauw, C. Castresana, M. Van Montagu, and J. Vandekerckhove, *Proceedings of the National Academy of Sciences (USA), 86*: 2673 (1989).
15. J. F. Bol, H. J. M. Linthorst, and B. J. C. Cornelissen, *Ann. Rev. Phytopathol.*, *28*: 113 (1990).
16. R. L. Jones and D. G. Robinson, *New Phytol.*, *111*: 567 (1989).
17. G. I. Cassab and J. E. Varner, *Ann. Rev. Plant Physiol. Plant Mol. Biol.*, *39*: 321 (1988).
18. Z.-H. Ye and J. E. Varner, *Plant Cell, 3*: 23 (1991).
19. G. Blobel and D. Dobberstein, *J. Cell Biol.*, *67*: 835 (1975).
20. D. F. Cutler, *J. Cell Sci.*, *91*: 1 (1988).
21. J. K. Rose and R. W. Doms, *Ann. Rev. Cell Biol.*, *4*: 257 (1988).
22. R. D. Klausner, *Cell*, *57*: 703 (1989).
23. H. R. B. Pelham, *Ann. Rev. Cell Biol.*, *5*: 1 (1989).
24. J. E. Rothman, *Cell*, *50*: 521 (1987).
25. R. B. Kelly, *Science*, *230*: 25 (1985).
26. T. L. Burgess and R. B. Kelly, *Ann. Rev. Cell Biol.*, *3*: 243 (1987).
27. C. Dorel, T. A. Voelker, E. M. Herman, and M. J. Chrispeels, *J. Cell Biol.*, *108*: 327 (1989).

28. D. C. Hunt and M. J. Chrispeels, *Plant Physiol.*, *96*: 18 (1991).
29. G. Iturriaga, R. A. Jefferson, and M. V. Bevan, *Plant Cell*, *1*: 381 (1989).
30. J. Deneche, J. Botterman, and R. Deblaere, *Plant Cell*, *2*: 51 (1990).
31. F. T. Wieland, M. L. Gleason, T. A. Serafini, and J. E. Rothman, *Cell*, *50*: 289 (1987).
32. A. Gonzalez-Noriega, J. H. Grubb, V. Talkad, and W. S. Sly, *J. Cell Biol.*, *85*: 839 (1980).
33. A. Hasilik and E. F. Neufeld, *J. Biol. Chem.*, *255*: 4937 (1980).
34. T. H. Stevens, J. H. Rothman, G. S. Payne, and R. Schekman, *J. Cell Biol.*, *102*: 1551 (1986).
35. J. H. Rothman, C. P. Hunter, L. A. Valls, and T. H. Stevens, *Proceedings of the National Academy of Sciences (USA)*, *83*: 3248 (1986).
36. S. Munro and H. R. B. Pelham, *Cell*, *48*: 988 (1987).
37. M. S. Poruchynsky, C. Tyndall, G. W. Both, F. Sato, A. R. Bellamy, and P. A. Atkinson, *J. Cell Biol.*, *101*: 2199 (1985).
38. S. Paabo, B. M. Bhat, W. S. M. Wold, and P. A. Peterson, *Cell*, *50*: 311 (1987).
39. S. M. Hurtley and A. Helenius, *Ann. Rev. Cell Biol.*, *5*: 277 (1989).
40. M.-J. Gething and J. Sambrook, *Biochem. Soc. Symp.*, *55*: 155 (1989).
41. H.-P. H. Moore, B. Gumbiner, and R. B. Kelly, *Nature*, *302*: 434 (1983).
42. D. D. Wagner, T. Mayadas, and V. J. Marder, *J. Cell Biol.*, *102*: 1320 (1986).
43. A. M. Tartakoff, *Cell*, *32*: 1026 (1983).
44. S. Craig and D. J. Goodchild, *Protoplasma*, *122*: 91 (1984).
45. D. J. Bowles, S. E. Marcus, J. C. Pappin, J. B. C. Findlay, E. Eliopoulos, P. R. Maylox, and J. Burgess, *J. Cell Biol.*, *102*: 1284 (1986).
46. M. J. Chrispeels, *Planta*, *158*: 140 (1983).
47. T. A. Wilkins, S. Y. Bednarek, and N. V. Raikhel, *Plant Cell*, *2*: 301 (1990).
48. S. Y. Bednarek and N. V. Raikhel, *Plant Mol. Biol.*, *20*: 133 (1992).
49. S. Kornfeld and I. Mellman, *Ann. Rev. Cell Biol.*, *5*: 483 (1989).
50. M. G. Farquhar, *Ann. Rev. Cell Biol.*, *1*: 447 (1985).
51. A. M. Gebhart and R. W. Ruddon, *Bioessays*, *4*: 213 (1986).
52. G. E. Palade, *Science*, *189*: 347 (1975).
53. K.-N. Chung, P. Walter, G. W. Aponte, and H.-P. H. Moore, *Science*, *243*: 192 (1989).
54. D. T.-H. Ho, *Genome Organization and Expression in Plants* (C. J. Leaver, ed.), Plenum Press, New York, 1980, p. 147 (1980).
55. P. Simon and R. L. Jones, *Eur. J. Cell Biol.*, *47*: 213 (1988).
56. M. J. Buchanan, S. H. Iman, W. A. Eskue, and W. J. Snell, *J. Cell Biol.*, *108*: 199 (1989).
57. G. von Heijne, *J. Mol. Biol.*, *184*: 99 (1985).
58. G. von Heijne, *J. Mol. Biol.*, *189*: 239 (1986).
59. P. Walter, R. Gilmore, and G. Blobel, *Cell*, *38*: 5 (1984).
60. P. Walter and V. R. Lingappa, *Ann. Rev. Cell Biol.*, *2*: 499 (1986).
61. J. E. Rothman, *Nature*, *340*: 433 (1989).
62. S. J. Rothstein, C. M. Lazarus, W. E. Smith, D. C. Baulcombe, and A. A. Gatenby, *Nature*, *308*: 662 (1984).
63. S. J. Rothstein, K. N. Lahners, C. M. Lazarus, D. C. Baulcombe, and A. A. Gatenby, *Gene*, *55*: 353 (1987).

64. J. E. Rothman, *Cell*, *59*: 591 (1989).
65. R. J. Deshaies, B. D. Koch, M. Werner-Washburne, E. A. Craig, and R. Schekman, *Nature*, *332*: 800 (1988).
66. W. J. Chirico, M. G. Waters, and G. Blobel, *Nature*, *332*: 805 (1988).
67. R. J. Deshaies, S. L. Sanders, D. A. Feldheim, and R. Schekman, *Nature*, 349: 806 (1991).
68. J. F. Harper, T. K. Surowy, and M. R. Sussman, *Proceedings of the National Academy of Sciences (USA)*, *86*: 1234 (1989).
69. K. D. Johnson, H. Hofte, and M. J. Chrispeels, *Plant Cell*, *2*: 525 (1990).
70. R. A. Rea, J. Kim, V. Sarafian, R. J. Poole, J. M. Davies, and D. Sanders, *Trends Biochem. Sci.*, *17*: 348 (1992).
71. V. Sarafian, Y. Kim, R. J. Poole, and R. A. Rea, *Proceedings of the National Academy of Sciences (USA)*, *89*: 1775 (1992).
72. H. Hofte and M. J. Chrispeels, *Plant Cell*, *4*: 995 (1992).
73. D. L. Melroy and E. M. Herman, *Planta*, *184*: 113 (1991).
74. H. Sze, J. M. Ward, and L. Shoupeng, *J. Bioenerg. Biomemb.*, *24*: 371 (1992).
75. F. J. M. Maathuis, and D. Sanders, *Curr. Opin. Cell Biol.*, *4*: 661 (1992).
76. E. Martinoia, *Bot. Acta*, *105*: 232 (1992).
77. H. R. B. Pelham, K. G. Hardwick, and M. J. Lewis, *EMBO J.*, *7*: 1757 (1988).
78. R. B. Freedman, *Trends Biochem. Sci.*, *9*: 438 (1984).
79. M. Geetha-Habib, R. Noiva, H. A. Kaplan, and W. J. Lennarz, *Cell*, *54*: 1053 (1988).
80. S. Munro and H. R. B. Pelham, *Cell*, *46*: 291 (1986).
81. U. Tillmann, G. Viola, B. Kayser, G. Siemeister, T. Hesse, K. Palme, M. Lobler, and D. Klambt, *EMBO J.*, *8*: 2463 (1989).
82. N. Inhohara, S. Shimomura, T. Fukui, and M. Futai, *Proceedings of the National Academy of Sciences (USA)*, *86*: 3564 (1989).
83. T. Hesse, J. Feldwisch, D. Balshusemann, G. Bauw, M. Puype, J. Vandekerckhove, M. Lobler, D. Klambt, J. Schell, and K. Palme, *EMBO J.*, *8*: 2453 (1989).
84. D. J. Meyer and A. B. Bennett, *International Society Plant Molecular Biology, Third International Congress*, Abstract 990 (1991).
85. E. B. P. Fontes, B. B. Shank, R. L. Wrobel, S. P. Moose, G. R. O'Brian, E. T. Wurtzel, and R. S. Boston, *Plant Cell*, *3*: 483 (1991).
86. J. Denecke, M. H. S. Goldman, J. Seurinck, and J. Botterman, *Plant Cell*, *3*: 1025 (1991).
87. B. S. Shorrosh and R. A. Dixon, *Proceedings of the National Academy of Sciences (USA)*, *88*: 10941 (1991).
88. H. R. B. Pelham, *EMBO J.*, *7*: 913 (1988).
89. R. A. Mazzarella, M. Srinivasan, S. M. Haugejorden, and M. Green, *J. Biol. Chem.*, *265*: 1094 (1990).
90. A. Ceriotti and A. Colman, *EMBO J.*, *7*: 633 (1988).
91. D. A. Andres, I. M. Dickerson, and J. E. Dixon, *J. Biol. Chem.*, *265*: 5952 (1990).
92. P. Zagouras and J. K. Rose, *J. Cell Biol.*, *109*: 2633 (1989).
93. E. M. Herman, B. W. Tague, L. M. Hoffman, S. E. Kjemtrup, and M. J. Chrispeels, *Planta*, *182*: 305 (1990).
94. D. Vaux, J. Tooze, and S. Fuller, *Nature*, *345*: 495 (1990).

95. C. Wandelt, W. Knibb, H. E. Schroeder, R. I. Khan, D. Spencer, S. Craig, and T. J. V. Higgins, *Plant Molecular Biology* (R. Herrmann, and B. Larkins, eds.) Plenum Press, (in press) (1991).

96. C. I. Wandelt, M. R. I. Khan, S. Craig, H. E. Schroeder, D. Spencer, and T. J. V. Higgins, *Plant J.*, *2*: 181 (1992).

97. J. Denecke, R. De Rycke, and J. Botterman, *EMBO J.*, *11*: 2345 (1992).

98. J. Denecke, B. Ek, M. Caspers, K. M. C. Sinjorgo, and E. T. Palva, *J. Exp. Bot.*, *44*(Suppl.): 213 (1993).

99. N. Dean and H. R. B. Pelham, *EMBO J.*, *9*: 623 (1990).

100. J. C. Semenza, K. G. Hardwick, N. Dean, and H. R. B. Pelham, *Cell*, *61*: 1349 (1990).

101. K. G. Hardwick, M. J. Lewis, J. Semenza, N. Dean, and H. R. B. Pelham, *EMBO J.*, *9*: 623 (1990).

102. M. J. Lewis, D. J. Sweet, and H. R. B. Pelham, *Cell*, *61*: 1359 (1990).

103. J. Lippincott-Swartz, J. G. Donaldson, A. Schweizer, E. G. Berger, H.-P. Hauri, L. C. Yuan, and R. D. Klausner, *Cell*, *60*: 821 (1990).

104. J. Armstrong and G. Warren, *Nature*, *344*: 383 (1990).

105. C. Lee and L. B. Chen, *Cell*, *54*: 37 (1988).

106. A. Nakona, D. Brada, and R. Scheckman, *J. Cell Biol.*, *107*: 851 (1988).

107. S. C. Stirzaker and G. W. Both, *Cell*, *56*: 741 (1989).

108. C. E. Machamer and J. K. Rose, *J. Cell Biol.*, *105*: 1205 (1987).

109. J. Armstrong and S. Patel, *J. Cell Sci.*, *98*: 567 (1991).

110. S. M. Hurtley, *Trends Biochem. Sci.*, *17*: 2 (1992).

111. G. P. Kaushal and A. D. Elbein, *Meth. Enzymol.*, *179*: 452 (1989).

112. L. Faye, K. D. Johnson, A. Sturm, and M. J. Chrispeels, *Physiol. Plant.*, *75*: 309 (1989).

113. A. Sturm, J. A. van Kuik, J. F. G. Vliegenthart, and M. J. Chrispeels, *J. Biol. Chem.*, *262*: 13392 (1987).

114. A. D. Elbein, *Ann. Rev. Plant Physiol.*, *30*: 239 (1979).

115. C. B. Hirschberg and M. D. Snider, *Ann. Rev. Biochem.*, *56*: 63 (1987).

116. A. Sturm, K. D. Johnson, T. Szumilo, A. D. Elbein, and M. J. Chrispeels, *Plant Physiol.*, *85*: 741 (1987).

117. A. Sturm, M. J. Chrispeels, J. M. Wieruszeski, G. Strecker, and J. Montreuil, Glycoconjugates, Proceedings of the 9th International Symposium, *B28*: (1987).

118. L. Faye and M. J. Chrispeels, *Glycoconjugate J.*, *5*: 245 (1988).

119. L. Faye, A. Sturm, R. Bollini, A. Vitale, and M. J. Chrispeels, *Eur. J. Biochem.*, *158*: 655 (1986).

120. E. M. Herman, L. M. Shannon, and M. J. Chrispeels, *Planta*, *165*: 23 (1985).

121. M. A. Mansfield, W. J. Peumans, and N. V. Raihkel, *Planta*, *173*: 482 (1988).

122. H. Shinshi, H. Wenzlur, J. Neuhaus, G. Felix, J. F. Hofsteenge, and J. Meins, *Proceedings of the National Academy of Sciences (USA)*, *85*: 5541 (1988).

123. D. Duksin and W. C. Mahoney, *J. Biol. Chem.*, *257*: 3105 (1982).

124. M. J. Chrispeels, T. J. V. Higgins, S. Craig, and D. Spencer, *J. Cell Biol.*, *93*: 5 (1982).

125. R. Bollini, A. Ceriotti, M. G. Daminati, and A. Vitale, *Physiol. Plant.*, *65*: 15 (1985).

126. U. Sonnewald, A. von Schaewen, and L. Willmitzer, *Plant Cell*, *2*: 345 (1990).
127. T. A. Voelker, E. M. Herman, and M. J. Chrispeels, *Plant Cell*, *1*: 95 (1989).
128. H. Schwaiger, A. Hasilik, K. von Figura, A. Wiemken, and W. Tanner, *Biochem. Biophys. Res. Commun.*, *104*: 950 (1982).
129. A. Driouich, P. Gonnet, M. Makkie, A.-C. Laine, and L. Faye, *Planta*, *180*: 96 (1989).
130. H. Hori and A. D. Elbein, *Plant Physiol.*, *67*: 882 (1981).
131. K. Ravi, C. Hu, P. S. Reddi, and R. B. V. Huystee, *J. Exp. Bot.*, *37*: 1708 (1986).
132. L. Faye and M. J. Chrispeels, *Plant Physiol.*, *89*: 845 (1989).
133. M. M. Bustos, F. A. Kalkan, K. A. Van den Bosch, and T. C. Hall, *Plant Mol. Biol.*, *16*: 381 (1991).
134. C. D. Dickinson, L. A. Floener, G. G. Lilley, and N. C. Nielsen, *Proceedings of the National Academy of Sciences (USA)*, *84*: 5525 (1987).
135. C. D. Dickinson, E. H. A. Hussein, and N. C. Nielsen, *Plant Cell*, *1*: 459 (1989).
136. A. S. Lee, *Trends Biochem. Sci.*, *12*: 20 (1987).
137. Y. Kozutsumi, M. Segal, K. Normington, M.-J. Gething, and J. Sambrook, *Nature*, *332*: 462 (1988).
138. M.-J. Gethings, K. McCammon, and J. Sambrook, *Cell*, *46*: 939 (1986).
139. L. J. Rizzolo, J. Finidori, A. Gonzalez, M. Arpin, I. E. Ivanov, M. Adesnik, and D. D. Sabatini, *J. Cell Biol.*, *101*: 1351 (1985).
140. S. Sharma, L. Rogers, J. Brandsma, M.-J. Gething, and J. Sambrook, *EMBO J.*, *4*: 1479 (1985).
141. C. K. Kassenbrock, P. D. Garcia, P. Walter, and R. B. Kelly, *Nature*, *333*: 90 (1988).
142. H. R. B. Pelham, *Cell*, *46*: 959 (1986).
143. D. G. Bole, L. M. Hendershot, and J. F. Kearny, *J. Cell Biol.*, *102*: 1558 (1986).
144. M. D. Rose, L. M. Misra, and J. P. Vogel, *Cell*, *57*: 1211 (1989).
145. A. Müsch, M. Weidmann, and T. A. Rapoport, *Cell*, *69*: 343 (1992).
146. S. L. Sanders, K. M. Whitfield, J. P. Vogel, M. D. Rose, and R. W. Schekman, *Cell*, *69*: 353 (1992).
147. T. A. Rapoport, *Science*, *258*: 931 (1992).
148. L. D'Amico, B. Valsasina, M. G. Daminati, M. S. Fabbrini, G. Nitti, R. Bollini, A. Ceriotti, and A. Vitale, *Plant J.*, *2*: 443 (1992).
149. R. L. Jones and D. S. Bush, *Plant Physiol.*, *97*: 456 (1991).
150. R. S. Boston, E. B. P. Fontes, B. B. Shank, and R. L. Wrobel, *Plant Cell*, *3*: 497 (1991).
151. R. D. Klausner and R. Sitia, *Cell*, *62*: 611 (1990).
152. J. F. Dice, *Trends Biochem. Sci.*, *15*: 305 (1990).
153. J. Lippincott-Swartz, J. S. Bonifacino, L. C. Yuan, and R. D. Klausner, *Cell*, *54*: 209 (1988).
154. C. Chen, J. S. Bonifacino, L. C. Yuan, and R. D. Klausner, *J. Cell Biol.*, *107*: 2149 (1988).
155. A. Ceriotti, E. Pedrazzini, M. S. Fabrini, M. Zoppe, R. Bollini, and A. Vitale, *Eur. J. Biochem.*, *202*: 959 (1991).

156. R. B. Freedman, N. J. Bulleid, H. C. Hawkins, and J. L. Paver, *Biochem. Soc. Symp.*, *55*: 167 (1989).
157. M. A. Roth and S. B. Pierce, *Biochem.*, *26*: 4179 (1987).
158. N. J. Bulleid and R. B. Freedman, *Nature*, *335*: 649 (1988).
159. M. J. Chrispeels, *Biochem. Biophys. Res. Commun.*, *39*: 732 (1970).
160. M. Andreae, P. Blankenstein, Y.-H. Zhung, and D. G. Robinson, *Eur. J. Cell Biol.*, *47*: 181 (1988).
161. M. Gardiner and M. J. Chrispeels, *Plant Physiol.*, *55*: 536 (1975).
162. M. J. Chrispeels, and B. W. Tague, *Recent Advances in the Development and Germination of Seeds* (R. B. Taylorson, ed.), Plenum Publishing, New York, p. 139 (1990).
163. P. E. Staswick, M. A. Hermodson, and N. C. Nielsen, *J. Biol. Chem.*, *259*: 13431 (1984).
164. C. D. Dickinson, P. M. Scott, E. H. A. Hussein, P. Argos, and N. C. Nielsen, *Plant Cell*, *2*: 403 (1990).
165. J. E. Rothman, *Scientific Amer.*, *253*: 85 (1985).
166. G. Griffiths and K. Simons, *Science*, *234*: 438 (1986).
167. S. A. Tooze and W. B. Huttner, *Cell*, *60*: 837 (1990).
168. R. A. Crowther, J. T. Finch, and B. M. F. Pearse, *J. Mol. Biol.*, *103*: 785 (1976).
169. M. S. Robinson, *J. Cell Biol.*, *204*: 887 (1987).
170. D. G. Robinson and H. Depta, *Ann. Rev. Plant Physiol. Plant Mol. Biol.*, *39*: 53 (1988).
171. F. M. Brodsky, *Science*, *242*: 1396 (1988).
172. B. M. F. Pearce, *EMBO J.*, *6*: 2507 (1987).
173. L. Orci, P. Halban, M. Amherdt, M. Ravazzola, J.-D. Vassalli, and A. Perrelet, *Cell*, *39*: 39 (1984).
174. L. Orci, B. S. Glick, and J.E. Rothman, *Cell*, *46*: 171 (1986).
175. G. Griffiths, S. Pfeiffer, K. Simons, and K. Matlin, *J. Cell Biol.*, *101*: 949 (1985).
176. G. F. Zhang and L. A. Staehelin, *Plant Physiol.*, *99*: 1070 (1992).
177. S. Hillmer, H. Freundt, and D. G. Robinson, *Eur. J. Cell Biol.*, *47*: 206 (1988).
178. D. G. Robinson, K. Balusek, and H. Freundt, *Protoplasma*, *150*: 79 (1989).
179. S. M. Harley and L. Beevers, *Plant Physiol.*, *91*: 674 (1989).
180. B. Hoh, G. Schauermann, and D. G. Robinson, *J. Plant Physiol.*, *138*: 309 (1991).
181. R. D. Record, and L. R. Griffing, *Planta*, *176*: 425 (1988).
182. S. Y. Bednarek, T. A. Wilkins, J. E. Dombrowski, and N. V. Raikhel, *Plant Cell*, *2*: 1145 (1990).
183. G. Hinz, B. Hoh, and D. G. Robinson, *J. Exp. Bot.*, *44*(Suppl.): 351 (1993).
184. E. Fries and J. E. Rothman, *Proceedings of the National Academy of Sciences (USA)*, *77*: 3870 (1980).
185. W. E. Balch, W. G. Dunphy, W. A. Braell, and J. E. Rothman, *Cell*, *39*: 405 (1984).
186. W. E. Balch and J. E. Rothman, *Arch. Biochem. Biophys.*, *240*: 413 (1985).

187. W. G. Dunphy, S. R. Pfeffer, D. O. Clary, B. W. Wattenberg, B. S. Glick, and J. E. Rothman, *Proceedings of the National Academy of Sciences (USA), 83*: 1622 (1986).
188. M. R. Paquet, S. R. Pfeffer, J. D. Burczak, B. S. Glick, and J. E. Rothman, *J. Biol. Chem., 261*: 4367 (1986).
189. L. Hellgren, D. J. Morre, G. Sellden, and A. S. Sandelius, *J. Exp. Bot., 44* (Suppl.): 197 (1993).
190. M. Barinaga, *Science, 260*: 487 (1993).
191. C. J. M. Beckers, D. S. Keller, and W. E. Balch, *Cell, 50*: 253 (1987).
192. K. Simons and H. Virta, *EMBO J., 6*: 2241 (1987).
193. D. Baker, L. Hicke, M. Rexach, M. Schleyer, and R. Schekman, *Cell, 54*: 335 (1988).
194. R. Deshaies and R. Schekman, *J. Cell Biol., 105*: 633 (1987).
195. S. Ferro-Novick, W. Hansen, I. Schauer, and R. Schekmen, *J. Cell Biol., 98*: 44 (1984).
196. P. Novick, C. Field, and R. Schekman, *Cell, 21*: 205 (1980).
197. R. Schekman, *Ann. Rev. Cell Biol., 1*: 115 (1985).
198. R. D. Burgoyne, *Trends Biochem. Sci., 13*: 241 (1988).
199. H. R. Bourne, *Cell, 53*: 669 (1988).
200. N. Segev, J. Mulholland, and D. Botstein, *Cell, 52*: 915 (1988).
201. H. Haubruck, C. Disela, P. Wagner, and D. Gallwitz, *EMBO J., 6*: 4049 (1987).
202. N. Touchot, P. Chardin, and A. Tavitian, *Proceedings of the National Academy of Sciences (USA), 84*: 8210 (1987).
203. K. Palme, T. Diefenthal, M. Vingron, C. Sander, and J. Schell, *Proceedings of the National Academy of Sciences (USA), 89*: 787 (1992).
204. K. Palme, T. Diefenthal, and I. Moore, *J. Exp. Bot., 44*(Suppl.): 183 (1993).
205. B. Goud, A. Salminen, N. C. Walworth, and P. J. Novick, *Cell, 53*: 753 (1988).
206. R. Bowser, H. Müller, B. Govindan, and P. Novick, *J. Cell Biol., 118*: 1041 (1992).
207. M. J. Chrispeels and N. V. Raikhel, *Cell, 68*: 613 (1992).
208. A. Vitale and M. J. Chrispeels, *Bioessays, 14*: 151 (1992).
209. K. Nakamura and K. Matsuoka, *Plant Physiol., 101*: 1 (1993).
210. A. B. Bennett, and K. W. Osteryoung, *Plant Genetic Engineering* (D. Grierson, ed.), Chapman and Hall, New York, p. 199 (1991).
211. A. R. Kermode, *Crit. Rev. Plant Sci., 9*: 155 (1990).
212. K. Muntz, *Biochem. Physiol. Pflanzen, 185*: 315 (1989).
213. N. Harris, *Ann. Rev. Plant Physiol., 37*: 73 (1986).
214. C. R. Lending and B. A. Larkins, *Plant Cell, 1*: 1011 (1989).
215. T. Boller and A. Wiemken, *Ann. Rev. Plant Physiol., 37*: 137 (1986).
216. W. T. Kim, V. R. Franseschi, H. B. Krishnan, and T. W. Okita, *Planta, 176*: 173 (1988).
217. E. M. Herman, C. N. Hankins, and L. M. Shannon, *Plant Physiol., 86*: 1027 (1988).

218. R. Simon, Y. Altschuler, R. Rubin, and G. Galili, *Plant Cell*, *2*: 941 (1990).

219. A. Vitale, A. Sturm, and R. Bollini, *Planta*, *169*: 108 (1986).

220. T. A. Voelker, R. Z. Florkiewicz, and M. J. Chrispeels, *Eur. J. Cell Biol.*, *42*: 218 (1986).

221. M. M. Bustos, V. A. Luckow, L. R. Griffing, M. D. Summers, and T. C. Hall, *Plant Mol. Biol.*, *10*: 475 (1988).

222. B. A. Larkins, K. Pedersen, A. K. Handa, W. J. Hurkman, and L. D. Smith, *Proceedings of the National Academy of Sciences (USA)*, *76*: 6448 (1979).

223. W. J. Hurkman, L. D. Smith, J. Richter, and B. A. Larkins, *J. Cell Biol.*, *89*: 292 (1981).

224. Y. Altschuler, N. Rosenberg, R. Harel, and G. Galili, *Plant Cell*, *5*: 443 (1993).

225. M. Kries, P. R. Shewry, B. G. Ford, J. Ford, and B. J. Miflin, *Oxford Surveys Plant Mol. Cell Biol.*, Vol. 2 (B. J. Miflin, ed.), Oxford University Press, Oxford, England, p. 253 (1985).

226. P. R. Shewry and A. S. Tatham, *Biochem. J.*, *267*: 1 (1990).

227. B. W. Tague and M. J. Chrispeels, *J. Cell Biol.*, *105*: 1971 (1987).

228. C. A. Woolford, L. B. Daniels, F. J. Park, E. W. Jones, J. N. van Arsdell, and M. A. Innis, *Mol. Cell. Biol.*, *6*: 2500 (1986).

229. J. H. Rothman, C. T. Yamashiro, P. M. Kane, and T. H. Stevens, *Trends Biochem. Sci.*, *14*: 347 (1989).

230. L. A. Valls, C. P. Hunter, J. H. Rothman, and T. H. Stevens, *Cell*, *48*: 887 (1987).

231. L. A. Valls, *Ph.D. Dissertation*, University of Oregon, Eugene, Oregon (1988).

232. L. M. Johnson, V. A. Bankaitis, and S. D. Emr, *Cell*, *48*: 875 (1987).

233. L. A. Valls, J. R. Winther, and T. H. Stevens, *J. Cell Biol.*, *111*: 361 (1990).

234. D. J. Klionsky, L. M. Banta, and S. D. Emr, *Mol. Cell Biol.*, *8*: 2105 (1988).

235. C. M. Moehle, C. K. Dixon, and R. B. Jones, *J. Cell Biol.*, *108*: 309 (1989).

236. D. J. Klionsky, and S. D. Emr, *EMBO J.*, *8*: 2241 (1989).

237. C. J. Roberts, G. Pohlig, J. H. Rothman, and T. H. Stevens, *J. Cell Biol.*, *108*: 1363 (1989).

238. V. A. Bankaitis, L. M. Johnson, and S. D. Emr, *Proceedings of the National Academy of Sciences (USA)*, *83*: 9075 (1986).

239. J. H. Rothman and T. H. Stevens, *Cell*, *47*: 1041 (1986).

240. J. S. Robinson, D. J. Klionsky, L. M. Banta, and S. D. Emr, *Mol. Cell Biol.*, *8*: 4936 (1988).

241. P. K. Herman, J. H. Stack, J. A. De Modena, and S. D. Emr, *Cell*, *64*: 425 (1991).

242. S. Felder, K. Miller, G. Moehren, A. Ullrich, J. Schlessinger, and C. R. Hopkins, *Cell*, *61*: 623 (1990).

243. A. M. Honegger, A. Schmidt, A. Ullrich, and J. Schlessinger, *J. Cell Biol.*, *110*: 1541 (1990).

244. P. K. Herman and S. D. Emr, *Mol. Cell Biol.*, *10*: 6742 (1990).

245. L. M. Banta, J. S. Robinson, D. J. Klionsky, and S. D. Emr, *J. Cell Biol.*, *107*: 1369 (1988).

246. J. H. Rothman, I. Howald, and T. H. Stevens, *EMBO J.*, *8*: 2057 (1989).

247. T. A. Vida, T. R. Graham, and S. D. Emr, *J. Cell Biol.*, *111*: 2871 (1990).
248. B. W. Tague and M. J. Chrispeels, *J. Cell Biochem. Suppl.*, *13D*: 230 (1989).
249. B. W. Tague, C. D. Dickinson, and M. J. Chrispeels, *Plant Cell*, 2: 533 (1990).
250. A. Von Schaewen and M. J. Chrispeels, *J. Exp. Bot.*, *44*(Suppl.): 339 (1993).
251. G. Saalbach, R. Jung, G. Kunze, I. Saalbach, K. Alder, and K. Muntz, *Plant Cell*, *3*: 695 (1991).
252. S. Bednarek and N. V. Raikhel, *Plant Cell*, *3*: 1195 (1991).
253. M. R. Schroeder, D. E. Dombrowski, S. Y. Bednarek, O. N. Borkhsenious, and N. V. Raikhel, *J. Exp. Bot.*, *44*(Suppl.): 315 (1993).
254. D. R. Lerner and N. V. Raikhel, *Plant Physiol.*, *91*: 124 (1989).
255. N. V. Raikhel and D. R. Lerner, *Devel. Genet.*, *12*: 255 (1991).
256. T. Hattori, S. Ichihara, and K. Nakamura, *Eur. J. Biochem.*, *166*: 533 (1987).
257. K. Matsuoka and K. Nakamura, *Proceedings of the National Academy of Sciences (USA)*, *88*: 834 (1991).
258. T. A Wilkins and N. V. Raikhel, *Plant Cell*, *1*: 541 (1989).
259. K. Nakamura, K. Matsuoka, F. Mukumoto, and N. Watanabe, *J. Exp. Bot.*, *44*(Suppl.): 331 (1993).
260. T. Boller and U. Vogeli, *Plant Physiol.*, *74*: 442 (1984).
261. F. Mauch and L. A. Staehelin, *Plant Cell*, *1*: 447 (1989).
262. J. F. Bol, J. H. M. Linthhorst, and B. J. C. Cornelissen, *Ann. Rev. Phytopathol.*, *28*: 113 (1990).
263. J. M. Neuhaus, L. Sticher, F. Meins, Jr., and T. Boller, *Proceedings of the National Academy of Sciences (USA)*, *88*: 10362.
264. B. C. Holwerda, N. J. Galvin, T. J. Baranski, and J. C. Rogers, *Plant Cell*, 2: 1091 (1990).
265. S. Koehler and T.-H. D. Ho, *Plant Cell*, 2: 769 (1990).
266. B. C. Holwerda, H. S. Padgett, and J. C. Rogers, *Plant Cell*, *4*: 307 (1992).
267. B. C. Holwerda and J. C. Rogers, *J. Exp. Bot.*, *44*(Suppl.): 321 (1993).
268. P. Eckes, P. Schmitt, W. Daub, and F. Wengenmayer, *Mol. Gen. Genet.*, *217*: 263 (1989).
269. V. A. Hilder, A. M. R. Gatehouse, S. E. Sheerman, R. F. Barker, and D. Boulter, *Nature*, *300*: 160 (1987).
270. A. Hiatt, R. Cafferkey, and K. Bowdish, *Nature*, *342*: 76 (1989).
271. R. Johnson, J. Narvaez, G. An, and C. Ryan, *Proceedings of the National Academy of Sciences (USA)*, *86*: 9871 (1989).
272. L. M. Hoffman, D. D. Donaldson, and E. M. Herman, *Plant Mol. Biol.*, *11*: 717 (1988).
273. M. S. Yang, N. O. Espinoza, P. G. Nagpola, J. H. Dodds, F. F. White, K. L. Schnorr, and J. M. Jaynes, *Plant Sci.*, *64*: 99 (1989).
274. A. R. Kermode, S. Craig, D. Spencer, and T. J. V. Higgins, International Society of Plant Molecular Biology, Third International Congress, Abstract 971 (1991).
275. J. U. Bowie, J. F. Reidhaar-Olson, A. L. Wendell, and R. T. Sauer, *Science*, *247*: 1306 (1990).
276. S. J. Anthony-Cahill, M. C. Griffith, C. J. Noren, D. J. Suich, and P. G. Schultz, *Trends Biochem. Sci.*, *14*: 400 (1989).

277. M. J. Zoller, *Curr. Opin. Biotechnol.*, *3*: 348 (1992).
278. D. Medynski, *Biotechnol.*, *10*: 1002 (1992).
279. J. C. Wallace, G. Galili, E. E. Kuwata, R. E. Cuellar, M. A. Shotwell, and B. A. Larkins, *Science*, *240*: 662 (1988).
280. S. Bagga, D. Sutton, J. D. Kemp, and C. Sengupta-Gopalan, *Plant Mol. Biol.*, *19*: 951 (1992).
281. M. A. Lawton, M. A. Tierney, I. Nakamura, P. Dube, N. Hoffman, R. T. Fraley, and R. N. Beachy, *Plant Mol. Biol.*, *9*: 315 (1987).
282. T. J. V. Higgins and D. Spencer, *Plant Sci.*, *74*: 89 (1991).
283. T. J. V. Higgins, personal communication.
284. I. Coraggio, E. Martegani, C. Compagno, D. Porro, L. Alberghina, L. Bernard, F. Faora, and A. Viotti, *Eur. J. Cell Biol.*, *47*: 165 (1988).
285. H. Hellebust, M. Murby, L. Abrahmsen, M. Uhlen, and S. O. Enfors, *Biotechn.*, *7*: 165 (1989).
286. M. Rechsteiner, S. Rogers, and K. Rote, *Trends Biochem. Sci.*, *12*: 390 (1987).
287. M. Mutter, *Trends Biochem. Sci.*, *13*: 260 (1988).
288. M. Matsumura, G. Signor, and B. W. Matthews, *Nature*, *342*: 291 (1989).
289. S. Mahadevan, J. D. Erfle, and F. D. Sauer, *J. Animal Sci.*, *50*: 723 (1980).
290. A. Hershko, *J. Biol. Chem.*, *263*: 15237 (1988).
291. B. P. Monia, D. J. Ecker, and S. T. Crooke, *Biotechn.*, *8*: 209 (1990).
292. T. R. Butt, S. Jonnalagadda, B. P. Monia, E. J. Sternberg, J. Marsh, J. M. Stadel, D. J. Ecker, and S. T. Crooke, *Proceedings of the National Academy of Sciences (USA)*, *86*: 2540 (1989).
293. E. A. Sabin, C. T. Lee-Ng, J. R. Shuster, and P. J. Barr, *Biotechn.*, *7*: 705 (1989).
294. D. J. Ecker, J. M. Stadel, T. R. Butt, J. A. Marsh, B. P. Monia, D. A. Powers, J. A. Gorman, P. E. Clark, F. Warren, A. Shatzman, and S. T. Crooke, *J. Biol. Chem.*, *264*: 7715 (1989).
295. S. Smeekens, P. Weisbeek, and C. Robinson, *Trends Biochem. Sci.*, *15*: 73 (1990).
296. K. Keegstra, L. J. Olsen, and S. M. Theg, *Ann. Rev. Plant Physiol. Plant Mol. Biol.*, *40*: 471 (1989).
297. C. Robinson, *Plant Genetic Engineering*, (D. Grierson, ed.), Chapman and Hall, New York, p. 179 (1991).
298. P. Weisbeek, J. Hageman, D. De Boer, and S. Smeekens, *Isr. J. Bot.*, *40*: 123 (1991).
299. R. J. Ellis, *Nature*, *328*: 378 (1987).
300. J. Yuan, K. Cline, and S. M. Theg, *Plant Physiol.*, *95*: 1259 (1991).
301. C. Robinson and R. J. Ellis, *Eur. J. Biochem.*, *142*: 337 (1984).
302. C. Robinson and R. J. Ellis, *Eur. J. Biochem.*, *142*: 343 (1984).
303. M. L. Mishkind, S. R. Wessler, and G. W. Schmidt, *J. Cell Biol.*, *100*: 226 (1985).
304. S. Smeekens, H. van Steeg, C. Bauerle, H. Bettenbroek, K. Keegstra, and P. Weisbeek, *Plant Mol. Biol.*, *9*: 377 (1987).
305. G. Van den Broeck, M. P. Timko, A. P. Kausch, A. R. Cashmore, M. Van Montagu, and L. Herrera-Estrella, *Nature*, *313*: 358 (1985).

306. P. H. Schreier, E. A. Seftor, J. Schell, and H. J. Bohnert, *EMBO J.*, *4*: 25 (1985).
307. T. H. Lubben, A. A. Gatenby, P. Ahlquist, and K. Keegstra, *Plant Mol. Biol.*, *12*: 13 (1989).
308. L. Comai, N. Larson-Kelly, J. Kiser, C. J. D. Mau, A. R. Pokalsky, C. K. Skewmaker, K. McBride, A. Jones, and D. M. Stalker, *J. Cell Biol.*, *263*: 15104 (1988).
309. T. H. Lubben and K. Keegstra, *Proceedings of the National Academy of Sciences (USA)*, *83*: 5502 (1986).
310. T. A. Kavanagh, R. A. Jefferson, and M. W. Bevan, *Mol. Gen. Genet.*, *215*: 38 (1988).
311. J. W. Meadows, J. B. Shackleton, A. Hudford, and C. Robinson, *FEBS LETT.*, *253*: 244 (1989).
312. K. Ko and A. R. Cashmore, *EMBO J.*, *8*: 3187 (1989).
313. C. C. Wasmann, B. Reiss, S. G. Bartlett, and H. J. Bohnert, *Mol. Gen. Genet.*, *205*: 446 (1986).
314. G. Della-Cioppa, S. C. Bauer, M. L. Taylor, D. E. Rochester, B. K. Klein, D. M. Shah, R. T. Fraley, and G. M. Kishore, *Biotechn.*, *5*: 579 (1987).
315. G. Von Heijne, J. Stepphun, and R. G. Herrmann, *Eur. J. Biochem.*, *180*: 535 (1989).
316. C. C. Wasmann, B. Reiss, and H. J. Bohnert, *J. Biol. Chem.*, *263*: 617 (1988).
317. B. Reiss, C. C. Wasmann, and H. J. Bohnert, *Mol. Gen. Genet.*, *209*: 116 (1987).
318. B. Reiss, C. C. Wasmann, J. Schell, and H. J. Bohnert, *Proceedings of the National Academy of Sciences (USA)*, *86*: 886 (1989).
319. S. Smeekens, D. Geerts, C. Bauerle, and P. Weisbeek, *Mol. Gen. Genet.*, *216*: 178 (1989).
320. M. Kuntz, A. Simons, J. Schell, and P. H. Schreier, *Mol. Gen. Genet.*, *205*: 454 (1986).
321. D. Roise, F. Theiler, S. J. Horvath, J. M. Tomich, J. H. Richards, D. S. Allison, and G. Shatz, *EMBO J.*, *7*: 649 (1988).
322. G. Von Heijne, *EMBO J.*, *5*: 1335 (1986).
323. G. Von Heijne and K. Nishikawa, *FEBS LETT.*, *278 1*: 1 (1991).
324. J. Hageman, C. Baecke, M. Ebskamp, R. Pilon, S. Smeekens, and P. Weisbeek, *Plant Cell*, *2*: 479 (1990).
325. S. Smeekens, C. Bauerle, J. Hageman, P. Keegstra, and P. Weisbeek, *Cell*, *46*: 365 (1986).
326. B. D. Kohorn and E. M. Tobin, *Plant Cell*, *1*: 159 (1989).
327. G. Van den Broeck, A. Van Houtven, M. Van Montagu, and L. Herrera-Estrella, *Plant Sci.*, *58*: 171 (1988).
328. L. Huang, Z. Adam, and N. E. Hoffman, *Plant Physiol.*, *99*: 247 (1992).
329. P. R. Chitnis, D. T. Morishige, R. Nechushtai, and J. P. Thornber, *Plant Mol. Biol.*, *11*: 95 (1988).
330. S. Smeekens and P. Weisbeek, *Photosyn. Res.*, *16*: 177 (1988).
331. J. Hageman, C. Robinson, S. Smeekens, and P. Weisbeek, *Nature*, *324*: 567 (1986).

332. P. M. Kirwin, P. D. Elderfield, and C. Robinson, *J. Biol. Chem.*, *262*: 16386 (1987).

333. P. M. Kirwin, P. E. Elderfield, R. S. Williams, and C. Robinson, *J. Biol. Chem.*, *263*: 18128 (1988).

334. H. E. James, D. Bartling, J. E. Musgrove, P. M. Kirwin, R. G. Herrmann, and C. Robinson, *J. Biol. Chem.*, *264*: 19573 (1989).

335. D. L. Willey, A. D. Auffret, and J. C. Gray, *Cell*, *36*: 555 (1984).

336. M. Salomon, K. Fischer, U.-I. Flugge, and J. Soll, *Proceedings of the National Academy of Sciences (USA)*, *87*: 5778 (1990).

337. H.-M. Li, T. Moore and K. Keegstra, *Plant Cell*, *3*: 709 (1991).

338. U. I. Flügge, K. Fischer, A. Gross, W. Sebold, F. Lottspeich, and C. Eckerskorn, *EMBO J.*, *8*: 39 (1989).

339. U. Dreses-Werringloer, K. Fischer, E. Wachter, T. A. Link, and U.-I. Flügge, *Eur. J. Biochem.*, *195*: 361 (1991).

340. H.-M. LI, T. D. Sullivan and K. Keegstra, *J. Biol. Chem.*, *267*: 18999 (1992).

341. S. M. Theg, C. Bauerle, L. J. Olsen, B. R. Selman, and K. Keegstra, *J. Biol. Chem.*, *264*: 6730 (1989).

342. L. J. Olsen, S. M. Theg, B. R. Selman, and K. Keegstra, *J. Biol. Chem.*, *264*: 6724 (1989).

343. G. Della-Cioppa and G. M. Kishore, *EMBO J.*, *7*: 1299 (1988).

344. K. Cline, *J. Biol. Chem.*, *261*: 14804 (1986).

345. P. Viitanen, E. Doran, and P. Dunsmuir, *J. Biol. Chem.*, *263*: 15000 (1988).

346. P. M. Kirwin, J. W. Meadows, J. B. Shackleton, J. E. Musgrove, J. E. Elderfield, R. Mould, N. A. Hay, and C. Robinson, *EMBO J.*, *8*: 2251 (1989).

347. K. Cline, M. Werner-Washburne, T. H. Lubben, and K. Keegstra, *J. Biol. Chem.*, *260*: 3691 (1985).

348. A. L. Friedman and K. Keegstra, *Plant Physiol.*, *89*: 993 (1989).

349. J. Pfisterer, P. Lachmann, and K. Kloppstech, *Eur. J. Biochem.*, *126*: 143 (1982).

350. K. L. Cornwell and K. Keegstra, *Plant Physiol.*, *85*: 780 (1987).

351. D. Pain, Y. S. Kanwar, and G. Blobel, *Nature*, *331*: 232 (1988).

352. J. Joyard and R. Douce, *Nature*, *333*: 306 (1988).

353. D. I. Meyer, *Nature*, *347*: 424 (1990).

354. G. Hinz and U. I. Flugge, *Eur. J. Biochem.*, *175*: 649 (1988).

355. R. J. Ellis, *Ann. Rev. Plant Physiol.*, *32*: 111 (1981).

356. S. M. Hemmingsen, C. Woolford, S. M. Van der Vies, K. Tilly, D. T. Dennis, C. P. Georgopoulos, R. W. Hendrix, and J. R. Ellis, *Nature*, *333*: 330 (1988).

357. H. Roy, *Plant Cell*, *1*: 1035 (1989).

358. R. J. Ellis and S. M. van der Vies, *Ann. Rev. Biochem.*, *60*: 321 (1991).

359. O. Fayet, T. Ziegelhoffer, and C. P. Georgopoulos, *Bacteriol.*, *171*: 1379 (1989).

360. M. Zweig and D. J. Cummings, *J. Mol. Biol.*, *80*: 505 (1973).

361. P. Goloubinoff, A. A. Gatenby, and G. H. Lorimer, *Nature*, *337*; 44 (1989).

362. E. S. Bochkareva, N. M. Lissin, and A. S. Girshovich, *Nature*, *336*: 254 (1988).

363. S. Lecker, R. Lill, T. Ziegelhoffer, C. Georgopoulos, P. J. Bassford, C. A. Kumamoto, and W. Wickner, *EMBO J.*, *8*: 2703 (1989).
364. E. Rintamaki, *Plant Physiol. Biochem.*, *29*: 1 (1991).
365. T. H. Lubben, G. K. Donaldson, P. V. Viitanem, and A. A. Gatenby, *Plant Cell*, *1*: 1223 (1989).
366. B. Dujon, *Molecular Biology of the Yeast Saccharomyces: Metabolism and Gene Expression* (J. Strathern, E. Jones, and J. Broach, eds.), Cold Spring Harbor Laboratory, Cold Spring Harbor, New York, p. 505 (1982).
367. G. Attardi and G. Schatz, *Ann. Rev. Cell Biol.*, *4*: 289 (1988).
368. F.-U. Hartl, N. Pfanner, D. W. Nicholson, and W. Neupert, *Biochem. Biophys. Acta*, *988*: 1(1989).
369. M. G. Douglas, M. T. McCammon, and A. Vassarotti, *Microbiol. Rev.*, *50*: 166 (1986).
370. F.-U. Hartl and W. Neupert, *Science*, *247*: 930 (1990).
371. R. J. Ellis and C. Robinson, *Adv. Bot. Res.*, *14*: 1 (1987).
372. B. S. Glick, E. M. Beasley, and G. Schatz, *Trends Biochem. Sci.*, *17*: 453 (1992).
373. R. Zimmermann, U. Paluch, M. Sprinzl, and W. Neupert, *Eur. J. Biochem.*, *99*: 247 (1979).
374. R. Zimmermann, U. Paluch, and W. Neupert, *FEBS LETT. 108*: 141 (1979).
375. A. Viebrock, A. Perz, and W. Sebald, *EMBO J.*, *1*: 565 (1982).
376. J. Kaput, S. Goltz, and G. Blobel, *J. Biol. Chem.*, *257*: 15054 (1982).
377. E. C. Hurt, B. Pesold-Hurt, and G. Schatz, *FEBS LETT.*, *178*: 306 (1984).
378. A. L. Horwich, F. Kalousek, I. Mellman, and L. E. Rosenberg, *EMBO J.*, *4*: 1129 (1985).
379. M. Boutry, F, Nagy, C. Poulsen, K. Aoyagi, and N.-H. Chua, *Nature*, *328*: 340 (1987).
380. J. A. Baum, J. M. Chandlee, and J. G. Scandalios, *Plant Physiol.*, *73*: 31 (1983).
381. J. A. White and J. G. Scandalios, *Biochim. Biophys. Acta.*, *926*: 16 (1987).
382. J. A. White and J. G. Scandalios, *Proceedings of the National Academy of Sciences (USA), 86*: 3534 (1989).
383. U. K. Schmitz and D. M. Lonsdale, *Plant Cell*, *1*: 783 (1989).
384. J. Huang, E. Hack, R. W. Thornburg, and A. M. Myers, *Plant Cell*, *2*: 1249 (1990).
385. D. Roise and G. Shatz, *J. Biol. Chem.*, *263*: 4509 (1988).
386. D. Eisenberg, R. M. Weiss, and T. C. Terwilliger, *Proceedings of the National Academy of Sciences (USA), 81*: 140 (1984).
387. E. T. Kaiser and F. J. Kezdy, *Ann. Rev. Biophys. Chem.*, *16*: 561 (1987).
388. M. Schleyer and W. Neupert, *Cell*, *43*: 339 (1985).
389. N. Pfanner, F.-U. Hartl, B. Guiard, and W. Neupert, *Eur. J. Biochem.*, *169*: 289 (1987).
390. B. Glick, C. Wachter, and G. Schatz, *Trends Cell Biol.*, *1*: 99 (1991).
391. N. Pfanner, J. Rassow, I. van der Klei, and W. Neupert, *Cell*, *68*: 999 (1992).

392. P. Bohni, S. Gasser, C. Leaver, and G. Schatz, *The Organization and Expression of the Mitochondrial Genome* (A. M. Kroon and C. Saccone, eds.), Elsevier, North-Holland, Amsterdam, p. 323 (1980).
393. C. Witte, R. E. Jensen, M. P. Yaffe, and G. Schatz, *EMBO J.*, *7*: 1439 (1988).
394. G. Hawlitschek, H. Schneider, B. Schmidt, M. Tropschug, F.-U. Hartl, and W. Neupert, *Cell*, *53*: 795 (1988).
395. M. Yang, R. E. Jensen, M. P. Yaffe, W. Oppliger, and G. Schatz, *EMBO J.*, *7*: 3857 (1988).
396. R. E. Jensen and M. P. Yaffe, *EMBO J.*, *7*: 3863 (1988).
397. B. Guiard, *EMBO J.*, *4*: 3265 (1985).
398. J. Romisch, M. Tropschug, W. Sebald, and H. Weiss, *Eur. J. Biochem.*, *164*: 111 (1987).
399. I. Sadler, K. Suda, G. Schatz, F. Kaudewitz, and A. Haid, *EMBO J.*, *3*: 2137 (1984).
400. F.-U. Hartl, J. Ostermann, B. Guiard, and W. Neupert, *Cell*, *51*: 1027 (1987).
401. A. P. G. M. Van Loon, A. W. Brandli, B. Pesold-Hurt, D. Blank, and G. Schatz, *EMBO J.*, *6*: 2433 (1987).
402. A. P. G. M. Van Loon and G. Schatz, *EMBO J.*, *6*: 2441 (1987).
403. S. Gasser, A. Ohashi, G. Daum, P. Bohni, J. Gibson, G. Reid, T. Yonetani, and G. Schatz, *Proceedings of the National Academy of Sciences (USA)*, *79*: 267 (1982).
404. M. Teintze, M. Slaughter, H. Weiss, and W. Neupert, *J. Biol. Chem.*, *257*: 10364 (1982).
405. H. P. Braun, M. Emmermann, V. Kruft, and U. K. Schmitz, *Mol. Gen. Genet.*, *231*: 217 (1992).
406. Y. A. Hatefi, *Ann. Rev. Biochem.*, *54*: 1015 (1985).
407. F.-U. Hartl, B. Schmidt, E. Wachter, H. Weiss, and W. Neupert, *Cell*, *47*: 939 (1986).
408. U. Harnisch, H. Weiss, and W. Sebald, *Eur. J. Biochem.*, *149*: 95 (1985).
409. J. D. Beckman, P. O. Ljungdahl, J. L. Lopez, and B. L. Trumpower, *J. Biol. Chem.*, *262*: 8901 (1987).
410. N. Pfanner, P. Hoeben, M. Tropschug, and W. Neupert, *J. Biol. Chem.*, *262*: 14851 (1987).
411. L. Xingquan, K. B. Freeman, and G. C. Shore, *J. Biol. Chem.*, *265*: 1 (1990).
412. D. W. Nicholson and W. Neupert, *Proceedings of the National Academy of Sciences (USA)*, *86*: 4340 (1989).
413. R. A. Stuart, D. W. Nicholson, and W. Neupert, *Cell*, *60*: 31 (1990).
414. M. Schwaiger, V. Herzog, and W. Neupert, *J. Cell Biol.*, *105*: 235 (1987).
415. D. Vestweber, and G. Schatz, *EMBO J.*, *7*: 1147 (1988).
416. J. Rassow, B. Guiard, U. Wienhues, V. Herzog, F.-U. Hartl, and W. Neupert, *J. Cell Biol.*, *109*: 1412 (1989).
417. R. Zimmermann, M. Sagstetter, M. J. Lewis, and H. R. B. Pelham. *EMBO J.*, *7*: 2875 (1988).
418. H. R. B. Pelham, *Nature*, *332*: 776 (1988).
419. H. Murakami, D. Pain, and G. Blobel, *J. Cell Biol.*, *107*: 2051 (1988).

420. S. K. Randall and G. C. Shore, *FEBS LETT.*, *250*: 561 (1989).
421. R. Pfaller and W. Neupert, *EMBO J.*, *6*: 2635 (1987).
422. R. Pfaller, N. Pfanner, and W. Neupert, *J. Biol. Chem.*, *264*: 34 (1989).
423. T. Sollner, G. Griffith, R. Pfaller, N. Pfanner, and W. Neupert, *Cell*, *59*: 1061 (1989).
424. D. Vestweber, J. Brunner, A. Baker, and G. Shatz, *Nature*, *341*: 205 (1989).
425. M. Ohba, and G. Schatz, *EMBO J.*, *6*: 2109 (1987).
426. M. Y. Cheng, F.-U. Hartl, R. A. Pollock, F. Kalousek, W. Neupert, E. M. Hallberg, R. L. Hallberg, and A. L. Horwich, *Nature*, *337*: 620 (1989).
427. J. Ostermann, A. L. Horwich, W. Neupert, and F.-U. Hartl, *Nature*, *341*: 125 (1989).
428. T. W. McMullin and R. L. Hallberg, *Mol. Cell Biol.*, *8*: 371 (1988).
429. D. S. Reading, R. L. Hallberg, and A. M. Myers, *Nature*, *337*: 655 (1989).
430. M. Eilers, W. Oppliger, and G. Schatz, *EMBO J.*, *6*: 1073 (1987).
431. J. E. Franz, *The Herbicide Glyphosate* (E. Grossman, and D. Atkinson, eds.), Butterworths, London, England, p. 3 (1985).
432. H. C. Steinrucken and N. Amrhein, *Biochem. Biophys. Res. Commun.*, *94*: 1207 (1980).
433. E. G. Jaworski, *J. Agric. Food Chem.*, *20*: 1195 (1972).
434. D. M. Shah, R. B. Horsch, H. J. Klee, G. M. Kishore, J. A. Winter, N. E. Tumer, C. M. Hironaka, P. R. Sanders, C. S. Gasser, S. Aykent, N. R. Siegal, S. G. Rogers, and R. T. Fraley, *Science*, *233*: 478 (1986).
435. P. Nagley and R. J. Devenish, *Trends Biochem. Sci.*, *14*: 31 (1989).
436. W. F. J. Vermaas, C. J. Arntzen, L.-Q. Gu, and C. A. Yu, *Biochim. Biophys. Acta*, *723*; 266 (1983).
437. J. Hirshberg and L. McIntosh, *Science*, *222*: 1346 (1984).
438. A. Y. Cheung, L. Bogorad, M. Van Montagu, and J. Schell, *Proceedings of the National Academy of Sciences (USA)*, *85*: 391 (1988).
439. R. J. Ellis and A. A. Gatenby, *Ann. Proc. Phytochem. Soc. Eur.*, *23*: 41 (1984).
440. D. Bradley, S. M. Van der Vies, and A. A. Gatenby, *Philosophical Transactions of the Royal Society of London B*, *313*: 447 (1986).
441. P. Winter and R. G. Herrmann, *Bot. Acta*, *101*: 42 (1987).
442. P. Lagley, L. B. Farrell, D. P. Gearing, D. Nero, S. Meltzer, and R. J. Devenish, *Proceedings of the National Academy of Sciences (USA)*, *85*: 2091 (1988).
443. J. Banroques, A. Delahodde, and C. Jacq, *Cell*, *46*: 837 (1986).
444. C. J. Howe, *Trends Genet.*, 4: 150 (1988).
445. D. Vestweber and G. Schatz, *Nature*, *338*: 170 (1989).
446. S. J. Gould, G.-A. Keller, M. Schneider, S. H. Howell, L. J. Garrard, J. M. Goodman, B. Distel, H. Tabak, and S. Subramani, *EMBO J.*, *9*: 85 (1990).
447. M. A. Restrepo-Hartwig and J. C. Carrington, *J. Virol.*, *66*: 5662 (1992).
448. M. A. Restrepo, D. D. Freed, and J. C. Carrington, *Plant Cell*, *2*: 987 (1990).
449. B. Tinland, Z. Koukolikova-Nicola, M. N. Hall, and B. Hohn, *Proceedings of the National Academy of Sciences (USA)*, *89*: 7442 (1992).

450. M. J. Varagona, R. J. Schmitz, and N. V. Raihkel, *Plant Cell*, *4*: 1213 (1992).
451. L. J. Olsen and J. J. Harada, International Society of Plant Molecular Biology, Third International Congress, Abstract 973 (1991).
452. A. R. van der Krol and N.-H. Chua, *Plant Cell*, *3*: 667 (1991).
453. H. Kindl, *Cell Biochem. Function*, *10*: 153 (1992).
454. G. A. Keller, S. Krisans, S. J. Gould, J. M. Sommer, C. C. Wang, W. Schliebs, W. Kunau, S. Brody, and S. Subramani, *J. Cell Biol.*, *114*: 893 (1991).

11

The Induction of Gene Expression in Response to Pathogenic Microbes

David B. Collinge, Per L. Gregersen, and Hans Thordal-Christensen

The Royal Veterinary and Agricultural University
Frederiksberg, Denmark

INTRODUCTION

Plants in nature are subject to constant attack by plant pathogenic microbes. Nevertheless, only a very limited number of these succeed in infecting the plant. This lack of success is caused by, in part, plant defense, which can be of either structural or chemical nature. Plants possess both defense systems which are established permanently, i.e., in place before the pathogen attacks [1,2,3], and those which are triggered by the presence of the pathogen. The former type is largely neglected due to the difficulty in providing critical proof of a role in defense. The latter type of defense is associated with physiological changes which can be observed as a hypersensitive response, alterations to the cell wall, as well as the production of novel metabolites and proteins [4,5]. It can also be visualized as "induced resistance," where a plant pretreated with a pathogen shows increased resistance to subsequent infection attempts [4,6,7,8]. The first event in the inducible defenses is recognition of the pathogen by the plant, or rather of elicitors produced either by or through the action of the pathogen [2,4,5,6,9]. Through supposed signal transduction pathways, this recognition activates a set of structural plant genes, called "defense response genes," "defense-related genes," or "resistance response genes" (or various related terms). The products of these genes, directly or indirectly, form the barrier which arrests the pathogen [2,9]. Studies of defense response genes are leading to an understanding of the mechanisms underlying resistance and are paving the way for practical application of such genes in disease control. In this review, we focus on the inducible biochemical mechanisms underlying resistance and the limited knowledge of the means by which plants regulate these responses to pathogens, and we discuss the main strategies used to obtain this knowledge. We conclude this chapter with a brief discussion of the perspectives for using the knowledge gained.

Inducible defense mechanisms are important against both nonhost pathogens and pathogens [4,5]. The concept "nonhost pathogen" results from coevolution of plants and microbes for aeons, and thus particular pathogen species are often specialized to particular plant taxa. Seen from the point of view of a particular pathogen, there are "hosts" and "nonhosts." Within the interaction of a pathogen and its host, the plant can possess specific resistance genes (R-genes), and the pathogen possess corresponding avirulence genes. The latter encode production of specific elicitors which are fortuitously recognized by the R-gene product [2,10,11]. A specific elicitor from this avirulent pathogen race will trigger resistance when the plant possesses a corresponding R-gene. Such an interaction is termed *incompatible*. In addition to interactions involving R-genes, a certain level of resistance induction can be observed in interactions between susceptible plants and virulent races. Although this is not normally enough to entirely prevent infection, nevertheless the result can be that not all infection attempts in compatible interactions succeed [12]. Likewise, differences in the rate or extent of disease development can be observed in, e.g., the interactions of a specific isolate with different host varieties or genotypes [13].

In order to study resistance, it is necessary to understand the infection strategies used by microbial pathogens. Three types of pathogens are recognized: biotrophic, hemibiotrophic, and necrotrophic [14,15]. Biotrophic pathogens (or "confidence men") live in a type of equilibrium with living host tissue, e.g., the obligate powdery mildew (Ascomycete) and rust (Basidiomycete) fungi. Necrotrophs (or "thugs") kill the tissue they infect and live on the dead tissue. These include bacteria, e.g., *Erwinia carotovora* and Ascomycetes such as *Botrytis cinerea*. Hemibiotrophs are intermediate, starting biotrophically, and ending necrotrophically. Examples include bacteria, e.g., *Xanthomonas* spp. and Oomycetes such as *Phytophthora* spp. Pathogenicity factors are defined as functions necessary for successful infection of the host, but not growth in vitro. Different types of pathogenicity factor, encoded by specific pathogenicity genes, are used under the different infection strategies [2,16,17]. Thus toxins and large amounts of hydrolytic enzymes such as pectinases are used by necrotrophs, whereas the supposedly more sophisticated mechanisms used by biotrophs are less understood.

THE NATURE OF INDUCIBLE RESISTANCE

Inducible resistance is exhibited in various ways, and involves a complex set of plant processes manifested after the arrival of the pathogen. Furthermore, resistance can be locally or systemically induced, often termed systemically (SAR) or locally acquired resistance [7,18,19]. The physiological processes can be considered at various levels: at the cytological level observed through the microscope as a hypersensitive response or changes in cell wall structure, or at the biochemical and molecular levels, e.g., as induced accumulation of particular proteins or their mRNAs. The induced accumulation of biosynthetic enzymes can result in the production of toxic secondary metabolites whereas other host proteins arrest pathogen development directly, e.g., by cell wall fortification. In addition, there are a few examples of pathogen-induced activation of preexisting proteins, e.g., callose synthase [20]. There is now a large body of data describing the nature of gene expression induced by pathogen attack, and an increasing body of data concerning the tissue distribution of their transcripts and products relative to the site of inoculation [e.g., 21]. However, there is relatively little evidence supporting a direct role for the individual components induced during the resistance response: most data describes correlations. Indeed, it is now clear that there is a general response within a species comprising many components of which a number are presumably ineffective against particular pathogens: e.g., it is hard to imagine a direct role for chitinase in inhibiting viral development (see pg. 402). Furthermore, for technical cloning reasons, there is a distinct bias toward the study of those responses for which activation is dependent on the accumulation of specific transcript in inoculated tissues, i.e., studies involving differential screening of cDNA libraries give results more rapidly than those involving mutagenesis.

We have tried to emphasize the critical evidence and approaches taken rather than merely catalog the responses observed in different interactions. A number of general reviews have appeared which describe the response of plants to pathogen attack [2,4,5,9,17,22–29].

The Hypersensitive Response

At the cytological level, the most striking mechanism in resistance is the hypersensitive response (HR), in which host cells coming in direct contact with the microbial pathogen undergo a programed cell death with the consequence that the pathogen also dies [13,30–33]. Little is known about the physiological mechanism underlying HR, and indeed, the phenomenon is defined differently by different authors, with the result that the literature can be difficult to interpret [4,32]. The consensus view is that HR involves irreversible membrane damage associated with necrosis localized to very few cells in direct contact with the invading pathogen. On the basis of observed early disturbances in ionic balance across the plasma membrane at the onset of the HR, measured as electrolyte leakage, it has been proposed that alterations in membrane structure occur [34]. HR is associated with an increase in enzymatic or nonenzymatic lipid peroxidation [32,35–37]. The latter, however, is perhaps unlikely due to the absence of the predicted autocatalytic products in bean plants undergoing HR [38]. Furthermore, increases in lipoxygenase activity occurred in several host pathogen interactions [39] and following methyl jasmonate treatment [40]. Slusarenko et al. [32] proposed that lipoxygenases oxidize certain free fatty acids released by enzymes such as lipid acyl hydrolases. The peroxidation also releases oxygen radicals, which, in contrast to lipoxygenases, can oxidize membrane bound lipids. Thus, it has been predicted that the oxygen radicals released cause an autocatalytic peroxidation of membrane lipids, initiating a chain reaction which irreversibly damages the integrity of the membrane. The fact that the application of free radical scavangers such as glutathione and ascorbic acid delays the onset of HR is consistent with a role for the observed increases in oxygen radicals in the development of HR [35,36,41–45]. As the reaction is potentially a chain reaction involving the production of oxygen radicals which act both in catalysis and possibly as signals (see below), a mechanism to contain the phenomenon is required. It has been proposed that superoxide dismutase (SOD) and/or peroxidase might act as protectants against such a chain reaction [46]. Indeed, there is evidence of SOD induction as part of the HR and other forms of stress [47]. In contrast, whereas an oxidative burst occurred within 10 minutes of elicitation of HR in clover leaves and tobacco cells, pre- or cotreatment of the eliciting cells with exogenous catalase or SOD quenched the oxidative burst without affecting the development of HR or subsequent phytoalexin accumulation [41].

An interesting feature of HR is that the induction of other defense responses occurs in cells outside the HR area. Thus in incompatible interactions of *Phytophthora infestans* with potato and of *P. megasperma* with parsley, transcripts associated with phenylpropanoid phytoalexin production accumulate in the adjacent zone [21,48,49]. It has been suggested that O_2^- radicals released may act as signal molecules for, e.g., phytoalexin biosynthesis [41]. Furthermore, it has been proposed that fatty acid residues released by lipid acyl hydrolases might activate specific signal transduction pathways, perhaps through the production of jasmonate. In contrast, methyl jasmonate has been shown to stimulate the production of lipoxygenase [39, 40]. A major problem for such studies is that, on an organ basis, the process of cell death occurs concomitantly with the induction of other responses such as the production of phytoalexins (pg. 400) and PR-proteins (pg. 402).

The main biological interest of HR lies in its effectiveness: most of the other identifiable cytological changes are also observed in compatible inter-actions, albeit later and to a lesser extent (e.g., 48,50]. Viable pathogen ma-terial cannot be extracted or spread from tissue which has undergone HR, with the exceptions of certain interactions with viruses [51] and *P. infestans* [49]. The mechanism underlying the effectiveness of HR is unknown. It has generally been considered that the production of phytoalexins in adjacent cells was responsible (see [4] and pg. 400). However, the pathogen appar-ently dies within the HR lesion, which suggests another mechanism. *Trans*-2-hexanal, which is highly toxic to *Pseudomonas syringae* pv. *phaseolicola* and various fungi, is produced in the bean plant as a product of lipid peroxi-dation during HR against *P. s.* pv. *phaseolicola*. Furthermore, it is pro-duced at levels expected to be toxic in vivo at a stage prior to the accumula-tion of the phytoalexin when the growth rate of the bacteria decreases [38]. This is highly indicative of a suitable lethal mechanism of HR, at least in this interaction.

Cell Wall Modifications

Alterations in host secondary cell wall structures are among the cytologi-cally most striking effects of pathogen attack. These can take the form of generally thickened and chemically modified walls, papillae, and tyloses. Hemispherical papillae and surrounding haloes are often formed at the in-ner side of the cell wall at sites of penetration attempts by fungal pathogens. Papillae occur widely in the plant kingdom, but are best described from the Gramineae [3]. Papillae are often effective barriers against direct penetra-tion, and their study is largely neglected at the molecular level. Papillae and haloes, in common with other types of secondary cell wall structure, are be-lieved to contain a variety of fortifying components including callose (β-1,3-glucans), polyphenolic compounds resembling lignins and suberin, proteins and silicon in addition to components typical of primary and secondary walls [3,12]. A major problem of studying the composition of secondary walls is that the components appear to be polymerized as heterogeneous polymers; and it is believed that peroxidases are at least partly responsible for this (see [52] and pp. 396 and 406).

It is not absolutely clear whether papillae actually work as a barrier to penetration. Nevertheless, their effectiveness is indicated in interactions be-tween barley and *Erysiphe graminis* where the fungus is prevented from penetrating as the papillae are formed, even in compatible interactions [53]. In several cases, a correlation has been found between papilla size and their resistance to penetration [12]. For example, papilla produced during the second (challenger) inoculation of barley were larger as a consequence of a pre-treatment with an inducer inoculum [54]. Furthermore, larger papillae, which were less frequently penetrated, were observed in plants possessing the *ml-o* gene [55].

Tyloses are another type of cell wall modification involved in defense. Tyloses are carbohydrate-filled protrusions of cell walls particularly into vascular tissue, where they are believed to inhibit pathogen spread. Tyloses are also associated with the development of secondary cell walls and other induced defenses [14].

Lignin and Suberin

Lignin and suberin are rather diffuse terms describing heterogeneous phenyl-propanoid polymers. Lignins are polymers of phenylpropanoids, whereas suberins are polymers of lignin monomers which also contain aliphatic fatty acid residues [56]. The aromatic lignin and suberin monomers are produced intracellularly from products of the phenylpropanoid, arginine and fatty acid pathways [56,57]. In addition to enzymes of the core phenylpropanoid pathway (see Figure 1 and p. 401), genes encoding several enzymes believed to be involved specifically in the production of the phenylpropanoid lignin and suberin monomers (but not the aliphatic monomers) have been cloned and characterized. These include caffeic acid 3-O-methyltransferase (OMT) from several species [58–60], and cinnamyl-alcohol dehydrogenase (CAD) from tobacco [61]. The identity of a putative CAD sequence from bean has been questioned recently [61–64]: the putative CAD sequence was shown to exhibit extensive sequence identity to malic enzyme from maize [64]. Genes encoding a third enzyme, cinnamoyl-CoA reductase, have not been isolated.

Virtually nothing is known of the mechanism of export from the cell to the site of polymerization, although it is thought that the monolignols are glycosylated prior to transport [57].

Peroxidases are believed to be responsible for the extracellular poly-merization of the monomers to give lignin and suberin, as well as to other cell wall components such as polysaccharides [see e.g., 65,66] and possibly cell wall proteins (p. 398). Lignification of wounded tobacco tissue is more rapid in plants expressing the gene encoding a peroxidase known to utilize lignin monomers [67,68], and peroxidase activity staining develops concom-itantly with lignification in bamboo [69]. A potato peroxidase transcript accumulates during suberinization of wounded tuber tissue [70], and a re-lated transcript is induced earlier following wounding in tomato fruits and elicitor-treated suspension cultures resistant to *Verticilium albo-atrium* than in susceptible tissues [71]. This tomato gene, and a related gene, are induced in transgenic tobacco by wounding and during a hypersensitive response to *Fusarium solani* f. sp. *pisi* [72]. Furthermore, it has been reported that over-expression of this gene in transgenic tobacco gave substantial protection

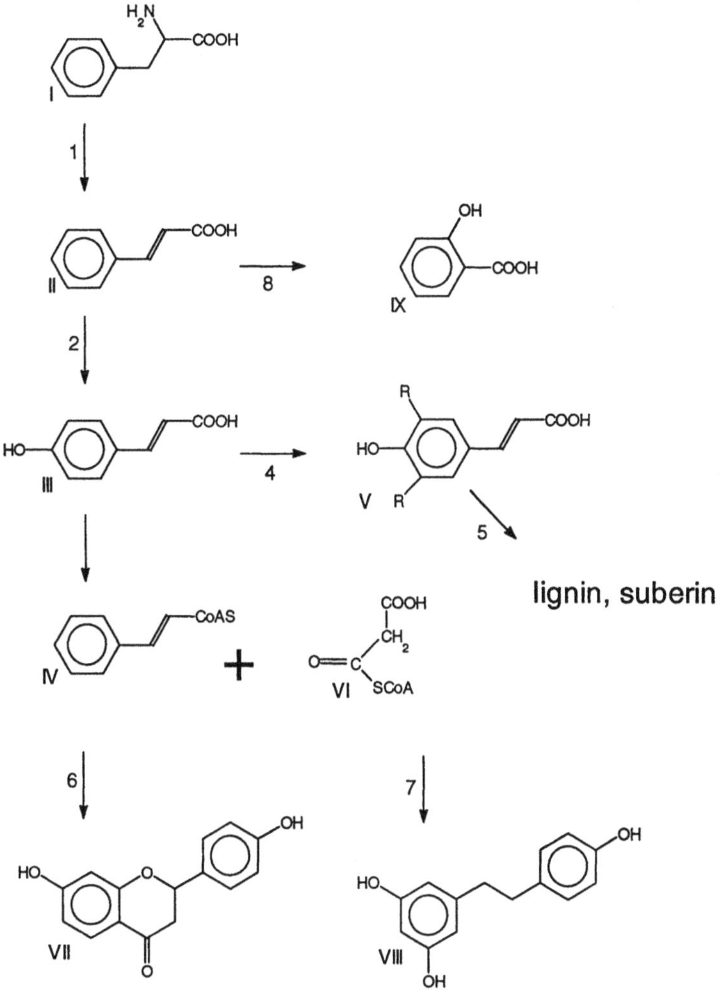

Figure 1 The central phenylpropanoid biosynthetic pathway. Compounds: Roman numerals – I. phenylalanine; II. *trans*-cinnamic acid; III. *p*-coumaric acid (= 4-hydroxycinnamic acid); IV. cinnamyl CoA; V. lignin precursors where R = H, OH or CH$_2$O; VI. malonyl-CoA; VII. naringenin; VIII. resveratrol (3,4′,5-trihydroxy-stilbene); IX. salicyclic acid. Enzymes: Arabic numerals. 1. phenylalanine ammonia lyase (PAL); 2. cinnamate-4 hydrolase (C4H); 3. 4-coumarate:CoA ligase (4CL); 5. cinnamate 3 and 5 hydroxylase, cinnamate *O*-methyl transferase (COMT), followed by polymerization by peroxidase; 6. flavonoid biosynthesis: chalcone synthase (CHS) followed by chalcone isomerase (CHI). See [118] for pathway from naringenin to isoflavonoid phytoalexin biosynthesis. Naringenin is also an intermediate in anthocyanin biosynthesis; 7. stilbene synthase (SS); 8. precise pathway undetermined [see 6].

against *Peronospora parasitica* [73]. Pathogen-induced peroxidase accumulation is discussed on p. 406.

Thionins

Thionins are cysteine-rich proteins best characterized from barley where they are known as seed storage proteins, leaf cell wall proteins and vacuolar proteins [74–76]. At least 50 genes are present which encode very similar c. 15 kDa precursors which are processed to 5 kDa mature cysteine-rich polypeptides [74,75]. Transcripts corresponding to leaf cell wall specific thionins accumulate in response to many external stimuli including infection with the powdery mildew fungus [77], heavy metals [78], jasmonic acid [79], and in etiolated seedlings [78]. However, the response to the powdery mildew fungus [77] is very slow in comparison to the accumulation to other transcripts [80]. The 5 kDa cysteine-rich pathogen-induced proteins of several dicots appear to resemble thionins (p. 406), and a thionin sequence has been obtained recently from tobacco flowers [81]. However, no serological cross-reaction was obtained with extracellular tomato proteins using antiserum raised against a barley prethionin [78]. Both leaf cell wall and soluble vacuolar thionins are toxic to fungi in in vitro tests [77,82], and, indeed, thionins were originally discovered in crude extracts of wheat seeds as being highly toxic to yeast [83,84]. It is thought that the toxicity is related to their amphiatic structure causing membrane disruption [74,75]. Interestingly, weak binding of antisera raised against barley leaf thionins was detected by immunogold labelling of wheat rachis and stems at penetration sites of *Fusarium culmorum* [85].

HRGP, PRP and GRP

Of several structural proteins present in cell walls, certain HRGP (= hydroxyproline-rich glycoprotein or extensin), proline-rich proteins (PRP), and glycine-rich proteins (GRP) accumulate in plants in response to pathogens and wounding [86–92]. The 50–60 kDa HRGPs are highly glycosylated (as much as 90% sugar: arabinose and galactose residues in the ratio 1:2) and contain high proportions of proline, lysine, and tyrosine residues [91]. It is believed that both glycosylation, and hydroxylation of the proline to give hydroxyproline, occur in the endoplasmic reticulum prior to export [93,93a]. HRGP and GRP (and possibly other developmentally regulated cell wall proteins) are thought to be cross-linked to lignin in the cell wall by peroxidase [94]. Immunocytological techniques demonstrated HRGP accumulation in cell walls of tomato infected with *Fusarium oxysporum* f.sp. *radicis-lycopersici* [95,96]. In melon and bean, accumulation occurred in the walls of cells adjacent to those exhibiting HR, as well as in response to a saprophyte [97]. Of several classes of HRGP and GRP sequences of tomato, only some accumulated following wounding, whereas levels of others decreased [98,99]. GRP transcripts accumulate in tobacco following TMV infection (and salicylate treatment [100,101]). Transcripts for both GRP and HRGP accumulate following elicitor treatment and pathogen infection, but at a later stage than for transcripts associated with phytoalexin accumulation [88,102]. HRGPs are minor constituents of monocot cell walls, and, in contrast to dicots, are encoded by single genes rather than by small gene families [103]. Overexpression of tobacco GRP did not result in increased resistance to TMV of transgenic tobacco plants [104]. The conclusion, in the absence of critical proof, is that HRGP and GRP probably have a role in physically containing pathogens, and prevent them from spreading from necrotic HR lesions.

Polysaccharides

Polysaccharides are components of both primary and secondary cell walls [105]. These include cellulose, callose, xyloglucans, and pectins (arabinogalactans and arabinans). Increases in activities of two enzymes (and/or the corresponding transcripts) associated with carbohydrate metabolism have been detected as part of the response to pathogen attack. Thus a sucrose synthase transcript accumulated weakly in barley inoculated with *Erysiphe graminis* [8,106], and increased levels of invertase (β-fructosidase) activity have been observed for carrot infected with the powderly mildew fungus, *E. heraclei* [107,108]. Invertase proteins accumulated in papillae, intercellular spaces, and the cell walls of tomato roots infected by *F. oxysporum* f.sp. *racidis-lycopersici* [109]. It is therefore plausible that these alterations in enzyme activity or transcript amount reflect a role in cell wall polysaccharide deposition, but it is also possible that they have a more general role associated with increased metabolic activity associated with the defense response [8,106,110].

Callose is proposed to function as a structural barrier to infection. Callose comprises β-1,3-glucans, whereas cellulose comprises β-1,4-glucan, both formed from UDP-glucose. It is generally believed that callose synthase and cellulose synthase are the same enzyme complex, comprising at least four and possibly five subunits [111,112]. This complex exhibits callose synthase activity at millimolar Ca^{2+} concentrations and cellulose synthase activity at micromolar Ca^{2+} concentrations [see 113,114]. Whereas callose deposition can be induced by elicitor treatment in several species, it is believed that this primarily occurs through the effects of changing Ca^{2+} levels in synergy with, e.g., β-furfuryl-β-glucoside, polyol and polyamine on preformed enzyme, and not be de novo transcription [20,115,116].

Secondary Metabolism

Histological staining techniques reveal the accumulation of phenolic compounds in many plants following infection with pathogens. These include the complex and ill-defined phenolic metabolites variously described as suberin, lignin, ligninlike compounds (see p. 396), tannins and melanins, in addition to the structurally discrete antimicrobial phytoalexins. Some of these compounds are associated with papillae or the primary cell wall, others seem to be more widely distributed in the reacting tissue. Phytoalexins are produced in many interactions, often in association with the hypersensitive response. They are soluble antimicrobial compounds representing a wide range of secondary metabolites from steroids to products of the phenylpropanoid pathway [117–121]. The term *phytoalexin*, though useful, needs to be used with care. The antimicrobial activity of many forms is not investigated. There is also the question as to the level of toxicity required of a compound for it to be considered antimicrobial either in vitro or in planta. Other antimicrobial metabolites are produced constitutively or are developmentally regulated, but their study is largely neglected [2,15,33,122,123]. Exceptions include the alkaloid tomatine, for which genetically controlled variation in the sensitivity of *Fusarium solani* to the alkaloid correlates with pathogenicity [124], and *cis,cis*-1-acetoxy-2-hydroxy-4-oxo-heneicosa-12,15-diene of avocado, the lipoxygenase mediated inactivation of which can result in triggering of a latent infection by *Colletotrichum gloesporioides* during ripening [125,126].

In situ hybridization often facilitates the localization of transcripts at the organ level, and has now been applied to several phytoalexin producing systems, i.e., potato [48,49] and parsley [21]. Less data has been obtained using immunocytological techniques, and that nearly exclusively concerns the immunogold localization of proteins at the cellular level [e.g., 95,107, 127], although some studies consider protein localization at the organ level [128]. Very little is known of the distribution of antimicrobial compounds at the cellular level [27]. In spite of the fact that antisera can be raised against many other and smaller molecules and parts of molecules, only limited cytological data is available on phytoalexin localization [e.g., 129,130]. The limited localization data available primarily concerns in situ localization of transcripts corresponding to the biosynthetic enzymes (see above), which is not necessarily the site of utilization or final localization of the product. Furthermore, the value of fluorescence microscopy and cytohistochemical techniques are often limited since they detect an active group rather than a whole structure, although in combination such techniques can be invaluable in the analysis of interactions. Various other techniques such as HPLC and laser microprobe mass spectral analysis has been applied in order to localize phytoalexins [27]. From the results available [27], it seems that lethal concentrations of several phytoalexins are strictly localized to the HR reacting cells and the immediate surrounding cells, although examples of more dispersed accumulation exists.

Phenylpropanoid Metabolism

Of the many types of secondary compounds, phenylpropanoid derivatives have received the most attention at the molecular level. The core phenylpropanoid pathway is presented in Figure 1. Various classes of defense-related compounds, the isoflavonoid, furanocoumarin, and related phytoalexins, the stilbene phytoalexins and lignin, as well as the signal molecule salicyclic acid (see p. 415) are all synthesized through this pathway, as are anthocyanin pigments and other compounds of unknown function. Genes encoding many of the enzymes involved in both the core phenylpropanoid and some of the distal branch pathways have been cloned from several species, and much is known about their structure and regulation. These studies have been exhaustively reviewed elsewhere [see e.g., 9,120,122,131]. There are three good pieces of critical evidence supporting a role for this pathway in defense: (1) α-aminooxyphenylpropionic acid, a specific inhibitor of the first enzyme, phenylalanine ammonia lyase (PAL), inhibits glyceollin production in soybean and makes the plants susceptible to avirulent races of *P. megasperma* f.sp. *glycinea* [132]; (2) phytoalexin detoxification [133, 134] and (3) expression of a vine stilbene synthase in tobacco [135] (see p. 402).

Some of the enzymes involved have multiple forms which can exhibit different regulation patterns [5]. Thus several species possess at least three PAL genes, of which some forms are induced by elicitors of resistance responses, whereas others are involved in normal plant development [136–138]. Similarly, of three chalcone synthase (CHS) genes characterized from soybean, only one transcript species accumulated following elicitor or UV treatment of cotyledons [139]. Likewise, french bean possesses at least seven CHS genes [138]. Promotors from two of these were fused to *uid*A (which encodes β-glucuronidase - GUS from *E. coli* [140]) and used to transform tobacco. The two showed differential regulation: whereas both were induced by $HgCl_2$ and UV light, only one was induced by oxalate and *Pseudomonas syringae*, during the induction of systemic induced resistance [138]. Transcriptional activation of response genes is treated in more detail on p. 417.

Only few studies have considered the effects of pathogens on the production of substrates for the phenylpropanoid pathway. There is, as a consequence, little data at the molecular level on these related pathways, although single enzymes have been cloned [141]. 3-hydroxy-D-arabino-heptosonate-7-phosphate synthase of *Arabidopsis thaliana*, the first enzyme in the shikimate pathway, is induced following infection by phytopathogenic bacteria or wounding of leaves [142]. The shikimate pathway is responsible for the production of phenylalanine and tyrosine, the immediate precursors of phenylpropanoid biosynthesis.

Detoxification of Phytoalexins

The fact that a compound extracted from plant material is able to kill or inhibit growth of pathogens in in vitro culture is only indicative data for the importance of such a compound in resistance. However, there are several examples of phytoalexin detoxification mechanisms possessed by fungal pathogens, of which the best characterized concerns pisatin demethylase of *Nectria haematocca* [134]. Indeed, it has been demonstrated that transfer of the key detoxification mechanism from a pea pathogen to a maize pathogen enabled the latter to form albeit limited lesions on pea [133]. This study is the best critical evidence for the importance of a phytoalexin in natural defense, and it demonstrates that the possession of a particular phytoalexin can limit the number of pathogens of a host. The adaptation of a particular pathogen to the specific defense mechanisms of its host also illustrates coevolution of host and pathogen. It has therefore been suggested that transfer of key enzymes from one species to another might allow the production of novel phytoalexins against which the natural pathogens of the recipient have no means of inactivating [17]. Indeed, transfer of stilbene synthase from vine to tobacco resulted in the synthesis of the stilbene phytoalexin resveratrol in transgenic tobacco plants with reduced susceptibility to *Botrytis cinerea* [135].

The production of the enzymes of phenylpropanoid biosynthesis in certain tissues remains unexplained and points to unsuspected roles [5]. One such role for flavonoids is as signals for the establishment of symbiosis with *Rhizobium* in legumes [143,144]. This is reflected by CHS expression in nodules [145], where the distribution of transcript during development does not indicate a role in defense. In contrast, other phenolic compounds, such as acetylsyringone, act as signals to phytopathogenic *Agrobacterium tumefaciens* [see 144].

Pathogenesis-Related (PR) Proteins, and Other Defense-Related Proteins

The term PR proteins covers a heterogeneous group of proteins induced in many plant species during infection. As PR proteins can represent up to 10% of the protein content of TMV infected tobacco leaves [146], and since their presence correlates with induced resistance against all types of pathogens, they have been subject to considerable attention in recent years. The original definition considered PR proteins to be extracellular and of extreme pI [see 147-149]. In this respect, PR-proteins are distinct from the structural cell wall proteins and enzymes of secondary metabolism (see pp. 395 and 402). On the basis of serological relationships and sequence data, essentially five families of PR proteins have been recognised with reference to tobacco, the

best characterized species. Under the simplest and currently approved nomenclature, these are designated PR-1 to 5 [149]. Members of these families
are apparently present in many species [147,150–152], including monocots
(e.g., 153–155], and all five are encoded by small gene families which include
both acidic and basic forms. Biochemical activities have been proven for
PR-2 (β-1,3-glucanase) and PR-3 (chitinase), but not the other families. As
it is now known that intracellular forms exist for many PR proteins, the
definition also encompasses related forms. In addition to the induced forms,
developmentally regulated forms are now known to exist for several PR
proteins: it has been proposed that these should be termed PR-like proteins
[149]. Some extend the definition to include any protein induced by pathogen
attack. Thus pathogen-induced, but not constitutive, peroxidases or HRGPs
are sometimes considered to be PR proteins. We will use the term *major PR
proteins* as a collective term for the five families originally described in tobacco, thus including both acidic and basic forms. Other protein and transcript families are being identified in a number of species, and to add to the
confusion, other species may have their own nomenclature: for example
PR-1 in potato [156] and parsley [157] are entirely unrelated to each other
or to the tobacco PR-1 family (see p. 411).

The structure and regulation of the major PR proteins have been exhaustively reviewed recently [6,25,26,147,150–152,158–160] and will be treated
briefly here. It is assumed that PR protein expression is transcriptionally
regulated, and this is indeed supported by run-off transcription studies with
bean chitinase [161] and wheat PR-5 [162,163]. Thus the accumulation of
PR protein transcript has been studied intensively. For example, in tobacco,
the regulation of acidic and basic PR proteins differs in terms of their induction by different chemicals [see 6,147,163,164]. Constitutive expression
of some PR proteins is known from certain tissues such as flower parts, roots,
and embryos which are particularly exposed to pathogen attack. Thus acidic
PR-1 protein of tobacco accumulated in leaves and is constitutively present
in sepals [165], whereas transcript encoding an inducible basic form is constitutively present in roots [166]. Similarly, expression of a basic maize PR-1
gene is constitutive in embryos and induced in excised germinating embryos
by infection [167]. However, it has not been determined whether these differently expressed PR-1 messengers are transcribed from the same genes or
from closely related genes.

The general view is that the chief role of PR proteins is in resistance,
although other roles cannot be excluded, particularly for those tissues where
the genes are constitutively expressed. Indeed, other roles have been proposed for PR-2 and PR-3, i.e., chitinases and β-1,3-glucanases [168].

PR-1

PR-1 proteins are c. 15 kDa proteins known from many species [169,170]. In tobacco, three closely related acidic forms have been purified, and a single basic form is known to exist. Whereas no biochemical function has been identified, transgenic tobacco plants expressing a tobacco PR-1 gene under the control of the 35S promotor of CaMV were found to be less susceptible to infection by *Peronospora tabacina*, although the same and similar plants were unaffected in their susceptibility to TMV and insect attack [104,171]. In vitro inhibition using a tomato PR-1 protein has been demonstrated against *P. infestans* [172]. In spite of the fact that proteins of the dominant PR-1 family have been known for more than two decades [152], and are believed to be present in most plant species, only a limited number of cDNA or gene sequences have been published: three represent acidic PR-1s and one represents a basic PR-1 from tobacco [6,150], two represent basic PR-1s from tomato [173], two represent apparently relatively neutral PR-1s from *Arabidopsis* [174,175], one represents a basic PR-1 from maize [176] and two represent basic PR-1s from barley [80].

PR-2 and PR-3

PR-2 (PR-N, PR-O), and PR-3 (PR-P, PR-Q) proteins possess β-1,3-glucanase and endochitinase activity, respectively [177,178]. These proteins occur widely in the plant kingdom [179], e.g., at least five distinct forms of β-1,3-glucanase are present in tobacco [177,180]. At least 50 cDNA and gene sequences from about 20 plant species representing 4 classes of endochitinase have been obtained [168,181] in addition to a combined chitinase/lysozyme [182], and chitosanases [183]. In combination with endochitinase, β-1,3-glucanases can inhibit the growth of several Ascomycetes (but not Oomycetes) species in in vitro tests since the cell wall of these fungal groups contains the substrates for these enzymes [168,181,184]. Transgenic plants expressing various classes of chitinases constitutively under the control of the strong 35S promotor of CaMV were more resistant to some pathogens [6, 185] but not to others [104,186]. However, the published studies are not strictly comparable since different chitinases and pathogens are used. Abolition of the natural level of a vacuolar β-1,3-glucanase by antisense transformation in tobacco did not affect the plant nor did it reduce its resistance to *Cercospora nicotianae*, which indicates that this β-1,3-glucanase is not essential for the plant [187].

In addition to the induced chitinases and β-1,3-glucanases present in plants, some are constitutively present in specific organs such as certain flower parts and roots [168]. It has been suggested that their presence in developing tissues reflects a role in organogenesis [188], and indeed there is evidence for potential endogenous substrates for chitinases in plants [189]. It can also be argued that their presence in tissues which are poorly protected by pre-formed physical barriers reflects a greater need for alternative defense mech-anisms. As β-1,3-glucans act as elicitors of phytoalexin biosynthesis, β-1,3-glucanase has been implicated in the development of the defense response [190,191,192]. Similarly, chitosan, a derivative of chitin, possesses elicitor activity in several systems [193].

PR-4

PR-4 proteins are extracellular 13 to 14.5 kDa proteins which were originally characterized in tobacco as TMV induced 13 or 14.5 kDa proteins and related wound-induced proteins [150]. Gene sequences have been obtained from to-bacco [194,195], tomato [195], potato [196], and barley [80]. An interesting structural feature of the PR-4 protein from potato tuber [196] is a cysteine rich chitin-binding domain, similar to the lectin domain of class I and class IV chitinases and the wound-induced lectin, hevein [197]. This domain is not present in PR-4 proteins of tobacco, tomato, or barley [194,195,198], and in this respect, the structures of different PR-3 and PR-4 proteins are analogous. Recently, the PR-4 protein of barley has been shown to inhibit growth in vitro of the Ascomycete *Trichoderma harzianum* [198]. Further-more, similar proteins [198] and transcripts [80] accumulate in barley leaves following infection with the powdery mildew fungus. No biochemical func-tion of PR-4 proteins has been identified.

PR-5

Genes encoding sequences corresponding to the 21 to 26 kDa PR-5 (PR-R, PR-S), acidic extracellular thaumatinlike or basic vacuolar osmotin proteins have been described from a number of species, e.g., *Thaumatococcus dan-iellii* [199], *Arabidopsis* [175], tobacco [200], potato [201], soybean [202], barley [80,203], maize [204], and wheat [162]. Osmotin of tobacco [205] possesses antifungal activity in vitro to the Oomycete *P. infestans*. Barley seed PR-5 inhibited the growth of the Ascomycetes *Candida albicans, F. oxysporum,* and *Trichoderma viridae* in in vitro tests, an effect enhanced in the presence of seed chitinase but not ribosome inhibiting protein (RIP) [206]. Similarly, the maize seed PR-5 inhibited the growth of *C. albicans, Neurospora crassa,* and *T. reesei*, and caused cell lysis in *C. albicans* and *N. crassa* [207].

It has been suggested that this group of proteins act as enzyme inhibitors as they possess some sequence similarity to a probable α-amylase/protease inhibitor (PAPI) of maize [208], which seems to be virtually identical to zeamatin, a maize antifungal seed protein, which lacks inhibitor activity [209]. Although barley leaf [153,203] and barley seed [206] PR-5 sequences show some similarity to a seed PAPI, no inhibitory activity has been detected [206]. Structurally unrelated proteinase inhibitors have been found to be pathogen-induced in tomato leaves [210].

Peroxidase

The possible role of peroxidases in lignification, suberinization, cross-linking of HRGP in secondary cell walls, and as possible scavengers of peroxides produced during HR have been discussed on pp. 394, 396, and 398 [see also 23,24]. Native electrophoresis gels stained for peroxidase activity reveal many different isozymes in plants, e.g., 23 isozymes have been demonstrated in barley [211]. Due to glycosylation and other posttranscriptional modifications, it is considered unlikely that all isoforms observed represent separate gene products [212,213].

Some forms, especially extracellular forms, accumulate following infection with pathogens [See 23,24,214,215]. There is firm evidence for seven different peroxidase genes in barley [216]. Transcripts for two of these, which accumulate following inoculation with *E. graminis*, encode peroxidases of less than 40% amino acid identity [217]. The observed transcript accumulation patterns show differences in timing, which is believed to reflect different roles in defense response. In addition, expression of one of these sequences is associated with papilla formation in response to the fungal germ tubes. Peroxidase transcript accumulation has also been observed for tobacco inoculated with TMV and *Fusarium solani* f.sp. *pisi* [72,218], wheat inoculated with the barley powdery mildew fungus [163] and elicitor-treated cell cultures of *Arabidopsis thaliana* and parsley [219].

Other Antifungal Proteins

In addition to the PR proteins, several plant proteins have been isolated recently which possess antimicrobial activity. It is not always clear whether these are induced following infection with pathogens. Two classes have been isolated from radish seeds [220], the first class comprises two 5 kDa highly basic proteins with potent antifungal activity. These resemble γ thionins, certain α-amylase inhibitors and two pea proteins which are induced by fungal attack [221]. The SAR8.2 proteins of tobacco [222] are cysteine rich, in common with this class, but do not otherwise share homology. The SAR8.2 protein is also highly fungitoxic, and induced later during SAR by both TMV and SA than the major PR-proteins.

The second class of radish seed protein, 2S seed albumins, show sequence identity to non-specific lipid transfer proteins, and also inhibit the growth of both phytopathogenic fungi and bacteria, an effect which is strongly antagonized by cations [223].

It has also been proposed that lectins have a role in defense in addition to being domains within other resistance response proteins such as class I chitinases [23,224]. Most of the evidence concerning a defense role is related to protection against herbivory, however, it seems likely that at least chitin-binding lectins, such as hevein, have a role in binding hyphae acting syn-geristically with other antifungal proteins. There is no evidence of pathogen-induced lectin gene expression, although hevein is induced by wounding [197].

Export and Organelle Targeting

Many components of the induced defenses are extracellular, either as apo-plastic soluble components or as part of a modified cell wall. Thus the cell needs to transport these components across the plasma membrane. Whereas little is known of the mechanisms of export of e.g., lignin monomers to the site of polymerization (see p. 396), rather more is known of the mechanisms of protein export and intracellular targeting [26,225]. In particular, it has been shown that many PR proteins [150] and peroxidases [217] possess leader sequences, and many PR proteins and peroxidases are indeed extracellular. However, the original characterization of PR proteins as being extracellular has proven too strict. Within several of the five major PR protein families in tobacco, the basic forms seem to be vacuolar. These vacuolar proteins have C-terminal extensions, relative to the acidic forms, which are believed to be responsible for targeting to the vacuole. For example, in addition to a hydrophobic N-terminal leader sequence, the C-terminal region of the basic β-1,3-glucanase of tobacco is first glycosylated and subsequently cleaved during vacuolar targeting [180,226]. Transformation studies, utilizing various chitinases genes with or without a C-terminal extension from a basic tobacco chitinase, demonstrated that vacuolar targeting depended on the presence of this extension [227].

It is likely that at least some peroxidase activity is required extracellu-larly for polymerization of phenolic compounds (see p. 396), and indeed, there are examples of peroxidases which, like certain major PR proteins [150], seem to be exported [215], and this is reflected by the presence of leader sequences in a number of the induced species [70,217,228].

The export and compartmentalization (e.g., to the vacuole) of proteins occurs through the endoplasmic reticulum, a process which involves specific groups of proteins [225]. It has been proposed that molecular chaperones are required in this process, and these include endoplasmic forms of certain large heat shock proteins [26,225,229,230]. HvGRP94 is a barley member of the HSP90 heat shock protein family, which, on the basis of the C-terminal extension "KDEL," is believed to be retained in the lumen of the endoplas-matic reticulum [231]. Indeed the GRP94 of tobacco has been isolated from microsomal fractions [232]. We suggest that HvGRP94 acts as chaperone for proteins bound for export, such as PR proteins, since the HvGRP94 transcript accummulates during the defense response.

STRATEGIES FOR STUDYING INDUCED RESPONSES

Two strategies are being used to study induced defense mechanisms; these have been dubbed the shotgun and targeted approaches [4]. The targeted approach relies on available physiological or cytological clues, or even guess-work to determine the course of study. Thus genes encoding enzymes of the phenylpropanoid biosynthetic pathway for production of certain phyto-alexins [9,120] and PR-proteins, e.g., chitinases [168] have been cloned using this approach. The approach is, however, intrinsically conservative since the choice of subject investigated is made on the basis of prior knowledge. Ths shotgun approaches for protein studies (i.e., the study of PR proteins, see p. 402) [147,148,150] and cDNA cloning allow the isolation of new in-duced components of response mechanisms without the conservative bias to known components. The rationale being that induced components are likely to participate in defense. This approach is increasingly fruitful now that sequence data bases are sufficiently large to indicate an identification of the majority of the sequences obtained. Indeed, a number of diverse hitherto unsuspected components of the defense response have been identified, in-cluding, e.g., alcohol dehydrogenase [233], glycine rich proteins [101], two heat shock proteins [156,231], and the thaumatinlike proteins [200], as well as to a number of novel sequence families for which no biochemical func-tion has been attributed (see p. 411). The two approaches are complemen-tary, and their combination for studying particular interactions has led to the identification of a number of physiologically independent processes in-duced during defense. Nevertheless, it is clear that components in defense remain to be discovered, and once it has been determined that the expres-sion of a gene discovered is induced during the induction of resistance, there can be a long way before a role in resistance is established. Thus despite their discovery in 1970, no biochemical function was attributed to a PR-protein until 1987 [177].

Model Systems

The information obtained arises from study of many different experimental systems of variable biological integrity. It is biologically most meaningful to study an intact interaction of a pathogen with its host. However, resistance is affected by developmental stage and environmental factors such as light, temperature, and nutrients [14]. The environment therefore needs to be under control, and greenhouse or growth chamber grown plants will differ physio-logically from plants grown in the field. The latter are themselves physiologi-cally variable to a high degree. Furthermore, in vitro studies of a response in which accumulation of a specific protein or mRNA is involved requires a synchronized response in a fairly large amount of the host tissue in order to extract sufficient material. This can be achieved when the pathogen develops in synchrony as, e.g., bacteria injected as a suspension. Such strategies are often more difficult with fungi [234], or those viruses which are dependent on insect vectors. This can be illustrated from tests with pathogens where the influence of specific factors on the outcome of an interaction is dependent on the assay used to study the interaction. For example, whereas strains of *Xanthomonas campestris* pv. *campestris* mutated in specific putative pathogenicity genes (e.g., protease), unlike the wild type *X.c.c.*, were unable to invade mature turnip leaves from the margin [235–237]. The same mutants appeared unaffected in pathogenicity toward seedlings.

Elicitors are components of diverse chemical structures derived from pathogens. Elicitors mimic the pathogen by inducing the resistance response in the host, either as a whole or in part (see p. 412 and [2,17,193,238]). Where elicitors are used to induce a response in the whole plant, the response can be compared to that induced by the intact pathogen, but, like any in vivo treatment of plants with, e.g., inhibitors or metabolic intermediates, the distribution and effect of the elicitor in the plant tissue can hardly be the same as for the pathogen.

A third level of study is to take cell suspension cultures and treat these with elicitors. Although such model systems bear limited resemblance to the intact interaction, results obtained by this type of study have facilitated studies with intact systems not least by providing tools, e.g., cDNA probes and specific antisera [e.g., 9,239]. Advantages of such model systems include the immediacy of the response and the possibility of uncoupling different parts of the response: the response of the plant to the pathogen may be dependent on a prior response of the pathogen to the presence of the host in order to stimulate attack. Thus cutinase, believed to be necessary for penetration of leaves by *Fusarium solanum*, is induced only in the presence of cutin monomers [240]. The analysis of the symbiotic interaction between *Rhizobium* and legumes shows that nodule development depends on complex mutual signaling where the perception of the host by the pathogen preceeds perception of the pathogen by the host [143,241,242].

Assay for Induced Gene Expression Using In Vitro Translation

The shotgun approach for studying induced gene expression utilizes cDNA libraries to obtain clones representing transcripts which accumulate in tissue responding to the pathogen (or elicitor). As plants respond rapidly and transiently to external stimuli by de novo expression of specific genes, it is necessary to take great care when choosing the starting material from which a cDNA library is prepared. Two-dimensional polyacrylamide gel electrophoretic analysis (2D-PAGE) of in vitro translation products obtained using mRNA extracted from a particular tissue gives a view of the overall pattern of gene expression in that tissue. By comparing the pattern obtained using mRNA samples harvested from plants at different time points after inoculation with a pathogen [24,243–249] or treatment with an elicitor [250, 251], it is possible to choose the best time point for making a cDNA library fulfilling the requirements. It is vital to incorporate adequate controls, e.g., mock inoculation, in such studies, as gene expression is affected by many external factors including diurnal rhythm, light, wounding, and fluctuations in temperature. For example, bean [247] and turnip [243] gene expression is greatly affected by the infusion of buffer or culture medium into mature leaves. The technique has two important limitations. First, in vitro translation does not measure gene expression, but the translatability of mRNA. Although translatable mRNA reflects altered gene expression patterns in practice, differences can arise through alterations in processing of mRNA. Secondly, it is very difficult to compare results obtained from different laboratories as precise reproducibility is difficult to achieve.

Shotgun Screening of cDNA Libraries

Two techniques have been used to identify clones in cDNA libraries which correspond to sequences that accumulate specifically as a response to pathogen attack. In differential hybridization, two separate probes (usually radioactively labeled first strand cDNA) are prepared from mRNA isolated from inoculated and control (uninoculated) tissues. These are used to screen replicate filters carrying cDNA clones from the library. Clones corresponding to mRNAs accumulating specifically in response to the pathogen give hybridization signal to the probe prepared from inoculated but not from control tissues (or only weakly), since the latter probe contains low amounts or none of these sequences. This technique has been used to identify plant cDNAs associated with the development of HR in the interaction between tobacco and *Pseudomonas solanacearum* [252,253], tobacco mosaic virus [100], salicyclic acid [6], and cytokinin-treated cell cultures [254], yielding many of the same sequences. Interactions involving other species include CaMV in turnip [255], *F. solani* f. sp. *phaseolicola* on pea [256], *E. graminis* on barley [80,217,257] and wheat [163], *P. infestans* on potato leaves [156], potato tubers treated with the elicitor arachidonic acid [258], and parsley cell suspensions treated with *P. megasperma* derived elicitor [259,260].

Differential hybridization, although relatively easy to perform, is not sufficiently sensitive to detect cDNAs corresponding to rare mRNAs, or to differentiate among poorly induced species [4]. Probe enrichment by subtractive hybridization provides an increase in screening sensitivity of perhaps an order of magnitude. Radioactively labelled cDNA probes prepared from mRNA extracted from inoculated plants is hybridized to control mRNA. The hybrid sequences are removed, leaving a probe solution enriched for sequences corresponding to transcripts induced in the response [261]. This method is technically more demanding, and requires the use of very hot radioactive probes to obtain sufficient sensitivity. Subtractive hybridization has led to the identification of cDNA clones in barley induced by *Erysiphe graminis* which would probably not have been found otherwise [80,231, 262]. The chief limitation with this approach lies in its inherent ability to detect only accumulating transcripts. Indeed there are several examples of physiological changes which are believed to be at least in part independent of de novo gene expression, e.g., callose formation [5].

It is also possible to use a shotgun approach to study responses at the protein level. Proteins can be purified using noninoculated controls for comparison. The purified proteins can be used both to raise antisera or for microsequencing. Antisera can be used to screen expression cDNA libraries; examples being PR proteins of tomato including a β-1,3-glucanase [173] and PR-1 of tobacco [263]. Alternatively, the microsequences can be used to design oligonucleotides which can be used either for PCR amplification from cDNA, e.g., chitinase of rape [264] and PR-4 of tobacco and tomato [195], or for direct screening of gene libraries, e.g., the antiviral protein from *Phytolacca americana* [265].

Shotgun studies have the fascinating tendency of producing unexpected results, as the following list of defense response-related sequences shows:

Alcohol dehydrogenase of potato [235].
14-3-3 protein—a putative protein kinase regulator or phospholipase A_2 of barley [262] (see p. 415).
GRP94—a HSP90 protein of barley [231] (see p. 407).
Glutathione-S-transferases of wheat [266].
Sucrose synthase of barley [8,24] (p. 399).
"PR-1" of potato—a homologue of HSP26 of soybean [156].

Other sequences identified bear no resemblance to published sequences, but homologues have nevertheless been identified independently in several species of plants. For example, homologues to "PR-1" from parsley [157, 259] have been isolated from potato [267,268], pea [269,270], asparagus [271,272], and birch [273]. The elicitor-induced transcript ELI3 of parsley cell cultures has been used to isolate a related sequence from *Arabidopsis thaliana*. These sequences are predicted to encode 45 kDa proteins [274].

Other transcript species have been cloned from a single species. These include pRP1 of barley [275] for which a related transcript also accumulates in wheat following inoculation of fungal pathogens [276]. Similarly, the

WIR1 transcript, which has been cloned from wheat inoculated with *Erysiphe graminis*, was also identified in barley [277]. "PvPR3" from bean [278] represents a new 15 kDa putative cytosolic protein present as a gene family (15 sequences).

THE MOLECULAR MECHANISMS UNDERLYING DEFENSE RESPONE GENE INDUCTION

In order to initiate a response, a host needs to perceive the presence of the pathogen. It is believed that the host takes advantage of the fact that the pathogen presents molecules or elicitors to it, which are then sensed by hypothetical receptors. The activated receptors are believed to be the first step in signal transduction processes, leading to defense gene induction and/or protein/enzyme activation. The current widely accepted concept for signal transduction in relation to the perception of pathogens by plants, summarized in Figure 2, has been discussed widely in the recent years [see 2,5,193,238,279–283].

Elicitors

A particular substance produced by the pathogen acts fortuitously as an elicitor, i.e., it is perceived by the plant with the result that the complex defense response described above is induced. Elicitors are a wide range of substances ranging from small molecules such as oligosaccharides and fatty acids to proteins and polysaccharides [193,238,280]. The ability of a molecule to function as an elicitor is not an intrinsic property of the molecule as such but is determined by the ability of the host plant to recognize the molecule in question, e.g., a cell wall glycoprotein of *P. megasperma* f.sp. *glycinea* acts as an elicitor of phytoalexin production in soybean and parsley. Pure peptide fragments of this protein possess elicitor activity in parsley but not soybean, whereas pure glucan from the same protein acts as an elicitor in soybean, but not parsley [284]. Pathogen-produced hydrolytic enzymes can also have elicitorlike activity as they liberate monomers of plant origin from cell wall polymers. These monomers subsequently act as endogenous elicitors. Indeed, it has been suggested that the production of β-1,3-glucanase (p. 405) results in the release of elicitors, i.e., β-1,3-glucans from fungal pathogens, reinforcing the response [190–192,280]. Elicitors can be specific or non-specific. Specific elicitors are characterized by being active only on certain plant genotypes (see p. 392). An example of a nonspecific elicitor is a *Colletotricum lindemuthianum* produced heptaglucan active on soybean. Only one specific structure of many possible heptaglucans has this elicitor activity [285]. Furthermore, certain elicitors act synergistically [286,287],

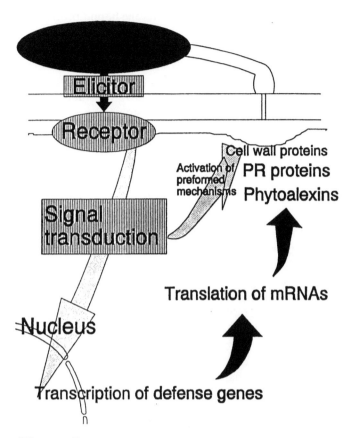

Figure 2   A simple model for induced gene expression using a plasma membrane receptor and secondary messengers. See [281,283] for more sophisticated models.

and others may only be responsible for activating a certain part of the response [5,17].

Receptors

The elicitor is believed to be perceived by a receptor, and for at least some elicitors inducing phytoalexin biosynthesis, there is evidence that their receptors are membrane-found [280,285,288,289]. Thus the soybean receptor of the heptaglucan from *Colletotricum lindemuthianum* described above has been studied using radiolabeled analogs of the elicitor [290,291]. Receptor-active membrane fractions have been isolated and partially characterized revealing binding of the elicitor to a 70 kDa membrane bound protein [288, 292].

It should be noted that many animal receptors for hormones use cytoplasmatic receptors, although there is currently no evidence for this in plants. Studies involving cell cultures and purified elicitors imply that the perception mechanism is constitutively present. Thus de novo resistance response gene expression can be detected in bean cell suspension cultures within five minutes of elicitor treatment [161], and the response to the elicitor appears to mimic the intact interaction qualitatively. However, in the absence of biochemical characterization of any receptor, conclusive proof that a receptor is indeed constitutively present is lacking. Indeed, it is conceivable that the production of such a receptor is induced.

Race Specificity

A fascinating aspect of plant resistance concerns genetically controlled variation in which certain isolates or races of a particular pathogen are only capable of attacking specific genotypes of their host species. In general, specific dominant resistance genes (R-genes) in the host are matched by specific dominant avirulence genes in the pathogen which leads to a unsuccessful infection attempt [10,11,293]. It is widely believed that this specificity reflects the ability of the host to recognize a specific elicitor produced by the pathogen [281]. A number of avirulence genes, primarily from phytopathogenic bacteria, have been characterized [2,10]. Furthermore, in three cases, avirulence genes have been shown to control the production of specific elicitors: the *P. syringae* pv. *tomato avr*D gene controls the production of a low molecular weight metabolite elicitor [294,295], the *Cladosporium fulvum avr*9 gene encodes a peptide elicitor [296,297], and an *avr*-gene of TMV encodes the coat protein [298]. The NIP 1 peptide of *Rhyncosporium secalis* also appears to represent an *avr*-gene product [203,299]. Although plasma membrane bound receptors for some nonspecific elicitors have been identified (see above), no receptors for specific elicitors have been identified, with the possible exception of the putative victorin receptor [300]. As plants apparently possess many different specific resistance genes (see [301] for examples of barley), it has been pointed out that the R-genes need not be involved in the primary detection as a receptor, as is current dogma, but encode some other component [279]. Several groups are apparently close to cloning specific R-genes by chromosome walking from linked RFLP markers [302] or transposon tagging [303].

Signal Transduction

The details of plant signal transduction processes are largely unknown. A number of components of signal transduction systems known from animals and lower eukaryotes have been identified in plants [304,305]. These include protein kinases [e.g., 306,307,308], G-proteins [309], phospholipases [310] and the enigmatic 14-3-3 protein [262,311,312], some of which possess protein kinase C inhibitor and phospholipase A_2 activities [313] and as well as apparently participating in transcription complexes [312]. Indications of the participation of a few of these components in signaling of the defense response has been presented. For example, the transcript of the barley 14-3-3 protein accumulates during the early stages of the response to the powdery mildew fungus [262]. Evidence that protein kinase/phosphatases are involved in the response to elicitors is presented as changes in protein phosphorylation pattern [314,315] and as abolition of recognized defense responses by the use of protein kinase inhibitors [316]. Furthermore, increases in inositol triphosphate levels preceded phytoalexin accumulation in cultured carrot cells [317]. Ca^{2+} is a potential signal transduction component for defense responses, and, with the calcium-binding protein calmodulin, has been implicated in the response of parsley to elicitors [289]. Plant cells have significant extracellular as well as intracellular stores of Ca^{2+} [304], and a Ca^{2+}-binding protein kinase has been found [307]. Finally, it has been suggested that activated oxygen species have a role in signal transduction in at least some interactions [45]: this should not be confused with the physiological role that activated oxygen species might play in HR (see p. 394).

Much work needs to be done in this area in order to clarify the nature of the signaling mechanism which is predicted to be constitutively present. Several current studies are using mutation of *Arabidopsis thaliana* to study signal transduction mechanisms. For example, a gene fusion using a modified rice basic chitinase promotor with alcohol dehydrogenase as the selective marker has been used to transform *A. thaliana*, and homozygous transformants are being selfed, mutagenized, and the selfed progeny will be screened for susceptibility to allyl alcohol following elicitor treatment [318].

Various compounds specific to plants are considered to be involved in short- and long-distance signaling of the defense response. These include salicyclic acid and ethylene (see below) as well as jasmonic acid [319–321], free radicals [41], and even electric signals [322]. Thus resistance cannot only be induced at the site of the infection attempt, but also systemically, in a phenomenon called systemic induced resistance or systemic acquired resistance (SAR).

Salicyclic Acid

Salicyclic acid (SA) and its role in plant disease resistance has recently been reviewed [323,324]. SA concentrations have been observed to increase systemically in the phloem sap of cucumber following inoculation with *Colletotrichum lagenarium* [325] and *Pseudomonas syringae* pv. *syringae* [326] as well as in tobacco leaves [327] and phloem exudates [328] after inoculation of a single leaf with TMV. Concomitant with the increases in SA levels, SAR was observed, and it was therefore suggested that SA is the signal of the SAR, possibly in the glucosylated form [329]. SA and acetyl salicyclic acid have long been known to be capable of inducing resistance [330]. As indicated above (p. 401), it is suggested that SA is derived from phenylalanine [6] via transcinnamic acid (see Figure 1). If this is the case, then the production of SA is dependent on PAL activity, which is a plausible explanation for its spread from HR lesions. Examples of defense responses caused by exogenously applied SA are peroxidase activity in cucumber [326], and accumulation of mRNAs for members of each of the PR-protein families 1 to 5 and of SAR8.2 in tobacco [6,164,222]. In general, the same set of transcripts accumulates in SAR-reacting plants, supporting the proposal that SA is responsible for SAR [6]. Furthermore, the expression of a bacterial salicyclic acid decarboxylase in transgenic tobacco plants deprived these plants of the ability to become systemically resistant [331]. Finally, a potential SA receptor has been identified [332], perhaps opening the way to studies on the molecular mechanisms of induction of SAR by SA.

Ethylene

Ethylene, a plant hormone, has long been known to be produced along with certain PR-proteins [333], however, in addition, treatment of plants with ethylene or ethephon, which spontaneously liberates ethylene, induced production of certain PR-proteins as well as other defense responses [103,334–337], which suggests that ethylene might be involved in regulation of PR-protein accumulation. An inhibitor of ethylene production, aminoethyoxylvinyl-glycine, has reduced both ethylene production and PR-protein accumulation in some inoculation experiments [338], while in other studies, it only reduced ethylene production [334]. Studies of *Arabidopsis* mutants affected in ethylene biosynthesis have shown that ethylene is not essential for specific resistance [339]. An explanation of these conflicting results might lie in the presence of at least two independent signal transduction pathways, one involving ethylene and Ca^{2+} and another which can be activated by, e.g., xylanase [340,341].

Suppressors

Many pathogens spread without visible effect in the living plant tissue before symptoms develop. This suggests either that these biotrophs or hemibiotrophs are capable of suppressing host defense responses or that they do not produce elicitors in planta. A number of structurally diverse macromolecules of pathogen origin have been identified which inhibit the effect of elicitors [342–344]. As for elicitors, their mode of action is unknown, although at least two are believed to act at later stages than the elicitor in the interaction since they delay rather than inhibit the response [344]. However, in a model system using tomato cell culture and a yeast glycopeptide elicitor, it has been demonstrated that the carbohydrate moiety acts as a suppressor of the elicitor activity. These data have suggested that a suppressor might act as a competitive inhibitor of elicitor binding to a specific receptor [345,346]. Suppressors would seem to have evolved as part of the adaptation of certain pathogens to specific hosts and have a role in preventing successful defense. The distinction between suppressors and toxins is subtle, in as much as toxins also block plant defense. However, in contrast to toxins, suppressors cannot cause disease symptoms. However, many examples of suppressors are based on model in vitro systems, and their role in interactions remains to be genetically proven.

Defense Response Gene Promotors

As indicated throughout this review, much is known about the genes activated in the defense response. Northern blotting of the entire messenger population in the tissue is the technique generally used to assay transcription, although it actually measures transcript accumulation. Only a few studies have confirmed that, e.g., genes encoding enzymes of phenylpropanoid metabolism [347,348] and PR-proteins [349,350] are transcriptionally activated in specific systems. It is, however, generally believed that defense response genes are transcriptionally regulated, and consequently, the last step in the signal transduction pathway is to be found in the promotor sequence of the specific response genes. Genomic sequences have been characterized for many response genes, and attempts are being made to localize promotor sequence elements capable of binding *trans*-acting factors. Thus, e.g., an element conferring inducibility by TMV and SA has been identified in the PR-1a and GRP promotors from tobacco [350]. The current status of promotor studies of the genes involved in phytoalexin production has been reviewed recently [131,351]. For example, various nuclear protein binding elements of a bean CHS promotor responsive to fungal elicitor treatment have been identified which contain sequences homologous to parts of the promotor of a coregulated PAL gene [352]. Furthermore, it appears that some of these elements are involved in elicitor responsiveness and proteins which are able to bind to the elements have been purified [351]. It will be interesting to see whether the elicitor responsiveness involves these purified proteins.

DISCUSSION

An overwhelming body of molecular and biochemical data has been presented in the field of plant-pathogen interactions over the last 10 years, most of which has dealt with the plant response to pathogen attack. Despite this significant increase in knowledge concerning the communication between plants and microbes, there is still much to be learned.

We have tried to scan the important data concerning the plant side of the interactions, and focused on some areas where more information is needed. The rate at which new types of defense responses are being discovered is by no means diminishing (see p. 410). This indicates that many genes involved are currently unknown. Furthermore, much remains to be done to demonstrate the importance of the responses mentioned in this review. So far, the main strategies have involved looking for antimicrobial activities in vitro (see p. 402) and overexpressing the genes in either homologous or heterologous systems, and indeed the importance of several genes in defense has been confirmed [6,353–355]. However, inhibition (or abolition) of the natural expression level of response genes by, e.g., mutation or antisense expression will be necessary in order to determine the importance of specific components in natural defense. A particularly difficult area is the hypersensitive response (p. 393): progress is being made, but many unanswered questions remain concerning the relation of this phenomenon to, e.g., phytoalexin (p. 401) and PR proteins production (p. 402).

The events taking place after an elicitor has been released from the pathogen and until transcription of the response genes starts, i.e., the signal transduction pathways, is still a "black box" (see Figure 2). This is the place where it all started with the definition of the specific resistance genes or R-genes. It looked so simple 40 years ago! However, no R-gene has been found, although at least one is localized to a cloned piece of DNA [302]. There are now so many alternative strategies for making artificial resistance [354,355] that R-genes are perhaps primarily of scientific interest, particularly since the resistance provided by incorporating R-genes by classical plant breeding has not proven durable [293].

As we have mentioned in the introduction, the area of preformed resistance components has largely been neglected, presumably due to the difficulties in obtaining conclusive data. Genetic data would often be required here, or in other words, time-consuming crossings and segregations of mutants or genetic variation achieved otherwise is necessary. Preformed resistance, however, should not be underestimated it is probably important in, e.g., nonhost resistance.

An observation which can be made from differential hybridization experiments (see p. 410), is that some transcripts appear to be down-regulated as opposed to the defense response transcripts. We have taken this a little further and demonstrated on northern blots that a couple of barley photosynthesis transcripts are suppressed very rapidly after inoculation with the powdery mildew fungus (Gregersen et al., unpublished). This phenomenon is perhaps essential in susceptibility, another "low input" research field, which is undoubtedly important, particularly in interactions involving biotrophic pathogens.

Plant diseases are not only the cause of great economic loss to Western agriculture with the subsequent need for pesticide use, they are a scourge of Third World agriculture where certain diseases can cause the total loss of specific crops. A major priority for the application of molecular biology for the Third World lies in producing durable low energy solutions for pathogen problems [356]. A combination of altered defenses will therefore be necessary. One application for molecular techniques lies in producing transgenic plants more resistant to specific pathogens. Cloned resistance response genes provide the material for altered expression of these genes, and inducible promotors can be applied in front of any gene which might lead to resistance. Recent examples include the introduction of T4 lysozyme into potato [357] and a bacterial phaseolotoxin-resistant ornithine carbamoyltransferase from *Pseudomonas syringae* pv. *phaseolicola* into tobacco. In the latter case, the transgenic plants developed a hypersensitive response to the pathogen [258]. The gene introduced made the plant insensitive to the toxin, and subsequently, the plant's own natural defense mechanisms successfully dealt with the pathogen. This type of strategy avoids the problem of coevolution: the pathogens of one species are not necessarily adapted to the defenses of another, or to gene products originating from other sources.

Possibilities, reaching far beyond pathogen defense, arise once receptors for elicitors have been identified and promotors can be activated by spraying the plants with inducer solutions.

ACKNOWLEDGMENTS

Our research is funded both directly and indirectly by The Nordic Council of Ministers and The Danish Agricultural and Veterinary Research Council (SJVF) directly and via the Biotechnology Framework Programme. Per L. Gregersen is supported by a Senior Research Fellowship from the Royal Veterinary and Agricultural University. We are grateful to Drs. Karsten M. Kragh, Peter Stephensen Lübeck, Rob Mellor, and Alan J. Slusarenko for their critical discussion and contributions to this review.

REFERENCES

1. A. J. Anderson, *Phytopathogenic Prokaryotes 2* (M. S. Mount and G. H. Lacy, eds.), Academic Press, London, p. 119 (1982).
2. N. T. Keen, *Plant Mol. Biol., 19*: 109 (1992).

3. M. G. Smart, *The Fungal Spore and Disease Initiation in Plants and Animals* (G. T. Cole and H. C. Hoch, eds.), Plenum Press, New York and London, p. 47 (1991).
4. D. B. Collinge and A. J. Slusarenko, *Plant Mol. Biol., 9*: 389 (1987).
5. R. A. Dixon and C. J. Lamb, *Ann. Rev. Plant Physiol. Plant Mol. Biol., 41*: 339 (1990).
6. J. Ryals, E. Ward, P. Ahl-Goy, and J. P. Métreaux, *Inducible Plant Proteins* (J. L. Wray, ed.), Soc. Exp. Biol. Sem. Ser., *49*, Cambridge University Press, Cambridge, England, p. 205 (1992).
7. L. Sequeira, *Ann. Rev. Microbiol., 37*: 51 (1983).
8. V. Smedegaard-Petersen, J. Brandt, B. H. Cho, D. B. Collinge, P. L. Gregersen, and H. Thordal-Christensen, Proceedings International Symposium Biotic Stress of Barley in Arid and Semi-Arid Environments (J. H. Riesselman, ed.), ICARDA, University of Montana, p. 123 (1992).
9. R. A. Dixon and M. J. Harrison, *Adv. Genet., 28*: 165 (1990).
10. P. J. G. M. de Wit, *Genes Involved in Plant Defense* (T. Boller and F. Meins Jr., eds.), Springer Verlag, New York, p. 25 (1992).
11. N. T. Keen, *Ann. Rev. Genet., 24*: 447 (1990).
12. J. R. Aist and W. R. Bushnell, *The Fungal Spore and Disease Initiation in Plants and Animals* (G. T. Cole and H. C. Hoch, eds.), Plenum Press, New York and London, p. 321 (1991).
13. M. C. Heath, *The Rust Fungi* (K. J. Scott and A. J. Chakravorty, eds.), Academic Press, New York, p. 223 (1982).
14. A. A. Bell, *Ann. Rev. Plant Physiol., 32*: 21 (1981).
15. N. T. Keen, *Recognition in Microbe–Plant Symbiotic and Pathogenic Interactions* (B. Lugtenberg, ed.), Springer Verlag, Berlin, Germany, p. 171 (1986).
16. M. J. Daniels, J. M. Dow, and A. E. Osbourn, *Ann. Rev. Phytopathol., 26*: 285 (1988).
17. C. J. Lamb, M. A. Lawton, M. Dron, and R. A. Dixon, *Cell, 56*: 215 (1989).
18. A. Ross, *Virology, 14*: 329 (1961).
19. A. Ross, *Virology, 14*: 340 (1961).
20. H. Köhle, W. Jeblick, F. Poten, W. Blaschek, and H. Kauss, *Plant Physio., 77*: 544 (1985).
21. E. Schmelzer, S. Krüger-Lebus, and K. Hahlbrock, *Plant Cell, 1*: 993 (1989).
22. T. Boller and F. Meins Jr., *Genes Involved in Plant Defense*, Springer Verlag, Vienna, Austria and New York (1992).
23. D. J. Bowles, *Ann. Rev. Biochem., 59*: 873 (1990).
24. T. Bryngelsson and D. B. Collinge, *Barley: Genetics, Molecular Biology and Biotechnology* (P. R. Shewry, ed.), C.A.B. International, Wallingford, England, p. 459 (1992).
25. J. P. Carr and D. F. Klessig, *Genetic Engineering, Principles and Methods*, Vol. 11 (J. K. Setlow, ed.), Plenum Press, New York, p. 65 (1989).
26. Y. Eyal and R. Fluhr, *Oxford Survey of Plant Mol. Cell Biol., 7*: 223 (1991).
27. T. L. Graham and M. Y. Graham, *Mol. Plant-Micr. Interact., 4*: 415 (1991).
28. W. Knogge, *Z. Naturforsch, 46c*: 969 (1991).
29. R. L. Nicholson and R. Hammerschmidt, *Ann. Rev. Phytopathol., 30*: 369 (1992).

30. Z. Klement, *Phytopathogenic Prokaryotes*, Vol. 2 (M. S. Mount and G. H. Lacy, eds.), Academic Press, London, p. 149 (1982).

31. J. Mansfield, *Recognition in Microbe-Plant Symbiotic and Pathogenic Interactions* (B. Lugtenberg, ed.), Springer Verlag, Berlin, Germany, p. 433 (1986).

32. J. Mansfield, *Recognition and Response in Plant–Virus Interactions* (R. S. S. Fraser, ed.), Springer Verlag, Berlin, Germany, p. 31 (1990).

32a. A. J. Slusarenko, K. P. Croft, and C. R. Voisey, *Biochemistry and Molecular Biology of Plant-Pathogen Interactions* (C. J. Smith, ed.), Oxford Science Publications, p. 126 (1991).

33. R. N. Strange, *Pests and Pathogens* (P. G. Ayres, ed.), ßios Scientific Publishers, Oxford, p. 39 (1992).

34. J. Pavlovkin, A. Novacky, and C. I. Ullrich-Eberius, *Physiol. Plant Pathol.*, *27*: 125 (1986).

35. A. Ádám, T. Farkas, G. Somlyai, M. Hevesi, and Z. Kíraly, *Physiol. Mol. Plant Pathol.*, *34*: 13 (1989).

36. L. D. Keppler and A. Novacky, *Phytopathol.*, *76*: 104 (1986).

37. L. D. Keppler and A. Novacky, *Phytopathol.*, *79*: 705 (1989).

38. K. P. C. Croft, F. Jüttner, and A. J. Slusarenko, *Plant Physiol.*, *101*: 13 (1993).

39. A. J. Slusarenko, B. Meier, K. P. C. Croft, and H. G. Eiben, *Mechanisms of Plant Defence Responses* (B. Fritig and M. Legrand, eds.), Symposium Proceedings, Kluwer Academic Publishing, Dordrecht, The Netherlands, p. 211 (1993).

40. H. D. Grimes, D. S. Koetje, and V. R. Franceschi, *Plant Physiol.*, *100*: 433 (1992).

41. W. S. Devlin and D. L. Gustine, *Plant Physiol.*, *100*: 1189 (1992).

42. L. D. Keppler and C. J. Baker, *Phytopathol.*, *79*: 555 (1989).

43. L. D. Keppler and A. Novacky, *Physiol. Mol. Plant Pathol.*, *30*: 233 (1987).

44. L. D. Keppler, M. M. Atkinson, and C. J. Baker, *Phytopathol.*, *79*: 974 (1989).

45. R. Vera-Estrella, E. Blumwald, and V. J. Higgins, *Plant Physiol.*, *99*: 1208 (1992).

46. K. P. C. Croft, C. R. Voisey, and A. J. Slusarenko, *Physiol. Mol. Plant Pathol.*, *36*: 49 (1990).

47. C. Bowler, T. Alliotte, M. de Loose, M. van Montagu, and D. Inzé, *EMBO J.*, *8*: 31 (1989).

48. B. Cuypers, E. Schmelzer, and K. Hahlbrock, *Mol. Plant-Micr. Interact.*, *1*: 157 (1988).

49. M. Schröder, K. Hahlbrock, and E. Kombrink, *Plant J.*, *2*: 161 (1992).

50. C. L. Cramer, J. N. Bell, T. B. Ryder, J. A. Bailey, W. Schuch, G. P. Bolwell, M. P. Robbins, R. A. Dixon, and C. J. Lamb, *EMBO J.*, *4*: 285 (1985).

51. R. N. Goodman, Z. Király, and K. R. Wood, *The Biochemistry and Physiology of Plant Disease*, Columbia, Missouri (1986).

52. S. C. Fry, *Ann. Rev. Plant Physiol.*, *37*: 165 (1986).

53. W. R. Bushnell, *Plant Cell/Cell Interactions* (I. Sussex, A. Ellingboe, M. Crouch, and R. Malmberg, eds.), Cold Spring Harbor Press, New York, p. 83 (1985).

54. H. Thordal-Christensen and V. Smedegaard-Petersen, *J. Phytopathol.*, *123*: 34 (1988).

55. J. P. Skou, *Phytopat. Z., 104*: 90 (1982).
56. P. E. Kolattukudy and K. E. Espelie, *Biosynthesis and Degradation of Wood Components* (T. Higuchi, ed.), Academic Press, New York, p. 161 (1985).
57. N. G. Lewis and E. Yamamoto, *Ann. Rev. Plant Physiol. Plant Mol. Biol., 41*: 455 (1990).
58. G. Gowri, R. C. Bugos, W. H. Campbell, C. A. Maxwell, and R. A. Dixon, *Plant Physiol., 97*: 7 (1991).
59. E. Jaeck, B. Dumas, P. Geoffroy, N. Favet, D. Inzé, M. van Montagu, B. Fritig, and M. Legrand, *Mol. Plant-Micr. Interact., 5*: 294 (1992).
60. D. Schmitt, A.-E. Pausch, and U. Mattern, *J. Biol. Chem., 266*: 17416 (1991).
61. M. E. Knight, C. Halpin, and W. Schuch, *Plant Mol. Biol., 19*: 793 (1992).
62. W. Schuch, *Inducible Plant Proteins* (J. L. Wray, ed.), *Soc. Exp. Biol. Sem. Ser., 49*, Cambridge University Press, Cambridge, England, p. 97 (1992).
63. M. H. Walter, L. Grima-Pettenati, C. Grand, A. M. Boudet, and C. J. Lamb, Proceedings National Academy of Sciences (U.S.), *85*, 5546 (1988).
64. M. H. Walter, L. Grima-Pettenati, C. Grand, A. M. Boudet, and C. J. Lamb, *Plant Mol. Biol., 15*: 525 (1990).
65. S. C. Fry, *Planta, 171*: 205 (1987).
66. J. Ralph, R. F. Helm, S. Quideau, and R. D. Hatfield, *J. Chem. Soc. Perkin Trans., 1*: 2961 (1992).
67. L. M. Lagrimini, *Plant Physiol., 96*: 577 (1991).
68. L. M. Lagrimini, W. Burkhart, M. Moyer, and S. Rothstein, Proceedings National Academy of Sciences (U.S.), *84*: 7542 (1987).
69. N. Yoshizawa, I. Satoh, S. Yokota, and T. Idei, *Holzforschung, 45*: 169 (1991).
70. E. Roberts, T. Kutchan, and P. E. Kolattukudy, *Plant Mol. Biol., 11*: 15 (1988).
71. R. Mohan and P. E. Kolattukudy, *Plant Physiol., 921*: 276 (1990).
72. R. Mohan, A. M. Bajar, and P. E. Kolattukudy, *Plant Mol. Biol., 21*: 341 (1993).
73. P. E. Kolattukudy, *Molecular Signals in Plant-Microbe Communication* (D. P. S. Verma, ed.), CRC Press, Boca Raton, Florida, p. 65 (1992).
74. K. Apel, H. Bohlmann, and U. Reimann-Philipp, *Physiol. Plant., 80*: 315 (1990).
75. H. Bohlmann and K. Apel, *Ann. Rev. Plant Physiol. Plant Mol. Biol., 42*: 227 (1991).
76. F. García-Olmedo, G. Salcedo, R. Sanchez-Monge, C. Hernandez-Lucas, M. J. Carmina, J. J. Lopez-Fando, J. Fernandez, L. Gomez, L. Royo, F. García-Maroto, A. Castagnaro, and P. Carbonero, *Barley: Genetics, Molecular Biology and Biotechnology* (P. R. Shewry, ed.), C.A.B. International, Wallingford, England, p. 335 (1992).
77. H. Bohlmann, S. Clausen, S. Behnke, H. Giese, C. Hiller, U. Reimann-Philipp, G. Schrader, V. Barkholt, and K. Apel, *EMBO J., 7*: 1559 (1988).
78. R. Fischer, S. Behnke, and K. Apel, *Planta, 178*: 61 (1989).
79. I. Andreasen, W. Becker, K. Schlüter, K. J. Burges, B. Parthier, and K. Apel, *Plant Mol. Biol., 19*: 193 (1992).
80. P. L. Gregersen, J. Brandt, H. Thordal-Christensen, and D. B. Collinge, *Mechanisms of Plant Defence Responses* (B. Fritig and M. Legrand, eds.), Symposium Proceedings, Kluwer Academic Publishing, Dordrecht, The Netherlands, p. 304 (1993).

81. Q. Gu, E. E. Kawata, M.-J. Morse, H.-M. Wu, and A. Y. Cheung, *Mol. Gen. Genet., 234*: 89 (1992).
82. U. Reimann-Philipp, G. Schrader, E. Martinoia, V. Barkholt, and K. Apel, *J. Biol. Chem., 264*: 8978 (1989).
83. A. K. Balls and W. S. Hale, *Cereal Chem., 17*: 243 (1940).
84. L. S. Stuart and T. H. Harris, *Cereal Chem., 19*: 288 (1942).
85. F. Ebrahim-Nesbat, K. Apel, and R. Heitefuss, *J. Phytopathol., 131*: 259 (1991).
86. G. I. Cassab and J. E. Varner, *Ann. Rev. Plant Physiol. Plant Mol. Biol., 39*: 321 (1988).
87. J. B. Cooper, J. A. Chen, G. J. van Holst, and J. E. Varner, *Trends Biochem. Sci., 12*: 24 (1987).
88. D. R. Corbin, N. Sauer, and C. J. Lamb, *Mol. Cell Biol., 7*: 4337 (1987).
89. P. Knox, *J. Cell Sci., 96*: 557 (1990).
90. D. Mazau and M. T. Esquerré-Tugayé, *Physiol. Mol. Plant Pathol., 29*: 147 (1986).
91. D. Rumeau, D. Mazau, and M. T. Esquerré-Tugayé, *Recognition in Microbe-Plant Symbiotic and Pathogenic Interactions* (B. Lugtenberg, ed.), Springer Verlag, Berlin, Germany, p. 377 (1986).
92. J. Sheng, R. D'Ovidio, and M. C. Mehdy, *Plant J., 1*: 345 (1991).
93. G. P. Bolwell, M. P. Robbins, and R. A. Dixon, *Biochem. J., 229*: 693 (1985).
93a. G. P. Bolwell, M. P. Robbins, and R. A. Dixon, *Eur. J. Biochem., 148*: 571 (1985).
94. J. B. Cooper and J. E. Varner, *Plant Physiol., 76*: 414 (1984).
95. N. Benhamou, D. Mazau, and M. T. Esquerré-Tugayé, *Phytopathol., 80*: 163 (1990).
96. N Benhamou, D. Mazau, J. Grenier, and M. T. Esquerré-Tugayé, *Planta, 184*: 196 (1991).
97. R. J. O'Connell, I. R. Brown, J. W. Mansfield, J. A. Bailey, D. Mazau, D. Rumeau, and M. T. Esquerré-Tugayé, *Mol. Plant-Micr. Interact., 3*: 33 (1990).
98. A. M. Showalter, A. D. Butt, and S. Kim, *Plant Mol. Biol., 19*: 205 (1992).
99. A. M. Showalter, J. Zhou, D. Rumeau, S. G. Worst, and J. E. Varner, *Plant Mol. Biol., 16*: 547 (1991).
100. R. A. M. Hooft van Huijsuijnen, L. C. van Loon, and J. F. Bol, *EMBO J., 5*: 2057 (1986).
101. J. A. L. van Kan, B. Cornelissen, and J. F. Bol, *Mol. Plant-Micr. Interact., 1*: 107 (1988).
102. A. M. Showalter, J. N. Bell, C. L. Cramer, J. A. Bailey, J. E. Varner, and C. J. Lamb, Proceedings National Academy of Sciences (U.S.), *82*: 6551 (1985).
103. D. Tagu, N. Walker, L. Ruis-Avila, S. Burgess, J. A. Martinez-Izquierdo, J.-J. Leguay, P. Netter, and P. Puigdomènech, *Plant Mol. Biol., 20*: 529 (1992).
104. H. J. M. Linthorst, R. L. J. Meuwissen, S. Kauffmann, and J. F. Bol, *Plant Cell, 1*: 285 (1989).

105. N. C. Carpita and D. M. Gibeaut, *Plant J., 3*: 1 (1993).
106. V. Smedegaard-Petersen, D. B. Collinge, H. Thordal-Christensen, J. Brandt, P. L. Gregersen, B. H. Cho, H. Walther-Larsen, H. J. Kristensen, and K. Vad, *Biological Control of Plant Diseases: Progress and Challenges for the Future* (E. C. Tjamos, G. Papavizas, and R. J. Cook, eds.), NATO-ASI Plenum Press, New York, p. 321 (1991).
107. N. Benhamou, J. Grenier, and M. J. Chrispeels, *Plant Physiol., 97*: 739 (1991).
108. A. Sturm and M. J. Chrispeels, *Plant Cell, 2*: 1107 (1990).
109. N. Benhamou, J. Grenier, A. Asselin, and M. Legrand, *Plant Cell, 1*: 1209 (1989).
110. V. Smedegaard-Petersen and K. Tolstrup, *Ann. Rev. Phytopathol., 23*: 475 (1985).
111. K. Frederikson, P. Kjellbom, and C. Larsson, *Physiol. Plant., 81*: 289 (1991).
112. R. M. Slay, A. E. Watada, D. J. Frost, and B. P. Wasserman, *Plant. Sci., 86*: 125 (1992).
113. H. Kauss, *Ann. Rev. Plant Physiol., 38*: 47 (1987).
114. H. Kauss, *The Plant Plasma Membrane* (C. Larsson and I. M. Møller, eds.), Springer Verlag, Berlin, Germany, p. 321 (1990).
115. D. P. Delmer, *Ann. Rev. Plant Physiol., 38*: 259 (1990).
116. P. Ohana, D. P. Delmer, J. C. Steffens, D. E. Matthews, R. Meyer, and M. Benziman, *J. Biol. Chem., 266*: 13742 (1991).
117. R. A. Dixon, *Biol. Rev., 61*: 239 (1986).
118. R. A. Dixon, A. D. Choudhary, M. J. Harrison, B. A. Stermer, L. Wu, S. A. Jenkins, C. J. Lamb, and M. A. Lawton, *Biochemistry and Molecular Biology of Plant-Pathogen Interactions* (C. J. Smith, ed.), Clarendon Press, Oxford, England, p. 271 (1991).
119. R. A. Dixon, P. A. Dey, and C. J. Lamb, *Adv. Enz., 55*: 1 (1983).
120. K. Hahlbrock and D. Scheel, *Ann. Rev. Plant Physiol. Plant Mol. Biol., 40*: 347 (1989).
121. D. A. Smith and S. W. Banks, *Phytochem., 25*: 979 (1986).
122. P. J. Edwards, *Pests and Pathogens* (P. G. Ayres, ed.), ßios Scientific Publishers, Oxford, England, p. 69 (1992).
123. P. J. Kuhn and J. A. Hargreaves, *Fungal Infection of Plants* (G. F. Pegg and P. G. Ayers, eds.), Cambridge University Press, Cambridge, England, p. 194 (1987).
124. D. Défago, H. Kern, and L. Sedlar, *Physiol. Plant Pathol., 22*: 39 (1983).
125. L. Karni, D. Prusky, I. Kobiler, E. Bar-Shira, and D. Kobiler, *Physiol. Mol. Plant Pathol., 35*: 367 (1989).
126. D. Prusky, N. T. Keen, and I. Eaks, *Physiol. Plant Pathol., 22*: 189 (1983).
127. J. P. Wubben, M. H. A. J. Joosten, J. A. L. van Kan, and P. J. G. M. De Wit, *Physiol. Mol. Plant Pathol., 41*: 23 (1992).
128. W. Jahnen and K. Hahlbrock, *Planta, 173*: 197 (1988).
129. M. G. Hahn, A. Bonhoff, and H. Grisebach, *Plant Physiol., 77*: 591 (1985).
130. P. Moesta, U. Seydel, B. Linder, and H. Grisebach, *Z. Naturforsch., 37c*: 748 (1982).
131. J. L. Dangl, *Genes Involved in Plant Defense* (T. Boller and F. Meins Jr., eds.), Springer Verlag, Vienna, Austria and New York, p. 303 (1992).

132. P. Moesta and H. Grisebach, *Physiol. Plant Pathol., 21*: 65 (1982).
133. W. Schäffer, D. Straney, L. Ciuffetti, H. D. VanEtten, and O. Yoder, *Science, 241*: 247 (1989).
134. H. D. VanEtten, D. E. Matthews, and P. S. Matthews, *Ann. Rev. Phytopathol., 27*: 143 (1989).
135. R. Hain, H.-J. Reif, E. Krause, R. Langeebartels, H. Kindl. B. Vornam, W. Wiese, E. Schmelzer, P. Schreier, R. H. Stöcker, and K. Stenzel, *Nature, 361*: 153 (1993).
136. A. Leyva, X. Liang, J. A. Pintor-Toro, R. A. Dixon, and C. J. Lamb, *Plant Cell, 4*: 263 (1992).
137. X. Liang, M. Dron, C. L. Cramer, R. A. Dixon, and C. J. Lamb, *J. Biol. Chem., 264*: 14486 (1989).
138. B. A. Stermer, J. Schmid, C. J. Lamb, and R. A. Dixon, *Mol. Plant Micr. Interact., 3*: 381 (1990).
139. R. Wingender, H. Röhrig, C. Höricke, D. Wing, and J. Schell, *Mol. Gen. Genet., 218*: 315 (1988).
140. R. A. Jefferson, S. M. Burgess, and D. Hirsch, Proceedings National Academy of Sciences (U.S.), *83*: 8447 (1986).
141. E. Bell and R. L. Malmberg, *Mol. Gen. Genet., 224*: 431 (1990).
142. B. Keith, X. Dong, F. M. Ausubel, and G. R. Fink, Proceedings National Academy of Sciences (U.S.), *88*: 8821 (1991).
143. H. J. Franssen, I. Vijn, W. C. Yang, T. Bisseling, *Plant Mol. Biol., 19*: 89 (1992).
144. N. K. Peters and D. P. S. Verma, *Mol. Plant-Micr. Interact., 3*: 4 (1990).
145. W.-C. Yang, H. C. J. C. Cremers, P. Hogendijk, P. Katinakis, C. A. Wijffelman, H. Franssen, A. Van Kammen, and T. Bisseling, *Plant J., 2*: 143 (1992).
146. L. C. van Loon and J. A. Callow, *Biochemical Plant Pathology* (J. A. Callow, ed.), John Wiley and Sons, Chichester, England, p. 385 (1983).
147. J. F. Bol, H. J. Lindhorst, and B. J. C. Cornelissen, *Ann. Rev. Phytopathol., 28*: 11 (1990).
148. L. C. van Loon, *Plant Mol. Biol., 4*: 111 (1985).
149. L. C. van Loon, *Physiol. Mol. Plant Pathol., 37*: 229 (1990).
150. H. J. M. Linthorst, *CRC Crit. Rev. Plant Sci., 10*: 123 (1991).
151. O. P. Sehgal and F. Mohamed, *Plant Viruses*, Vol. 2, *Pathology* (C. L. Mandahar, ed.), CRC Press, Boca Raton, Florida, p. 65 (1990).
152. R. F. White and J. F. Antoniw, *CRC Crit. Rev. Plant Sci., 9*: 443 (1991).
153. T. Bryngelsson and B. Gréen, *Physiol. Mol. Plant Pathol., 35*: 45 (1989).
154. T. Bryngelsson, M. Gustafsson, M. R. Leal, and E. Bartonek, *J. Phytopathol., 123*: 193 (1988).
155. W. Nasser, M. De Tapia, S. Kaufmann, S. Montasser-Kouhsari, and G. Burkhard, *Plant Mol. Biol., 11*: 529 (1988).
156. J. L. Taylor, K.-H. Fritzmeier, I. Häuser, E. Kombrink, F. Rohwer, G. Strittmatter, and K. Hahlbrock, *Mol. Plant-Micr. Inter., 3*: 72 (1990).
157. I. E. Somssich, E. Schmelzer, P. Kawalleck, and K. Hahlbrock, *Mol. Gen. Genet., 213*: 93 (1988).
158. J. R. Cutt and D. F. Klessig, *Genes Involved in Plant Defense* (T. Boller and F. Meins Jr., eds.), Springer Verlag, New York, p. 209 (1992).

159. B. Dumas, E. Jaeck, A. Stintzi, J. Rouster, S. Kaufmann, P. Geoffroy, M. Kopp, M. Legrand, and B. Fritig, *Plant Molecular Biology*, Vol. 2 (R. G. Herrmann and B. A. Larkins, eds.), NATO ASI, Series A, Vol. 212, Plenum Press, New York and London, p. 153 (1991).

160. Y. Ohashi and M. Ohshima, *Plant Cell Physiol., 33*: 819 (1992).

161. M. A. Lawton and C. J. Lamb, *Mol. Cell Biol., 7*: 335 (1987).

162. G. Rebmann, F. Mauch, and R. Dudler, *Plant Mol. Biol., 17*: 283 (1991).

163. F. T. Brederode, H. J. M. Linthorst, and J. F. Bol, *Plant Mol. Biol., 17*: 1117 (1991).

163a. P. Schweizer, W. Hunziker, and E. Mösinger, *Plant Mol. Biol., 12*: 643 (1989).

164. E. R. Ward, S. J. Uknes, S. C. Williams, S. S. Dincher, D. L. Wiederhold, D. C. Alexander, P. Ahl-Goy, J.-P. Métraux, and J. A. Ryals, *Plant Cell, 3*: 1085 (1991).

165. T. Lotan, N. Ori, and R. Fluhr, *Plant Cell, 1*: 881 (1989).

166. J. Memelink, H. J. M. Linthorst, R. A. Schilperoort, and J. H. C. Hoge, *Plant Mol. Biol., 14*: 119 (1990).

167. J. M. Casacuberta, D. Raventós, P. Puigdomènech, and B. San Segundo, *Mol. Gen. Genet., 234*: 97 (1992).

168. D. B. Collinge, K. M. Kragh, J. D. Mikkelsen, K. K. Nielsen, U. Rasmussen, and K. Vad, *Plant J., 3*: 31 (1993).

169. A. Nassuth and H. L. Sänger, *Virus Res., 4*: 229 (1986).

170. R. F. White, E. P. Rybicki, M. B. Von Wechmar, J. L. Dekker, and J. F. Antoniw, *J. Gen. Virol., 68*: 2043 (1987).

171. D. Alexander, C. Glascock, J. Pear, J. Stinson, P. Ahl-Goy, M. Gut-Rella, E. Ward, R. M. Goodman, and J. Ryals, *Advances in Molecular Genetics of Plant Microbe Interactions*, Vol. 2 (E. W. Nester and D. P. S. Verma, eds.), Kluwer Academic Publishing, Dordrecht, The Netherlands, p. 527 (1993).

172. T. Niderman, T. Bruyère, K. Gügler, and E. Mösinger, *3rd International Workshop on Pathogenesis-Related Proteins in Plants*, Arolla, Switzerland, August ;16–20 (Abstract) (1992).

173. J. A. L. Van Kan, M. H. A. J. Joosten, C. A. M. Wagemakers, G. C. M. van den Berg-Velthuis, and P. J. G. M. De Wit, *Plant Mol. Biol., 20*: 513 (1992).

174. M. C. Metzler, J. R. Cutt, and D. F. Klessig, *Plant Physiol., 96*: 346 (1991).

175. S. Uknes, B. Mauch-Mani, M. Moyer, S. Potter, S. Williams, S. Dincher, D. Chandler, A. Slusarenko, E. Ward, and J. Ryals, *Plant Cell, 4*: 645 (1992).

176. J. M. Casacuberta, P. Puigdomènech, and B. San Segundo, *Plant Mol. Biol., 16*: 527 (1991).

177. S. Kauffmann, M. Legrand, P. Geoffroy, and B. Fritig, *EMBO J., 6*: 3209 (1987).

178. M. Legrand, S. Kauffmann, P. Geoffroy, and B. Fritig, Proceedings National Academy of Sciences (U.S.), *84*: 6750 (1987).

179. T. Boller, *Oxford Surveys of Plant Mol. Cell. Biol., 5*: 145 (1988).

180. M. Van den Bulcke, C. Bauw, C. Castresana, M. Van Montagu, and J. Vandekerckhove, Proceedings National Academy of Sciences (U.S.), *86*: 2673 (1989).

181. J. Flach, P.-E. Pilet, and P. Jollès, *Experientia, 48*: 701 (1992).
182. M. B. Ary, M. Richardson, and P. R. Shewry, *Biochim. Biophys. A., 993*: 260 (1989).
183. J. Grenier and A. Asselin, *Mol. Plant–Micr. Interact., 3*: 401 (1990).
184. F. Mauch, L. A. Hadwiger, and T. Boller, *Plant Physiol., 87*: 325 (1988).
185. K. Groglie, I. Chet, M. Holliday, R. Cressman, P. Biddle, S. Knowlton, C. J. Mauvais, and R. Broglie, *Science, 254*: 1194 (1991).
186. J.-M. Neuhaus, P. Ahl-Goy, U. Hinz, S. Flores, and F. Meins Jr., *Plant Mol. Biol., 16*: 141 (1991).
187. J.-M. Neuhaus, S. Flores, D. Keefe, P. Ahl-Goy, and F. Meins Jr., *Plant Mol. Biol., 19*: 803 (1992).
188. A. de Jong, J. Cordewener, F. Lo Schiavo, M. Terzi, J. Vandekerckhove, A. Van Kammen, and S. C. De Vries, *Plant Cell, 4*: 425 (1992).
189. N. Benhamou and A. Asselin, *Biol. Cell, 67*: 341 (1989).
190. N. T. Keen and M. Yoshikawa, *Plant Physiol., 71*: 460 (1983).
191. F. Mauch and L. A. Staehelin, *Plant Cell, 1*: 447 (1989).
192. K.-S. Ham, S. Kauffmann, P. Albersheim, and A. G. Darvill, *Mol. Plant- Micr. Interact., 4*: 545 (1991).
193. M. G. Hahn and J.-J. Cheong, *Advances in Molecular Genetics of Plant- Microbe Interactions. Vol. 1* (H. Hennecke and D. P. S. Verma, eds.), Kluwer Academic Publishing, Dordrecht, The Netherlands, p. 403 (1991).
194. L. Friedrich, M. Moyer, E. Ward, and J. Ryals, *Mol. Gen. Genet., 230*: 113 (1991).
195. H. J. M. Linthorst, N. Danahash, F. T. Brederode, J. A. L. Van Kan, P. J. G. M. De Wit, and J. F. Bol, *Mol. Plant-Micr. Interact., 4*: 586 (1991).
196. A. Standford, M. Bevan, and D. Northcote, *Mol. Gen. Genet., 215*: 200 (1989).
197. W. F. Broekaert, H. I. Lee, A. Kush, N. H. Chua, and N. Raikhel, Proceedings National Academy of Sciences (U.S.), *87*: 7633 (1990).
198. J. Hejgaard, S. Jacobsen, S. E. Bjørn, and K. M. Kragh, *FEBS Lett., 307*: 389 (1992).
199. L. Edens, L. Heslinga, R. Klock, A. M. Ledeboer, J. Maat, M. Y. Toonen, C. Visser, and C. T. Verrips, *Gene, 18*: 1 (1982).
200. B. J. C. Cornelissen, R. A. M. Hooft van Huijsuijnen, and J. F. Bol, *Nature, 321*: 531 (1986).
201. W. S. Pierpoint, P. J. Jackson, and R. M. Evans, *Physiol. Mol. Plant Path- ol., 36*: 325 (1990).
202. J. S. Graham, W. Burkhart, J. Xiong, and J. W. Gillikin, *Plant Physiol., 98*: 163 (1992).
203. M. Hahn, S. Jüngling, and W. Knogge, personal communication.
204. P. Frendo, L. Didierjean, E. Passelegue, and G. Burkard, *Plant Sci., 85*: 61 (1992).
205. C. P. Woloshuk, J. S. Meulenhoff, M. Sela-Buurlage, P. J. M. van den Elzen, and B. J. Cornelissen, *Plant Cell, 3*: 619 (1991).
206. J. Hejgaard, S. Jacobsen, and I. Svendsen, *FEBS Lett., 291*: 127 (1991).
207. W. K. Roberts and C. P. Selitrennikoff, *J. Gen. Microbiol., 136*: 1771 (1990).
208. M. Richardson, S. Valdes-Rodriguez, and A. Blanco-Labra, *Nature, 327*: 432 (1987).

209. A. J. Vigers, W. K. Roberts, and C. P. Selitrennikoff, *Mol. Plant-Micr. Interact., 4*: 315 (1991).
210. V. Pautot, F. M. Holzer, and L. L. Walling, *Mol. Plant-Micr. Interact., 4*: 284 (1991).
211. L. Coppens and D. Dewitte, *Plant Sci., 67*: 97 (1990).
212. K. Saeki, O. Ishikawa, T. Fukuoka, H. Nakagawa, Y. Kai, T. Kakuno, J. Yamashita, N. Kasai, and T. Horio, *J. Biochem., 99*: 485 (1986).
213. R. B. van Huystee, *Ann. Rev. Plant Physiol., 38*: 205 (1987).
214. K. Kerby and S. Somerville, *Physiol. Mol. Plant Pathol., 35*: 323 (1989).
215. K. Kerby and S. Somerville, *Plant Physiol., 100*: 397 (1992).
216. B. Theilade, *Cloning and Characterization of two Peroxidase Genes from Barley*, Ph.D. thesis, University of Aarhus, Denmark (1993).
217. H. Thordal-Christensen, J. Brandt, B. H. Cho, P. L. Gregersen, S. K. Rasmussen, V. Smedegaard-Petersen, and D. B. Collinge, *Physiol. Mol. Plant Pathol., 40*: 395 (1992).
218. L. M. Lagrimini and S. Rothstein, *Plant Physiol., 84*: 438 (1987).
219. G. F. Trezzini, A. Horrichs, and I. E. Somssich, *Plant Mol. Biol., 21*: 385 (1993).
220. F. R. G. Terras, H. M. E. Schoofs, M. F. C. De Bolle, F. Van Leuven, S. B. Rees, J. Vanderleyden, B. P. A. Cammue, and W. F. Broekaert, *J. Biol. Chem., 267*: 15301 (1992).
221. C. C. Chiang and L. A. Hadwiger, *Mol. Plant-Micr. Interact., 4*: 324 (1991).
222. D. Alexander, J. Stinson, J. Pear, C. Glascock, E. Ward, R. M. Goodman, and J. Ryals, *Mol. Plant-Micr. Interact., 5*: 513 (1992).
223. F. R. G. Terras, I. J. Goderis, F. Van Leuven, J. Vanderleyden, B. P. A. Cammue, and W. F. Broekaert, *Plant Physiol., 100*: 1055 (1992).
224. M. J. Chrispeels and N. V. Raikhel, *Plant Cell, 3*: 1 (1991).
225. M. J. Chrispeels, *Ann. Rev. Plant Physiol. Plant Mol. Biol., 42*: 21 (1991).
226. H. Shinshi, H. Wenzler, J.-M. Neuhaus, G. Felix, J. Hofsteenge, and F. Meins Jr., Proceedings National Academy of Sciences (U.S.), *85*: 5541 (1988).
227. J.-M. Neuhaus, L. Sticher, F. Meins Jr., and T. Boller, Proceedings National Academy of Sciences (U.S.), *88*: 10362 (1991).
228. G. Rebmann, C. Hertig, J. Bull, F. Mauch, and R. Dudler, *Plant Mol. Biol., 16*: 329 (1991).
229. J. Ellis, *Sem. Cell Biol., 1*: 1 (1990).
230. M. Gething and J. Sambrook, *Nature, 355*: 33 (1992).
231. H. Walther-Larsen, J. Brandt, D. B. Collinge, and H. Thordal-Christensen, *Plant Mol. Biol., 21*: 1097 (1993).
232. J. Denecke, B. Ek, M. Caspers, K. M. C. Sinjorgo, and E. T. Palva, *J. Exp. Bot., 44(suppl.)*: 213 (1993).
233. D. P. Matton, P. Constabel, and N. Brisson, *Plant Mol. Biol., 14*: 775 (1990).
234. O. C. Yoder and B. G. Turgeon, *Gene Manipulations in Fungi* (J. W. Bennet and L. L. Lasure, eds.), Academic Press, New York, p. 417 (1985).
235. J. M. Dow, B. R. Clarke, D. E. Milligan, J.-L. Tang, and M. J. Daniels, *Appl. Env. Microbiol., 56*: 2994 (1990).
236. C. L. Gough, J. M. Dow, C. E. Barber, and M. J. Daniels, *Mol. Plant–Micr. Interact., 1*: 275 (1988).

237. A. E. Osbourn, B. R. Clarke, and M. J. Daniels, *Mol. Plant–Micr. Interact., 1*: 275 (1988).
238. D. Scheel and J. Parker, *Naturforsch, 45c*: 569 (1990).
239. K. Hahlbrock, N. Arabatzis, M. Becker-Andre, H.-J. Joos, E. Kombrink, M. Schröder, G. Styrittmatter, and J. Taylor, *Signal Molecules in Plants and Plant-Microbe Interactions* (B. J. J. Lugtenberg, ed.), NATO ASI series H: Cell Biology, Vol. 36. Springer Verlag, Berlin, p. 241 (1989).
240. P. E. Kolattukudy, G. K. Podila, B. A. Sherf, M. A. Bajar, and R. Mohan, *Advances in Molecular Genetics of Plant-Microbe Interactions* (H. Hennecke and D. P. S. Verma, eds.), Kluwer Academic Publishing, Dordrecht, The Netherlands, p. 242 (1991).
241. R. F. Fisher and S. R. Long, *Nature, 357*: 655 (1992).
242. H. Spaink, *Plant Mol. Biol., 20*: 977 (1992).
243. D. B. Collinge, D. E. Milligan, J. M. Dow, G. Scofield, and M. J. Daniels, *Plant Mol. Biol., 8*: 405 (1987).
244. A. D. Davidson, J. M. Manners, R. S. Simpson, and K. J. Scott, *Physiol. Mol. Plant Pathol., 32*: 127 (1988).
245. P. L. Gregersen, D. B. Collinge, and V. Smedegaard-Petersen, *Physiol. Mol. Plant Pathol., 36*: 471 (1990).
246. L. A. Hadwiger and W. Wagoner, *Physiol. Plant Pathol., 23*: 153 (1983).
247. A. J. Slusarenko and A. Longland, *Physiol. Mol. Plant Pathol., 29*: 79 (1986).
248. T. E. Smart, D. D. Dunigan, and M. Zaitlin, *Virology, 158*: 461 (1987).
249. R. Stratford and S. N. Covey, *Mol. Plant–Micr. Interact., 1*: 243 (1988).
250. D. J. Bowles, J. Hogg, and H. Small, *Physiol. Mol. Plant Pathol., 34*: 463 (1989).
251. M. A. M. S. Hamdan and R. A. Dixon, *Physiol. Mol. Plant Pathol., 31*: 105 (1986).
252. L. Godiard, D. Froissard, J. Fournier, M. Axelos, and Y. Marco, *Plant Mol. Biol., 17*: 409 (1991).
253. Y. Marco, F. Ragueh, L. Godiard, and D. Froissard, *Plant Mol. Biol., 15*: 145 (1990).
254. J. Memelink, J. H. C. Hoge, and R. A. Schilperoort, *EMBO J., 6*: 3579 (1987).
255. K. Saunders, A. P. Lucy, and S. N. Covey, *Physiol. Mol. Plant Pathol., 35*: 339 (1989).
256. R. C. Riggleman, B. Fristensky, and L. A. Hadwiger, *Plant Mol. Biol., 4*: 81 (1985).
257. A. D. Davidson, J. M. Manners, R. S. Simpson, and K. J. Scott, *Plant Mol. Biol., 8*: 77 (1987).
258. C. Marineau, D. P. Matton, and N. Brisson, *Plant Mol. Biol., 9*: 335 (1987).
259. I. E. Sommssich, J. Bohlmann, K. Hahlbrock, E. Kombrink, and W. Schulz, *Plant Mol. Biol., 12*: 227 (1989).
260. I. E. Somssich, E. Schmelzer, J. Bohlmann, and K. Hahlbrock, Proceedings National Academy of Sciences (U.S.), *83*: 2427 (1986).
261. T. D. Sargent, *Meth. Enz., 152*: 423 (1987).
262. J. Brandt, H. Thordal-Christensen, K. Vad, P. L. Gregersen, and D. B. Collinge, *Plant J., 2*: 815 (1992).

263. J. R. Cutt, D. C. Dixon, J. P. Carr, and D. F. Klessig, *Nucl. Acid. Res., 16*: 9861 (1988).
264. U. Rasmussen, K. Bojsen, and D. B. Collinge, *Plant Mol. Biol., 20*: 277 (1992).
265. Q. Lin, Z. C. Chen, J. F. Antoniw, and R. F. White, *Plant Mol. Biol., 17*: 609 (1991).
266. R. Dudler, C. Hertig, G. Rebmann, J. Bull, and F. Mauch, *Mol. Plant-Micr. Interact., 4*: 14 (1991).
267. D. P. Matton and N. Brisson, *Mol. Plant-Micr. Interact., 2*: 325 (1989).
268. D. P. Matton, B. Bell, and N. Brisson, *Plant Mol. Biol., 14*: 963 (1990).
269. C. C. Chiang and L. A. Hadwiger, *Mol. Plant-Micr. Interact., 3*: 78 (1990).
270. B. Fristensky, D. Horovitz, and L. A. Hadwiger, *Plant Mol. Biol., 11*: 713 (1988).
271. S. A. J. Warner, R. Scott, and J. Draper, *Plant Mol. Biol., 19*: 555 (1992).
272. S. A. J. Warner, R. Scott, and J. Draper, *Plant J., 3*: 191 (1993).
273. H. Breiteneder, K. Pettenburger, A. Bito, R. Valenta, D. Kraft, H. Rumpold, O. Scheiner, and M. Brieteneder, *EMBO J., 8*: 1935 (1989).
274. S. Kiedrowski, K. Hahlbrock, P. Kawalleck, I. E. Somssich, and J. L. Dangl, *EMBO J., 11*: 4677 (1992).
275. W. Jutidamrongphan, G. Mackinnon, J. M. Manners, and K. J. Scott, *Nucleic Acids Res., 17*: 9478 (1989).
276. K. J. Scott, A. D. Davidson, W. Jutidamrongphan, G. Mackinnon, and J. M. Manners, *Austral. J. Plant Physiol., 17*: 229 (1990).
277. J. Bull, F. Mauch, C. Hertig, G. Rebmann, and R. Dudler, *Mol. Plant-Micr. Interact., 5*: 516 (1992).
278. Y. K. Sharma, C. M. Hinojos, and M. C. Mehdy, *Mol. Plant-Micr. Interact., 5*: 89 (1992).
279. J. L. Dangl, *Plant J., 2*: 3 (1992).
280. J. Ebel and D. Scheel, *Genes Involved in Plant Defense* (T. Boller and F. Meins Jr., eds.), Springer Verlag, Vienna, Austria, p. 183 (1992).
281. D. W. Gabriel and B. G. Rolfe, *Ann. Rev. Phytopathol., 28*: 365 (1990).
282. M. G. Hahn, *Signal Molecules in Plants and Plant–Microbe Interactions* (B. J. J. Lugtenberg, ed.), NATO ASI series H: Cell biology, Vol. 36, Springer Verlag, Berlin, Germany, p. 1 (1989).
283. C. A. West, R. Bruce, and Y.-Y. Ren, *Signal Molecules in Plants and Plant–Microbe Interactions* (B. J. J. Lugtenberg, ed.), NATO ASI series H: Cell biology, Vol. 36, Springer Verlag, Berlin, Germany, p. 27 (1989).
284. J. E. Parker, W. Schulte, K. Hahlbrock, and D. Scheel, *Mol. Plant-Micr. Interact., 4*: 19 (1991).
285. A. Darvill, C. Augur, C. Bergmann, R. W. Carlson, J.-J. Cheong, S. Eberhard, M. G. Hahn, V.-M. Ló, V. Marfà, B. Meyer, D. Mohen, M. A. O'Neill, M. D. Spiro, H. van Halbeek, W. S. York, and P. Albersheim, *Glycobiol., 2*: 181 (1992).
286. K. R. Davis, A. G. Darvill, and P. Albersheim, *Plant Mol. Biol., 6*: 23 (1986).
287. K. R. Davis and K. Hahlbrock, *Plant Physiol., 85*: 1286 (1987).
288. J. Ebel, E. G. Cosio, M. Feger, T. Frey, U. Kissel, S. Reinold, and T. Waldmüller, *Advances in Molecular Genetics of Plant Microbe Interactions*, Vol. 2

(E. W. Nester and D. P. S. Verma, eds.), Kluwer Academic Publishing, Dordrecht, The Netherlands, p. 477 (1993).

289. W. R. Sacks, P. Ferreira, K. Hahlbrock, T. Jabs, T. Nümberger, A. Renelt, and D. Scheel, *Advances in Molecular Genetics of Plant Microbe Interactions*, Vol. 2 (E. W. Nester and D. P. S. Verma, eds.), Kluwer Academic Publishing, Dordrecht, The Netherlands, p. 485 (1993).
290. J.-J. Cheong and M. G. Hahn, *Plant Cell, 3*: 137 (1991).
291. E. G. Cosio, T. Frey, R. Verduyn, J. van Boom, and J. Ebel, *FEBS Lett., 271*: 223 (1990).
292. E. G. Cosio, T. Frey, and J. Ebel, *Eur. J. Biochem., 204*: 1115 (1992).
293. J. Barrett, *Ecology and Genetics of Host-Parasite Interactions* (D. Rollinson and R. M. Anderson, eds.), Linnean Society Symposium, No. 11, Academic Press, London, p. 215 (1985).
294. N. T. Keen, J. J. Sims, S. Midland, M. Yoder, F. Jurnak, H. Shen, C. Boyd, I. Yucel, J. Lorang, and J. Murillo, *Advances in Molecular Genetics of Plant Microbe Interactions*, Vol. 2 (E. W. Nester and D. P. S. Verma, eds.), Kluwer Academic Publishing, Dordrecht, The Netherlands, p. 211 (1993).
295. N. T. Keen, S. Tamaki, D. Kobayashi, D. Gerhold, M. Stayton, H. Shen, S. Gold, J. Lorang, H. Thordal-Christensen, D. Dahlbeck, and B. Staskawicz, *Mol. Plant-Micr. Interact., 3*: 112 (1990).
296. I. M. J. Scholtens-Toma and P. J. G. M. De Wit, *Physiol. Mol. Plant Pathol., 33*: 59 (1988).
297. G. F. J. M. Van den Ackerveken, J. A. L. Van Kan, and P. J. G. M. De Wit, *Plant J., 2*: 359 (1992).
298. J. N. Culver and W. O. Dawson, *Mol. Plant-Micr. Interact., 4*: 458 (1991).
299. L. Wevelsiep, K. H. Kogel, and W. Knogge, *Physiol. Mol. Plant Pathol., 39*: 471 (1992).
300. T. J. Wolpert and V. Macko, Proceedings National Academy of Sciences (U.S.), *86*: 4092 (1989).
301. J. H. Jørgensen, *Barley: Genetics, Molecular Biology and Biotechnology* (P. R. Shewry, ed.), C.A.B. International, Wallingford, England, p. 441 (1992).
302. J. L. Dangl, T. Debner, M. Gerwin, S. Kiedrowski, C. Ritter, A. Bendahmane, H. Liedgens, and J. Lewald, *Advances in Molecular Genetics of Plant Microbe Interactions*, Vol. 2 (E. W. Nester and D. P. S. Verma, eds.), Kluwer Academic Publishing, Dordrecht, The Netherlands, p. 405 (1993).
303. E. A. B. Aitken, J. A. Callow, and H. J. Newbury, *Plant J., 2*: 775 (1992).
304. G. F. E. Scherer, *Hormone Perception and Signal Transduction in Animals and Plants* (J. Roberts, C. Kirk, and M. Venis, eds.), S. E. B. Symposium 44, Company of Biologists, Cambridge, p. 257 (1990).
305. A. Trewavas and S. Gilroy, *Trends Biochem. Sci., 7*: 356 (1991).
306. A. Alderson, P. O. Sabelli, J. R. Dickinson, D. Cole, M. Richardson, M. Kreis, P. R. Shewry, M. G. Halford, Proceedings National Academy of Sciences (U.S.), *88*: 8602 (1991).
307. J. F. Harper, M. R. Sussman, G. E. Schaller, C. Putman-Evans, H. Charbonneau, and A. C. Harmon, *Science, 252*: 951 (1991).
308. M. A. Lawton, R. T. Yamamoto, S. K. Hanks, and C. J. Lamb, Proceedings National Academy of Sciences (U.S.), *86*: 3140 (1989).

309. S. Aruntalabhochai, N. Terryn, M. van Montagu, and D. Inzé, *Plant J., 1*: 167 (1991).
310. M. G. Palmgren, M. Sommarin, P. Ulvskov, P. L. Jørgensen, *Plant Physiol., 74*: 11 (1988).
311. S. Hirsch, A. Aitken, U. Bretsch, and J. Söll, *FEBS Lett., 296*: 222 (1992).
312. G. Lu, A. J. DeLisle, N. C. de Vetten, and R. J. Ferl, Proceedings National Academy of Sciences (U.S.), *89*: 11490 (1992).
313. A. Aitken, D. B. Collinge, G. P. H. van Heusden, T. Isobe, P. H. Roseboom, G. Rosenfeld, and J. Soll, *Trends Biochem. Sci., 17*: 498 (1992).
314. A. Dietrich, J. E. Mayer, and K. Hahlbrock, *J. Biol. Chem., 265*: 6360 (1990).
315. D. Grab, M. Feger, and J. Ebel, *Planta, 179*: 340 (1989).
316. G. Felix, D. G. Grosskopf, M. Regenass, and T. Boller, Proceeding National Academy of Sciences (U.S.), *88*: 8831 (1991).
317. F. Kurosaki, Y. Tsurusawa, and A. Nishi, *Plant Physiol., 85*: 601 (1987).
318. R. A. de Maagd, R. K. Cameron, R. A. Dixon, and C. J. Lamb, *Advances in Molecular Genetics of Plant Microbe Interactions*, Vol. 2 (E. W. Nester and D. P. S. Verma, eds.), Kluwer Academic Publishing, Dordrecht, The Netherlands, p. 445 (1993).
319. E. E. Farmer and C. A. Ryan, Proceedings National Academy of Sciences (U.S.), *87*: 7713 (1990).
320. H. Gundlach, M. J. Müller, T. M. Kutchan, and M. H. Zenk, Proceedings National Academy of Sciences (U.S.), *89*: 2389 (1992).
321. B. Parthier, *J. Plant Growth Reg., 9*: 57 (1990).
322. D. C. Wildon, J. F. Thain, P. E. H. Minchin, I. R. Gubb, Y. D. Skipper, H. M. Doherty, P. J. O'Donnell, and D. J. Bowles, *Nature, 360*: 62 (1992).
323. J. Malamy and D. F. Klessig, *Plant J., 2*: 643 (1992).
324. I. Raskin, *Plant Physiol., 99*: 799 (1992).
325. J. P. Métreaux, H. Signer, J. Ryals, E. Ward, M. Wyss-Benz, J. Gaudin, K. Raschdorf, E. Scmid, W. Blum, and B. Inveradi, *Science, 250*: 1004 (1990).
326. J. B. Rasmussen, R. Hammerschmidt, and M. N. Zook, *Plant Physiol., 97*: 1342 (1991).
327. J. Malamy, J. P. Carr, D. F. Klessig, and I. Raskin, *Science, 250*: 1002 (1990).
328. N. Yalpani, P. Silverman, T. M. A. Wilson, D. A. Kleier, and I. Raskin, *Plant Cell, 3*: 809 (1991).
329. A. J. Enyedi, N. Yalpani, P. Silverman, and I. Raskin, Proceedings National Academy of Sciences (U.S.), *89*: 2480 (1992).
330. R. F. White, *Virology, 99*: 410 (1979).
331. J. Ryals, E. Ward, S. Uknes, K. Lawton, D. Alexander, R. Goodman, J.-P. Métreaux, H. Kessmann, and P. Ahl-Goy, 3rd International Workshop on Pathogenesis-Related Proteins in Plants, Arolla, Switzerland, Aug. 16–20 (Abstract) (1992).
332. Z. Chen and D. F. Klessig, Proceedings National Academy of Sciences (U.S.), *88*: 8179 (1991).
333. G. F. Pegg, *Encyclopedia of Plant Pathology* New Series (R. Heitefuss and P. A. Williams, eds.), Springer Verlag, Heidelberg, Germany, p. 582 (1976).
334. F. Mauch, L. A. Hadwiger, and T. Boller, *Plant Physiol., 76*: 607 (1984).

335. E. Mauch, J. B. Meehl, and L. A. Staehelin, *Planta, 186*: 367 (1992).
336. C. R. Simmons, J. C. Litts, N. Huang, and R. L. Rodriguez, *Plant Mol. Biol., 18*: 33 (1992).
337. L. C. van Loon and J. F. Antoniw, *Neth. J. Plant Pathol., 88*: 237 (1982).
338. T. Lotan and R. Fluhr, *Plant Physiol., 93*: 811 (1990).
339. A. F. Bent, R. W. Innes, J. R. Ecker, and B. J. Staskawicz, *Mol. Plant-Micr. Interact., 5*: 372 (1992).
340. T. Lotan and R. Fluhr, *Symbiosis, 8*: 33 (1990).
341. V. Raz and R. Fluhr, *Plant Cell, 4*: 1123 (1992).
342. N. Doke, N. A. Garas, and J. Kuć, *Physiol. Plant Pathol., 15*: 127 (1979).
343. G. Maniara, R. Laine, and J. Kuć, *Physiol. Plant Pathol., 24*: 177 (1984).
344. T. Yamada, H. Hashimoto, T. Shiraishi, and H. Oku, *Mol. Plant–Micr. Interact., 2*: 256 (1989).
345. C. W. Basse, K. Bock, and T. Boller, *J. Biol. Chem., 267*: 10258 (1992).
346. C. W. Basse and T. Boller, *Plant Physiol., 98*: 1239 (1992).
347. J. Chappell and K. Hahlbrock, *Nature, 311*: 76 (1984).
348. C. L. Cramer, T. B. Ryder, J. N. Bell, and C. J. Lamb, *Science, 227*: 1240 (1985).
349. D. A. Samac and D. M. Shah, *Plant Cell, 3*: 1063 (1991).
350. M. D. van de Rhee, J. A. L. van Kan, M. T. Gonzáles-Jaén, and J. Bol, *Plant Cell, 2*: 357 (1990).
351. R. A. Dixon, M. K. Bhattacharyya, M. J. Harrison, O. Faktor, C. J. Lamb, G. J. Loake, W. Ni, A. Oommen, N. Paiva, B. Stermer, and L. M. Yo, *Advances in Molecular Genetics of Plant Microbe Interactions*, Vol. 2 (E. W. Nester and D. P. S. Verma, eds.), Kluwer Academic Publishing, Dordrecht, The Netherlands, p. 497 (1993).
352. M. A. Lawton, S. M. Dean, M. Dron, J. M. Kooter, K. M. Kragh, M. J. Harrison, L. Yu, L. Tanguay, R. A. Dixon, and C. J. Lamb, *Plant Mol. Biol., 16*: 235 (1991).
353. B. J. C. Cornelissen and L. S. Melchers, *Plant Physiol.,* in press (1993).
354. C. T. Harms, *Crop Protection, 11*: 291 (1992).
355. C. J. Lamb, J. A. Ryals, E. R. Ward, and R. A. Dixon, *Bio/Technol., 10*: 1436 (1992).
356. L. Sequeira, *Advances in Molecular Genetics of Plant Microbe Interactions*, Vol. 2 (E. W. Nester and D. P. S. Verma, eds.), Kluwer Academic Publishing, Dordrecht, The Netherlands, p. 1 (1993).
357. K. Düring, M. Fladung, and H. Lörz, *Advances in Molecular Genetics of Plant Microbe Interactions*, Vol. 2 (E. W. Nester and D. P. S. Verma, eds.), Kluwer Academic Publishing, Dordrecht, The Netherlands, p. 573 (1993).
358. J. M. de la Fuente-Martínez, G. Mosqueda-Cano, A. Alvarez-Morales, and L. Herrera-Estrella, *Bio/Technol., 10*: 905 (1992).

12

Engineering Stress-Resistant Plants Through Biotechnological Approaches

Marc Van den Bulcke

Universiteit Gent
Ghent, Belgium

INTRODUCTION

Plants are constantly exposed to changes in their environment which require permanent adjustment of their cellular metabolism. Diurnal light cycles, temperature and humidity changes, and fluctuating nutrient resources are just a few factors that influence the growth and development of plants. External changes monitor the expression of specific gene patterns in such a way that the plant can cope with the altered situation. Extreme environmental changes impose a stress on the organism and can lead to drastic limitations on normal plant development. Stress-mediated damage ranges from barely detectable to reduced growth and even death of the plant.

In agricultural terms environmental stress is an important factor in yield reduction. Plant breeding programs have often succeeded in selecting suitable, more resistant crops. Inherent problems with this strategy are multiple: breeding programs are labor-intensive, time-consuming, and, as a consequence, expensive. Moreover, selected genetic resistance traits are only transmissible between compatible species. An approach that could overcome these drawbacks and present a simple, widely applicable strategy to develop stress-resistant varieties would have a major impact on crop agriculture.

Biotechnology represents the most promising way to increase stress resistance of plants. There are many different aspects as to how agriculture can benefit from biotechnological processes. Here, we will only discuss the possibilities that genetic engineering and recombinant DNA technology offer in improving stress resistance. A major breakthrough in this fascinating research field has been the development of the molecular tools to efficiently introduce new genetic traits into plants. We will describe some major successful applications in engineering pest resistance and herbicide tolerance, and discuss prospectives in biocontrolling temperature, salt, heavy metal, and oxidative stress. It will become apparent that often research is only at the foot of the hill (or should we say mountain). However, the established knowledge in plant physiology and ecology, combined with an ever increasing effort in the biochemical and molecular analysis of plant environment relations, should lead to stress-resistant crops in the near future.

TRANSGENIC PLANTS: A BREAKTHROUGH IN CROP ENGINEERING

Apart from identifying potential genetic traits that can influence stress resistance, a methodology allowing efficient introduction and stable inheritance of the desired characteristic, is essential. Several strategies have been devised to stably transform plants [1,2]. Here, we will discuss two well-established methods: *Agrobacterium tumefaciens*-mediated DNA transfer and naked DNA electroporation. Both methods have been applied in many different species (both monocotyledonous and dicotyledonous plants), are rapid, relatively simple, and do not require sophisticated laboratory equipment, which makes them generally applicable all over the world.

Agrobacterium tumefaciens is a Gram-negative bacterium, which causes crown gall tumor formation on a large number of plants, especially *Solanaceae*. Tumor-inducing functions are localized on a large plasmid, the Ti or tumor-inducing plasmid (for reviews, see [3–5]). A particular DNA fragment, the T-DNA, bordered by 25 bp repeats, was shown to be specifically transferred to the plant and responsible for tumor formation [6,7]. Nononcogenic strains have been generated and it was shown that any DNA sequence put between the 25 bp borders can be transferred to and is stably inherited in the transgenic plant [6]. A second important Ti-linked region, the *vir*(ulence) region, controls a number of processes essential at distinct stages of the transformation [8]. It was demonstrated that *vir* functions when supplied *in trans*, can drive transfer of any DNA sequence bordered by 25-bp repeats [9,10]. These properties formed the basis for the development of two types of transformation vectors: (1) the cointegrate vectors where the DNA fragment is inserted between the borders through a recombinational event, and (2) a binary vector system, where the DNA fragment is cloned between the borders in any suitable vector and the *vir* functions are localized on a different replicon (for more details, see [2]). Both vector systems have been successfully used to transform different dicotyledonous and monocotyledonous plant species. Recent progress in transforming forest trees demonstrates the general importance of *Agrobacterium tumefaciens* as a tool to introduce new genetic traits into plants [11–13].

A second important line of gene transfer techniques is based upon naked DNA introduction into vegetative cells. The most widely applied method is DNA transfer through electroporation [14,15]. In short, pure linearized or circular DNA molecules are introduced directly into plants cells by subjecting the latter to a short, high-voltage electric pulse. In this way, pores are generated into the plasma membrane allowing large molecules, such as DNA, to diffuse freely into the cell. Voltage strength and pulse duration depend mainly on the source of the protoplasts and their diameter. The major prerequisite for successful application of the electroporation technique are the efficient generation of protoplasts, cells (partially) stripped from their cell wall, without losing the capacity to regenerate into entire, fertile plants. Protoplasting plant tissue, using macerating (fungal) enzymes, is widely applied, but the regeneration into healthy plants is for a large number of species still complicated or even impossible. An interesting feature of the electroporation technique is the possibility to directly introduce naked DNA into viable tissue [16]. This approach might be particularly interesting in obtaining transgenic plants from species with only limited numbers of regenerable cells (e.g., *Oryza sativa*). Next to electroporation, other naked DNA transfer methods have been described [1,17]. Apart from the high sophisticated requirements of some of these techniques, only a limited number of experiments have been described using these methods and their usefulness and potential in becoming general tools to transform plants is still awaited.

HOW TO TACKLE STRESS RESISTANCE IN PLANTS?

Having the tools at hand to introduce new genetic traits into plants, the next important step is to identify which genes can mediate increased stress resistance of the host. The final success of any applied strategy to limit stress effects will rely on a thorough and precise insight into how stress mediated damage occurs and into the mechanism(s) plants have evolved to reduce these adverse effects. Some general approaches currently pursued which promise to produce some interesting lines of research to identify stress resistance genes are discussed.

Stress Proteins: Potential Plant Stress Downers

A typical feature of the response of plants to (severe) changes in their environment is the selective biosynthesis of specific sets of proteins. Different

sets of "stress" proteins have been identified upon exposure of plants to high
and low temperature, to high salt concentrations, to drought, to heavy metals,
to UV light, to pathogen infection, to oxidative environment, to anaerobiosis.
The precise function of these stress proteins in the protective response is still
poorly understood. The fact that these polypeptides are rapidly induced upon
stress exposure and accumulate only transiently in the tissue, might indicate
that they play a role in the immediate limitation of the damaging effects
caused by the imposed stress.

At the moment, representative genes of the different classes of stress
proteins have already been isolated and characterized. In most cases, these
genes have been isolated by differential screening of cDNA or genomic libraries and only represent strongly induced members within the respective
sets of stress proteins. Alternative approaches identified a number of stress
proteins which are expressed at lower levels, through tedious subcellular
fractionation, protein chemical analysis and microsequencing [10]. Molecular analyses provided a vast amount of information on how their expression and accumulation is regulated upon and during the stress response.
Detailed analysis of the 5'-regulatory promoter region has allowed identification of stress-responsive enhancer and silencer sequences and the search
for stress-regulating DNA-binding factors continues [19–21]. Transgenic
plants harboring chimeric fusions between 5'-regulatory sequences of stress
genes and a reporter gene, together with immunochemical analyses, documented tissue and organelle specificity of the respective gene products. In
spite of all these data, the precise role of most of these polypeptides in the
plant stress response is still not clear.

Two major approaches are currently being followed to answer the above
question: (1) by constitutively overexpressing one or several members of the
respective classes of stress proteins, and (2) by introducing "antisense" constructions which modulate the inactivation of a particular stress protein itself. If these proteins play an important role in limiting damage, a positive
effect is expected in the first case, while the second approach should result
in more pronounced damage to the plant. Although apparently very simple,
enhancing or reducing the presence of only one member within a class of
stress proteins seems to have little impact on the observed response [22].
More pronounced effects will probably be detected if the expression of a
whole set of stress proteins is enhanced or repressed in a concerted way.
Major efforts are currently made to identify the signal transduction pathways that control and activate the expression of the stress proteins [23]. Approaches using transgenic plants which harbored chimeric gene fusions between 5'-regulatory promotor sequences of stress proteins linked to selectable

markers or reporter genes, in combination with random or directed mutagenesis, will provide an elegant "stress regulatory gene" retrieval system.

Plant Mutants as a Source for
Stress Resistance Genes

A second field of major interest for isolating stress resistant traits is the broad
variety of plant mutants available for most environmental stress situations
in different species. A simple gene isolation technique, such as shot-gun
cloning, is not yet feasible in isolating the corresponding plant genes due to
the large genome size and the lack of an efficient, simple assay system. In
some cases, direct complementation of mutant *Escherichia coli* or yeast strains
with plant cDNA clones has been successfully used to identify a particular
functional plant gene [24]. However, a selectable phenotype for certain resistance traits is often not expressed in these organisms. The most widely
applied strategies to identify these particular genes will be gene-tagging techniques (either transposon or T-DNA mediated), and physical mapping of
the respective traits by linkage analysis (aided through RFLP mapping and
chromosome walking) [25,26].

Gene-tagging techniques have been extensively described for maize,
basically as a result of the detailed description of its transposable elements.
In this way, up to a dozen different genes have been isolated. This method
has two limitations: (1) the high copy number of the Tn elements, and (2) the
high rate of spontaneous mutations [27]. Several strategies have been worked
out to use transposons as a gene-tagging tool and promising results are likely
in the near future [28,29]. T-DNA-mediated gene tagging could provide an
elegant alternative to the above method. The high transformation rate and
the usual low copy number of inserted T-DNAs makes it a highly potential
gene-tagging system, especially in combination with host plants with a small
genome (e.g., *Arabidopsis thaliana, Oryza sativa*) [30]. Already a number
of interesting plant developmental genes have been isolated through T-DNA
tagging strategies [31].

Some stress resistance traits have already been mapped in great detail
and in some cases these loci are closely linked to genetic markers that have
been isolated and characterized [32]. When a high-density map of that species is available, suitable hybridization probes can be designed and efforts
can be made to identify the resistance gene through chromosome walking.
The isolation of these genes will be greatly facilitated through developing
yeast artificial chromosomes (YACs). When applied to the model plant *Arabidopsis thaliana* and having the molecular and genetic tools at hand (including a detailed physical and genetic map and the availability of a com-

plete Contig library), these gene isolation techniques will prove to represent an important aid in defining and isolating stress resistance traits [33,34].

Increasing Stress Resistance Applying "Alien" Genes

Both the above strategies initiate the quest for stress resistance traits through biochemical and genetic analysis of stress responsive elements of the plant. Alternatively, one could envisage the introduction of stress-limiting elements or even pathways, originating from other organisms, particularly bacteria and fungi. Whilst the idea might sound like science fiction, some basic concepts in establishing efficient transformation procedures of plants owe everything to these organisms. The commonly used selectable or screenable markers in transformation experiments are all of bacterial origin (e.g., NPTII, CAT, β-Gal, β-GUS). These reporter genes are not only used to identify transgenic plants; NPTII, GUS amongst others, are currently applied as molecular tools to study gene expression at multiple levels, to isolate novel plant genes, to study protein trafficking, etc.

A number of bacterial genes have already been used in approaches dealing with such agriculturally important issues as pest and herbicide resistance, male sterility, etc. (p. 445). The multitude of microbial organisms that colonize awkward, detrimental environments will constitute a major source of stress resistance or stress adaptation genes. Just think of the microorganisms that withstand extreme temperature and pressure, tolerate high salt concentrations, and cope easily with astonishingly high concentrations of heavy metals. Unravelling the pathways which mediate survival under these adverse conditions and identifying the key enzymes and/or the regulatory proteins, compatible with plant metabolic pathways, will remain a very exciting research area. This is especially the case when combined with cell specific regulatory sequences; in such circumstances "alien" genes can provide a rich pool of stress resistance elements.

PEST RESISTANCE: A GOAL WITHIN REACH OF THE POTENTIALS

Plant-pathogenic microorganisms are an important cause of severe crop losses. A lot of effort has been put in the development of pest-resistant varieties, but until now pathogens kept up the pace remarkably well. Fortunately, plants are susceptible to only a minority of existing microorganisms. However, when cultivated under agricultural conditions, the subtle balance between plant and surrounding bacteria, fungi, viruses, nematodes, and insects may easily be disturbed. Monoculture has been recognized as a major cause of epidemic pest explosions. Culturing plants in nonnative localizations can result in a dramatic burst of new diseases (e.g., potato infection by the Colorado beetle). Another result of human impact on crop resistance has been the prolonged selection for an increased crop yield, neglecting interest in retaining pest resistance.

At the moment, pest-controlling strategies mainly involve applying chemical compounds, but these approaches are under severe constraints. Not only are several of these pesticides toxic to the environment (including to animals and humans), these products are getting less effective due to the appearance of pesticide resistant species. Alternative protection strategies are urgently needed and developing pest resistant crops is a major goal in plant biotechnology. Two classics in tailoring pest resistance (*Bacillus thuringiensis* endotoxin and viral coat protein-mediated resistance) are discussed and we will point out some lines of research that can form the platform for engineering broad pest resistance.

Insect Resistant Plants: A Dual Success Story

Bacillus thuringiensis *Toxins,*
A Powerful Bacterial Protector

The application of *Bacillus thuringiensis* (Bt) to monitor insect pests (especially Lepidoptera) has been extensively used for a long time [35]. Two important inconveniences of this method are high production costs and the need of carefully timed repeated treatments. The usefulness of developing transgenic Bt crops has been recognized very early and a number of successful applications have already been reported [36–38]. Insect toxicity of Bt toxins relies on the presence of the δ-endotoxin in the parasporal crystals [39]. Different Bt strains contain distinct toxins which monitor specificity toward different classes of insects. The active compound is synthesized as a larger precursor (molecular mass, 130 kDa) and processing occurs in the insect gut [40]. Thus, only a small part of the protein ($\pm$ 68 kDa) is needed to obtain full toxicity, which resulted in the construction of truncated Bt toxin genes, deleting any redundant domains of the precursor [37,38].

Major research has been conducted using the lepidopteran specific δ-endotoxins of the Bt strains *Berliner* and *Kurstaki*. Their respective δ-endotoxin encoding genes have been cloned, and truncated toxin genes were expressed in a number of plants including tobacco and tomato [37,38]. The transgenic plants harbored high resistance toward larvae of the lepidopteran *Manduca sexta*, at toxin concentrations similar to the ones routinely applied by spraying commercial dry-powder Bt toxin [41]. Resistance was not only found in leaf tissue but also tomato fruit were protected [37]. A technical drawback in using Bt genes has been the low expression level of the toxin

(<0.02% of total leaf protein). Increasing the Bt expression level is one potential way to broaden the protective range and designing an artificial Bt toxin gene optimized to plant codon usage could be an interesting option. Also, the isolation of toxin genes from *Bacillus thuringiensis* strains with specificity toward other insect classes is being pursued. Finally, having elucidated the three-dimensional structure of the Bt toxin [42], a better understanding of the cytotoxicity of the Bt toxin and especially the identification of the Bt receptor protein, would allow us to develop new strategies in engineering insect resistance using this bacterial toxin.

Proteinase Inhibitors: Endogenous Phytoprotectors

Proteinase inhibitors (PI) are common to all classes of organisms. In plants, PI are mainly restricted to storage organs (tubers, seeds, etc.). Their function would be twofold: (1) to prevent uncontrolled proteolysis, and (2) to protect plant tissues against foreign proteases. It has become clear that the defense response of plants to pathogens is an extremely complex process and PI represents only one aspect amongst many others. Studies with resistant and sensitive cowpea seeds to bruchid beetle, *Callosobruchus maculatus*, provided initial evidence for the role of trypsin inhibitory activity in the resistance [43]. Purified cowpea trypsin inhibitor (CpTI), when added to artificial diets, had detrimental effects on larval development, only when applied at concentrations found in resistant seeds. Although, these data pointed out a potential role of CpTI in resistance, the importance of additional factors in resistance has been demonstrated [44,45].

Nevertheless, CpTI was an interesting target to apply in pest resistance. A wide range of insects were shown to be susceptible to CpTI, including members of lepidoptera and Coleoptera [46]. Thus, CpTI could function as a broad species toxin. The corresponding gene was cloned, constructions driving constitutive high concentrations of the CpTI in all plant tissues were made and transferred to plants [47]. CpTI accumulated to levels as high as 1% of soluble total leaf protein and was fully active as measured by inactivating bovine trypsin. Toxicity assays demonstrated that high levels of CpTI are necessary to obtain insecticidal toxicity, but its effectiveness applied to different species. CpTI does not stand alone; recently transgenic tobacco harboring PI of tomato or potato were shown to be fully resistant to *Manduca sexta*, demonstrating the general applicability of PI in developing insect resistant crops [48]. In contrast to the highly efficient, but narrow range Bt toxin, PI are effective to lesser extend but have the advantage of a broader toxicity spectrum. An interesting question is how double Bt/PI transgenic crops would perform. Indeed, cotransforming a pea lectin gene (*lec*A) and CpTI generated transgenic plants with increased resistance compared to plants which harbored only either of both genes [49,50].

A Key as to How to Eliminate
Your Virus of Preference

Virus infections cause severe plant diseases all over the world. Through classical breeding programs virus resistant plants have been obtained, but unfortunately often a severe quality and/or yield penalty had to be paid. Nowadays three distinct ways of controlling viral infection are common: The first implies establishing virus-free seed or plantlet stocks. The second implicates the elimination of virus transmission vectors, essentially through extensive pesticide usage. The third way applies molecular biology technology to develop virus resistant crops. The latter will become more and more important because this method limits environmental damage and is in theory applicable to any stably transformable plant species. Several strategies have been worked out to date and we refer to a recent review for more detailed discussion [51].

One method applies the viral coat proteins (CP) and is known as "coat protein-mediated" virus resistance. Transgenic plants expressing viral CP, were found to be highly protected toward viral infection [52,53]. The actual protection mechanism is still largely unclear, but it has been suggested that the presence of high concentrations of CP prior to infection would interfere with uncoating of the virus [53–55]. A clear correlation between resistance and CP concentration could be drawn, suggesting that resistance is linked to CP only. Protection was not only obtained to the original virus, but could also be demonstrated against other (more or less) related strains. A threshold value of homology between the CP of the respective viruses is in general needed, although interference with infection of unrelated viruses has been reported [56]. In some cases, field trials have already being performed and initial results are promising; yield increases of 20–30% were obtained [57, 58]. Although this method has been widely applied, a number of questions still remain. Does CP-mediated resistance work against all types of viruses (DNA viruses, double stranded RNA viruses)? In the case of viruses with multiple CP, does overexpressing only one CP give resistance or are all of the CPs necessary? Can CP-mediated protection sustain well environmental constraints such as temperature changes, drought, and poor soil conditions, etc.

A second approach in the pursuit of virus resistant crops applies expressing sense or antisense viral RNA. Antisense RNA applied to the 3' region of different CP has been introduced into host plants and conferred resistance toward the respective viruses [59–61]. The level of effectiveness was, however, significantly lower than in CP-mediated protection. Probably, the lower effectiveness is due to the inherent leakiness of this strategy, inhibiting only 90–95% of gene expression [62]. Possible improvements could include increasing the expression of the antisense construct, through applying amplification vectors [63], or carefully choosing the region to be inactivated [64]. The sense RNA strategy has allowed to generate tobacco mosaic virus (TMV) resistant lines, expressing the P54 gene at constitutive high levels [65].

Several other strategies have been used to develop virus resistance. Complete protection was obtained through introducing benign satellite viruses [66,67]. No cross-protection against other viruses could be detected, and more importantly, often reversion from benign to malign satellite virus only requires a few bp mutations [68], leaving the usefulness of this method open to debate. The recently discovered "ribozymes," RNA molecules with enzymatic activity, could also be modulated to attack viral genomes [69,70]. The potential use of ribozymes will, however, depend on their in vivo stability and expression level, cleavage efficiency, and recycling ability [51].

General Pest Resistance: Dream or Reality?

The appropriate choice of "molecular blockers" [51] might provide an efficient tool in developing pest resistance. The above described protection strategies only apply to a restricted number of pathogens and the development of a general resistance strategy remains a major goal. In view of the presence of common defensive responses of plants to viruses, bacteria, fungi, and even some nematodes, we favor the idea that it will become technically feasible to establish a general resistance strategy. The plant defense response is a complex process, controlled at multiple levels and probably integrating a large number of protective pathways. Unravelling the mechanism of a common general active defense response of the plant would greatly facilitate the progress toward broad pest resistance. In our opinion, the best suited candidate is the hypersensitive response (HR), which, by definition, results from an incompatible interaction between plant and pathogen and is characterized by a rapid dying of the infected region. The generality of this response is determined at multiple levels: (1) HR occurs both in monocotyledonous and dicotyledonous plants; (2) HR has been described for viruses, bacteria, and fungi; (3) the respective responses occurring during HR triggered by different organisms are highly similar. Thus, a clear insight into how HR is regulated and controlled should allow to develop a common resistance strategy to a large number of important pathogens.

Hypersensitive response has been extensively studied in many plant-pathogen interactions. Genetically, it was defined through the interaction of a pathogen avirulence gene product and a plant resistance gene [71]. Recently, a number of avirulence genes have been isolated and cloned, but no clear function could be assigned to them [72–74]. To date, no plant resistance genes have been identified. A nematode resistance gene of tomato is the most likely candidate for cloning and characterization in the near future [32]. Thus, although precisely defined genetically, a firm molecular understanding of HR is still awaited.

On the other hand, many distinct biochemical and physiological changes associated with HR have been described, and genes involved in these processes have been isolated (for recent reviews, see [75,76]). In vitro experiments demonstrated a potential role of some of these defense-related genes in directly attacking or interfering with pathogen growth [77,78]. Their in vivo effectiveness is, however, still unclear [22]. Basically, it has been suggested that a concerted activation of the different defense pathways, might be essential for establishing increased resistance.

Several groups have reported that treatment of plant cells with certain chemical elicitors protects them from pathogen induced cytotoxic effects [79,80]. Also, applying the heptoglucan elicitor of *Phytophthora megasperma* induced resistance toward a large number of viruses [81]. Whereas salicyclic acid treatment results in the induction of defense pathways similar to the ones induced during HR [82], the heptoglucan elicitor mediated protection did not result in triggering several of these defense genes, including phytoalexin biosynthesis and the production of pathogenesis-related proteins. These data indicate that an "artificial" treatment can trigger a fully effective plant defense response.

In view of the extremely rapid induction of a number of the defense responses, it is likely that only few intermediate steps are present between the initial recognition and the activation of the defense response [23]. Major efforts have been made to identify and isolate the carrier genes involved in the defense signal transduction pathway. To date, the overall scheme is starting to emerge and shows considerable analogy with signal transduction pathways in other organisms (comprising phosphorylated membrane proteins, plasma membrane elicitor-binding proteins, etc.). Combined with the quest for *trans*-acting factors controlling the transcription of defense genes upon infection, in the near future all tools necessary to modulate allaround pest resistance will be available. Next to these defense controlling molecules, genes involved in systemic acquired resistance (SAR) might be interesting targets. Recent evidence for the possible in vivo role of salicyclic acid in SAR, makes the isolation and cloning of genes involved in salicyclic acid synthesis one of the most intriguing aspects of plant protection [83–85].

HERBICIDE RESISTANCE:
THE ECOLOGICAL ANSWER

Uncontrolled growth of weeds in crop fields is an important issue in agriculture. For years, herbicide treatment was the only cheap and effective solution to this nuisance. We wish to include some thoughts on this aspect in plant engineering as herbicide resistance might drastically reduce environmental stress caused through extensive culturing of crops. High selectivity, eliminating only the weed not the crop is a major prerequisite for the usefulness of any compound as a herbicide. At the moment, selectivity is monitored at three levels: (1) the differential uptake capacity between weed and crop; (2) a precise timing of applying the herbicide; and (3) the resistance of the crop plants [86]. The general public and especially the many ecological organizations have stressed the necessity for highly effective, biodegradable, nontoxic (at least for humans) herbicides. A lot of research has been investigating possibilities to develop environment-compatible, highly efficient, and less expensive herbicides. The combination with genetically engineering crops, resistant or (more) tolerant to herbicides, provides the actual ultimate step toward rational weed control.

A considerable variability in herbicide resistance exists in nature. Several weed species are resistant to triazines and some species developed paraquat resistance [87,88]. Through extensive selection, mutants of tobacco and maize were obtained, resistant toward sylfonylureas and imidazoles [89, 90]. Often resistance was due to a single dominant mutation. In such cases, genetic engineering can offer a straightforward, valuable tool in developing herbicide resistant crops. Two strategies are being followed; the first by altering the sensitivity and/or concentration of the plant target enzyme, the second by introducing an "alien" gene encoding a herbicide detoxifying enzyme.

Increasing the expression level of herbicide target enzymes has been successful in a number of cases. Glyphosate resistant petunia have been obtained through overexpressing a plant 5-enolpyruvylshikimate 3-phosphate synthase (EPSPS), the natural target for glyphosate [91,92]. Also, enhanced activity of glutamine synthase, the target for the amino acid biosynthesis blocker phosphinothricin, resulted in resistant alfalfa lines [93]. Expressing a mutant Salmonella EPSPS gene or a mutant alfalfa glutamine synthase gene resulted in increased resistance towards the respective herbicides [94–96]. Some of these herbicide targets are retained in specific organelles (e.g., EPSPS in the chloroplasts) and fusing the appropriate targeting sequences to the corresponding bacterial genes can enhance their efficiency considerably [97].

The ability of a number of herbicides to detoxify enzymes from plants and bacteria has been identified. Strategies applying plant detoxifying enzymes have had little success, but engineering herbicide resistance using bacterial enzymes has been achieved in several cases.

The *Streptomyces hygroscopicus bar* gene, encoding a phosphinothricin acyltransferase, was introduced into tobacco, potato, and tomato, giving fully phosphinothricin-resistant transgenic plants [98,99]. A *Klebsiella oraenae* nitrilase gene (*bxn*) can confer increased tolerance of tobacco and tomato to the photosynthesis inhibitor, bromoxylim [100]. Also, the *tfd*A gene of *Alialigenes eutrophus*, encoding a 2,4-dichlorophenoxyacetate mono-oxygenase, was capable of generating 2,4-dichlorophenol-resistant tobacco [101, 102]. In general, strategies applying detoxifying enzymes are considered the more straightforward. Modifying the specificity or expression level of host target enzymes could affect normal cell metabolism and turn out to be detrimental. Of course, applying detoxifying enzymes can only be used when suitable and effective candidates are available.

Herbicide resistance genes have become a new asset in selection procedures during transformation and regeneration experiments. However, plant breeding programs could also benefit. Resistance genes could be physically linked to other agricultural traits, which are not easily detectable phenotypically and greatly facilitate tracing the "cryptic" trait at early stages in the crossing to other cultivars through selecting for herbicide resistance [86].

THE HEAVY METAL PROBLEM:
A PLANT SOLUTION

Although heavy metals such as Cu and Zn, etc., are essential cofactors for normal growth and development, at excess concentration these micronutrients and related compounds such as Cd, Hg, Ni, and Pb become highly toxic. Lethal concentrations of heavy metals accumulate near mining pits and industrial wastelands, but also in some natural soils. Some plant species evolved a naturally increased tolerance toward high metal concentrations. The temporal but rapid adaptation to heavy metal exposure has been well documented, but the biochemical and molecular mechanism of this response is still only poorly documented [103]. In order for any metal ion to become toxic the following conditions should be met. The metal ions should accumulate to toxic concentrations at the appropriate subcellular localization, have free accessibility toward the ligand, and bind strongly to the latter [104]. Obviously, data accumulated by in vitro experiments have to be carefully evaluated when extrapolating to the in planta situation. Although not of direct use in any food agricultural program, plants could be used as heavy metal bioconcentration device. Just imagine heavy metal accumulating grasses that extract toxic metals from the soil, are easily harvested, and disposed of; clearly, in this way detoxification of polluted soils and leakage of these compounds into the environment could be reduced.

High metal concentrations are toxic to several photosynthetic pathways. Chlorophyll biosynthesis, photosynthetic electron transport, and photorespiration are all highly sensitive to increased heavy metal concentrations

[105,106]. Some of the enzymes or proteins involved in these pathways have been identified and could represent potential targets to increase tolerance. The metal uptake mechanism into chloroplasts in particular deserves major attention because interference with chloroplast function mainly depends on the subcellular metal concentration [107]. Increased heavy metal ion accumulation affects peroxidase and lipoxygenase isozyme patterns, two classes of enzymes which can act as scavengers of highly toxic oxidative reagents, whose synthesis is often catalyzed by heavy metals [108–110]. The recent purification and cloning of different peroxidases will allow a more accurate evaluation of a potential function of this enzyme in heavy metal protection [111, 112].

Upon exposure of cells to high heavy metal concentrations, cells rapidly accumulate high amounts of heavy metal binding proteins [103]. In plants, two distinct classes are present: the phytochelatins and the metallothionins. Both classes of low molecular weight polypeptides contain high cysteine levels, but differ in structure and biosynthesis. While metallothionins are direct gene products, phytochelatins are nonprotein metal binding peptides in which the peptide bond is mediated through the γ-carboxylic acid of glutamic acid. Metallothionins are found in all species and the corresponding genes have been cloned from mouse, rat, human, and a number of other organisms [113]. Some of these exogenous metallothionins have been transferred to plants and increased tolerance to at least Cd was observed [114–116]. The recent cloning of plant metallothionin genes will allow us to further pursue the use of this class of metal binding proteins in engineering heavy metal resistance [117]. Indeed, applying endogenous metallothionins might not only solve tissue specificity problems [114], but also increase the effectiveness and eventually provide broader ion compatibility. The structure of phytochelatins has been studied in detail and their biosynthesis parallels the glutathione synthesis pathway, involving gamma-glutamyl-cysteine synthase (GCS) and glutamine synthase [118]. Phytochelatin synthase, the key enzyme, is constitutively expressed but highly induced upon heavy metal exposure [119,120], and forms an interesting target for increasing heavy metal tolerance. Cd tolerance in tomato paralleled the increased phytochelatin levels and GCS levels [121]. Comparison of resistant and sensitive *Datura innoxia* cell cultures did not demonstrate significant differences in phytochelatin concentrations [122]. However, the rate of metal complexation in the resistant line was much higher, suggesting the involvement of additional factors in controlling metal resistance. Interestingly, heavy metals also induce the expression of some heat-shock proteins (HSP). Some HSP contain chaperoning activity, an energy dependent assistance in protein assembling processes [123]. Heat-shock proteins might account for the increased metal tolerance in the resistant Datura cultures, and provide an elegant asset to engineering metal tolerance.

FISHING FOR HEAT AND COLD RESISTANCE TRAITS

Temperature changes are probably the most frequent and common environmental fluctuations plants are exposed to. While species are normally well adapted to temperature variations within their native range, once beyond it, severe adverse effects can be observed. Generating crops with ampler temperature tolerance could become very important in view of the predicted rapidly increasing global temperature and the never ending need to exploit less advantageous areas for crop cultivation. Stress effects imposed by high or low temperature are very specific, but in both cases water plays a key role. Heat-shock leads to increased transpiration and water loss, resulting in rapid wilting; low temperature induces extrusion of intracellular water accompanied with intracellular freezing and tissue injury. In contrast to this simple homology, the adaptive processes accompanying heat and cold stress are extremely complex and we refer to some reviews for detailed discussion of the physiological, biochemical, and molecular aspects of both [124–128]. We will describe some lines of research which could form the basis for engineering temperature tolerant crops.

The response of plants to increased temperature is very well characterized. Shifting the environmental temperature a few degrees above optimal growth temperature, results in a repression of normal protein and MRNA synthesis but induces the accumulation of a small set of heat-shock proteins (HSP) (for a recent review, see [128]). The heat-shock response is not unique to plants but occurs in members of all species and apparently the respective HSPs are highly conserved. Organisms can be preconditioned to tolerate normally lethal temperatures by preincubation at high, nonlethal temperatures. Evidence has been presented for the role of HSP in this short-term thermotolerance [126,127]. In maize and wheat varieties adapted to high temperature, HSP expression has been found constitutively [127]. In mammalian cells, overexpressing a small number of HSP successfully improved their thermotolerance [129]. In yeast, thermotolerance could be enhanced by altering the expression of only one HSP [130]. Thus, HSPs seem a good target for engineering increased thermotolerance. However, genetic data disprove this hypothesis as back-crossing thermosensitive with thermoresistant varieties, and did not reveal a clear linkage between thermotolerance and HSP expression [131]. More detailed studies overexpressing plant HSP should provide a precise answer as to the involvement of HSP in thermotolerance of plants.

Cold acclimation of plants is an adaptive response to low temperature resulting in elevated frost resistance and an increased tolerance to freeze-induced dehydration. Cold acclimatization influences a large number of processes including membrane lipid composition, increased sugar and soluble protein content, and the induction of a set of cold stress proteins [132,133]. Genetic studies in wheat and barley demonstrated that winter hardiness is controlled both by dominant and recessive alleles [134,135]. In pea, this process might be controlled by as little as three genes (or linkage groups) [136], which opens possibilities for isolating the respective genes through gene tagging strategies. Levitt stated that plants can be hardened by pretreatment for short periods at low temperatures [133]. The inducible, transient nature of cold acclimatization has been demonstrated in several species [137,138] and has allowed us to identify genes which specifically accumulate upon low temperature treatment [139,140]. In vivo and in vitro labeling studies have provided some evidence for an altered expression of at least 300 proteins at low temperature exposure [124]. It is clear that a lot more work is needed to explain the role of any of these genes in the cold acclimatization process [141,142].

Several interesting targets to monitor cold resistance are, however, already at hand. Low-molecular weight compounds such as sugars, glycine, betaine, proline, and polyamines accumulate to high concentrations upon low temperature exposure and are thought to act as cryoprotectants. A number of them are also increased upon drought and salt stress what could indicate their role as general protective components in processes mediated through water sequestration or loss (see p. 451). A number of genes involved in the biosynthesis of these cryoprotectants have already been characterized (sucrose synthase, sucrose phosphate synthase) and could represent valuable tools in monitoring cold resistance [143,144].

A number of polypeptides have the capacity to alter the freezing of water [145]. Insect hysteresis proteins [146], fish antifreeze proteins [147], and bacterial ice nucleation proteins [148,149] have either the potential to reduce or increase freezing. Ice nucleation protein has already been successfully applied in increasing frost resistance of tomato root tissue, although not through engineering transgenic plants. Applications of the other two classes of peptide cryoprotectors are still awaited. The recent identification of homologous proteins to fish antifreeze proteins in plants [150] strengthens the hypothesis that expressing non planta protectors in trangenic plants may attribute to some extend to cold resistance of crops.

SALINE SOILS: A TEMPTING WASTELAND

Water represents the most critical growth element of terrestrial plants. Although apparently present in excess, water uptake and storage by plants are complex, energy consuming processes. Plants face major problems, especially when growing in (semi) arid climates. In addition to water loss through extensive transpiration, soils in these climate zones often have high osmotic potentials causing increased water withdrawal. In terms of agriculture, these regions contain land that is promising for cultivation; soils are relatively un-leached and often fertile, the growing season is long with high temperatures and light intensities while humidity is low. All these properties could result in high productivity if only salinity problems can be overcome.

Plant survival on saline soils or in dry climates is basically mediated through two processes: water stress avoidance and water stress tolerance. Water stress avoidance evolved through developing specialized structural adaptations in root and shoot tissues [151] and sometimes by altering photo-synthetic pathways [152]. Water stress tolerance is accompanied by the ac-cumulation of osmoprotective solutes, and the biosynthesis of a large number of proteins [153]. Developmental adaptation is probably controlled by a complex mechanism through the interaction of several gene products. Straight-forward strategies to engineer any of these developmental processes are not available at the moment. Biochemical adaptory pathways involving only a limited number of steps could, however, represent an interesting target.

A general response to osmotic stress is the intracellular accumulation of common solutes such as sucrose, polyamines, and proline which are thought to act as osmoprotectants. These molecules would not only play a role in adjusting the osmotic potential of the cell but moreover help in preserving the overall cell structure. Although increased solute accumulation has been reported for many species, a direct correlation between solute accumulation and osmoprotection still has to be demonstrated. In adapted tobacco suspen-sion cultures, the accumulated proline only accounts for less than 5% of the intracellular osmotic potential [154,155]. Sometimes proline synthesis is only induced as part of a late response to osmotic stress [155], in stressed callus tissue high proline concentrations remain present even after transferring to normal growth conditions [156]. Other organic sugars such as sucrose, polyols, etc., contribute to similarly extend proline to the osmotic potential of the cell [154], but their contribution is not always consistent with stress toler-ance [157]. Polyamines have been proposed as osmoprotectants, stabilizing as diverse structures as nucleotides and membranes [158]. Genes involved in the biosynthesis of these molecules have been cloned or their isolation is well underway. Transgenic lines overexpressing and accumulating one or several of these respective osmoprotectants will provide an essential tool in designat-ing the role and potential of them in high salt resistance.

Exposure to high salt concentrations or adaptation to high osmotic po-tential, results in many changes in plant gene expression [159], and often distinct changes are observed in different tissues [160–162]. Altogether, the importance of any of these changes in osmotic adaptation is not clear. The most studied member of these osmotic stress proteins has been called "osmo-tin"; this protein was originally isolated from adapted tomato cell cultures but homologous proteins have been identified in other species [163,164]. At the sequence level, osmotin shows homology with one of the pathogenesis related proteins with the sweet tasting fruit protein thaumatin and with a bifunctional α-amylase/trypsin inhibitor [165–167]. No functional similarity with any of these classes of proteins could be determined, but recently it was shown that, under very specific in vitro conditions, osmotin has an antifungicidal activity [78]. A second important group are the late embryogenic abundant (LEA) proteins [168]. These proteins were initially identified during investigations of embryogenesis, and later studies demonstrated that at least some LEA proteins are inducible by salt stress and drought. In general, these polypeptides are highly hydrophilic and not particularly associated with any organelle. The dehydrin family has been found in rice, barley, cotton, and maize is extremely rapidly induced and it has been proposed that dehydrin could be part of the primary adaptive response [169]. Glycine-rich Proteins inducible by water stress have been found in maize and *Arabidopsis* [170,171] and some of them contain a RNA-binding domain. Also, the induction of the chaperoning HSPs [172] in response to drought or high salts could indicate that at least part of the protective response is dependent on stabilizing present macromolecules such as nucleotides and polypeptides. In rice, a small protein called SalT is rapidly induced and its distribution throughout the plant parallels Na^+-concentration in the respective tissue [162]. It was proposed that SalT would be involved in establishing a Na^+-gradient within the plant or participate in processes that restrict damage invoked by such gradient.

All these osmotic stress-influenced genes could represent targets for altering salt stress resistance of crops. In view of the complex physiological impact salts and drought have on plant cells, it seems unlikely that changing the expression of only one of these osmotic stress genes might be beneficial. Rather an integrated induction of several groups of them could be more efficient. For this a detailed understanding of the transcriptional regulation of the different osmotic stress genes is needed. To date no reports have been presented describing the isolation of osmotic stress regulating *trans*-acting factors or signal transduction carrier molecules. A lot more research will be needed to clarify these mysteries and successful engineering of drought- or salt-stress resistant crops is not on the horizon.

OXIDATIVE STRESS RESISTANCE: SUPEROXIDE DISMUTASES SHOW THE WAY

Plants are constantly exposed to toxic oxygen species such as hydrogen peroxide (H_2O_2), superoxide radicals (O_2^-), hydroxyl radicals ($OH\cdot$), and ozone (O_3). These oxidative molecules originate as by-products of oxidative cellular metabolism [173] or from environmental conditions, e.g., increased temperature combined with high light intensities [174–176]. Electron transport chains within the respective cellular compartments, especially mitochondria and chloroplasts, are well-defined sources of oxidating molecules [177]. In excess, these oxidantia can inhibit photosynthesis and result in photoinhibition, invoking major damage to pigments and membranes [178]. The most reactive hydroxyl radicals can react with almost all cellular constituents, including nucleotides, proteins, and lipids [177,179,180].

Plants have evolved a battery of mechanisms to detoxify and eliminate these highly reactive oxygen species. Most of these antioxidantia are common to other organisms, suggesting that all species deal with oxidative stress in a similar way. Three major classes can be distinguished: (1) antioxidantia which act as oxygen scavengers (glutathione, ascorbate, α-tocopherol); (2) peroxidases and catalases which remove hydrogen peroxide; and (3) superoxide dismutases which catalyze the conversion of superoxides to hydrogen peroxide. In respect to engineering oxidative stress-resistant crops, each of these classes can represent interesting targets. Altering the endogenous levels of any of these oxidative stress protective pathways must be carefully considered. In general, oxidative metabolism is highly integrated; eliminating only one oxygen species can easily drive equilibrium toward an excess of another toxic oxygen species, resulting in a null protection or, in the worst cases, in even more severe damage. To modulate efficiently the accumulation of deleterious oxygen species, a clear knowledge of the role of each of them and their participation in oxidative metabolism is essential. Here, we will only describe recent progress made in limiting oxidative stress by using plant superoxide dismutases (SODs).

Three classes of plant SODs can be distinguished, according to their metal cofactor: copper/zinc (Cu/Zn), iron (Fe), and manganese (Mn) superoxide dismutases. FeSOD accumulates mainly in the chloroplasts, MnSOD is found in the mitochondria, whereas Cu/ZnSOD is either cytoplasmic or chloroplastic (for reviews, see [181,182]). Members of each class have been cloned allowing detailed transcriptional and functional studies of the contribution of the each class in oxidative detoxification [24,183–187]. Initial studies demonstrated that MnSOD expression parallels mitochondrial metabolic activity [186]. Chloroplastic FeSOD expression was highly induced upon paraquat treatment, a herbicide mainly invoking oxidative stress in the chloroplasts. Also it was demonstrated that paraquat treatment also induces mitochondrial MnSOD and the cytoplasmic Cu/ZnSOD [187]. These results indicate that a concerted activation of multiple oxidative stress protective pathways might be needed to obtain significant protection.

Data on overproducing plant SODs in order to obtain oxidative stress resistance are somewhat conflicting. Initial resports demonstrated that a high overproduction of a petunia chloroplastic Cu/ZnSOD in tobacco did not alter sensitivity of the transgenic plants towards paraquat [188]. It was concluded that detoxification of paraquat-induced toxins requires more than just increasing chloroplastic SOD. However, SOD transgenics did not become hypersensitive what indicates that the H_2O_2-detoxifying chloroplastic systems can cope with the increased H_2O_2 accumulation resulting from the superoxide dismutation. A detailed study of the effects of overproducing and recompartimentalization of a *Nicotiana plumbaginifolia* MnSOD gave very different results [189]. Increased levels of chloroplast-directed MnSOD drastically reduced cellular damage upon oxidative stress invoked by paraquat. Increasing the level of mitochondrial MnSOD was only poorly effective. Unexpectedly, small increases in chloroplastic MnSOD were deleterious when paraquat treatments were performed in the dark. These data were explained by the synergistic action of SOD and H_2O_2-scavenging pathways under light conditions in eliminating or detoxifying the oxidative compounds [182,189]. Thus, it seems that SOD can be applied in generating resistance towards oxidative stress. The degree of protection will, however, be dependent on the level of overproduction, the type of SOD used, and the endogenous scavenging systems of the host [182].

Other targets for increasing oxidative stress resistance are peroxidases and catalases. cDNA clones encoding members of both classes of H_2O_2-detoxifying enzymes have already been cloned [140,141,192,193]. Multiple isoforms of peroxidases and catalases are present in plants and their expression is controlled by many factors during development or environmental stress [194]. It might be that overproducing or recompartmentalizing any of these enzymes results in "weird" phenotypes. Indeed, overproducing an extracellular peroxidase in tobacco did not interfere with normal development until plants started flowering; then, rapid wilting occurred [140]. One might also expect interference with cell wall assembly processes [195] and even pathogen resistance [196].

Thus, developing oxidative stress resistance might be more difficult than originally thought. As indicated by Shaaltiel and Gressel [197], successful resistance might only be obtained by modulating the whole oxidative protection response. Manipulating the overall control of this response would be most effective; in *Conyza bonariensis*, one nuclear gene is responsible for the activation of all the genes involved in paraquat resistance [198]. In analogy, also in crop plants a limited number of controlling genes might be present. Alternatively, one might attempt to broaden the applicability of SOD. Especially, a concerted elimination of the excess H_2O_2 produced by SOD action, through overproduction and relocalization of H_2O_2-eliminating enzymes could be attempted. Catalases would be the most suitable as they do not need any substrate for catalysis.

CONCLUSION

The advent of recombinant DNA technology and the establishment of efficient genetic transformation techniques has allowed us to study plant responses to environmental stress in detail at the molecular level. Some successful applications toward engineering stress-resistant crops have demonstrated the usefulness of molecular approaches in crop improvement. Our knowledge of how the adaptive or protective response of plants to adverse situations is organized, is increasing at dazzling speed, although we lack of clear insight into the concerted regulation of the different reactions to plants to a particular stress. Gradually, it is becoming clear that engineering only one characteristic into transgenic plants, is often not sufficient to obtain resistance. Future major efforts should, therefore, concentrate on identifying the mechanisms that control the protective response of plants to environmental stress.

ACKNOWLEDGMENTS

The author wishes to thank Dr. D. Inzé and Dr. C. Simoens for critical reading; Professor M. Van Montagu and Dr. D. de Oliveira for encouragement, and Dr. M. De Cock for help with the manuscript.

REFERENCES

1. I. Potrykus, *Ann. Rev. Plant Physiol. Plant Mol. Biol., 42*: 205 (1991).
2. J. Draper and R. Scott, *Plant Genetic Engineering* (Plant Biotechnology, Vol. 1) (D. Grierson, ed.), Blackie, Glasgow, Scotland, p. 38 (1991).
3. G. Gheysen, P. Dhaese, M. Van Montagu, and J. Schell, *Genetic Flux in Plants* (Advances in Plant Gene Research, Vol. 2) (B. Hohn and E. S. Dennis, eds.), Springer Verlag, Vienna, Austria, p. 11 (1985).
4. H. Klee, R. Horsch, and S. Rogers, *Ann. Rev. Plant Physiol., 38*: 467 (1987).
5. P. Zambryski, J. Tempé, and J. Schell, *Cell, 56*: 193 (1989).
6. P. Zambryski, H. Joos, C. Genetello, J. Leemans, M. Van Montagu, and J. Schell, *EMBO J., 2*: 2143 (1983).
7. K. Wang, L. Herrera-Estrella, M. Van Montagu, and P. Zambryski, *Cell, 38*: 455 (1984).
8. S. E. Stachel and P. Zambryski, *Cell, 47*: 155 (1986).
9. A. Koekema, P. R. Hirsch, P. J. J. Hooykaas, and R. A. Schilperoort, *Nature, 303*: 179 (1983).
10. M. Bevan, *Nucl. Acids Res., 12*: 8711 (1984).
11. J. J. Fillatti, J. Kiser, R. Rose, and L. Comai, *Bio/technology, 5*: 726 (1987).
12. G. H. McGranahan, C. A. Leslie, S. L. Uratsu, L. A. Martin, and A. M. Dandekar, *Bio/technol., 6*: 800 (1988).
13. S. Mante, P. H. Morgens, R. Scorza, J. M. Cordts, and A. M. Callahan, *Bio/technol., 9*: 853 (1991).

14. R. D. Shillito, M. W. Saul, J. Paszkowski, M. Müller, and I. Potrykus, *Bio/technol., 3*: 1099 (1985).
15. M. E. Fromm, L. P. Taylor, and V. Walbot, *Nature, 319*: 791 (1986).
16. R. A. Dekeyser, B. Claes, R. M. U. De Rycke, M. E. Habets, M. Van Montagu, and A. B. Caplan, *Plant Cell, 2*: 591 (1990).
17. S. G. Rogers, *Curr. Opin. Biotechnol., 2*: 153 (1991).
18. G. Bauw, M. De Loose, D. Inzé, M. Van Montagu, and J. Vandekerckhove, Proceedings National Academy of Sciences (U.S.), *84*: 4806 (1987).
19. D. Roby, K. Broglie, R. Cressman, P. Biddle, I. Chet, and R. Broglie, *Plant Cell, 2*: 999 (1990).
20. M. D. Van de Rhee, J. A. L. Van Kan, M. T. González-Jaén, and J. F. Bol, *Plant Cell, 2*: 357 (1990).
21. R. Wingender, H. Röhrig, C. Höricke, and J. Schell, *Plant Cell, 2*: 1019 (1990).
22. H. J. M. Linthorst, R. L. J. Meuwissen, S. Kauffmann, and J. F. Bol, *Plant Cell, 1*: 285 (1989).
23. C. J. Lamb, M. A. Lawton, M. Dron, and R. A. Dixon, *Cell, 56*: 215 (1989).
24. W. Van Camp, C. Bowler, R. Villarroel, E. W. T. Tsang, M. Van Montagu, and D. Inzé, Proceedings National Academy of Sciences (U.S.), *87*: 9903 (1990).
25. N. V. Fedoroff, D. B. Furtek, and O. E. Nelson Jr., Proceedings National Academy of Sciences (U.S.), *81*: 3825 (1984).
26. S. D. Tanksley, N. D. Young, A. H. Paterson, and M. W. Bonierbale, *Bio/technol., 7*: 257 (1989).
27. J. I. Yoder, *Plant Cell, 2*: 723 (1990).
28. R. Hehl and B. Baker, *Plant Cell, 2*: 709 (1990).
29. J. D. G. Jones, F. Carland, E. Lim, E. Ralston, and H. K. Dooner, *Plant Cell, 2*: 701 (1990).
30. C. Koncz, N. Martini, R. Mayerhofer, Z. Koncz-Kalman, H. Körber, G. P. Redei, and J. Schell, Proceedings National Academy of Sciences (U.S.), *86*: 8467 (1989).
31. Z. Schwarz-Sommer, P. Huijser, W. Nacken, H. Saedler, and H. Sommer, *Science, 250*: 931 (1990).
32. J. M. M. J. G. Aarts, J. G. J. Hontelez, P. Fischer, R. Verkerk, A. van Kammen, and P. Zabel, *Plant Mol. Biol., 16*: 647 (1991).
33. C. Chang and E. M. Meyerowitz, *Curr. Opin. Biotechnol., 1*: 178 (1991).
34. E. M. Meyerowitz, *Cell, 56*: 263 (1989).
35. H. T. Dulmage, *J. Invert. Pathol., 15*: 232 (1970).
36. K. A. Barton, H. R. Whiteley, and N.-S. Yang, *Plant Physiol., 85*: 1103 (1987).
37. D. A. Fischhoff, K. S. Bowdish, F. J. Perlak, P. G. Marrone, S. M. McCormick, J. G. Niedermeyer, D. A. Dean, K. Kusano-Kretzmer, E. J. Mayer, D. E. Rochester, S. G. Rogers, and R. T. Fraley, *Bio/technol., 5*: 807 (1987).
38. M. Vaeck, A. Reynaerts, H. Höfte, S. Jansens, M. De Beuckeleer, C. Dean, M. Zabeau, M. Van Montagu, and J. Leemans, *Nature, 328*: 33 (1987).
39. T. A. Angus, *Nature, 173*: 545 (1954).
40. V. F. Sacchi, P. Parenti, G. M. Hanozet, B. Giordana, P. Lütly, and M. G. Wolfersberger, *FEBS Lett., 204*: 213 (1986).

41. H. Van Mellaert, H. Höfte, A. Reynaerts, and M. Vaeck, *Plant–Microbe Interactions, Molecular and Genetic Perspectives*, Vol. 3 (T. Kosuge and E. W. Nester, eds.), McGraw-Hill, New York, p. 3 (1989).
42. J. Li, J. Carroll, and D. J. Ellar, *Nature,* 815 (1991).
43. A. M. R. Gatehouse, J. A. Gatehouse, P. Dobie, A. M. Kilminster, and D. Boulter, *J. Sci. Food Agric., 30*: 948 (1979).
44. R. J. Redden, P. Dobie, and A. M. R. Gatehouse, *Austral. J. Agric. Res., 34*: 681 (1983).
45. A. M. R. Gatehouse, K. J. Butler, K. A. Fenton, and J. A. Gatehouse, *Entomol. Exp. Appl., 39*: 279 (1985).
46. A. M. R. Gatehouse and D. Boulter, *J. Sci. Food Agric., 34*: 345 (1983).
47. V. A. Hilder, A. M. R. Gatehouse, S. E. Sheerman, R. F. Barker, and D. Boulter, *Nature, 330*: 160 (1987).
48. R. Johnson, J. Narvaez, G. An, and C. Ryan, Proceedings National Academy of Sciences (U.S.), *86*: 9871 (1989).
49. D. Boulter, G. A. Edwards, A. M. R. Gatehouse, J. A. Gatehouse, and V. A. Hilder, *Crop Protect., 9*: 351 (1990).
50. J. A. Gatehouse, V. A. Hilder, and A. M. R. Gatehouse, *Plant Genetic Engineering* (Plant Biotechnology, Vol. 1) (D. Grierson, ed.), Blackie, Glasgow, Scotland, p. 105 (1991).
51. K. W. Buck, *Plant Genetic Engineering* (Plant Biotechnology, Vol. 1) (D. Grierson, ed.), Blackie, Glasgow, Scotland, p. 136 (1991).
52. P. Powell Abel, R. S. Nelson, B. De, N. Hoffmann, S. G. Rogers, R. T. Fraley, and R. N. Beachy, *Science, 232*: 738 (1986).
53. R. S. Nelson, P. Powell Abel, and R. N. Beachy, *Virology, 158*: 126 (1987).
54. D. R. Gallie, D. E. Sleat, J. W. Watts, P. C. Turner, and T. M. A. Wilson, *Science, 236*: 1122 (1987).
55. J. C. Register, III and R. N. Beachy, *Virology, 166*: 524 (1988).
56. E. J. Anderson, D. M. Stark, R. S. Nelson, P. A. Powell, N. E. Tumer, and R. N. Beachy, *Phytopathol., 79*: 1284 (1989).
57. R. S. Nelson, S. M. McCormick, X. Delannay, P. Dubé, J. Layton, E. J. Anderson, M. Kaniewska, R. K. Proksch, R. B. Horsch, S. G. Rogers, R. T. Fraley, and R. N. Beachy, *Bio/technol., 6*: 403 (1988).
58. W. Kaniewski, C. Lawson, B. Sammons, L. Haley, J. Hart, X. Delannay, and N. E. Tumer, *Bio/technology, 8*: 750 (1990).
59. M. Cuozzo, K. M. O'Connell, W. Kaniewski, R.-X. Fang, N.-H. Chua, and N. E. Tumer, *Bio/technol., 6*: 549 (1988).
60. C. Hemenway, R.-X. Fang, W. K. Kaniewski, N.-H. Chua, and N. E. Tumer, *EMBO J., 7*: 1273 (1988).
61. P. A. Powell, D. M. Stark, P. R. Sanders, and R. N. Beachy, Proceedings National Academy of Sciences (U.S.), *86*: 6949 (1989).
62. L. S. Robert, P. A. Donaldson, C. Ladaique, I. Altosaar, P. G. Arnison, and S. F. Fabijanski, *Plant Mol. Biol., 13*: 399 (1989).
63. R. J. Hayes, R. H. A. Coutts, and K. W. Buck, *Nucl. Acids Res., 17*: 2391 (1989).

64. A. Hirashima, S. Sawaki, Y. Inokuchi, and M. Inouye, Proceedings National Academy of Sciences (U.S.), *83*: 7726 (1986).
65. Y. Inokuchi and A. Hirashima, *J. Virol., 61*: 3946 (1987).
66. D. C. Baulcombe, G. R. Saunders, M. W. Bevan, M. A. Mayo, and B. D. Harrison, *Nature, 321*: 446 (1986).
67. B. D. Harrison, M. A. Mayo, and D. C. Baulcombe, *Nature, 328*: 799 (1987).
68. P. Palukaitis, *Mol. Plant-Microbe Interact., 1*: 175 (1988).
69. K. Kruger, P. J. Grabowski, A. J. Zaug, J. Sands, D. E. Gottschling, and T. R. Cech, *Cell, 31*: 147 (1982).
70. A. J. Zaug, C. A. Grosshans, and T. R. Cech, *Biochemistry, 27*: 8924 (1988).
71. H. H. Flor, *Phytopathol., 32*: 653 (1942).
72. S. Kelemu and J. E. Leach, *Mol. Plant–Microbe Interact., 3*: 59 (1990).
73. D. Y. Kobayashi, S. J. Tamaki, D. J. Trollinger, S. Gold, and N. T. Keen, *Mol. Plant-Microbe Interact., 3*: 103 (1990).
74. D. Y. Kobayashi, S. J. Tamaki, and N. T. Keen, *Mol. Plant-Microbe Interactions, 3*: 94 (1990).
75. J. P. Carr and D. F. Klessig, *Genetic Engineering, Principles and Methods*, Vol. 11 (J. K. Setlow, ed.), Plenum Press, New York, p. 65 (1990).
76. D. J. Bowles, *Ann. Rev. Biochem., 59*: 873 (1990).
77. F. Mauch, B. Mauch-Mani, and T. Boller, *Plant Physiol., 88*: 936 (1988).
78. C. P. Woloshuk, J. S. Meulenhoff, M. Sela-Buurlage, P. J. M. van den Elzen, and B. J. C. Cornelissen, *Plant Cell, 3*: 619 (1991).
79. R. A. M. Hooft van Huijsduijnen, S. W. Alblas, R. H. De Rijk, and J. F. Bol, *J. Gen. Virol., 67*: 2135 (1986).
80. C. Masuta, M. Van den Bulcke, G. Bauw, M. Van Montagu, and A. B. Caplan, *Plant Physiol., 97*: 619 (1991).
81. M. Kopp, J. Rouster, B. Fritig, A. Darvill, and P. Albersheim, *Plant Physiol., 90*: 208 (1989).
82. R. F. White, *Virology, 99*: 410 (1979).
83. J. Malamy, J. P. Carr, D. F. Klessig, and I. Raskin, *Science, 250*: 1002 (1990).
84. J. P. Métraux, H. Signer, J. Ryals, E. Ward, M. Wyss-Benz, J. Gaudin, K. raschdorf, E. Schmid, W. Blum, and B. Inverardi, *Science, 250*: 1004 (1990).
85. E. R. Ward, S. J. Uknes, S. C. Williams, S. S. Dincher, D. L. Wiederhold, D. C. Alexander, P. Ahl-Goy, J.-P. Métraux, and J. A. Ryals, *Plant Cell, 3*: 1085 (1991).
86. D. M. Stalker, *Plant Genetic Engineering* (Plant Biotechnology, Vol. 1) (D. Grierson, ed.), Blackie, Glasgow, Scotland, p. 82 (1991).
87. J. Gressel and G. Ben-Sinai, G, *Plant Science, 38*: 29 (1985).
88. L. G. Hickok and O. J. Schwarz, *Plant Science, 47*: 153 (1986).
89. W. J. Netzer, *Bio/technol., 2*: 939 (1984).
90. G. W. Haughn and C. Somerville, *Mol. Gen. Genet., 204*: 430 (1986).
91. N. Amrhein, D. Johänning, J. Schab, and A. Schulz, *FEBS Lett., 157*: 191 (1983).
92. H. C. Steinruecken, A. Schulz, N. Amrhein, C. A. Porter, and R. T. Fraley, *Arch. Biochem. Biophys., 244*: 169 (1986).

93. L. Comai, D. Facciotti, W. R. Hiatt, G. Thompson, R. E. Rose, and D. M. Stalker, *Nature, 317*: 741 (1985).
94. G. Donn, E. Tischer, J. A. Smith, and H. M. Goodman, *J. Mol. Appl. Genet., 2*: 621 (1984).
95. J. J. Fillatti, J. Sellmer, B. McCown, B. Haissig, and L. Comai, *Mol. Gen. Genet., 206*: 192 (1987).
96. E. Tischer, S. DasSarma, and H. M. Goodman, *Mol. Gen. Genet., 203*: 221 (1986).
97. G. della-Cioppa, S. C. Bauer, M. L. Taylor, D. E. Rochester, B. K. Klein, D. M. Shah, R. T. Fraley, and G. M. Kishore, *Bio/technol., 5*: 579 (1987).
98. T. Murakami, H. Anzai, S. Imai, A. Satoh, K. Nagaoka, and C. J. Thompson, *Mol. Gen. Genet., 205*: 42 (1986).
99. M. De Block, J. Botterman, M. Vandewiele, J. Dockx, C. Thoen, V. Gosselé, R. Movva, C. Thompson, M. Van Montagu, and J. Leemans, *EMBO J., 6*: 2513 (1987).
100. D. M. Stalker and K. E. McBride, *J. Bacteriol., 169*: 955 (1987).
101. W. R. Streber, K. N. Timmis, and M. H. Zenk, *J. Bacteriol., 169*: 2950 (1987).
102. W. R. Streber and L. Willmitzer, *Bio/technol., 7*: 811 (1989).
103. J. C. Steffens, *Ann. Rev. Plant Physiol. Plant Mol. Biol., 41*: 553 (1990).
104. F. Van Assche and H. Clijsters, *Plant Cell Environm., 13*: 195 (1990).
105. H. Clijsters and F. Van Assche, *Photosynthesis Res., 7*: 31 (1985).
106. D. D. K. Prasad and A. R. K. Prasad, *J. Plant Physiol., 127*: 241 (1987).
107. W. H. O. Ernst, *Cadmium in the Environment* (J. O. Nriagu, ed.), John Wiley, New York, p. 639 (1980).
108. F. Van Assche, C. Put, and H. Clijsters, *Arch. Internat. Physiol. Biochim., 94*: 60 (1986).
109. L. Gora and H. Clijsters, *Biochemical and Physiological Aspects Ethylene Production in Lower and Higher Plants* (H. Clijsters, M. De Proft, R. Marcelle, and M. Van Poucke, eds.), Kluwer Academic Pubishing, Dordrecht, The Netherlands, p. 219 (1989).
110. E. F. Elstner, G. A. Wagner, and W. Schutz, *Curr. Topics Plant Biochem. Physiol., 7*: 159 (1988).
111. D. A. Dalton, F. J. Hanus, S. A. Russell, and H. J. Evans, *Plant Physiol., 83*: 789 (1987).
112. L. M. Lagrimini, S. Bradford, and S. Rothstein, *Plant Cell, 2*: 7 (1990).
113. D. H. Hamer, *Ann. Rev. Biochem., 55*: 913 (1986).
114. I. B. Maiti, G. J. Wagner, R. Yeargan, and A. G. Hunt, *Plant Physiol., 91*: 1020 (1989).
115. I. B. Maiti, G. J. Wagner, and A. G. Hunt, *Plant Science, 76*: 99 (1991).
116. S. Misra and L. Gedamu, *Theor. Appl. Genet., 78*: 161 (1989).
117. B. Lane, R. Kajioka, and T. Kennedy, *Biochem. Cell Biol., 65*: 1001–1005 (1987).
118. H. Rennenberg, *Plant Molecular Biology* (D. von Wettstein and N.-H. Chua, eds.), Plenum Press, New York, p. 279 (1987).
119. H. V. Scheller, B. Huang, E. Hatch, and P. B. Goldsbrough, *Plant Physiol., 85*: 1031 (1987).

120. E. Grill, S. Löffler, E.-L. Winnacker, and M. H. Zenk, Proceedings National Academy of Sciences (U.S.), *86*: 6838 (1989).
121. J. C. Steffens, D. F. Hunt, and B. G. Williams, *J. Biol. Chem., 261*: 13879 (1986).
122. E. Delhaize, P. J. Jackson, L. D. Lujan, and N. J. Robinson, *Plant Physiol., 89*: 700 (1989).
123. R. J. Ellis, *Science, 250*: 954 (1990).
124. C. L. Guy, *Ann. Rev. Plant Physiol. Plant Mol. Biol., 41*: 187 (1990).
125. P. L. Steponkus, *Ann. Rev. Plant Physiol., 35*: 543 (1984).
126. S. Lindquist, *Ann. Rev. Biochem., 55*: 1151 (1986).
127. R. T. Nagao, J. A. Kimpel, E. Vierling, and J. L. Key, *Oxford Surveys Plant Mol. Cell. Biol., 3*: 384 (1986).
128. E. Vierling, *Ann. Rev. Plant Physiol. Plant Mol. Biol., 42*: 579 (1991).
129. J. Landry, P. Chrétien, H. Lambert, E. Hickey, and L. A. Weber, *J. Cell Biol., 109*: 7 (1989).
130. Y. Sanchez and S. L. Lindquist, *Science, 248*: 1112 (1990).
131. S. E. Fender and M. A. O'Connell, *Plant Cell Reports, 8*: 37 (1989).
132. A. Sakai and W. Larcher, *Frost Survival of Plants. Responses and Adaptations to Freezing Stress.* Springer Verlag, Berlin, Germany (1987).
133. J. Levitt, *Responses of Plants to Environmental Stresses*, Vol. 1. *Chilling, Freezing and High Temperature Sresses*, 2nd ed., Academic Press, New York (1980).
134. C. R. Rohde and C. F. Pulham, *Agronomy J., 52*: 584 (1960).
135. C. N. Law and G. Jenkins, *Genet. Res., 15*: 197 (1970).
136. D. R. Liesenfeld, D. L. Auld, G. A. Murray, and J. B. Swensen, *Crop Science, 26*: 49 (1986).
137. P. L. Steponkus and F. O. Lanphear, *Physiol. Plant., 21*: 777 (1968).
138. A. Fennell and P. H. Li, *Acta Hort., 168*: 179 (1985).
139. M. A. Schaffer and R. L. Fischer, *Plant Physiol., 87*: 431 (1988).
140. S. S. Mohapatra, L. Wolfraim, R. J. Poole, and R. S. Dhindsa, *Plant Physiol., 89*: 375 (1989).
141. R. K. Hajela, D. P. Horvath, S. J. Gilmour, and M. F. Thomashow, *Plant Physiol., 93*: 1246 (1990).
142. K. Nordin, P. Heino, and E. T. Palva, *Plant Mol. Biol., 16*: 1061 (1991).
143. P. Calderón and H. G. Pontis, *Plant Science, 42*: 173 (1985).
144. G. L. Salerno and H. G. Pontis, *Plant Physiol., 89*: 648 (1989).
145. J. A. Raymond, P. Wilson, and A. L. DeVries, Proceedings National Academy of Sciences (U.S.), *86*: 881 (1989).
146. J. G. Duman, *J. Insect Physiol., 25*: 805 (1979).
147. R. E. Feeney and Y. Yeh, *Adv. Protein Chem., 32*: 191 (1978).
148. C. Orser, B. J. Staskawicz, N. J. Panopoulos, D. Dahlbeck, and S. E. Lindow, *J. Bacteriol., 164*: 359 (1985).
149. P. K. Wolber, C. A. Deininger, M. W. Southworth, J. Vandekerckhove, M. Van Montagu, and G. J. Warren, Proceedings National Academy of Sciences (U.S.), *83*: 7256 (1986).
150. S. Kurkela and M. Franck, *Plant Mol. Biol., 15*: 137 (1990).

151. A. Blum, *Plant Breeding for Stress Environment.* CRC Press, Boca Raton, Florida (1988).
152. L. G. Paleg and D. Aspinall, *The Physiology and Biochemistry of Drought Resistance in Plants.* Academic Press, New York (1981).
153. D. Rhodes, *Physiology of Metabolism* (The Biochemistry of Plants: A Comprehensive Treatise, Vol. 12) (D. D. Davies, ed.), Academic Press, New York, p. 201 (1987).
154. M. L. Binzel, P. M. Hasegawa, D. Rhodes, S. Handa, A. K. Handa, and R. A. Bressan, *Plant Physiol., 84*: 1408 (1987).
155. A. E. Moftah and B. E. Michel, *Plant Physiol., 83*: 238 (1987).
156. S. F. Chandler and T. A. Thorpe, *Plant Cell Reports, 6*: 176 (1987).
157. S. Bhaskaran, R. H. Smith, and R. J. Newton, *Plant Physiol., 79*: 266 (1985).
158. H. E. Flores, N. D. Young, and A. W. Galston, *Cellular and Molecular Biology Plant Stress* (UCLA Symposia on Molecular and Cellular Biology, New Series Vol. 22) (J. L. Key and T. Kosuge, eds.), Alan R. Liss, New York, p. 93 (1985).
159. N. K. Singh, A. K. Handa, P. M. Hasegawa, and R. A. Bressan, *Plant Physiol., 79*: 126 (1985).
160. S. Ramagopal, Proceedings National Academy of Sciences (U.S.), *84*: 94 (1987).
161. W. J. Hurkman and C. K. Tanaka, *Plant Physiol., 83*: 517 (1987).
162. B. Claes, R. Dekeyser, R. Villarroel, M. Van den Bulcke, G. Bauw, M. Van Montagu, and A. Caplan, *Plant Cell, 2*: 19 (1990).
163. N. K. Singh, C. A. Bracker, P. M. Hasegawa, A. K. Handa, S. Buckel, M. A. Hermodson, E. Pfankoch, F. E. Regnier, and R. A. Bressan, *Plant Physiol., 85*: 529 (1987).
164. G. J. King, V. A. Turner, C. E. Hussey Jr, E. S. Wurtele, and S. M. Lee, *Plant Mol. Biol., 10*: 401 (1988).
165. B. J. C. Cornelissen, R. A. M. Hooft van Huijsduijnen, and J. F. Bol, *Nature, 321*: 531 (1986).
166. W. S. Pierpoint, A. S. Tatham, and D. J. C. Pappin, *Physiol. Mol. Plant Pathol., 31*: 291 (1987).
167. M. Richardson, S. Valdes-Rodriguez, and A. Blanco-Labra, *Nature, 327*: 432 (1987).
168. L. Dure III, M. Crouch, J. Harada, T.-H. D. Ho, J. Mundy, R. Quatrano, T. Thomas, and Z. R. Sung, *Plant Mol. Biol., 12*: 475 (1989).
169. J. Mundy and N.-H. Chua, *EMBO J., 7*: 2279 (1988).
170. E. Mortonson and G. Dreyfuss, *Nature, 337*: 312 (1989).
171. D. E. de Oliveira, J. Seurinck, D. Inzé, M. Van Montagu, and J. Botterman, *Plant Cell, 2*: 427 (1990).
172. J. J. Heikkila, J. E. T. Papp, G. A. Schultz, and J. D. Bewley, *Plant Physiol., 76*: 270 (1984).
173. I. Fridovich, *Arch. Biochem. Biophys., 247*: 1-11 (1986).
174. B. Barényi and G. H. Krause, *Planta, 163*: 218 (1985).
175. R. R. Wise and A. W. Naylor, *Plant Physiol., 83*: 272 (1987).
176. R. R. Wise and A. W. Naylor, *Plant Physiol., 83*: 278 (1987).
177. E. Cadenas, *Ann. Rev. Biochem., 58*: 79 (1989).

178. S. B. Powles, *Ann. Rev. Plant Physiol., 35*: 15 (1984).
179. B. Halliwell, *Chem. Physics Lipids, 44*: 327 (1987).
180. I. Fridovich, *J. Biol. Chem., 264*: 7761 (1989).
181. J. V. Bannister, W. H. Bannister, and G. Rotilio, *CRC Crit. Rev. Biochem., 22*: 111 (1987).
182. C. Bowler, M. Van Montagu, and D. Inzé, *Ann. Rev. Plant Physiol. Plant Mol. Biol., 43*: 83 (1992).
183. R. Perl-Treves, B. Nacmias, D. Aviv, E. P. Zeelon, and E. Galun, *Plant Mol. Biol., 11*: 609 (1988).
184. J. Tepperman, C. Katayama, and P. Dunsmuir, *Plant Mol. Biol., 11*: 871 (1988).
185. R. E. Cannon and J. G. Scandalios, *Mol. Gen. Genet., 219*: 1 (1989).
186. C. Bowler, T. Alliotte, M. De Loose, M. Van Montagu, and D. Inzé, *EMBO J., 8*: 31 (1989).
187. E. W. T. Tsang, C. Bowler, D. Hérouart, W. Van Camp, R. Villarroel, C. Genetello, M. Van Montagu, and D. Inzé, *Plant Cell, 3*: 783 (1991).
188. J. M. Tepperman and P. Dunsmuir, *Plant Mol. Biol., 14*: 501 (1990).
189. C. Bowler, L. Slooten, S. Vandenbranden, R. De Rycke, J. Botterman, C. Sybesma, M. Van Montagu, and D. Inzé, *EMBO J., 10*: 1723 (1991).
190. L. M. Lagrimini, W. Burkhart, M. Moyer, and S. Rothstein, Proceedings National Academy of Sciences (U.S.), *84*: 7542 (1987).
191. S. Sakajo, K. Nakamura, and T. Asahi, *Eur. J. Biochem., 165*: 437 (1987).
192. E. Roberts, T. Kutchan, and P. E. Kolattukudy, *Plant Mol. Biol., 11*: 15 (1988).
193. E. Roberts and P. E. Kolattukudy, *Mol. Gen. Genet., 217*: 223 (1989).
194. R. B. van Huystee, *Ann. Rev. Plant Physiol., 38*: 205 (1987).
195. S. C. Fry, *Ann. Rev. Plant Physiol., 37*: 165 (1986).
196. I. Apostol, P. F. Heinstein, and P. S. Low, *Plant Physiol., 90*: 109 (1989).
197. Y. Shaaltiel and J. Gressel, *Pest. Biochem. Physiol., 26*: 22 (1986).
198. Y. Shaaltiel, N.-H. Chua, S. Gepstein, and J. Gressel, *Theor. Appl. Genet., 75*: 850 (1988).

Index

Abiotic stress-induced genes, 295–298
  anaerobic stress, 295
  genes related to water, salt, and
    cold stress, 295–298
Abscinins, 174
Abscisic acid (ABA), 14, 174, 184,
  185
Acer, as source of xylem sap, 5–7
Acetylene, 174
Agrobacterium-mediated DNA trans-
  formation, 259–260
Agronomically useful genes, trans-
  genic plants carrying, 262–263,
  264
Alien gene introgression, somaclonal
  variation to enhance, 250
Aluminum (Al) effect on root growth,
  156
Amino acids, diurnal production of,
  25
Ammonium. *See* Nitrogen (N)
Anaerobic stress, genes related to, 295

Anther culture (plant improvement
  technique), 231, 244–247
Antifungal proteins, 406–407
  See also Pathogenesis-related (PR)
  proteins
Apoplastic versus symplastic pathways:
  for loading of photoassimilates,
  17–19
  for unloading of photoassimilates,
  20–23
Auxins, 14, 180–183
  inhibitors of auxin transport, 190–
  191
Azorhizobium species, 74–75, 78

Bacillus thuringiensis (bacterial pro-
  tector), 441–442
Balanced nutrition, maintenance of,
  55–64
  ionic balance in plants, 55–58
  nutrient transport in plants, 58–
  60